T0076181

A HISTORY
OF THEIR OWN

Women in Europe From Prehistory to the Present

VOLUME II
REVISED EDITION

BONNIE S. ANDERSON
JUDITH P. ZINSSER

NEW YORK   OXFORD
OXFORD UNIVERSITY PRESS
2000

Oxford University Press

Oxford   New York
Athens   Auckland   Bangkok   Bogotá   Buenos Aires   Calcutta
Cape Town   Chennai   Dar es Salaam   Delhi   Florence   Hong Kong   Istanbul
Karachi   Kuala Lumpur   Madrid   Melbourne   Mexico City   Mumbai
Nairobi   Paris   São Paulo   Singapore   Taipei   Tokyo   Toronto   Warsaw

and associated companies in

Berlin   Ibadan

Introduction copyright © 2000 by Bonnie S. Anderson and Judith P. Zinsser
A *History of Their Own, Volume II*. Copyright © 1988 by Bonnie S. Anderson and Judith P. Zinsser
Published by arrangement with HarperCollins Publishers, Inc. All rights reserved.
Illustration credits follow the Index.

Published by Oxford University Press, Inc.,
198 Madison Avenue, New York, New York 10016

Oxford is a registered trademark of Oxford University Press

All rights reserved. No part of this publication
may be reproduced, stored in a retrieval system, or transmitted
in any form or by any means, electronic, mechanical,
photocopying, recording, or otherwise, without the prior
permission of Oxford University Press.

Library of Congress Cataloging-in-Publication Data

Anderson, Bonnie S.
    A history of their own : women in Europe from prehistory to the present / Bonnie S. Anderson,
Judith P. Zinsser. — Rev. ed.
       p.   cm.
    Includes bibliographical references and index.

    ISBN-13 978-0-19-512839-0

    1. Women—Europe—History.   2. Feminism—Europe—History.   I. Zinsser, Judith P.   II. Title.
HQ1587.A53   1999
305.4'094—dc21   98-46743
                                                                                                              CIP

Printed in the United States of America

Contents

Illustrated sections follow pages 8, 106, 230, and 338.

Acknowledgments

With the publication of this second edition of A *History of Their Own* we are pleased to acknowledge the individuals and institutions who helped to make it possible. First, thank you to all those who, through their letters and e-mail messages, eloquently expressed their support for this project. Second, our thanks to Gioia Stevens, our editor at Oxford University Press, whose cheerfulness, competence, and calm efficiency made our revisions a pleasure to complete. Third, thank you to Laura Kitch, Acting Provost of Brooklyn College, CUNY; John H. Skillings, Associate Dean, and Charlotte Newman Goldy, Chair of the History Department at Miami University, who gave us the time and resources we needed. Fourth, the task of surveying and assessing the vast literature of recent European women's history would have been impossible without the tireless efforts of Jennifer M. Morris and Jeri L. Schaner. Jenny L. Presnell was never too busy to answer a bibliographic question; Elizabeth A. Smith helped us meet last minute deadlines. Finally, we wish to thank our colleagues and friends, who continue to give so generously of their time, affection, and encouragement—in particular, Mary E. Frederickson and Roger J. Millar.

Introduction

For the first edition of A *History of Their Own* in 1988 we stated goals remarkably similar to those of forward-looking women in mid-nineteenth-century Europe and the United States. "It would be really fine if someone would undertake to write a historical work on the position of women in society—how it has been developed from the earliest times up to today," wrote the German feminist Louise Otto in 1844,

> The lack of such a work is always strongly felt, but even more now—to put lessons from the past on a firm historical foundation, to seek prospects for the future, even more to have women begin to feel that they . . . are not just wives and mothers of the people, but half of this people themselves. What has passed for the history of women until now is only biographies of famous women, saints, princesses, heroines, etc.[1]

In the 1980s, we also wanted to legitimize the writing of women's history, to include lives and accomplishments long forgotten, to compensate for the absence of half of humanity from the historical record.

In many ways this project has been successful. During the last decade, scholarship on women has grown dramatically for every region and in every historical era. Because of this growth, the very existence of the history of women no longer needs to be justified as it did in Louise Otto's time. Few would now dare suggest, as they did twenty years ago when we first began the research for A *History of Their Own*, that women "had no history," or that they had achieved little worthy of inclusion in the historical record. Today, the study of women's past has become an accepted field within the discipline, and its practitioners have risen to the top ranks of the profession. Women's history courses are an accepted part of high school, college, and university curricula.[2]

Academic series of books in women's history, rich monographs, specialized encyclopedias, dictionaries, and bibliographic guides, new journals in women's history, internet groups, and web sites have proliferated. Growing networks of scholars, both in this country and throughout Europe, attest to the vitality of the subject. Topics that had to be pieced together in the 1980s are now recognized subfields in their own right: the history of the family, of sexuality, of violence against women, of laws and customs governing their lives. Subsequent research has confirmed analyses we assembled in order to write a work of synthesis: from the patterns of women's religious participation to the gendered dynamics of the witchcraft persecutions, from the shift in male roles when commerce made "providers" out of warriors to the connection between feminist demands and women's participation in political revolutions. Almost all of the principal figures singled out in our narrative history now have their own biographies and often scholarly editions of their writings. Talks that we cited in our notes, ones given at the Berkshire Conference on the History of Women, have become full-length monographs; long-range, general studies have finally been published—for example, an essential survey history of Catholic nuns, and analyses of women's lives in nineteenth-century Russia.[3]

In other ways, success has eluded feminist historians. Most works in women's history are found not in the history sections of libraries and book stores, but instead are catalogued under the impossibly vast category of Women's Studies. Scholarship in the field is read primarily by other women's historians or by feminists. Too many traditional historians make only the most cursory changes to their old narratives. European women now have their own histories, but this knowledge of their past has not significantly altered general accounts of European history. The question with which Louise Otto closed her thoughts on women's history in 1844 still applies today: "Shouldn't the female sex be given more attention in a general history of civilization than they were before?"

Women's historians of all regions of the world, not just Europe, have critiqued the male-centered nature, the subjectivity, the inadequacies of these supposedly gender-neutral prevailing accounts of the past. But however clearly the political nature of historical narratives and the choices and exclusions made in them are demonstrated, the full integration of women's history into all kinds of historical writing remains, in the words of the U.S. historian Anne Firor Scott, an "unfinished business."[4] The burden of proof still lies with advocates for the inclusion of women and the dynamics of gender, not with those who oppose it. Most traditional historians continue to insist that their version of the past, which suppresses women's history and ignores gender, presents an "objective," "apolitical" account of all that is significant.[5]

Nowhere is this resistance more obvious than in European history textbooks. Although much lip service is paid in explanatory introductions to the need for a new, more inclusive history, the canons of the male-focused narrative still govern periodization, the organization of sections and chapters, the choice of heroes and the occasional heroine. At best, women are subsumed under "social history," implying that their contributions to politics, the economy, or intellectual and spiritual life have been negligible. Too often, women then appear only as victims, the subjects of laws and customs that constrained and denigrated them. Some textbooks tell far more about men's views of appropriate female behavior than women's actual lives.[6] At worst, too many European textbook authors have simply added a few paragraphs about women to the ends of pre-existing chapters, pigeonholing the entire female sex as a "minority," an afterthought, an inconsequential and marginalized group whose lives exist outside the main story of Western Civilization.

None of the leading European history texts offers an innovative framework that views women as active participants with men in all areas of human endeavor. None analyzes men as a group or explicitly identifies what men as men have done. None examines the creation of historical definitions of masculinity, none explores how male-centered customs and practices became institutionalized and codified, none demonstrates how concepts of sexual difference have operated as social forces. Instead, the male has been universalized, so that accounts of men's achievements are assumed to be a complete history of the European past. What women accomplished despite constraints, what they made of their lives around and about the traditional men's narratives, has been omitted. There is no history of Europe that adequately describes the past experiences of women and men.[7]

As a result, women still need a history of their own that provides a continuous, female-centered narrative of Europe's past. A number of fine histories of European women now exist, but they either cover only a few centuries or particular groups and nationalities. Anthologies of articles present case studies across time but offer only brief transitional essays to link their disparate parts. Oxford University Press decided to publish a new edition of A *History of Their Own* because it provides a true synthesis, an unbroken, readable account of the European past from prehistory to the present that consistently puts women at the center of the narrative. Freed by its unique organization from the forced exclusions of traditional periodization, this two-volume work tells the familiar stories of the aristocrats and queens, but from a female perspective. In addition, by focusing on women previously ignored, like peasants and domestic servants, A *History of Their Own* provides a truly feminist account of European women's and men's collective history.

Choices that we made in the writing of these two volumes have become models for the field. Many historians have repeatedly called for reorganization of the standard divisions that shape our views of the past, chronological designations like "Middle Ages," "Renaissance," and "Industrial Revolution," but few have actually done this.[8] Instead, the traditional units continue to shape histories of women, however exclusionary or distorting they may be. Freeing European women's history from what the literary critic Jane Marcus calls "the yoke of male periodization" was the most liberating aspect of our collaboration.[9] By forcing ourselves to go against our training, we could re-examine standard historical events and thus reconceptualize the past from women's perspectives. In our reconceptualization, however, we did not abandon chronology. Each chapter is internally chronological, but the same event may appear more than once, as it affects different groups of women. Industrialization, for instance, had an entirely different impact on elite women, peasant women, and women of the modern cities, so it appears in each of the three chapters that focus on these categories. The Renaissance, the Enlightenment, the French Revolution, the World Wars are all dealt with in this way. As traditional historical periods and events receded in significance, others grew in importance. Factors often ignored in histories of men, whether contraception or clothing, diseases or the design of houses, proved crucial in women's lives.

Equally liberating was our decision to change standard historical words and phrases. All historians are trained to examine documents critically, to distinguish between actual reality and what is being said about that reality by a specific group of people. In English, supposedly gender-neutral terms like "peasant" or "revolutionary" appear inclusive but the descriptions that follow never mention women. Other nomenclature based on the unspoken assumption that male experience is everyone's experience also distorts the reality of women's lives. "Working class," for instance, implies both that women in other categories did not labor and that women's lives in this social stratum mirrored those of men. We used descriptive phrases like "women of the people" and "women who earned income outside the home" to convey the reality of poor urban women's lives more accurately. In addition, we reversed traditional modes of expression, writing of "women and men," "queens and kings," "mothers and fathers" in order to make women the focus of our narrative and to counter the weight of a male-oriented past and male-dominant forms of expression.

Rejecting the male-centered biases of our discipline, we turned to other fields for methodological and analytical tools to break what the feminist historian and theorist Joan Wallach Scott calls the "epistemological frame of orthodox history."[10] Anthropology proved the most fruitful. The beginning of each section of this work uses the anthropological technique of "thick description." Women within a particular category in specific geographic localities are described in "cross-cultural montages" generated from all kinds of sources—from folklore, archaeology, and art history to sociology and economics.[11] This technique also allows us to highlight a single woman's life or production—her weaving, her basketwork, her painting, her diary. We then used this "text" to provide insight into the overall social structures and cultural contexts in which she lived. Thus, well-known heroines illumine the lives of unnamed women in similar circumstances. Joan of Arc sheds light on other peasant women's place in society, while Anne Frank's brief life illustrates the tragedies of the Nazi genocide.

In addition to giving us descriptive and narrative techniques, anthropology provided new categories of organization. The concepts of "place" and "function" allowed us to abandon the inadequate and inappropriate periodization of men's histories. Placing women in specific geographic and institutional contexts, identifying them according to their broad functions within European society, revealed the unity of certain groups over time and across the continent. Peasant women, usually rendered invisible in historical narratives, emerged as a separate group whose similarities outweighed geographic, ethnic, or temporal differences. Women within the Christian churches constituted another category unified across time and secular boundaries by place and function. In the modern era, these concepts allowed us to distinguish between different types of experience in the same place and era. In the nineteenth century, the lives of poor urban women differed so markedly from those of peasants that they constituted separate categories, even though the same woman may have lived first in the countryside and then in the city.

These categories also enabled us to assess the long trajectories and patterns of European women's experience over time. We had hoped to find a "Golden Age" for women, a time when European women were not subordinated to and valued less than men. While the possibility of a matriarchal culture in prehistory cannot be ruled out completely, we discovered no era in the historical past in which women dominated.[12] This unequal relationship between women and men, present in the earliest written documents of European culture—the Hebrew Bible, Homer's epics, Roman law—intensified as time went on. The nineteenth century marked the nadir of European women's powers and opportunities. In earlier eras, alternative authorities and customs, as well as regional, governmental, and religious variations, created a range of circumstances that enabled some European women to achieve relative independence and relative dominance. Gradually, however, the growing centralization, rationalization, and uniformity imposed in government, law, the economy, and religion worked to erode these options and further limited women's lives.

The centuries from the Renaissance through the Enlightenment broadened possibilities for most men, giving them greater access to education and more choices of occupation. The opposite happened for women. New national law codes denied them control of their property and earnings, gave primary authority within the family to the husband alone, outlawed any efforts by women to control their fertility, and barred them from higher education and the newly defined professions. During these centuries, increasingly polarized images of the physical and psychological differences, both real and imagined, between women and men, between the "feminine" and the "masculine," justified these growing disparities. Female incapacity and male authority came to seem self-evident and natural.

The popular nineteenth-century ideal of the "angel in the house," a woman happily limited to the care of her household and children, offered a more restricted life to women than even the didactic treatises of previous centuries. The reality was always different from the imagined ideal. The majority of Europe's women continued to earn income; some "angels of the house" created paths out of the parlor and into the world. Even so, the concept of domesticity, with its constraining assumptions and definitions, remained for women of all classes and in all circumstances. The creation of "women's movements" in the nineteenth century occurred in part as a response to this narrowed view of women's capacities and activities.

In the twentieth century, women have changed laws and modified institutions. They, like European men, have benefited from prosperity, universal medical care, and technological progress. Most European women today enjoy full rights of citizenship, have access to education and employment, live longer, and face fewer risks from sexual activity and childbearing than women in earlier ages. While neither complete equality nor a realistic assessment of the value of women's contributions to European life has yet been realized, women's circumstances and opportunities have improved. In large part, change has come because of the effectiveness of the Women's Liberation Movement of the 1970s and 1980s. As the anthropologist Kathleen Gough observes, "It is not necessary to believe myths of a feminist Golden Age in order to plan for parity in the future."[13]

In writing about each category of women, certain questions guided our research. First, how had ordinary women lived? What tasks filled their days? What motivated their actions and determined their attitudes? Second, how to explain the startling contrasts between women's and men's lives in the same eras? Why had laws, economic systems, religion, and politics excluded European women from the most valued activities in life? How had cultural attitudes evolved that defined women, and qualities identified as "feminine," as innately inferior and placed all things female in a subordinate relationship to men and all things male? Why had men created or acquiesced in this inherently unequal system of social relationships? Perhaps more importantly, why had most women accepted or been forced to accept these limitations, which devalued their activities, denigrated their nature, and subordinated them to men?

Third, we looked at the exceptions—those women who achieved prominence and were included in traditional histories: St. Bridget of Sweden, Queen Isabella of Castile, Mme. de Pompadour, Florence Nightingale, Marie Curie. Why had these women gained recognition? Were they exceptions because of their character or historical circumstance? Finally, we studied those women, like Christine de Pizan, who first publicly questioned women's disadvantaged and denigrated status. Why did some women question all women's subordination? How had they come to identify with all women and to work for expanded opportunities for their sex? How and why did feminism begin and where might it lead, as it calls into question the basic values of European culture and society?

The answers to these questions led us to the central thesis of these volumes: that gender has been the most important factor in shaping the lives of European women. However, not all women's experiences are alike. Our narrative recognizes the gulf between a woman in medieval France and a woman in modern England, between a fifteenth-century female merchant and a twelfth-century day laborer, between a German Social Democrat and a Soviet Bolshevik. Our method of organizing women into separate categories graphically indicates the significance we accord these differences. However, underlying these differences are similarities decreed by gender. Throughout the centuries we found an awesome similarity in the effects of gender on European women's lives, in the continued power of the denigrating qualities classified as "feminine." Unlike men, who have been primarily identified by class, ethnic origin, or historical era, European women have traditionally been seen first as female, a separate category of being. As the French socialist Louise Michel wrote in 1885, it has been "painful" for us "to admit that we are a separate caste, made one across the ages," but as we compared our findings from studies of different eras, classes, and ethnic circumstances, no other conclusion was possible.[14]

Part I of A History of Their Own, "Traditions Inherited," speculates about women's and men's lives in prehistory and the origins of European culture's largely negative views of women and their subordinate status. It then examines the Greek, Roman, Hebrew, Celtic, Germanic, and Christian traditions about women and their relationships to men, traditions already in place when Europe emerges as a more recognizable entity in the ninth century. Part II, "Women of the Fields: Sustaining the Generations," surveys the lives of European peasant women into the 1980s. Because they make up the vast majority of Europe's women until well into the eighteenth century, we placed their narrative first, thus highlighting their numbers and affirming their significance. Our account gives priority to the constants in their experiences over local differences in geography, custom, patterns of landholding and trade. Part III, "Women of the Churches: The Power of the Faithful," shows how Christianity provided a unique environment for European women. From the early centuries of the religion's growth, through the Reformations of the sixteenth and seventeenth centuries, women could gain authority and relative autonomy not possible in other circumstances.

Part IV, "Women of the Castles and Manors," argues that the lives of Europe's noblewomen from the ninth to the seventeenth centuries are connected because of their elite status and their function as "custodians of land and lineage." While these women sometimes acquired power and acted in place of men, they remained vulnerable because of their gender. Part V, "Women of the Walled Towns: Providers and Partners," distinguishes urban women of the twelfth to seventeenth centuries from their rural counterparts. From the poorest day laborer to the wealthiest merchant's wife, townswomen participated in the significant economic developments of their era: the formation of guilds and the evolution of commercial capitalism. Neither, however, freed them from the constraints of circumstance and attitude that traditionally limited women's lives. These five sections comprise Volume I.

While our first volume focuses on the centuries before 1600 and our second on those after, this division is not rigid. Traditional chronologies are not the organizing principle of this work; the categories of place and function demarcating women's lives are. Thus although Part VI, "Women of the Courts: Rulers, Patrons, and Attendants," appears as the first section of Volume II, it describes court life from the fifteenth through the eighteenth centuries. We argue that the growth of dynastic monarchy created special circumstances in which some women had opportunities to become educated, to write, to exercise political influence, and, in a few instances, to rule. Part VII, "Women of the Salons and Parlors: Ladies, Housewives, and Professionals," examines the lives of economically privileged women from the late seventeenth century to the present. Ideals of domesticity and the realities of better standards of living distinguished these women's lives from those in other classes. Some women turned these conditions to advantage, using their moral and material authority to play active roles outside of their homes.

Part VIII, "Women of the Cities: Mothers, Workers, and Revolutionaries," deals with the lives of everyday urban women in the same centuries, focusing on their participation in economic, social, and political movements. Here we pioneered the thesis that urbanization was more important than industrialization in shaping these women's lives. The chapters grouped under the heading "Women of the Cities" parallel our earlier category, "Women of the Fields." Together, urban and country women comprise the two most numerous groups of women, and so are near the beginning and end of the two volumes.

Part IX, "Traditions Rejected: A History of Feminism in Europe," mirrors the first section of Volume I, "Traditions Inherited" when Europe coalesced. Beginning with the writings of the courtier Christine de Pizan in the fifteenth century, this final section views European feminism as a series of repudiations of the negative traditions that limited women's lives.[15] In this process, a women-centered view of the world, which is still being elucidated and realized today, evolved into feminism. This new edition of *A History of Their Own* concludes with an epilogue on developments in both Western and Eastern Europe since the collapse of communism in the late 1980s. The conservative shift in politics, the shrinkage of the welfare state, the rise of unregulated market economies, the continuance of primary responsibility for child-raising and housework still differentiate women's lives from those of men in the same nation, ethnic group, social stratum, or even family.

Aside from our central thesis about the significance of gender in European women's history, no other aspect of our book has been so controversial as our emphasis on the continuities in European women's lives throughout the centuries.[16] These two convictions are linked. There are variations in European women's lives across time, place, function, and circumstance. They were active participants in political, social, economic, and religious change. But these variations and this participation did not alter women's status relative to men. Despite dramatic transformations of European culture and society from the ninth to the early twentieth century, the meanings given to sexual difference and to "feminine" and "masculine" identities worked to maintain the disadvantaged status of women.[17] The medieval historian Judith Bennett describes a "patriarchal equilibrium" creating an overall pattern of European women's history that resembles a dance, in which the steps and rhythms, partners and groups may change, but the men always lead.[18]

Distinguishing variation from genuine transformation, we argue that a number of important aspects shaping women's lives have remained unchanged over time. Until the last decades of the twentieth century, all women were defined by their relationships to men. Many women—far many more than men—remain in the historical record only as men's women: the daughters of Priam, Lot's wife, and the mother of the Maccabees are but a few of the earliest examples. A woman is first identified as her father's daughter, her husband's wife or widow, her son's mother. No matter what the era in European history, what their class or social rank, what their nationality or ethnic group, most women have lived their lives as members of a male-dominated family. Even those who lived more autonomous lives as part of women's spiritual communities were defined by their rejection of earthly marriage. Nuns, as members of religious orders, were described as the "brides of Christ."

These definitions, as historians of sexuality and of the family have demonstrated, constrained women and men. The "family" protected by law and custom, the union of a woman with a lawful husband for the purpose of procreation, presumed the heterosexuality of both partners and dictated their primary functions and roles. In the male-headed family, child rearing and maintenance of the household have always been gendered, seen throughout Europe's history as women's preordained, biologically appropriate tasks.

Defining women's primary duties as care of the family and the home has not precluded other work. In all historical eras, the vast majority of European women have labored at other chores and assumed other responsibilities.[19] They have worked in the fields. They have earned wages. They have generated additional income for their families. Weeding, reaping, sewing, knitting, cleaning others' homes, raising others' children, working in factories or offices, women's labor has made the continuance of their families possible. This "double burden" of caring for a family and home and earning additional income has characterized the lives of most European women and differentiated them from men. It is women, not men, who have these multiple responsibilities and must find work compatible with these duties or arrange for substitutes to care for their children and their household while they earn income.

In addition, "women's work," whether in the home or outside of it, has traditionally been valued less and considered less important than men's work. Raising children and maintaining the home have been taken for granted and have never been valued as much as labor that men perform, whatever it may be. Paid labor available to women has usually been less prestigious than men's, has traditionally required less formal training, and has been more vulnerable to fluctuations in the economy. As a result, when they have been paid for their work, women have consistently received between one-half and two-thirds of what men earn. Sometimes connected by scholars to different economic systems, this factor has always been present in European history. In reckonings of female and male worth in the Old Testament, in the manor rolls of noble households, in account books of sixteenth-century merchants, in payrolls of nineteenth- and twentieth-century factories, women received less than men. The amount that they are paid may vary: labor shortages or economic regulations may raise women's wages, but, so far, they have rarely equaled those of men. As historians of European women's labor have demonstrated, there was no "Golden Age" of women's household production. Commercial capitalism brought different activities and relationships within the marketplace to women, but not the transformation of the underlying gendered patterns of social interaction and the institutions that protected them. All of these factors shaping women's work limited European women's lives by curtailing their opportunities and resources.[20]

Some women maneuvered around these limits or found ways of setting one institutional constraint against another—the aristocratic woman who sought the Church's support in her choice of a husband against the dictates of her family; the royal women who ruled as queens in their own right; the merchant's wives who managed a husband's fortune after his death; the successful court musicians, poets, and artists. Even they, however, were subject to the most damning aspects of gender: European culture's largely negative views of women. Considered innately flawed, less valuable, and thus inferior to men, all women were supposed to be subordinate to men. This subordination seemed part of the natural order. A woman who did rule over men, who held a dominant role, whether from a throne or within a family, was seen as "unwomanly," a danger to the universe's natural hierarchy, which made man come first.

These cultural views, expressed in the earliest writings of the Hebrews, Greeks, and Romans, changed remarkably little over time. The biblical injunction to Eve that "your desire shall be for your husband, and he shall rule over you" (Genesis 3:16) is repeated in every era and every European nation. The view that "the best woman is she who is silent"—first written down in ancient Greece—reappears often in European men's writings about women. The assumption that only men are truly human—that "a hen is not a bird and a woman is not a person," as the Russian proverb explains—echoes throughout European history. No woman could escape the impact of these views completely. Of all the factors that have limited women's lives, these negative cultural traditions, these negative constructions of what it means to be female, have proved the most powerful and the most resistant to change.

But they have never been all-powerful. Throughout the centuries is also scattered the evidence of European women's agency, of the multiplicity of ways in which they gave value, beauty, and power to their lives. Many took pleasure and pride in their reproductive and nurturing role, in their daily tasks, however mundane. Sadly, much of women's creation has been anonymous and evanescent. Yet it is evidence none the less: the basket of willow branches created to gather food, the weaving in hand-dyed wools which clothed Europeans in the early centuries, the lace tablecloth for a daughter's trousseau, the household objects and children's toys designed to make life easier and more pleasant.

Although most of Europe's women accepted the institution of the male-dominated family for its guarantee of subsistence, an approved partner for life, and a sense of being protected from forces beyond their control, they have not just been victims. Resistance can take many forms. Even when unable to see beyond their culture's attitudes, they mastered the strategies of those in subordinate positions, manipulating, pleasing, enduring, surviving.[21] Some claimed spiritual or moral authority as women, drawing on those religious or ethical traditions that empowered women rather than subordinated them. There is magnificence in the fragments of Sappho's poetry, in Hildegard of Bingen's visions, in Marie de Gournay's defense of women, in Paula Modersohn-Becker's self-portraits, in Mo Mowlam's negotiations in Ireland.

Our belief in women's abilities to create such excellence, to expand the boundaries of human creativity and endeavor, to transcend conditions that seek to limit and control them underlies every section of these volumes. We have never held that women are determined by "their essence" to remain in certain roles or to fulfill certain functions—a charge that our emphasis on continuity has sometimes prompted.[22] Instead we have affirmed women's diversity and applauded their accomplishments. Increasingly, women's varied contributions have affected the lives of all. In 1998, Mary Robinson, as the United Nation's High Commissioner for Human Rights, brought her feminist sensibility to international affairs. While she was president of Ireland from 1990 to 1997, Robinson once explained: "A society that is without the voice and vision of a woman is not less feminine. It is less human."[23]

Throughout our collaboration on A History of Their Own, we have taken heart from Virginia Woolf's vision in A Room of One's Own of a future in which a woman with the talent of Shakespeare could flourish. "My belief is that if we live another century or so, if we have the habit of freedom and courage to write exactly as we think," she predicted in 1929, a way for this genius can be prepared.[24] With these volumes we make our contribution to this collective enterprise. We look forward to the creation of a world in which women and men will acquire "the habit of freedom and courage," as well as the means and opportunity to succeed as they choose.

Notes

1. Louise Otto cited in Ruth-Ellen Boetcher Joeres, *Die Anfänge der deutschen Frauenbewegung: Louise Otto-Peters* (Frankfurt-am-Main: Fischer Taschenbuch Verlag, 1983), p. 82.

2. For a survey of the changes in the academy and the profession for women's history and women historians, see for example, Judith P. Zinsser, Part III of *History and Feminism: A Glass Half Full* (New York: Twayne Publishers, 1993).

3. Jo Ann McNamara, *Sisters in Arms: Catholic Nuns through Two Millennia* (Cambridge, Mass.: Harvard University Press, 1996); Barbara Alpern Engel, *Between the Fields and the City: Women, Work, and Family in Russia, 1861–1914* (New York: Cambridge University Press, 1994).

4. Anne Firor Scott, "Unfinished Business," *Journal of Women's History*, vol. 8, no. 2 (Summer 1996), pp. 111–20.

5. For discussions of this problem, see the special issue of *History and Theory: Studies in the Philosophy of History*, Ann-Louise Shapiro, ed., vol. 31, no. 4 (December 1992) published in an expanded version as *Feminists Revision History* (New Brunswick, N.J.: Rutgers University Press, 1994).

6. For discussion of this in specialized works as well, see Amanda Vickery, "Golden Age to Separate Spheres? A Review of the Categories and Chronology of Women's History," *Historical Journal*, vol. 36 (1993), pp. 413–14.

7. See for example, current editions of Lynn Hunt, Theodore R. Martin, Barbara H. Rosenwein, R. Po-chia Hsia, and Bonnie G. Smith, *The Challenge of the West: Peoples and Cultures from the Stone Age to the Global Age* (Lexington, Mass.: D. C. Heath and Co.); Donald Kagan, Steven Ozment, and Frank M. Turner, *The Western Heritage* (New York: Macmillan); Mark Kishlansky, Patrick Geary, and Patricia O'Brien, *Civilization in the West* (New York: West Publishing Co.); Jackson J. Spielvogel, *Western Civilization* (New York: West Publishing Co.).

8. Joan Kelly and Gerda Lerner were the first to articulate the need for new periodization in their now famous essays: "Did Women Have a Renaissance?" in *Women, History & Theory: The Essays of Joan Kelly* (Chicago: Chicago University Press, 1984 [1977]); "The Challenge of Women's History," in *The Majority Finds Its Past: Placing Women in History* (New York: Oxford University Press, 1979).

9. Jane Marcus, "The Asylums of Antaeus: Women, War, and Madness—Is There a Feminist Fetishism?" in H. Aram Veeser, ed., *The New Historicism* (New York: Routledge, 1989), p. 140.

10. Joan Scott has discussed this dilemma in many of her essays. See for example, "The Evidence of Experience," in James Chandler, Arnold I. Davidson, and Harry Harootunian, eds., *Questions of Evidence: Proof, Practice, and Persuasion across the Disciplines* (Chicago: University of Chicago Press, 1994), pp. 367–69, 372–73, 376, 378. See also, *Only Paradoxes to Offer: French Feminists and the Rights of Man* (New York: Cambridge University Press, 1996), p. 124.

11. Judith Lowder-Newton uses this phrase. See her essay "History as Usual? Feminism and the 'New Historicism,'" in Veeser, pp. 153–54.

12. Anthropologists have documented parity between women and men, and

societies in which women exercised significant direct power, but not equality as it has traditionally been defined in Western cultures. For essays exploring this question, see for example, Rayna R. Reiter's collection, *Toward an Anthropology of Women* (New York: Monthly Review Press, 1975) and Peggy Reeves Sanday and Ruth Gallagher Goodenough, eds., *Beyond the Second Sex: New Directions in the Anthropology of Gender* (Philadelphia: University of Pennsylvania Press, 1990).

13. Kathleen Gough, "The Origin of the Family," in Reiter, p. 54.
14. Louise Michel, *The Red Virgin: Memoirs of Louise Michel*, Bullitt Lowry, and Elizabeth Ellington Gunter, eds. and trans. (University: University of Alabama Press, 1981), p. 139.
15. Joan Kelly enunciated this idea in "Early Feminist Theory and the *Querelle des Femmes*, 1400–1789," in *Essays*, pp. 65–109. This is also a central thesis of Gerda Lerner's *The Creation of Feminist Consciousness from the Middle Ages to 1870* (New York: Oxford University Press, 1993).
16. On the difficulties of accounting for continuities and changes at the same time, see Sandra E. Greene, "A Perspective from African Women's History: Comment on 'Confronting Continuity,'" *Journal of Women's History* vol. 9, no. 3 (Autumn 1997), pp. 95–104.
17. For a clear formulation of questions about the meanings assigned to sexual difference, and the gendering of identities to women's disadvantage, see Ava Baron, "Gender and Labor History: Learning from the Past, Looking to the Future," in Ava Baron, ed., *Work Engendered: Toward a New History of American Labor* (Ithaca, N.Y.: Cornell University Press, 1991).
18. Judith M. Bennett, "Theoretical Issues: Confronting Continuity," *Journal of Women's History*, vol. 9, no. 3 (Autumn 1997), p. 86.
19. Even though Europe's prescriptive literature might endorse the idea of a "separate sphere" as an isolated, protected women's world, the concept had no reality in fact. A number of historians have written on this misconception. See for example, Linda K. Kerber, "Separate Spheres, Female Worlds, Woman's Place: The Rhetoric of Women's History," *Journal of American History*, vol. 75, no. 1 (June 1988), pp. 9–39; Dorothy O. Helly and Susan Reverby's introduction in *Gendered Domains: Rethinking Public and Private in Women's History* (Ithaca, N.Y.: Cornell University Press, 1992).
20. See Vickery, pp. 401–4; see also Bennett, "Theoretical Issues," p. 86. Note that Bennett, in her discussion of medieval English brewsters, does not see these constraints as specifically designed to affect women. Rather "[t]hese factors affected some women differently from others, but they affected all women to some extent. These factors shaped the lives of men as well as women, but they constrained most women more than most men. And these factors grew from fundamental institutions of English life at the time, institutions that were much more than mechanisms for the subordination of women," pp. 87, 86.

21. Anthropologists have made systematic studies of this phenomenon in contemporary cultures. See for example, James C. Scott, *Domination and the Arts of Resistance; Hidden Transcripts* (New Haven: Yale University Press, 1990).
22. For a discussion of the need to avoid the extremes of "essentialism" or "relativism," see Linda J. Nicholson, *Feminism/Postmodernism* (New York: Routledge, 1990), p. 9.
23. Cited in Alida Brill, ed., *A Rising Public Voice: Women in Politics Worldwide* (New York: The Feminist Press at the City University of New York, 1995), p. 155.
24. Virginia Woolf, *A Room of One's Own* (New York: Harcourt, Brace & World, 1957 [1929]), pp. 117–18.

VI

WOMEN OF THE COURTS

•

RULERS, PATRONS, AND ATTENDANTS

1

THE WORLD OF ABSOLUTE
MONARCHS FROM THE FIFTEENTH TO
THE EIGHTEENTH CENTURIES

✤

CATHERINE THE GREAT, Empress of Russia, personified the absolute monarch for her contemporaries and for subsequent generations. A late-eighteenth-century engraving portrays her in her early sixties ready for an afternoon walk. Though heavyset, almost dowdy in a lace cap and grosgrain silk coat with her pet whippet beside her, she looks the ruler. One is struck by her direct gaze, the determined set of her chin. She is clearly the matriarch of an empire. To maintain this power, to administer her vast territories, Catherine II (1729–1796) filled her days with a seemingly endless stream of activities. She rose at five in the morning to deal with the real business of government. In these early hours she considered and formulated policy, writing and signing the *ukazy* (imperial decrees) that determined everything from the design of a new imperial residence and a soldier's cloak to the wording of Russia's new law codes and a treaty annexing territory in the Crimea.

The rest of Catherine the Great's day consisted of the ceremonial duties that also filled the life of the absolute ruler. These events constituted the life of the court as well, the activities and obligations of the imperial aristocracy and the bureaucracy who were her courtiers. After dressing in private, assisted by her personal attendants, she joined other favored courtiers in an antechamber for the last decorative touches: a diamond and pearl pendant for her hair, a laced ruffle to soften the neckline of her heavy brocade gown, the sash of a military order. The empress was then ready for her first grand public appearance of the day. From nine until one she met with advisers, heard petitioners, and received visitors. At two o'clock she dined with ten to twenty guests and made her midday meal yet another ceremonial occasion; to eat in the imperial presence was an honor coveted by courtiers. Catherine II gave the afternoon over to more private activities: reading, sewing, knitting, seeing her grandchildren, or conversing with a special visitor like the French Encyclopedist Denis Diderot.

When she stayed at the Winter Palace in St. Petersburg, the rituals and formal appearances that characterized the life of the ruler and her courtiers resumed again in the early evening, at six o'clock. The empress attended a formal reception and then made a ceremonial progress from room to room, greeting the assembled members of her court. Servants produced supper at ten, but Catherine II rarely stayed, preferring instead to retire to her rooms. The only variation in the exacting routine of business and public appearances would be the occasional informal evening in her more private apartments with her most intimate attendants and perhaps her current lover. Then she played a game of cards, listened to a concert, or watched a play.[1]

Women participated in this opulent, ritualized world in a variety of ways. Like Catherine the Great some women ruled as absolute monarchs of powerful dynastic states. She had seized the right to rule. Others, like Elizabeth I in sixteenth-century England, inherited the throne in their own right. From the fifteenth to the eighteenth centuries women also exercised power on behalf of their families, acting as surrogate rulers and royal regents, as did Eleanora of Ferrara in fifteenth-century Italy and Marie de' Medici in seventeenth-century France.

Royal and aristocratic women might become rulers. The vast majority of privileged women, however, whether in Renaissance Italy or seventeenth-century France, found other opportunities offered by the world of the courts. Like Catherine the Great's personal attendants, noblewomen gave service and companionship to royal and aristocratic patrons. They became friends and unofficial advisers, held a privileged position at their benefactor's court, and gained the tangible favors of lands, money, offices, and titles for themselves and their families. At the end of the seventeenth century Sarah Jennings became the confidante of the future Queen Anne of England and made an advantageous marriage to one of the royal military commanders, John Churchill. Together, she and her husband, enriched and rewarded by their royal patron, established a new aristocratic dynasty as Duchess and Duke of Marlborough.

A skillful and gifted attendant could even win a king. Although her initial introduction to the court of the English monarch Henry VIII came through her father's efforts, Anne Boleyn played the role of courtier with supreme artistry. She used her youth, her beauty, and her courtly skills to excite and please the king. With the promise of her sexual favors and the children she would bear, she turned Henry VIII against his wife and gained marriage and the title it bestowed, Queen of England. In seventeenth- and eighteenth-century France, royal mistresses, like the Marquise de Montespan, achieved positions second only to the members of the royal family. They presided over their own courts and saw their children acknowledged as "royal bastards," with households, titles, and privileges almost equal to those of the king's legitimate offspring.

From the fifteenth to the eighteenth centuries women of less privileged birth also found opportunities for reward and preferment in this world of ceremonies and courtly gatherings. What had been patrician and aristocratic amusements came to be entertainments staged by professionals. Women trained in all the arts. They sang; they performed the music they had composed. They acted in plays and tableaus. They danced in pageants and the entr'actes of operas. They wrote poetry and prose, they painted grand decorative scenes and portraits. In the early fifteenth century the writer and courtier Christine de Pizan gained the patronage of the Dukes of Burgundy and the queen of France with her poetry and didactic writings, and thus supported her family. A Venetian lace designer, Rosalba Carriera, received commissions and earned her living by drawing pastel portraits of royal and aristocratic patrons at the courts throughout eighteenth-century Europe.

Privilege, service, influence, the chance to exercise indirect or direct royal power, all characterized the life open to enterprising women in this new world of the courts. Women, like men, seized the opportunities offered. Women, like men, had to adapt when this world passed away. At the end of the eighteenth century, revolutionary upheavals first in France and then across Europe curtailed this world of privilege and opulence. Royal and aristocratic women and men died on the scaffolds erected during the French Revolution. Enterprising adventurers also were killed. Mme. du Barry, the illegitimate daughter of an innkeeper, the last *maîtresse-en-titre* (official mistress) of Louis XV of France, went to the guillotine in 1793, just a year after the members of the royal family.

Though imperial and aristocratic families maintained their wealth and special status into the twentieth century, the courts over which they presided bore little relation to the world of the absolute monarch. Victoria was Queen of England in the nineteenth century, but governed only on the advice of cabinet ministers responsible to an elected legislature, the House of Commons. Though often portrayed as queen-empress, Victoria was just as likely to be photographed in her more traditional role of wife to her consort, Albert, and mother to their nine children.

Women, whether from privileged families or not, continued to gain favor and position at the courts of the nineteenth century, but the rituals became ever more elaborate and the roles more ceremonial than substantive. Elizabeth, the sixteen-year-old daughter to the King of Bavaria, captured the affection of Emperor Franz Joseph of Austria in 1853. She married him, became his empress, but refused to participate in his exacting life. Instead she spent her days nurturing her legendary beauty and lived most of her life away from court.[2] When Franz Joseph took a mistress, it was a companionable actress who lived separate from the court, who entertained him with morning coffee and gossip about his courtiers. Neither woman played a significant role in the government or influenced imperial policy.

Though an actress might become the emperor's mistress, women in the arts in the nineteenth and early twentieth centuries gained their greatest success outside the world of the courts. They found patrons and ways to earn a living elsewhere. Elizabeth Gaskell, the English novelist, wrote for a popular audience. Berthe Morisot exhibited her paintings in the major art salons of nineteenth-century Paris. Clara Schumann performed on the public concert stages of Europe's capitals and thus supported her family. Singers, dancers, and actresses became celebrities, their performances heralded in newspapers, their favors sought by the wealthy and the privileged.

Although the world of the courts had been eclipsed, Europe's privileged women adapted and accommodated to the changes this signified. They accepted the constraints of constitutions; they found new audiences for their talents. As in centuries past the vast majority found satisfaction in their lives, whatever the limitations.

2

THE LIFE OF THE COURTIER

❦

THE WORLD OF THE COURTS and the life it offered from the fifteenth to the eighteenth centuries emerged because of dramatic political changes in Europe. All power came to revolve around the monarch who in turn created the luxurious world of the palace and the array of opportunities for the enterprising courtier and her family.

This political transformation, this accumulation of authority in the hands of the ruler, came about gradually. By the early fifteenth century in northern Italy and in parts of France, then throughout Europe in the course of the next two centuries, the feudal pattern of government was superseded. Families like the Viscontis of Milan, the Valois of France, and the Hapsburgs of Austria established their hegemony by conquest, centralized their power, and governed by new methods. The old feudal kingdoms of noble fiefs and chartered towns—the alliances of overlords and vassals as in the time of Eleanor of Aquitaine in twelfth-century England and France, the federations of aristocratic and merchant interests as in Margrethe of Denmark's fourteenth-century Scandinavia—all lost authority.

Government of the realm remained the province of the privileged, but not automatically of the old aristocracy or the wealthy, urban patricians. Monarchs ruled through appointed bureaucracies. They maintained professional standing armies. With the passing of feudalism, the nobility no longer gained access to lands and revenues by giving military service. Few among the aristocracy served as warriors and few died violently.[1] In the new world of the courts and the absolute monarch, the members of the aristocracy found other kinds of service asked of them. They no longer hoped for glory in battle, but for other prizes, the rewards of the court: offices, exemptions from taxes, pensions, trading monopolies, lands, and titles.[2]

As in past centuries privileged women adapted to these changes and participated in these developments as members of families. They were the

wives, mothers, and daughters of the royal dynasties, of the landed aristocracy, of the ruling families of the great mercantile empires of northern Italy.
Thus, they functioned in their traditional female roles: bearing children,
supervising staff and the household, acting as surrogates for their husbands.
In addition, however, life in the palaces and châteaux of the well-born and
powerful offered women other roles to fulfill for their families. Where women
had traditionally been barred from direct participation in the warrior's feudal
world of tournaments, drawn swords, and hand-to-hand combat, they were
welcomed from the fifteenth to the eighteenth centuries in the world of the
courts. Families looked to all members to advance and to protect their
interests. Daughters as well as sons might become courtiers and gain favors
for themselves and their kin. For success in this new courtly world went to
the most talented, whether female or male, in a competition among families
as fierce as that in the era of feudal warriors. Only the field of battle and the
skills required had changed.

The competition among families arose in these centuries because so
few—2 percent or less of the population—had access to the new centers of
power. "Les grands," the highest nobility of eighteenth-century Bourbon
France, probably numbered one thousand families. Elsewhere, many might
possess titles, but only a few hundred landed families had any real power. For
instance, this was the situation in eighteenth-century Hungary. Where in
previous centuries families had established principles of inheritance favoring
sons that they believed would guarantee the continuation of privilege and
power over generations, nothing was so certain in the world of the absolute
monarch. New families found favor and rose, old families lost and fell from
favor. Much depended on the whim and the political and financial needs of
the rulers. In the seventeenth century an enterprising king like Louis XIV
of France sold estates and their titles, then revoked them, resold and revoked
them again. Other monarchs did the same.[3]

In these variable circumstances numerous privileged women became
courtiers. For women, too, cultivated the courtier's attributes—affability,
charm, circumspection—and used them as skillfully and successfully as men.
A woman, by services to her patron, by her beauty, by her wit, could earn
a lucrative position at court, make valuable friendships and alliances, and thus
raise her family to extraordinary levels of wealth and prestige. Imagination
and enterprise, not gender, brought success in this opulent new world of
Europe's great courts.

1. Christine de Pizan (1365– c.1430), the French writer and courtier, presents her manuscript to Queen Isabelle of France.

2. *The Court of Louis XV at Play in the Palace of Versailles,* mid-eighteenth century.

3. Anne Boleyn (1507–1536), the English courtier, queen, and mother of Elizabeth I.

4. Lady Arabella Stuart (1575–1615), a distant cousin of Queen Elizabeth I and daughter of the Earl of Lennox, at twenty-three months.

5. A *Women's Literary Gathering*, from a seventeenth-century engraving by Abraham Bosse.

6. A medal of Catherine II (Catherine the Great) of Russia (1729–1796) in 1762.

7. The Empress Maria Theresa of Austria (1717–1780) at age forty-eight with her husband and eleven of their children.

8. Isabella d'Este (1474–1539), the Renaissance model of a learned woman, in a drawing by Leonardo da Vinci.

9. Emilie du Châtelet (1706–1749), Fren scientist, mathematician, and courtier.

10. Marianne Mozart and her family, early 1760s.

The Courtier's Roles and Rewards

THE ROLES

The monarch, the aristocratic patron, and the patrician presided over the courtly world of the fifteenth to the eighteenth centuries. A few women held such power, ruling as regents for male members of their families, inheriting the throne from their fathers. In the gilded and tapestried rooms of the great palaces, however, the majority of elite women and men aspired to many other roles. They were the courtiers, the noble attendants serving a royal, aristocratic, or patrician patron. Though the descriptions of the ideal courtier envision men, the distinction of gender receded into the background in the courts. The role of courtier required qualities and skills that elite women had been trained to acquire for centuries. Early guides to behavior written for those eager to serve the new rulers of Renaissance Italy stress characteristics traditionally attributed to women. The ideal male courtier of Baldassare Castiglione's *Book of the Courtier* had "talent, beauty of countenance, comeliness of person, and that grace which will make him at first sight loveable to all." He was strong, "shapely of limb," supple and graceful in his movements, never exciting envy, "genial and discreet" with women, taking care to guard his reputation.[4]

The function of Castiglione's courtier also mirrored the duties traditionally assigned to women. Like the ideal wife described in the treatises instructing women on how to serve their husbands, the courtier was to spend his life "devoting his every desire and habit of manner to pleasing" his patron. His male courtiers must entertain, be attentive without being intrusive, obey all honorable requests, and when chosen to be an adviser keep the master from doing evil, correct without appearing insubordinate, and suggest without appearing superior.[5]

Just as there were guides to help the aspiring young man, so there were descriptions of what the ideal young woman of the courts should be like. Castiglione in his sixteenth-century treatise included advice for the aspiring female courtier. Like the man she was to be of "gentle birth," to have natural grace, cleverness, prudence, circumspection, and to take care to guard her reputation. She should exhibit "a certain pleasing affability," a "quick vivacity of spirit," and be able "to entertain graciously every kind of man with agreeable and comely conversation."[6]

When Charles Perrault wrote his fairy tales for the French court of the seventeenth century he gave all of these qualities to one of his heroines, the princess, "Sleeping Beauty." She was the "most beautiful person in the world," with the "temper of an angel" who did "everything with wonderful grace." In addition, she had acquired what the French of the seventeenth and eighteenth centuries called the *arts d'agrément*, the "arts of pleasing." She could "dance to perfection," "sing like a nightingale," and "play every kind of music with the utmost skill."[7]

Typically, sixteenth- and seventeenth-century aristocratic families arranged for private instruction for their daughters in music, dancing, and languages. In Italy and France, the little girls might be sent to a convent for this. Catherine de' Medici, the future Queen of France, studied with the nuns in Rome before coming to the court at thirteen after her betrothal.[8] After the minimal schooling at home or in the convent, daughters of aristocratic families often went to the household of another family, ideally one grander and wealthier than their own, where they acquired the instincts, skills, and polish of the successful courtier. Sixteenth-century English families sought out places with the French nobility. Honor, Lady Basset, wife to Viscount Lisle, sent her two daughters to French families. Thomas Boleyn, courtier to King Henry VIII of England, did even better. Both of his little girls, Mary and Anne, entered the service of the queen of France.[9] If the girl learned well, she might be allowed to remain and thus make her future. Catherine Willoughby succeeded admirably. Sent to learn from Mary Brandon, sister to King Henry VIII, she stayed on after her patron's death. At fourteen, she married Brandon's widower and became the new Duchess of Suffolk.

Among royalty the betrothal might dictate a daughter's education. In the sixteenth century Mary Stuart, the heir to the Scottish throne, went as a child to the French court after the arrangements for her marriage to one of the king's sons had been completed. In the eighteenth century Marie Antoinette of Austria, betrothed at thirteen to the dauphin (the royal heir to France), began to study French and dancing with a tutor sent to Vienna by the king.

Guides written for young women advised them on how best to ingratiate themselves with their patron. In 1405 Christine de Pizan dedicated her *Book of Three Virtues* to Marguerite, daughter of the Duke of Burgundy. She described how best to serve the "lady": cultivate the ability to please, advise her against misdeeds, and always exhibit the most extreme loyalty—defend her if she was discovered in adultery, and claim the infant of her illegitimate pregnancy as one's own.[10]

Others gave practical advice on how to succeed at court. In the sixteenth century the Italian Annibal Guasco sent his twelve-year-old daughter Lavinia to join the household of the Count of Savoy. The father made one absolute rule, obedience to the lady. He even wanted the little girl to learn to anticipate her patron's wishes so that she could please her all the more. He worried about other aspects of his daughter's behavior. She should look to her appearance, clean stains from her clothes, stand properly, and never look tired. She was to curb her appetite. She should trust no one and watch her words, especially with men, so as not to endanger her reputation.[11] French seventeenth-century books of manners cautioned women not to pull their skirts up to their thighs when they stood in front of the fire to warm their legs. They could, however, still clean their teeth at the table with a toothpick, and in intimate company kill lice or fleas that they found on themselves.[12] By the eighteenth century, rules elaborated. Before presenting her at court, Emilie du Châtelet's mother instructed her in the eight acceptable ways to eat a soft-boiled egg.[13]

Once admitted to this courtly world, everyone from a well-born noblewoman to a royal princess or a queen played her part in its rituals, whether as servant or as the one to be served. Patrons required the attention of their courtiers both for special occasions and for the most intimate activities of their daily lives. On the one hand, the sixteenth-century christening of the English prince, the future Edward VI, necessitated processions to the chapel of as many as four hundred courtiers.[14] On the other, Marguerite, Queen of Navarre, in her collection of tales, *The Heptameron*, indicated that one of her attendants even slept in the same room with her. For the claims to absolute power to rule required ceremonies and a way of life that made almost everything public.

All monarchs accepted this public aspect of their lives and used it to create the illusion of privilege for their courtiers. It was an honor to march in a royal procession, but the height of status and potential influence was to attend the "royal person." In the fifteenth century John Baker proudly bequeathed to his daughter the title of "grand sergeant to the monarch" with its right to hold the king's head if he were seasick crossing the English Channel.[15] Some courtiers were allowed to view their royal patrons at meals, others were allowed to serve them. The French and the Spanish created the most exacting rituals surrounding this and other aspects of court life. In the sixteenth century, when Philip II of Spain and his queen ate together on certain holy days, her ladies served her from their knees. In the next century, Philip IV and his queen required three attendants and a physician even when they ate alone. Different courtiers had the honor of bringing the wine glass and then the napkin so that they might wipe their lips.[16]

The design of the grand residences and palaces reflected the relationship between patron and attendant: a series of rooms, each more exclusive, each closer to the most private apartments, each signifying more access to favor.

The woman courtier admitted to the "cabinet" of France or the "withdraw-ing" chamber of England, the woman allowed to attend the monarch on waking or on going to bed (the *lever* and *coucher* of the French court), knew that she had achieved the height of status, privilege, and potential power.

Attendance in these hallowed sanctuaries could be exhausting. In the seventeenth century Marie-Anne de la Trémoille, Princesse des Ursins, went as the first attendant of the French wife chosen for King Philip V of Spain by Louis XIV. Maria Louisa of Savoy was just thirteen, and the princess wrote complaining of all she had to do, serving not only the little girl but the young king as well: everything from handing him his slippers to carrying out the chamber pot handed to her by his attendant. As "it would be sacrilege if anyone but myself went into the Queen's room when they were in bed," she had the responsibility for drawing the curtains and being the first to wake the couple in the morning.[17]

The duties might seem trivial, the personalities of the other courtiers distasteful. The eighteenth-century English queen, Charlotte, wife of George III, chose the novelist Fanny Burney for her entourage. Appoint-ment as Second Keeper of the Robes gave Burney an income of £200 a year but meant being the companion of the Senior Keeper, a silly woman who played cards all day and enjoyed humiliating her subordinates. Burney hated the routine and the intrigues among the attendants. She complained in her journal of "the mischiefs of jealousy, narrowness and selfishness" that characterized life in the queen's service.[18]

Even the members of the royal family did not escape service and lived according to the stringent rituals of the court. As a child in Spain, María Teresa, later the wife of King Louis XIV of France, went each day to a great, long hall and inquired of a functionary if her father would grant her an interview. King Philip IV would reply "yes." She approached him, received a ritual kiss, and then left his presence, with not a word spoken between them.[19] Louis XV of France took coffee with his daughter Adelaide every morning; she and her ladies returned the call in the afternoon. On Tuesdays and Thursdays, Caroline, wife of the eighteenth-century English king, George II, required her daughters to walk with her when they were at Hampton Court. On Wednesdays and Saturdays when the king hunted, they accompanied her in the carriage that followed the royal party.[20]

THE REWARDS

Mme. de La Fayette, in her novel of 1678, *The Princess of Clèves*, described the world of patronage and service: "The Court gravitated around ambition," she explained. "Nobody was tranquil or indifferent—everybody was busily trying to better their position by pleasing, by helping, or by hindering somebody else."[21] Women, like men, sought out the uncertainty and the challenge, for the most gifted practitioners of the courtier's arts won prizes beyond measure. The expansion of the monarch's role and the rituals of the court in the seventeenth and eighteenth centuries multiplied the possibilities and opportunities, the rewards of patience, persistence, and affability. The seventeenth-century courtier Mme. de Motteville explained, "The house of kings is like a large market-place where it is necessary to trade for the maintenance of life and for the interests of those to whom we are bound by duty or friendship."[22]

The court of France's Louis XIV set the pattern for the kinds of rewards a woman might aspire to. Already at the age of ten Françoise Bertaut, the daughter of successful courtiers, the future Mme. de Motteville, had an allowance of 600 livres from the royal family. At twenty-one she enjoyed a prestigious marriage to the eighty-year-old president of the royal Chambre des Comptes and had become the confidante of the regent, Anne of Austria, with an allowance now of 2,000 livres.[23] It was the Duchess of Saint Simon, not the Duke (the famous diarist of Louis XIV and Louis XV's courts), who served as a *dame d'honneur* to a member of the royal family. This position permitted her husband's attendance at court, his right to live in the palace of Versailles, and thus the opportunity to write his famous memoirs.

In the next century Emilie du Châtelet used her many attributes and talents to succeed as a courtier. She arrived at the French court in 1722, aged sixteen, attractive, ingenuous, and enthusiastic. By her dress, her manner, and her outrageous actions she appeared, as King Louis XV remarked, to take "joy in all things."[24] When courtiers were betting on who would seduce her first, she used her quick tongue to protect herself. Athletic, and tall for a woman (she was 5 feet 8 inches), she challenged Colonel Jacques de Brun, the commander of the household guard, to a fencing match. She lost but fenced so well that those who had talked of seduction feared they would have to fight her themselves. The king honored her with his praise: "one is never bored in her company."[25] In the end she won a husband (a marquis and general of France), and two rich and prominent courtiers as lovers: the Duke of Richelieu, a commander of the army and adviser to the king, and Voltaire, the famous playwright and man of letters.

The closer to the monarch a courtier came, the more the opportunity for preferment and influence, the more the need for tact and patience. In the eighteenth century, Maria Theresa, the ruler of Austria, had women in attendance all the time. Her ladies rose with her at five in the morning and read to her while her hair was dressed.[26] They accompanied her in all of her daily activities, even walks in the garden. The empress allowed them power in return. Her attendants took charge of her calendar and arranged who might and who might not have an audience. Thus, these courtiers determined who had access to the ruler of the Austrian Empire.

The French court offered wonderful prizes to the skilled courtier. At fifteen, Henriette Genet, daughter to a royal official, became "reader," or *lectrice,* to Louis XV's sisters. When Marie Antoinette arrived to become "dauphine" (wife to the heir), Genet passed easily to the post of *femme de chambre.* A dowry, marriage to M. Campan—a courtier in the service of another member of the royal family—and an income of 15,000 francs a year were her rewards. When Marie Antoinette chose her as confidante, Campan's salary and the perquisites multiplied. Others paid for access to the young queen. Other benefits of living in the unreal opulence of Versailles came to her; for example, all of the candles from the queen's apartments were hers by right—a considerable number, as they were changed every day whether they had been used or not.[27]

In 1744 the eighteenth-century courtier Emilie du Châtelet wrote her *Traité sur le bonheur,* literally, a treatise on happiness. Even she acknowledged that women's lives might be more restricted than men's. But in the world of the court she saw this as no deterrence. Happiness came from good health, from the privileges of wealth and position—the jewels, the clothes, the food and wine—and from study, the setting of goals and working for them.[28] For those women like Emilie du Châtelet who persevered at court and triumphed, there would be days, weeks, months filled with tasks to be performed, patrons to be attended, but also rewards to be won: lands, titles, an advantageous marriage—the prizes of the courtier's world.

The Court Setting, Costumes, and Activities

THE SETTING

The eighteenth-century noblewomen who attended Russia's Empress Catherine the Great, England's Queen Anne, or France's Marie Antoinette had earned special privileges and rewards for themselves and their families. In addition, those congenial and aristocratic women who became the personal attendants to their respective patrons enjoyed additional status and opportunity. They might help the empress dress, pour the queen's wine at the midday meal, or read to the monarch in the afternoon when she retired from the formal gathering of the court. These women had become the favored courtiers admitted to the empress's bedchamber, to the queen's private apartments. The less-favored had to be content with an audience in one of the great public halls where elite families jostled each other for a glimpse of the patron or ruler.

Such distinctions between members of the court would have been of little consequence in earlier centuries. In the twelfth century Eleanor of Aquitaine's and her daughter Marie of Champagne's retainers and attendants assembled in the great hall of a castle with no walls to divide public from private space. Most noblewomen slept in an enclosed bed in the same great hall where they received their vassals and ate their meals. As early as the fifteenth century in Renaissance Italy and in northern France the courts of the elite began to change. By the eighteenth century the palatial residences of the privileged had a multiplicity of rooms, many designated for special uses: bedchambers, studies and libraries, great audience halls, more intimate salons. Patrician, royal, and aristocratic rulers commanded new power and wealth. They used both to create settings for themselves which—like Versailles, Schönbrunn, and Tsarskoye Selo—not only gave them privacy and comfort but also mirrored their importance.

These grand new settings could never have been built without significant changes in the way the powerful fought their wars. The gradual shift of battles, first to the towns and then away from settled areas altogether, meant that the great house, the château, the palace, no longer had to be a fortress. Instead of trenches and moats, the lands surrounding the Hôtel de Saint Pol, the residence for one of the French kings of the early fifteenth century, had orchards, gardens (maintained by a woman gardener, Jeanne La Bouchère), an aviary with doves and nightingales, and a menagerie of ten lions.[29] All across Europe the elite remodeled and expanded old castles and manors, and built new ones. They summoned gardeners to landscape the grounds, architects to design the buildings, and craftsmen to embellish the interiors. By the eighteenth century the residences of royal and aristocratic families had become declarations of the taste, the sophistication, and the opulence of the courts and the rulers they housed. Empress Maria Theresa brought special craftsmen to Austria. She hired a Dutchman to design the surrounding parks and an Italian for the buildings when she remodeled her palace of Schönbrunn. All testified to her wealth and authority: the decorative exterior limestone, painted in pale yellow and pink; grand outside staircases, and glass windows across the entire facade; carved niches for life-size allegorical statues; 1,441 rooms for every kind of court activity.[30]

The rooms in such a vast palace ranged from public areas like Henry VIII of England's Great Hall at Hampton Court—where his courtiers watched him eat seated on a raised dais—to the most private, like his daughter Queen Elizabeth's Privy Chamber where she sat in the only chair with a few of her ladies in attendance. As the number of rooms proliferated, the uses became more specialized. Anne of Austria, the seventeenth-century Queen of France, had her own apartments, a long set of connecting rooms, each with their own function—from the "great cabinet" and the salon, used for audiences and gala balls, through to the bedroom where her family gathered, to her boudoir where she dressed and bathed. In addition, she had a special oratory built for her collection of holy relics.[31]

Royal and aristocratic families came to favor one or two palaces. They no longer moved from dwelling to dwelling taking their possessions with them. Louise of Savoy, mother to a future King of France, moved her household of about three hundred people to Cambrai in 1529. She used twelve carts for her clothes, furniture, paintings, and the kitchen utensils.[32] This changed during subsequent centuries as all thought of practicality and mobility gave way to luxury, display, and comfort. Monarchs hired artisans to create ever more sumptuous interiors: white stucco and gilt around the moldings of painted and plastered walls, mirrors, tapestries, hand-painted wallpapers from China, woven silks and brocades to set off large glassed windows, inlaid floors in intricate geometric patterns of differently colored woods. Catherine the Great's bedroom at Tsarskoye Selo had thin faience columns divided in three sections colored in a pattern of green and white with gold trim. In another room all of the decoration and furniture was silver plated.[33] Chandeliers and sconces held dozens of candles that regularly made day out of night so that the court's activities could continue into the small hours of the morning.

Furniture filled the rooms: each piece with one specific function, unlike the simpler, multipurpose pieces of earlier eras. There were armchairs, wing chairs, tapestried benches, upholstered sofas covered in satin with woven stripes or delicately embroidered flowers in repeating patterns. The simple chest became a "chest of drawers"; the bureau, the secretary—each had its special use. Everything was decorated with the greatest care and artistry. For example, Anne of Austria's seventeenth-century boudoir had blue enameled furniture. Aristocratic households boasted similar luxuries, like a large chest of drawers with brass handles in the shape of vines and a marble top with an inlaid design of playing cards strewn across the shiny surface as if the players had left for another amusement. Shelves in cabinets, "niches" in walls, and the tops of tables displayed other possessions: porcelain statuettes of dancing characters from the Commedia dell'Arte like Pierrot and Colombine; enameled copper or porcelain snuffboxes painted with miniatures of pastoral scenes or mottos like "love and live happy"; a gold bell to ring for assistance at the toilette. Whole manufacturing centers like Sèvres, Gobelin, and Beauvais grew up to supply the courts with the luxuries their inhabitants demanded.

Of all the royal palaces King Louis XIV of France's new residence, Versailles, under construction for decades during the seventeenth century, magnificently demonstrated the unlimited wealth and authority of a dynastic monarchy. Cultivated grounds stretched for acres; the Grand Canal, big enough to accommodate a sailing ship for the court entertainments, was almost a mile long. In the next century Mme. de Pompadour, Louis XV's mistress, and Marie Antoinette, Louis XVI's queen, made extensive changes and additions to the design of the gardens and parks. Mme. de Pompadour had one garden designed just to indulge the sense of smell. Each day the gardeners changed the plants to insure their freshness and the purity of their scents.

Everything about Versailles—the façade with its gilded balconies, row on row of windows, statuettes along the parapets, the hundreds of rooms— expressed grandeur, opulence, and privilege. The Hall of Mirrors, a long rectangular gallery, had an elaborately painted ceiling, carved and gilded decoration everywhere, marble facing on the walls. All was made even more impressive by the light from the multipaned glass windows along one wall and the reflections from the mirrors along the other. At night three thousand candles created endless reflections in the mirrors. The scale of Versailles had the desired effect and awed contemporaries. Other rulers copied its design and tried to create the same impression. They too hoped that a grand, extravagant setting would signify their power, their wealth, and the luxurious life they could command.

THE COSTUMES

Just as the setting for an aristocratic or royal woman changed dramatically from the simple and practical to the elaborate and luxurious, so too did the clothing she wore. Eleanor of Aquitaine, the late-twelfth-century queen (first of France and then of England), is depicted in her effigy dressed in a long, belted tunic with a cloak around her shoulders and draped across her legs. The effigy of her son Richard I shows him in clothes that are virtually the same. An early-fourteenth-century manuscript shows the wife of the Margrave of Brandenburg playing chess with her husband. Their dress is identical except for the difference in the neckline of their overtunics and the sleeves. This common style of clothing for women and men began to disappear first in the late-fourteenth-century courts of northern Italy and France and then gradually in the rest of Europe.

Designs first differentiated the male form, then the female. From the middle of the fourteenth century, men's clothes revealed their legs, accentuated the breadth of their shoulders, exaggerated the size of their chests with padded jackets, and the size of their genitals with "codpieces." Women's clothes began to outline and draw attention to the shape of their bodies as well. The manuscript illustrations for the fifteenth-century book of hours of the Duke of Berry show women in fitted undertunics with buttons down the side to pull the fabric tight across the torso and bust. The bodices are cut low, the sleeves fitted to the wrists.

Then in the fifteenth century the clothing of women and men of the courts rapidly displayed a sumptuousness and a grandeur unknown in Europe before. Margaret of Flanders, wife to the fifteenth-century Duke of Burgundy, Philip the Good, wore a dress of white velvet decorated with gold stars and a crimson silk cloak for their son's wedding. Her husband wore green and white silk with gold decoration. Both had other robes, including sets of gem-embroidered sleeves made of satin, velvet, and cloth of gold.[34]

From as early as the sixteenth century, courtiers and royalty began to accumulate whole wardrobes of clothes. They reserved the heavy silk brocades in silver, gold, and other metallic thread for ceremonial court functions. These costumes became more cumbersome and restrictive to wear. In 1698 Liselotte, King Louis XIV's sister-in-law, described her daughter's trousseau of fifteen dresses. She feared that the two designed for court occasions were too heavy. One in black velvet had a twelve-inch border of embroidered garlands in gold thread and an underskirt of cloth of silver oversewn in two different gold-colored threads.[35]

It was the Spanish fashion to have the skirt just long enough to make it appear as if a woman simply flowed across the floor. To create such impressions and to give exaggerated definition to the female form, court dress from the sixteenth century on included undergarments and apparatus that restricted women's movements even more than the weight of the decorated fabrics. There was a corset (sometimes made of wood splints covered in linen), called a "busc" in England, a *corps pique* in France. The "stomacher" flattened both the bust and the abdomen. The "farthingale," a roll of cloth tied around the waist under the skirt, made the hips appear larger, and thus accentuated the slimness of the waist. Frames and other metal devices gave unnatural shapes to the skirt and thus to the woman's body. The hook and eye, together with lacing, made the bodice and sleeves even more fitted and thus tighter. The French noblewoman Mme. d'Aulnoy, given a court appointment to King Carlos II of Spain in the seventeenth century, described the plight of the appropriately clad courtier:

I cannot imagine a more uncomfortable dress, your shoulders have to be so tightly bound that they hurt, you can't raise your arm, and you can hardly get it into a sleeve. I was put into an enormous bulky *guardingant.* . . . I didn't know what to do with this strange machine. You can't sit down.[36]

Such discomfort, however, meant little when compared to the life, the luxuries, and the rewards available to the successful courtier.

By the eighteenth century dressing for the court, or the "toilette" as it was called in France, could take as long as four hours. For courtiers, whether female or male, changing into the appropriate clothing had become an activity in and of itself. A young woman in attendance at Versailles in the eighteenth century had dozens of dresses, all designed for specific times of the day, all signifying the luxurious life and decorative role of the wearer. The French artist François Boucher painted such a courtier, young Mme. de Bergeret, in a white satin day dress, her skirt billowing with the yards of material. Other women would have "walking" or "morning" dresses in pale colors, the bodice decorated with braid, with ribbons, or with ruffles down the front. Linen and soft cotton dresses had small embroidered or delicately hand-painted flowers, vines, and butterflies, patterns copied from Chinese porcelain designs. For an afternoon ball, courtiers wore a fitted gown with extra fabric down the back making a short train, a robe *à la francaise.* In colder weather courtiers wore patterned silk and quilted velvet with the skirt and sleeves trimmed in fur.

Women of the courts also had a whole wardrobe of accessories. Mary Stuart, the sixteenth-century Queen of Scotland, had Italian pattern books and designs by two French craftsmen that she consulted when embroidering accessories like the soft, tightly fitted kid gloves she wore. Louis XIV's niece went to her Spanish marriage with lace-trimmed linens and chests full of ribbons and jewels. An eighteenth-century courtier like Mme. de Bergeret had painted silk fans to cool herself in the summer, muffs to warm her hands in the winter. She would also have had handkerchiefs, laces, belts, small decorated purses, tricornered hats for riding, and square-heeled shoes covered in brocade and satin. The French artist Boucher painted one young courtier sitting by the fire tying her garter while her maid shows her a lace-trimmed cap that she might wear for the morning.[37]

Most women of the courts, including the queens and the princesses, reveled in the excitement of dressing magnificently and in the spoken and unspoken competition of fashion. Anne of Denmark (James I of England's wife) and Anne of Austria (Louis XIV of France's mother) wore their predecessors' refurbished dresses, but most queens preferred to set new fashions.[38] In the 1670s Mme. de Montespan, Louis XIV of France's mistress, continuously pregnant by the king, made a new style of a draped bodice, with the back, not the front, of the dress fitted so as to hide the swelling of her body into the last months. Her successor as the king's favorite, Mme. de Maintenon, prided herself on her subdued taste and made a fashion of darker colors, covered shoulders, and high-necked bodices.[39] In the eighteenth century Marie Antoinette, queen to King Louis XVI, made softer fabrics and lines fashionable. The artist Elizabeth Vigée-Lebrun painted the queen in a gauze-like white dress, all loosely gathered with short ruffles at the sleeves and across the lowered neckline.[40]

Queen Catherine de' Medici of France is credited with introducing to the rest of Europe's courts the Italian fashion in cosmetics, another time-consuming aspect of the toilette. In Italy at the end of the fifteenth century the ideal was a high forehead, the palest eyebrows, white skin, and reddish blond hair. To create this illusion women plucked their foreheads and eyebrows, used bleaches and dyes to change the color of their hair, and added falls to give fullness. They whitened their skin with powders made of lead, reddened their cheeks with abrasives, and rouged their lips. They could find suggestions and formulas for cosmetic preparations in the treatise by Trotula, the twelfth-century woman doctor of Salerno, and in the fifteenth-century manuscript *Experimenti* by Caterina Sforza, one of the rulers of northern Italy. At the turn of the eighteenth century the courts of France and England created the fashion of adding patches to their makeup: a black spot, a star, a half-moon discreetly placed to draw attention to a cheek, to the breast. Russian noblewomen outlined their eyes in black and colored their lids in vivid blues.[41]

The finishing touch to a courtier's costume was a headdress in the earlier centuries, the arrangement of the hair or a wig in the seventeenth and eighteenth centuries. The "coif" ranged from the simple pearl-rimmed net worn at the court of Milan in late-fifteenth-century Italy to the conical hats (hennin) of fifteenth-century Burgundy, from the starched linen caps of sixteenth-century France to the elaborate squared-arch that framed the face at the sixteenth-century English and Spanish courts. Wigs could be even more complicated, involving wire frames, falls of curls, feathers, jewels, and even artificial creations or ornaments like stars, birds, and sailing ships.

The eighteenth-century French courtier Emilie du Châtelet enjoyed putting on makeup and dressing for the court. She wrote in a treatise in 1774 that jewelry was one of the great sources of happiness in her life. In the same treatise she wrote of other aspects of this courtly world. She described the pleasure she took in the vast array of activities planned to keep the courtiers and their patron amused and entertained: evenings of cards and dancing, week-long celebrations with hired performers and fireworks.

THE ACTIVITIES

Carefully dressed and coiffed to draw attention to their beauty, the young women admitted to the world of the courts found a never-ending program of events and celebrations. In the seventeenth century Mme. de Sévigné spoke of Versailles as if it were another country—"ce bons pays-là"—that she visited at least one evening a week.[42] From the fifteenth to the eighteenth centuries the entertainments became more elaborate. Courtiers played cards, danced, sang, participated in comedies, and watched others perform. In the sixteenth century Mary Tudor, as a princess at her father King Henry VIII's court, loved to act and dance in the "masques," an entertainment imported from the Italian courts in which the courtiers wore masks as they performed. When done at Versailles the theatrical entertainments had scenery, themes for the costumes of the guests, and might last as long as eight hours. Louis XIV of France created the model for the great ball with a master of ceremonies and the ritual partnering of the participants by rank. In the time of Louis XV and his mistress, Mme. de Pompadour, the "First Gentlemen of the Bedchamber" supervised the court entertainments, called "Les Menus Plaisirs." They included theater twice a week, with ballets, balls, and fireworks for special occasions. Pompadour planned events for the king as well. She had a small private theater built at Versailles over which she presided. She assigned the parts, set the rehearsals, and supervised the performances.[43] An eighteenth-century masked ball at one of the Austrian imperial palaces, the Belvedere, had seven thousand courtiers in attendance.[44]

Most elaborate of all were the succession of fêtes and banquets organized to commemorate or celebrate some event of significance to the reigning monarch. The feasting and entertainments went on for days, sometimes weeks. The Italian courts of the great Renaissance families and churchmen initially set the standard and the scale. The fifteenth-century court of Burgundy added the grandeur of chivalric themes. For the 1454 Feast of the Pheasant, the duke, Philip the Good, called for tournaments as part of the festivities, with challengers, judges, and pavilions for the spectators. Other courts carried on the traditions.

In the sixteenth century Queen Elizabeth I of England watched her courtiers jousting in the tiltyard at the palaces of Westminster or Whitehall each year as part of the celebrations on the anniversary day of her accession. Edward de Vere, Earl of Oxford and one of the queen's favorites, first came to her attention at the tournament of 1571. In 1585 she authorized over two months of pageants and banquets to entertain the French prince, the Duke of Alençon, and his entourage when he came to court her and to negotiate a marriage treaty.[45]

Such magnificence became commonplace at Louis XIV's and Louis XV's Versailles in the seventeenth and eighteenth centuries. One fête, "Les Plaisirs de l'île enchantée," was a week-long extravaganza for eight hundred guests using the palace, the gardens, and the canals and featured gondolas, fireworks, plays, and ballets.[46] Not to be outdone, the rulers of principalities and duchies organized similar events; at the beginning of the eighteenth century the Elector of Saxony staged the "Diversions of the Planets" with operas, fireworks, and a water pageant of Jason fighting dragons. At the Austrian court, "Fasching" (Lent) was a time of special celebration. One winter in the 1780s, seven hundred cartloads of snow were brought to the capital so that thirty carved and decorated sleighs could race with their noble passengers through the streets of Vienna.[47]

In the days between the great entertainments, in the hours given to ritual duties to the royal or aristocratic patron, courtiers spent their time in a variety of ways. Members of Marguerite of Navarre's sixteenth-century learned court attended mass and spent the day reading. Women courtiers also read romances and travel accounts. They amused themselves with conversation, with riddles and word games. At the end of the seventeenth century, Countess Marie-Catherine Le Jumel de Barneville D'Aulnoy made up stories to amuse the ladies that were published as a collection of fairy tales.[48]

Aristocratic women wrote letters to friends and family. In the eighteenth century Emilie du Châtelet commonly sent twenty pages a week to her friend and former lover, the Duke de Richelieu, and he was just one of her correspondents. Mme. de Sévigné, the seventeenth-century French courtier, wrote to her daughter every day, page after page, as if they were sitting talking together in her boudoir. She sent news of deaths, marriages, hairstyles, wars. She reported what the king, Louis XIV, said, what he did. She told the weather. She described the scandal of La Grande Mademoiselle and her lover Lauzun, of the death of Vatel, the Prince de Condé's majordomo, who impaled himself on his sword when not enough meat had been served at an official dinner and the fish for the next day's meal had not arrived. She loved mocking the latest whims of the court, like the rise and fall of chocolate:

> And now, my dear child, I must tell you that chocolate no longer holds the place in my esteem that it used to do; fashion has influenced me, as it always does; those who used to praise chocolate, now speak ill of it, revile it, and accuse it of all the disorders to which we are subject.[49]

Members of the seventeenth- and eighteenth-century French and English courts also spent time entertaining in the grand houses that they owned or leased in the capital. Meals served by the great aristocratic families were sumptuous. All the food came at the same time, with the main roast at the center of the table and the other dishes—fowl, salads, meats, pâtés—placed around so that the guests literally ate their way to the center, selecting from the nearest dishes first. Eight courses was common; it might take as long as six days to wash all of the china and tableware needed.[50] In her memoirs the eighteenth-century French artist Elizabeth Vigée-Lebrun remembered that the Duchess of Mazarin had planned a special dish for one supper party, a pie with one hundred live birds inside, that was to be the highlight of the evening. Everything went awry, however, when the birds panicked and flew into the faces and hair of the sixty guests.[51]

Sixteenth-century Venetian and French artists depicted the nobility gathered in the shade of trees in the countryside. The French writer Christine de Pizan remembered the pleasure of a day in 1400 at Poissy, the convent where her daughter was a nun. There had been a drive out, the stroll in the grounds, all a rest from the routine and the pressure of the court. In the sixteenth century King Francis I had elaborate hunts staged for as many as three or four thousand attendants, with the participants sleeping in tents. Craftsmen of all varieties serviced their needs. A more modest outing with Maria Theresa, the eighteenth-century ruler of Austria, meant beaters driving the animals to the hunters and then an elaborate picnic with chairs, wicker baskets of food, linen and silver, and a staff of servants to see to the guests.[52]

Queens and kings also liked to travel into the countryside to be honored and entertained. In the summer months Elizabeth I made her "progress" through the English countryside. She went in an open litter, in a procession of as many as five hundred courtiers. She enjoyed taking a leisured pace, ten or twelve miles a day, and stopped to talk to her subjects and to receive their tributes.[53] Special favorites arranged grand visits for their monarchs. In 1575 Queen Elizabeth spent three weeks with the Earl of Leicester at his estate of Kenilworth. He had a special tapestry woven for her room, the house enlarged to accommodate her retinue, and a pond with three islands created as a setting for the entertainments.[54] In the eighteenth century Prince Grigori Potemkin, adviser and former lover to Empress Catherine the Great of Russia, staged a grand birthday party for her and three thousand guests at the Tauride Palace, the country estate she had given him.[55]

Most courtiers owned lands in the country. As early as the fifteenth century the great Italian families like the Medicis retired to their villas in the

summer for a change of scene. By the 1590s the hard wooden traveling carts
and long rides on horseback that had characterized trips over long distances
had been replaced by private coaches with cushioned interiors and springs to
ease the battering from the uneven roads and tracks. Aristocratic families
visited each other. The seventeenth-century Duchess of Hamilton organized
hunting, hawking, and music to entertain her guests. Others, like the eighteenth-century Duchess of Maine, at her Château de Sceaux, staged grander
events: processions, fêtes, and elaborate entertainments rivaling those of the
royal court.

For some the migrations to the country estate came to be a welcome
change from the arduous duties, rituals, and intrigues of royal service. At first,
Anne-Marie-Louise d'Orléans, Louis XIV's cousin (known to the court as La
Grande Mademoiselle), exiled in 1653, "found it hard to conceive what kind
of mental exercise anyone accustomed to life at court, and born to the rank
I occupy could possibly engage in when obliged to live in the country."[56]
Soon, however, she came to enjoy the time at Saint Fargeau: reading, doing
needlework, seeing friends, and riding in the countryside. She supervised the
rebuilding of the château and the plantings for the gardens. She wrote her
memoirs, employed her own printer, brought in a troupe of actors. The time
away from Versailles—an exile of four and a half years—gave her a new
perspective. She explained to a friend:

> If bored with court, you can go to the country, to your estates, where you hold
> court. You can build; you can enjoy yourself. In short, when you are your own
> mistress, you are happy, because you do what you like.[57]

Mme. de Sévigné, the seventeenth-century French courtier, had the same
view of the time spent at her country estate, Les Rochers, in Brittany. Over
the door she had a quotation from the sixteenth-century French author
Rabelais: "Saint Liberty, or do what you like."[58] Widowed early, she had
come to think of herself as without a "lord and Master," as she explained to
a friend. Les Rochers gave her "perfect freedom." She could walk alone, or
have the occasional company.[59] She went to mass once a day and spent
money on the gardens, but otherwise she did not concern herself with the
estate or its inhabitants. When the visitors bored her she did her needlework.
She had time to read: books in French, Italian, Spanish, and Latin—the
works of the twelfth-century Byzantine historian Anna Comnena, the
Aeneid, religious books from the writers at Port Royal, others by Bossuet,
Montaigne, Corneille, and the Italian poet Tasso.[60]

Despite the real or imagined pleasures of country life, both La Grande
Mademoiselle and Mme. de Sévigné returned to the royal court, to its dis-
comforts, uncertainties, and restrictions. Neither would have given up the

excitement, the activities, and the opportunities. The vast marbled rooms of the palaces might be drafty and cold. Ceremonial dress might be expensive and difficult to wear, the queens and kings exacting and exasperating to serve. Still, this world of the courts was the center of authority, the source of honor and wealth. Women and men jealously guarded their access to its confines and the privilege of competing for its perquisites.

3

THE TRADITIONAL LIFE
IN A GRAND SETTING:
WIFE AND QUEEN CONSORT

❦

THOUGH A WOMAN of aristocratic or royal birth might have unusual opportunities in the courtly world of the fifteenth to the eighteenth centuries, like
her predecessors among Europe's elite she gained access to this world through
her family and was defined by her relationship to its men.[1] Even in the
elegantly frescoed and mirrored halls of aristocratic palazzos and royal palaces
privileged women and men assumed that their daughters would not only
fulfill their role as a courtier, patron, or ruler, but also as a daughter, wife,
and mother. Maria Theresa, heiress to the Hapsburg domain, married according to her father's wishes and dutifully presented her husband, Francis of
Lorraine, with numerous sons and daughters.

The Ideal

Just as families schooled their daughters in the arts and rituals of the
court, so they prepared them for their more traditional role and function. The
same authors and the same didactic treatises that pictured the young woman
courtier also presented an image of the ideal wife presiding over her assembled attendants or serving her husband and children. At the beginning of the
sixteenth century Baldassare Castiglione in his *Book of the Courtier* described
Elisabetta Gonzaga, the wife of his patron, the Duke of Mantua, as wise,
gracious, and congenial, and admired "that modesty and grandeur which
ruled over all the acts, words, and gestures of the Duchess." He believed her
to be an inspiration to her attendants both female and male, for "each one
strove to imitate her style, deriving, as it were, a rule of fine manners from
the presence of so great and virtuous a lady."[2] As *The Book of the Courtier*
went into edition after edition—forty in the sixteenth century alone—and
was translated into all the major European languages, the Duchess of Mantua
became as popular and desirable a model as that of the male courtier.[3] The

qualities valued in the royal or aristocratic wife did not change. Fanny Burney, one of the courtiers in attendance to Queen Charlotte, wife of England's George III, wrote in her diaries for 1785 of her royal patron. She was a paragon, "a most charming woman,"

> full of sense and graciousness, mingled with delicacy of mind and liveliness of temper. . . . Her manners have an easy dignity with a most engaging simplicity, and she has all that fine high breeding which the mind, not the station, gives, of carefully avoiding to distress those who converse with her, or studiously removing the embarrassment she cannot prevent.[4]

In their treatises the early-sixteenth-century Humanists like Sir Thomas More, Juan Luis Vives, and Desiderius Erasmus filled out every other detail of the image, describing the more private life of the well-born woman with her family. They delineated the wife's duties, the relationship with her husband, and the behavior that brought honor or dishonor to herself and the family she had joined. The didactic treatises made no real distinctions according to rank; the ideal image and the prescriptions about appropriate behavior held true for the wife of a minor court official as much as for the consort of a king.

To please her husband—her "lion," as Desiderius Erasmus, the Dutch Humanist, called her spouse—was the new bride's principal function. She accomplished this by the careful supervision of the household. Classical writers like Aristotle and Xenophon gave authority to the traditional descriptions of the wife's duties. Although the husband made all the major decisions as in previous ages, the wife had responsibility for their execution. She kept the accounts, she instructed the servants. To better supervise her staff, some suggested she also know how to perform their duties, such as cooking, sewing, spinning, and weaving.[5] The Spaniard Juan Luis Vives, in his *Instruction of a Christian Woman*, written at the suggestion of England's Queen Catherine of Aragon as a guide for the education of her daughter Mary Tudor, believed that such tasks performed by a rich woman would help her "to avoid idleness." With the ideal wife of the Old Testament Book of Proverbs in mind, he explained that "a woman well brought up" in this way "is fruitful and profitable unto her husband, for so shall his house be wisely governed."[6]

If pleasing her husband was her principal function, to accept his authority in all matters was her principal duty. Though Sir Thomas More, the English Humanist, advocated choosing a wife who might "be learned if possible or at least capable of being made so," he did not mean to imply equality in the family or in the marriage.[7] Rather, More believed that education made wives better companions to their husbands and better teachers of their children. Both More and Erasmus saw instruction of a young wife as a way of inculcating the proper attitude toward her husband. Erasmus imagined the response of such a favored young bride:

> O happy me who fell to such a husband. What a beast would I have been unless he had instructed me.[8]

Advocating that the wife accept the husband's tutelage led easily to descriptions of the wife's acceptance of his will in all things. From the reading of the Church Fathers and other religious writings, Vives expected her to learn "to love and honour her husband, whom she should take as a divine and holy thing, and obey his will as the law of God."[9] As it had in other ages, "love" meant a joining in which the wife lost her separate identity and any independent authority. The tradition continued into the seventeenth century when Paul Scarron, husband to the future Mme. de Maintenon, companion of Louis XIV of France, instructed women to "yield and accommodate themselves to the ways and moods of their husbands, like a good mirror which faithfully reflects the face, having no desire, love or thought of its own."[10]

The learned men of these centuries believed that wives reflected their husbands in another way. Ancient attitudes about women were reaffirmed in the new voices of the Humanists. Wives, by their actions, and in particular by their sexual behavior, brought honor or dishonor to the man and his family. Vives, the sixteenth-century Spanish churchman, presented the same narrow view of women reflected in the customs of the Greeks, Romans, Hebrews, and Germanic peoples. He described how females could maintain their reputation:

> As for a woman she hath no charge to see to, but her honesty and chastity. Wherefore when she is reformed of that, she is sufficiently appointed.[11]

In addition, he explained that the woman's body belonged to her husband. Even the suggestion of infidelity meant "such utter opprobrium and shame that any woman of whom ill is once spoken is disgraced forever, whether what is said be calumny or not," wrote the sixteenth-century Italian courtier Castiglione, echoing views commonplace in ancient times.[12]

In these writings the traditional differing standards of sexual behavior for women and men were restated. Castiglione repeated the distinction, excusing men: "we ourselves have set a rule that a dissolute life in us is not a vice, or fault, or disgrace."[13] The sixteenth-century French essayist Michel de Montaigne wrote with a veneer of courtly nonchalance, but presented the same attitude:

> There is hardly one among us who does not fear more the shame which comes to him from the vices of his wife than from his own; . . . who would not prefer

to be a thief and a blasphemer and that his wife were a murderer and heretic, than that she should be less chaste than her husband.[14]

To guard her reputation Castiglione wanted the wife to be careful of every aspect of her demeanor: to be pleasing but not urbane, to blush at inappropriate words and phrases, to eschew vigorous dancing or loud singing, and to characterize her whole behavior by a "gentle delicacy."[15]

Learned men expected this dutiful, chaste, and obedient behavior to continue after the husband's death. A widow maintained her reputation by her modesty, her chastity, and by her acceptance of the authority of others. From the fourteenth century on, male court writers like the English poet Geoffrey Chaucer and the Spaniard Juan Ruiz drew on traditional images like those of the Roman poet Ovid and mocked any older woman whose behavior was not circumspect as the wanton widow or the old hag.[16] Even as a widow a woman was to conform to the traditional male ideal of the female.

Education

With obedience and chastity the desired qualities in a wife, the young girl's education focused on suppressing what male scholars believed to be the negative aspects of her nature. In this, Humanists sharply differentiated girls' and boys' schooling. For while fifteenth- and sixteenth-century Humanists designed an education for boys and men that heightened their faculties, trained their reason, and thus expanded their expectations for themselves and their lives, they advocated the opposite for girls. Revolutionary in their views of the education of the male, when it came to women the Humanists reinforced the most traditional negative attitudes about the female nature and the way women should be treated. In every way the learned men of this new age intended to restrain and limit, not to encourage or expand, the young girl's or the woman's talents and the possibilities for her life. The traditional premises of secular and religious leaders from earlier ages found new advocates in men like the sixteenth-century Spaniard Juan Luis Vives, and in the seventeenth-century French churchman and educator François Fénelon. Even in the young girl, these learned men imagined the negative qualities delineated by classical and early Christian writers: the desire to dominate, a tendency to anger, pride, and idleness, and a propensity to sin and lustfulness.[17]

To instill the proper moral values, to keep her natural proclivities under control, and to accustom the young girl to circumspection and obedience, Humanists advised restricting every aspect of her behavior. Vives wanted the girl to eat bland food, to drink diluted wine, to wear coarse, simple clothing, and to sleep in a hard bed. Her mother must be diligent and strict, her father severe lest his caresses make her too familiar with other men. "The daughter should be handled without any cherishing," he explained. "For cherishing marreth sons, but it utterly destroys daughters."[18] Humanists also advocated restricting the girl's intellectual training. They prescribed specific texts and forbade others. Erasmus and Vives believed that the female could read the Epistles of Paul from the Bible, the Church Fathers like Jerome and Augustine, and carefully selected classical authors like Plato, Cicero, and Seneca. They favored exemplary tales of obedience and chastity like those of the classical heroines Penelope, Virginia, and Lucretia. The choice of reading was designed to avoid awakening the characteristics these men feared and to instill accommodation and reconciliation to the carefully defined future role.[19]

Women of the courts accepted these premises and this ideal. In the seventeenth century Mme. de Maintenon, companion to Louis XIV of France, came to personify the image of the chaste, accepting woman—active in both the public world of the court and the private world of the family. At St. Cyr, the school that she and the king founded in 1684, she hoped to educate poor but aristocratic young women to the virtues she tried to live by. As models of industry, duty, and piety, dowered by the king and suitably married, they would, she believed, counteract the degeneracy and idleness of Versailles. Like the men who had written before her, she made modesty and obedience the woman's principal virtues and extolled the ideal of the quiet, accepting, dependent wife. "Accustom yourself to the humor of others," she explained to the students at St. Cyr. A wife is subject to the whim of the husband, "not even in charge of closing a window." "Learn to obey," went one of her maxims, "for you will obey forever."[20] The school, with its limited curriculum and the moralistic writings of Mme. de Maintenon's principal adviser, Bishop Fénelon, was admired and copied throughout Europe. The eighteenth-century Russian empress, Catherine the Great, modeled the Smolny Institute after St. Cyr. Similar schools were established in Germany, Poland, Switzerland, and Denmark.

The traditional pronouncements and admonitions about the nature and education of young women survived in this extravagant world of queens, kings, and courtiers: the female by nature evil; the female of necessity consigned to the protective and subordinating care of her father or her husband. The preferred life for a young woman, even of the highest rank, was still defined in relation to the men of her family as the exemplary virtuous daughter, wife, or mother.

Marriage Arrangements

Just as traditions about the image and the behavior of the ideal wife survived into the new courtly world, so did the traditions about the reasons for marriage and the ways in which it should be arranged. As they had in previous centuries, royal and aristocratic mothers and fathers chose the marriage partners and set the terms of the union. Negotiated exchanges of property and money facilitated the young couple's life together and guaranteed the daughter's future. As in past centuries a young woman, by marrying, passed from the protection and care of her parents to the protection and care of her husband and his family. The"dowry" her parents contributed was matched by "a widow's portion," or "dower," a set amount of income to be given to her should her husband predecease her.

Marriage remained key because it meant the union of two families for mutual advantage. For the privileged, marriage was a means to an end. In the scramble for favors and influence that characterized the world of the courts, the exchange of money and lands, the alliance of two lineages with their political and social connections, their titles and bloodlines, could mean fortunes renewed, preferment, or increased privilege for one or both families.

In the eighteenth century, for instance, Emilie de Breteuil, the future Mme. du Châtelet, restored the family position by her marriage. Her father had been principal secretary and Chief of Protocol to King Louis XIV. He lost everything with the king's death. The daughter gained the attention of a marquis from an old Lorraine family. When she became his wife, her family once again had access to the court. Emilie du Châtelet in turn arranged for her own daughter to marry a duke and thus raised the family to an even more prestigious rank.

Often marriages among the elite and with royalty meant the exchange of money for lineage. Usually the groom had the requisite number of aristocratic forebears, called "quarterings," that differentiated one elite family from another and gave privileges like exemptions from taxes and access to offices reserved to the nobility.[21] In seventeenth-century France Marie de Rabutin-Chantal, the future Marquise de Sévigné, came from a family that made this kind of match. Her father had the title; her mother, the daughter of a tax collector, had the wealth. The marquise negotiated a similar marriage for her daughter: a dowry of 300,000 livres in exchange for a noble widower, the Comte de Grignan.[22] With enough wealth a family could aspire to marry royalty. When Catherine de' Medici (1519–1589), daughter of the Duke of Urbino, married the future King of France, Henry II, members of the royal circle described her as coming from a family of "tradesmen."[23] A foreign ambassador reported: "It is sufficient to say that she is a woman, a foreigner, and a Florentine to boot, born of a simple house, altogether beneath the dignity of the kingdom of France."[24] The King of France, her future father-in-law, disagreed. The fourteen-year-old bride brought to the French crown not only the towns of Pisa, Livorno, Reggio, Modena, Rubiera; she brought money as well—100,000 écus from her uncle, Pope Clement VII.[25] The union between Caterina Sforza, daughter of the Duke of Milan, and the King of Poland at the end of the sixteenth century brought a dowry large enough to pay off the Swedish national debt.[26]

Sometimes money, property, and title were less important than the political alliances signified by the union. Among royalty such unions created a network of international alliances between dynasties.[27] The English Tudor monarchs of the late fifteenth and early sixteenth centuries, Henry VII and his son Henry VIII, used marriage this way and so raised their new dynasty to a level comparable with other European royal families. Henry VII's oldest daughter, Margaret, married the King of Scotland; his youngest, Mary, the King of France; and his son Arthur married Catherine, the daughter of Isabella and Ferdinand of Spain (after Arthur's death Catherine married his second son, the future Henry VIII). During his reign, Henry VIII negotiated in the same way. As a little girl, his daughter Mary Tudor was betrothed first to Charles V, the Holy Roman Emperor, and then six years later when the political realities had changed, to Charles's rival, Francis I, the King of France.[28]

A careful choice of marriage partner could mean new lands, even whole kingdoms, added to the royal domain. The Hapsburgs of Austria came to rule most of Western Europe in this way. At the end of the fifteenth century Mary of Burgundy's marriage to Holy Roman Emperor Maximilian brought her duchy to the Hapsburg dominions. Their son's marriage to Juana of Castile, heir to Isabella and Ferdinand, united the Spanish and Austrian territories. Mary of Burgundy's great grandson, Philip II of Spain, by his marriage to Maria of Portugal brought her father's kingdom into the family. The French Bourbons proved almost as adept. King Louis XIV claimed the Netherlands, Franche-Comté, the German Palatinate, and Spain, all as a result of the terms of marriage alliances.

As in previous centuries noble and royal families sometimes found it difficult to provide the dowries that made such alliances possible. Paying out the widow's portion also strained the family's resources. At the end of the seventeenth century the English statesman Sir William Temple called "the giving of good portions to daughters" a "public grievance which has since ruined so many estates."[29] To guard against the "public grievance" between the thirteenth and the seventeenth centuries new principles of inheritance favoring sons evolved all across Europe as part of the practical and legal accommodation between the aristocracy and dynastic monarchs. The aristocracy hoped these new practices would protect their lands, the main source of the family's income, from division among numerous claimants, and reserve it to one heir, preferably a male who could best hold it against the claims of increasingly powerful kings. From the fifteenth to the seventeenth centuries European jurists wrote justifying primogeniture (inheritance in the male line) and entail (the indivisibility of the property). The seventeenth-century French jurist Laurens Boucalh wrote that these principles were meant to "preserve more enduringly the name, arms and property in the family and to avoid the property being lost and consumed by the prodigality, crime and disaster of daughters or undeserving heirs." In the sixteenth and seventeenth centuries queens and kings, empresses and emperors codified and made uniform the feudal patchwork of regional laws and customs of their realms. They created a paramount system of royal law. Together with decisions of judges in royal courts, these new laws defined and upheld the new inheritance practices favoring a single male heir as part of "private," or "family," law.[30]

In principle a daughter's inheritance and her dowry had become one and the same, a single payment made at the time of her betrothal. Usually the amount was less than the sum of the two payments she might have received in earlier ages, thus easing the financial burden on her family. In addition, kings offered other ways to protect the "patrimony," as the main family holdings were called. In their decrees of the sixteenth and seventeenth centuries, rulers tried to limit the size of dowries to a percentage of the rents on the family property or of its overall value, to half the amount given to the male heir, or to an amount commensurate with the daughter's position at court.[31]

Families found other strategems to safeguard the patrimony. They provided the dowry and negotiated the promise of the widow's portion by using peripheral lands and properties, and by mortgaging the entailed lands. A queen called on other resources. Catherine de' Medici, the Regent of France, negotiated a loan in 1567 to raise the 40,000 francs she needed for a dowry.[32] At their worst such strategems led to the situation of some English estates entailed to the eldest son, but which were so encumbered with debts incurred to provide for daughters and widows that the young lord had the title but not the wealth.[33] Sometimes a mother provided the lands for dowries. Though the laws of the sixteenth, seventeenth, and eighteenth centuries restricted a wife's rights over her property, marriage contracts could abrogate the prohibitions. While the father's lands went to the eldest son, the mother's could be assigned to daughters and to younger sons. Maria Theresa, the eighteenth-century ruler of Austria, followed this practice. Her husband, Francis Stephen, Duke of Lorraine, brought a private fortune to the marriage and increased both it and hers by his careful administration of their lands. He willed all of his property to Joseph, their eldest son. She provided for their other children.

As in earlier centuries, when faced with no sons to inherit or with the prospect of a particularly advantageous match for a daughter, families circumvented the very rules of inheritance through the male line they had fought to establish. Sometimes even the title was allowed to pass through the female line. The young woman took it with her as part of her marriage portion and passed it on to her child. In sixteenth-century Spain thirty-two years of court battle ended with the creation of the sixth Duchess of Infantado, a daughter holding the title in her own right. In this and the next century so many dispensations were made to the rules of inheritance that the Spanish law courts were filled with disputes, including daughters claiming lands over distant male cousins. In sixteenth-century France a similar increase in litigation led the government to restrict entail in hopes of minimizing the need for exceptions and thus the source of the disputes. England's courts adhered more strictly to the customary laws. In the late seventeenth century Lady Anne Clifford regained the family property but only after the death of the last possible male heir, a half-brother.[34]

Too often noble widows of this courtly world found that only by litigation could they assert their rights to their portion after their husband's death. At the end of the fourteenth century Christine de Pizan, the French courtier, turned to writing and sought patrons because of her difficulties as a widow. Her father, astrologer to Charles V of France, had no steady income at the time of his death. Her husband, Etienne de Castel, also a court official, had to take on responsibility for the family. At twenty-six, he too died, leaving Christine de Pizan with her mother, two brothers, and her own children to support. Questionable claims on her husband's salary and awards were her only hope of income. Years later she remembered herself "so confused with grief that I became a recluse, dull, sad, alone and weary."[35] To make matters worse, she had known nothing of their situation. As she explained, "It is a common habit of married men, not to inform their wives entirely of their affairs."[36] At one point she paid court to four different aristocrats, reminding them of their obligations. Household officials mocked her appearances, the lone woman in the anterooms. Creditors took her furniture; she had to borrow. It took fourteen years of wrangling to assert the family's claim to her father's lands in Italy, and her right to the moneys owed by various nobles to her husband. Christine de Pizan's experience was not extraordinary, and throughout these centuries widows of even the most privileged families continued to have difficulty collecting their "portion" after their husband's death.

With so much of the family's welfare and future at stake, both secular and religious law upheld the primary right of parents to consent to their daughters' and sons' marriages. Church decrees like those of the Council of Trent in the sixteenth century, royal codifications of laws, and the decrees of absolute monarchs required public declarations of the intent to marry and prescribed punishment for any independent action on the part of children. For example, sixteenth- and seventeenth-century English and French laws required publishing the "banns" (announcement of the intent to wed), and marriage in a church. Laws forbade elopements, made abduction of an heiress a secular crime, and allowed disinheritance for marriage without consent. A 1673 French ordinance gave the father the right to confine his daughter to a convent until the age of twenty-five, the legal age of consent for a female.[37]

For a young woman of the courts, as in previous eras, the approval of the monarch might also be needed. As in the past, the queen or king viewed the giving of an aristocratic daughter in marriage as a favor to bestow on a loyal follower. Or they might want to prevent a marriage because it created an alliance between two potentially threatening families. Elizabeth I, the sixteenth-century Queen of England, favored five young women with the title "maid of honor." An advantageous marriage arranged by the queen and a sinecure or privileges for a father or husband were the expected perquisites of such service. To marry against Elizabeth I's wishes, however, meant imprisonment in the Tower of London. Lady Catherine Grey (the sister of Lady Jane Grey, executed in 1554 for treason against Elizabeth's sister Mary Tudor) married without the queen's permission and ended bearing her two children in the Tower, one conceived on her wedding day, the other during a brief visit allowed with her husband.

Protective of their royal lineage, rulers prevented the marriages of their relatives to people they considered their inferiors. At the beginning of the seventeenth century Anne of Austria forbade the marriage of her adviser's niece, Marie de Mancini, to her son, the young King of France, Louis XIV. She berated the young woman's uncle, Cardinal Mazarin, saying, "I warn you that all of France would rebel against you and against him, that I myself would lead the rebels."[38] Louis XIV in turn forbade the marriage of his forty-three-year-old cousin, known to the court as La Grande Mademoiselle, to M. de Lauzun, a thirty-six-year-old Gascon who had risen at court to be captain of the guards.[39]

Just as royal families prevented marriages, so, as in previous ages, they forced them for their own ends and on their own terms. When Catherine de Bourbon (1559–1604) fell in love with the son of a rival family, her brother, King Henry IV of France, tore up the agreement the young people had signed.[40] In return for the protection for his lands, Elisabeth Charlotte of the Palatinate's father gave her to Louis XIV to marry the king's homosexual brother, Monsieur, the Duke of Orléans. Liselotte, as she came to be called, remembered that she had cried "all night long from Strasbourg to Chalons."[41]

Once the marriage arrangements had been completed, the wedding, as in the past, remained a way in which aristocratic and royal families displayed their wealth and thus indicated their power and importance. The ceremony and celebrations surrounding marriages in the fifteenth-century court of Burgundy rivaled the extravagance of the great Italian Renaissance families. In 1430 when Isabel of Portugal married Philip the Good in Bruges, one hundred carts of wine, fifty carts of furniture and gems, and fifteen carts of armor and arms for the tournament were brought for the festivities. As part of the decorations, carpenters constructed a stag, a unicorn, and a painted wooden lion with a paw that spouted wine.[42] At the 1468 banquet of Margaret of York's marriage to Charles the Bold of Burgundy contemporaries marveled at the white and blue silks billowing over their heads and the tapestries of Jason and the Golden Fleece and King Clovis that covered the walls.[43]

Agents of the Fuggers, the Austrian merchant family, described marriages among great sixteenth-century noble families of similar opulence. Jousts, tournaments, and a chamois hunt celebrated the wedding of the Duke of Mantua's daughter to an Austrian archduke. The wedding guests feasted on twelve tons of venison, nine tons of suckling pigs, 11,000 hares, 20,620 eggs, two tons of cheese, and six kinds of wine.[44]

Royal weddings in the eighteenth century became even more extravagant. In Russia in August of 1746 the festivities surrounding the marriage of the future Catherine the Great lasted ten days.[45] In May of 1770 Marie Antoinette, daughter of Maria Theresa, the ruler of Austria, left Vienna to go to France to marry the heir to the throne, the dauphin. Her mother's courtiers, both women and men, had already marked the occasion with plays and receptions. French ladies-in-waiting joined her at the border of the two kingdoms. Hundreds of thousands of courtiers and performers participated in the festivities. At Versailles, actresses, dancers, and musicians performed in three theaters. Three hundred thousand attended the marriage ball and marveled at the flotillas of boats in the canals and the fireworks set off for their entertainment.[46]

Childbearing

Just as in earlier centuries, by marrying, the young women of Europe's royal and aristocratic elite brought not only lands, money, and an alliance to a new family, but also the promise of heirs. Both families looked to the match for the continuation of their lineages. To insure this, treatises on marriage advised choosing a healthy young woman with a pleasing appearance.[47] Given the tacit understanding that one of the wife's principal duties was to produce legitimate offspring, all families preferred that she bear males. The noblewoman and the queen consort gained their reputations not only for the grace and congeniality of their manner, but also for the number and gender of their children. The English Duchess of Newcastle knew that her husband, "having but two sons proposed to marry me, a young woman that might prove fruitful to him and increase his posterity by a masculine offspring."[48] The French aristocrat Louise de Coligny in 1598 congratulated her stepdaughter Brabantine:

> My girl, it is a boy. I weep for joy. . . . God be praised that you have come through so well. . . . I am just dying to have a look at this tiny grandson and to see how you manage him with your little hands. . . . The duc de Bouillou is consumed with envy that his wife has not given him a boy.[49]

If a son brought joy, a daughter could bring dismay. When the princess Elizabeth and not the promised son was born to Queen Anne Boleyn on a September Sunday in 1533, King Henry VIII, the royal father, ordered the church bells rung and nothing more: no artillery salutes, no banquets or fêtes to mark the occasion.[50] The birth of the future Queen of Sweden, Christina, created havoc. Only Princess Katherine, King Gustavus Adolphus's sister, had the courage to bring him his daughter after the midwife had mistakenly announced a boy child. "I am content, dear sister, God protect the child," he told her, and unlike Henry VIII ordered all of the celebratory rites commonly performed for the birth of a male heir. Christina later wrote in her autobiography that they had waited still longer to tell her mother, "until she was able to bear such a disappointment."[51]

Noble and royal wives accepted their biological function and this important duty of their marriage. Eleanora of Ferrara, daughter of the King of Naples and wife to Ercole d'Este of the great fifteenth-century Roman family, produced daughters and sons almost every year between 1474 and 1481—six in all. Catherine de' Medici birthed ten children in twelve years for France's sixteenth-century king, Henry II. Henrietta Maria, the seventeenth-century queen of England's King Charles I, had nine children. When in the eighteenth century the regent for France's young king, Louis XV, chose a bride,

Marie Leszczyńska, the daughter of the exiled King of Poland, proved the perfect match. It was no alliance, there was no money, but she was young enough (in her early twenties), healthy, and able to bear. Ten children were born in the first twelve years of the marriage. She described herself as "always going to bed, always pregnant, always giving birth."[52] Even rulers in their own right accepted this obligation. Maria Theresa of Austria bore sixteen children from her marriage at nineteen in 1736 until 1756 when she was thirty-nine.

For the birth of her child an elite woman had the help of a midwife and the companionship and encouragement of her family and retainers. In 1670 Mme. de Sévigné stayed with her pregnant daughter for her confinement. A queen or a princess giving birth to potential heirs had many in attendance to ensure the identity and safe delivery of the infant. All kinds of rituals surrounded the occasion. In September of 1533 when her pains began, Anne Boleyn, second wife to England's King Henry VIII, walked in procession with her ladies to rooms specially prepared for the event. "Taking her Chamber," as it was called, had become a ceremony with a mass, communion, and the court official, the chamberlain, as her escort.[53] In France the queen had her lying-in in a special green-walled room. She delivered on a red velvet birthing stool, with Princes of the Blood and the the king in attendance. Two nuns prayed for her, and special relics of the royal family (like the Virgin Mary's belt) would be brought to the room to lend their protective powers. Successfully delivered, the sixteenth-century French queen, Marie de' Medici, moved to the state bed to receive the congratulations of the courtiers.[54]

In the case of the eighteenth-century French queen, Marie Antoinette, the rituals surrounding the birth of her first child almost killed her. With no pregnancy for the first seven years of her marriage, intense excitement filled the last hours of her labor in January of 1770. As the infant's head presented, the attending physician ran from the lying-in room yelling "The Queen is going to give birth" to the waiting courtiers and Princes of the Blood. Although screens had been set up around the bed to give her some privacy, there was still a crush of people. Two climbed up on the furniture so that they could see better. The sheer numbers, the noise, the closeness of the room with the windows sealed shut, and perhaps her own relief that she had managed it all without ever crying out, caused Marie Antoinette to have what contemporaries described as a "seizure." The king, fearing that she would die, forced open the windows himself. She was bled from the foot. She revived. After that only a few at the French court had the privilege of observing births. In gratitude and perhaps in anticipation of more children to come, Marie Antoinette's father-in-law, Louis XV, gave her 100,000 livres.[55]

As in previous ages the birth of more than one baby to aristocratic and royal mothers assumed significance because children died so easily. One of Marie Antoinette's two sons died in 1789. Smallpox decimated noble and royal families. Louis XIV of France lost his son and grandson in one outbreak. The last son of Queen Anne of England (1665–1714) died of the disease (though pregnant sixteen times, Anne saw none of her children survive to adulthood). Typhoid, typhus, malaria, sweating sicknesses, pleurisy, tuberculosis, and infections of all sorts claimed the offspring of even the most prolific mothers. As Liselotte, Louis XIV's sister-in-law, wrote to the children's governess after the birth of her third child, Elizabeth Charlotte:

> It is such a strain. If they survived it might be a different story, but just to see them die as I did earlier this year—truly, there is no joy in it.[56]

No matter how well-born or privileged, wives risked death in childbirth. A study of the English aristocracy from 1588–1641 estimated that a wife's chances of dying early were double those of her husband. About 25 percent of the wives who died before the age of fifty died in childbirth or from its complications.[57] The histories of the royal houses of Europe from the fifteenth to the eighteenth centuries are filled with the deaths of princesses and queens in childbirth.

Noble and royal wives knew these hazards, and once they had done their duty, they sometimes tried to withdraw their sexual favors. The seventeenth-century French courtier Mme. de La Fayette gave her husband two sons, but after subsequent miscarriages and stillbirths she moved to Paris, while he stayed on the estate in the country. Charlotte de Saumaize fought for a separation from her husband, the eighteenth-century Count of Brégy. She had borne four children and did not want to risk her life in another pregnancy. Queen Marie Leszczyńska, wife to Louis XV, abstained first on saints' days and gradually refused intercourse altogether.[58]

Everything about childbirth assumed such importance because a wife's inability to produce children, especially a son, had serious consequences for a family. A royal dynasty that had newly seized and consolidated its power needed sons to ensure order and continuity. Royal wives understood the significance of their potential fecundity. In the sixteenth century Catherine de' Medici, queen to Henry II of France, spent ten years as a barren wife. In 1539, when a peasant woman her husband had raped conceived and gave birth, she offered to retire to a convent so that he could take another wife.[59] In the mid-eighteenth century the future Catherine the Great of Russia admitted to her mother-in-law, Empress Elizabeth, that she had taken a lover when unable to produce an heir with her husband.[60]

Catherine of Aragon, the sixteenth-century Queen of England, is the most famous of the royal wives divorced for her inability to bear a son. Always behaving like contemporaries' ideal of a daughter and wife, she dealt patiently with adversity even before her husband, Henry VIII, fell in love with the courtier Anne Boleyn. Catherine had been a pawn in the marriage-alliance games of her parents. She first married Prince Arthur, heir to the English throne. After his death only six weeks later, she waited eight humiliating years with little money and no status while her father, Ferdinand of Spain, and England's King Henry VII haggled over another betrothal, this time to Arthur's younger brother Henry. Catherine married him in 1509, but only in 1516 did she carry a child to term and have it survive. (She was pregnant six times and had numerous miscarriages.) Even then, it was a daughter, the future Queen Mary Tudor. By the 1520s her husband, Henry VIII, knew that he could beget sons with other women; he wanted to marry Anne Boleyn and he talked of divorce. Catherine of Aragon refused him. Even Thomas Cromwell, her opponent and Henry's adviser, admired her courage: "But for her sex she would have surpassed all the heroes of history."[61] Her implacability and pious determination enraged the king, who in the end left the Catholic church, created his own church court, and became his own Pope to bring about the dissolution of their marriage.

The inability of Catherine's successor, Anne Boleyn, to produce a male child led Henry VIII to discard her even more ruthlessly with fabricated charges of treason that meant her execution. In the last resort, even a queen could be judged solely by her reproductive ability. Even in the world of the courts all other aspects of her character, of her function, counted for little when weighed against the number of sons she bore.[62]

The Relationship Between Wives and Husbands

For aristocratic or royal wives marriage could be more than the negotiated exchanges of inheritance, vows, and the performance of duties. In the seventeenth century, King Louis XIV's courtiers enjoyed the fairy tales of Charles Perrault in which Cinderella and Sleeping Beauty loved, married, and found happiness with their princes.[63] Such tales had their real counterparts in the world of the courts. Queens wrote of the happiness and love in their marriages. Henrietta Maria, daughter of Henry IV of France, married Charles I of England when she was fourteen. As the influence of her husband's male favorites came to an end, as they mourned the death of their young sons together, their devotion to each other grew. She wrote, "I was the happiest and most fortunate of queens; for not only had I every pleasure the heart could desire; I had a husband who adored me."[64] She wrote to her mother, Marie de' Medici, "The only dispute that now exists between us is that of conquering each other by affection, both esteeming ourselves victorious in following the will of the other."[65]

Maria Theresa, the eighteenth-century ruler of the Habsburg empire, continued to love her husband, Francis Stephen of Lorraine, even after she learned of his infidelities. When he died suddenly in August of 1765 after twenty-nine years of marriage, she helped to sew his shroud and paid the money he had promised to his last mistress, Countess Wilhelmina Auersperg. On the day of his death she calculated in her prayer book the span of his life down to the days and hours, and the length of their time together to the months and days, so as to know the numbers of appropriate prayers and alms. Five months later, in January of 1766, she still mourned him:

> I hardly know myself now, for I have become like an animal with no true life or reasoning power. I forget everything. I get up at five. I go to bed late, and the live long day I seem to do nothing. I do not even think. It is a terrible state to be in.[66]

To her son Leopold she confided, "I have lost everything, a tender husband, a perfect friend, my own support, to whom I owed everything."[67] She remained in semi-mourning for the rest of her life and customarily went into seclusion each August to pray and to commemorate her husband's death.[68]

Marriage might bring other satisfactions to an aristocratic or royal woman. As in past centuries a wife might become the principal cultural influence and source of patronage at the court over which she and her husband presided. The fifteenth- and sixteenth-century courts of the great Italian noble families set this pattern for subsequent generations. In the fifteenth century Eleanora of Ferrara belonged to the first of two generations of learned, cultured wives in great Italian families. She had her own library of religious and classical texts, and saw to it that her daughters Isabella and Beatrice d'Este had, like their brothers, an exemplary education including Latin, history, French, and music. Isabella d'Este became a famous patron of the artist Leonardo da Vinci. In the sixteenth century, Italian noble daughters sent as wives to other courts brought their Humanistic learning and their artistic tastes with them and began the transformation of these gatherings as well.[69] Valentina Visconti—the daughter of Gian Galeazzo Visconti, the ruler of Milan—and Catherine de' Medici, from the great Florentine family, are credited with bringing the Renaissance to France.

Juan Luis Vives, the sixteenth-century Spanish churchman, in the treatise he wrote for Princess Mary Tudor, described another kind of influence that a wife might have. He advised that a woman, by creating "pleasant conditions," by her "sweet speech," would "couple and bind her husband unto her."[70] Wives assumed that a husband bound to his wife by such exemplary service would be amenable to her advice and influence. A skilled wife could thus gain indirect access to powers usually reserved to aristocratic and royal men.

Queen Caroline of Brandenburg-Ansbach, wife of the eighteenth-century English king, George II, manipulated her husband in this way according to her courtier and principal companion, Lord Hervey. She became the ally of Prime Minister Horace Walpole, who explained to her how important she was in the management of the king:

> Your Majesty knows that this country is entirely in your hands, that the fondness the King has for you, the opinion he has of your affection, and the regard he has for your judgment, are the only reins by which it is possible to restrain the natural violences of his temper or to guide him through any part where he is wanted to go.[71]

To have such influence had its adverse consequences as well. The wife, like the courtier, knew a different kind of servitude and paid a price for her indirect power. Hervey explained:

> To him she sacrificed her time, to him she mortified her inclination; she looked, spake and breathed but for him . . . and governed him (if such influence so gained can bear the name of government) by being so great a slave to him thus ruled as any other wife could be to a man who ruled her.[72]

When Maria Theresa, the eighteenth-century Empress of Austria, sent her fifteen-year-old daughter Marie Antoinette to France, she expected the young wife to exercise influence both in her own right as dauphine (wife to the heir) and through careful management of Louis, her husband. Marie Antoinette proved a poor instrument for her mother's plans. As her mother's letters described it, with no formal education to such a subtle role, dazzled by the new world of her father-in-law Louis XV's court, Marie Antoinette continually displeased and disappointed with "her immaturity as well as her lack of application."[73] She created a virtual scandal in her first months at court when she refused to speak to the king's mistress, Mme. du Barry. Nothing changed when she became queen. She eagerly fell into the politics of court factions, and turned from her husband to pursuits that both her mother and her brother, Emperor Joseph II, thought frivolous and unworthy: gambling at cards, flirting, dressing up, and playing at the rustic life with her attendants. By the mid-1780s, despite the birth of children and a more matronly appearance and activities, Marie Antoinette's reputation was so bad that scandals about her irresponsibility and extravagance flourished.

Ironically, given her mother's hopes, the French queen was seen by contemporaries as having a tremendous influence—but for ill, not for good. Marie Antoinette was blamed for diverting the king from governing, for removing effective royal ministers like Turgot from office, for draining the treasury of France to buy jewelry and clothes. In 1785 few believed that she had not been involved in the negotiations to purchase a diamond necklace worth over one and a half million livres, and the populace assumed that she had been the lover of the courtier arrested for the incident.

To her contemporaries and to subsequent generations, Marie Antoinette came to personify all the evils of the "dilapidated" French monarchy, as Marie Theresa called it.[74] This identification of women with the weakness and corruption of government was an ancient and powerful European tradition. Like their predecessors in earlier centuries French male political theorists and moralists of the eighteenth century portrayed a people reveling in luxury, dissipation, and indulgence. As a result of female influence, "the entire nation, formerly full of courage, grows soft and becomes effeminate, and the love of pleasure and money succeeds that of virtue," explained Jacques Joseph Duguet, a French moralist, in 1750.[75] So strong was this tradition that women courtiers came to espouse the idea as well. Mme. de Motteville, a friend to Anne of Austria, the seventeenth-century Regent of France, expressed the contemporary view of the hazards of female influence and rule:

> Women are generally the prime cause of the greatest upheavals of states; and wars, which destroy kingdoms and empires, are almost invariably the effects produced by their beauty or their malice.[76]

Thus, the most ancient traditions about women's nature, about the dangers of women stepping outside the limited roles and functions of the family, passed on to yet another generation of Europe's elite.

4

WOMEN RULERS

�֍

The Combination of Circumstances

Though courtiers and political theorists criticized women's influence in government and revived the deprecating images of females exercising power, from the fifteenth to the eighteenth centuries circumstances still arose that required allowing women to rule like men. For positive as well as negative traditions survived into the world of the courts. In the absence of a male ruler, families turned to wives and mothers to act as their surrogates. In the absence of a male heir, families passed their claims to kingdoms and empires to their daughters. As in centuries past families turned to their women rather than lose all claim to power and position.

Just as they had in the eras of feudal homage, vassals, and lords, privileged women acted as trusted surrogates for their male relatives. The tradition personified by Eleanor of Aquitaine in the twelfth century—partner to her husband, King Henry II of England, and ally to her sons, Richard and John—continued with women of the courts like Isabella d'Este and Catherine de' Medici.

As in earlier centuries aristocratic and royal women also ruled in their own name. Rather than see the territories fall into civil war with the throne passing to young male heirs, fourteenth-century Scandinavian lords and townsmen had chosen a woman, Margrethe, to rule the kingdoms of Denmark, Norway, and Sweden. Daughter and wife to kings, she became queen in her own right of a unified Scandinavia. So again from the sixteenth to the eighteenth centuries the male elite of the aristocracy, the towns, and the royal bureaucracy supported the claims of women over men—those of Catherine the Great of Russia over those of her erratic husband and her young son.

Women came to be queens in other circumstances as well. When there was no son to inherit, royal fathers designated daughters as their heirs.

Powerful men accepted the unorthodox choice of a female heir from the established dynasty, rather than risk the alternative—contending claimants, disorder, and civil war. In such circumstances at the end of the fifteenth century the Spanish nobility had accepted Isabella of Castile as their queen. So again from the sixteenth to the eighteenth centuries in countries as diverse as Scotland and England, Sweden and Austria, women inherited and ruled. Kings and emperors tried to guarantee their daughters' rights to the throne. Parliaments, merchant leagues, powerful aristocrats, and members of the royal bureaucracy concurred. Strict inheritance, even to a female—such as Christina of Sweden—meant peace in a time of economic expansion when all were eager to enjoy the bounty of mercantile capitalism. Strict inheritance by Elizabeth Tudor discouraged rival claimants in a time of volatile religious controversy between Protestants and Catholics. Strict inheritance by Maria Theresa produced a ruler from the established dynasty able to rally support and maintain unity when challenged by monarchs of other nations.

Queen Mother and Regent

RULING FOR THE FAMILY

As in feudal times, so in the Renaissance, wives acted for their husbands among the aristocratic families of northern Italy. Isabella d'Este (1474–1539) supervised the defense of territories under the rule of her husband, Francesco Gonzaga, and managed to establish a lucrative cloth manufacturing industry in his absence.[1] When the Venetians besieged Ferrara from 1482–1484, the duchess, Eleanora, acted as commander in her husband's absence. Royal and aristocratic wives commonly performed this duty and accepted this responsibility in courts all across Europe. English queens acted as regents when their husbands left the kingdom, from Catherine of Aragon for Henry VIII to Queen Caroline for George II.

Some women were given their own lands to govern by their families. The Austrian and Spanish Hapsburgs of the late fifteenth and sixteenth centuries relied on their aunts and sisters to assist in the administration of their vast, diverse territories. It became almost a tradition for female members of the family to administer the northern provinces of Flanders and the Netherlands. While regent, Margaret of Austria (1480–1530) negotiated the Treaty of Cambrai for her nephew Charles V, the Holy Roman Emperor. Her niece Mary of Hungary acquired a reputation for toleration when she administered the Low Countries. Charles V's son Philip II of Spain appointed his stepsister Margaret of Parma (1522–1586) to rule the area for him from 1559–1567.[2]

As in earlier eras from the fifteenth to the eighteenth centuries aristocratic and royal women also gained power over their husbands' lands as widows. Just as in past centuries the death of a husband might mean the throne left to a male heir too young to reign. In the sixteenth century Isabella d'Este governed for her young son. For a royal family the Queen Mother was sometimes the obvious and most trustworthy candidate to rule until the young prince had reached his majority. In the sixteenth and seventeenth centuries accidents of early death repeatedly left a young boy as king of France. As a result, an almost unbroken line of women ruled in the name of their dynasty from Louise of Savoy (1476–1531) to Anne of Austria (1601–1666).[3]

Of all of these French regents, Catherine de' Medici (1519–1589) earned both the highest praise and the most severe criticism from her contemporaries and subsequent generations. A drawing by the court painter Clouet shows her in black mourning dress, white ruff, and simple, thin white cap. Her eyes are clear, the lips full and set. She makes no pretense to vanity; the heaviness shows in her cheeks and in the extra flesh under her chin. She is staid, solid, and imposing. She looks as she liked to describe herself, the "Queen Mother." Many contemporaries saw her differently and gave her other titles. Some vilified her as "the Sinister Queen," who, though regent for only a few years during the minority of her son Charles IX, made her influence felt for many decades nonetheless. Some accused her of debauching her children so that they would be too weak to rule. Subsequent generations accused her of causing the wars of religion that divided France from 1562 until 1596. Many believed that she alone had ordered the massacre of Protestant leaders in 1572 on St. Bartholomew's Day.

Unquestionably, Catherine de' Medici enjoyed exercising power. In 1563 the Venetian ambassador explained, "it pleases her to make one believe that everything stems from her authority."[4] On her husband's death in 1560 she skillfully took the regency away from the two aristocratic factions that had assumed they could rule in the young heir's name.[5] She used her power not for herself but, like Queen Mothers of previous ages, to ensure that her children would rule as their father had. While her weak, erratic, and unstable sons governed from 1559–1589 (Francis II, Charles IX, Henry III) she held the monarchy for the Valois dynasty.

Catherine de' Medici's history as a wife suggested the strengths she would exhibit as a ruler. She had been the model consort to Henry II. Married at fourteen, she bore him ten children from 1544–1556 and saw seven survive to adulthood.[6] She acted as regent for him from 1553–1559 when he left France to fight. After his death, she maintained his legacy for their sons. She used all of her influence to still the religious and political controversies that threatened her children's hold on the throne. She sought to avoid or to end confrontations between the great lords and Princes of the Blood (like Guise, Coligny, Montmorency, and Navarre) who might take the crown for themselves. She sought to keep the King of Spain and the Queen of England favorable to her family's rule, or at least neutral. Knowing the value of marriage alliances in matters of war and peace, Catherine de' Medici used one daughter to forestall Spanish intervention and another to create an alliance with the leader of a rival family, the King of Navarre.[7]

The Huguenots (the French Protestants) remained the biggest threat to her family's continued rule throughout the period of her influence. An alliance of aristocrats and patricians (many of them Protestant themselves) sympathetic to the Huguenots challenged the authority of the dynasty and threatened the dissolution of the kingdom. Catherine de' Medici hoped to reconcile the Catholic and Protestant interests and thus preserve peace.[8] She called a meeting on doctrine and made treaties acknowledging limited worship for Huguenots. When this failed to satisfy the dissenters, she wanted to follow the example of the German states and allow the two sects to live separately in their own territories.[9]

Catherine de' Medici was remembered by subsequent generations not for what she accomplished but for what she could not prevent. Contemporaries, especially her enemies, ascribed to her the sole responsibility for the horror of the St. Bartholomew's Day massacre of the Huguenot leaders gathered in Paris to celebrate the marriage of her daughter Margaret to Henry of Navarre in 1572. Most historians disagree. They argue that the situation had become so volatile that anything could have set off the carnage. Many must share the blame, including her son Henry III. It is unlikely that she favored such an extreme measure. Rather she was unable to prevent the impressionable king from participating in the machinations of leaders cleverer than himself.[10] The power of the monarchy had been upheld, the claims of the dynasty protected, but at a terrible price.

Catherine de' Medici acted as surrogate for her sons because they were too young or too weak to rule effectively. In a period of political uncertainty, the majority of great aristocratic and patrician families accepted her authority as an alternative to outright civil war.

RULING ALONE

Catherine de' Medici held and exercised power because of an unusual combination of circumstance and character. In a period of political uncertainty she retained the throne for the dynasty. Other women went one step further and seized the power on their own behalf. For example, Caterina Sforza in fifteenth-century Italy and Catherine the Great in eighteenth-century Russia made this choice and proved that a strong and ruthless woman could use power and govern as effectively as any man.

Caterina Sforza (1462–1509) lived in the chaos of contending aristocratic families in fifteenth-century Italy. The illegitimate daughter of Galeazzo Maria Sforza, the Duke of Milan, she spent the early years of her life fulfilling her duties as a wife to Gerolamo Riario, nephew to Pope Sixtus IV. She bore eight sons and daughters from 1478–1487. Then she took on roles usually reserved to men. She defended the estates from the attacks of other families. She hired mercenary troops and led them into battle. On one occasion, though seven months pregnant, she joined a military expedition to Rome to determine the choice of the next Pope. When her first husband was assassinated in 1488 she took absolute control of their properties, the towns of Imola and Forlì. In recognition of her unique activities, contemporaries called her "the Virago" (the Amazon, or woman warrior). Until her capture and defeat in 1500 at the hands of Cesare Borgia, she proved a ruthless and successful player in the warring competition between families and against the French after their invasion of Italy in 1494. Circumstances overwhelmed her as they did leaders of other Italian city-states. Surrender to Cesare Borgia brought humiliation and imprisonment. She never ruled again.[11]

Two centuries later similar circumstances—the practical dangers of contending claimants to power, the hazards of being a female in a male role— challenged Catherine the Great (1729–1796), Empress of Russia. She met each challenge successfully. From her arrival in Russia at the age of fourteen this exceptional woman turned every circumstance to her advantage: she played the dutiful wife to the heir, plotted his overthrow to become regent, and then consolidated the power of empress in her own hands.[12]

Because the descendants of Peter the Great failed to produce male offspring, Russia had a tradition of female regents and of women taking the imperial crown for themselves, as Peter's own daughter Elizabeth did from 1741–1762.[13] Empress Elizabeth chose Catherine, born to a Prussian general and a Danish noblewoman, to wed the heir, her nephew. The future Catherine II had been given little education to suit her to her new nation or powerful future position, but even in 1745, the year of her marriage, Catherine the Great was ambitious and determined to succeed. She quickly learned Russian, studied the Russian Orthodox religion, and adopted the customs of the Russian court. As she remembered later in her *Memoirs:* "I was set in Russia, and never lost sight for a moment: that I must (1) please the grand duke, (2) please the Empress, (3) please the nation."[14]

She worked to be a perfect wife and daughter-in-law. She joined her husband, Peter, in his games with toy soldiers. She ignored his temper and his drinking. She humbly cried if the empress reprimanded her. She accepted the strict supervision ordered for her household, including the censorship of her letters to her family. Even so her position as wife to the heir was not secure. After eight years of marriage she had not once been pregnant. To prove to the empress that she could conceive a child, she took a lover. For a royal wife adultery was treason. She gambled everything, however, on her ability to bear a son.[15] The birth of Paul in 1754 was the first of many victories. She wore her success well. At twenty-five, according to one of her admirers, she looked magnificent: "just risen from her first delivery" with black hair, blue eyes, dark lashes, "a kissable mouth, perfect hands and arms, slender figure, tall rather than small; she moved quickly yet with great nobility, and had an agreeable voice and a gay good-tempered laugh."[16]

The difficulties of her role as wife only increased with the death of the empress in December of 1761 and the accession of Peter. By this time he had taken his own mistress, Elizabeth Vorontosova, and talked publicly of marrying her. He had not named Catherine's son as heir. His rages and drunken scenes at night made him unpredictable. She survived and prospered in these adverse circumstances because Peter endangered and angered others as well. These men became her allies.

In July of 1762 Catherine plotted with a diverse group which included Nikita Panin, a diplomat and her son's tutor, and her latest lover, Grigory Orlov, an officer in the Russian army. Orlov and his four brothers planned the coup to overthrow the emperor and directed Catherine's first appearance in her new role as ruler of Russia. She was driven early in the morning to the principal barracks in Moscow. Wearing the imperial guardsman's uniform used before her husband's adoption of Prussian customs, she stepped from the carriage, her hair unpowdered, her man's uniform suggesting the authority of the ruler and showing her allegiance to the old Russian ways. The soldiers and their officers cheered. At Kazan Cathedral the churchmen proclaimed her empress.

From the beginning she skillfully maintained her independence. She orchestrated her own coronation in September of 1762. By 1764 the rival claimants to the throne were dead.[17] In the course of her reign she neutralized possible opposition and created a government—unlike those that had preceded hers—based on civilian, not military, authority, on persuasion, not force.[18] Gradually she fashioned absolute authority for herself as empress. She called together groups of advisers to study the issues that had made Peter unpopular: the secularization of the Church and the reform of the army. She allowed a legislative commission to discuss a new law code while she made the final decisions.[19] When Panin, her ally in the coup against Peter, suggested the advisory council have more power, she refused. She limited the authority of the nobility's senate by redefining and dividing its powers. She played factions and individuals against each other with consummate skill. As she explained to a new officer of her government in 1764, the "sensible policy from my point of view requires me to pay no attention to them [to their demands and suggestions] . . . so I merely watch them with an unsleeping eye and employ people in this or that affair according to their capacity."[20] She made church officials into salaried government employees and reorganized local government across the empire, bringing it under the control of the Procuracy, a department answering only to her. The nobility retained the right given by her predecessor, Peter the Great, of free choice of government service, but many chose to volunteer as the empress's expansionist foreign policy brought more commissions and new lands to administer.

As her son grew to maturity, she refused him any real participation in affairs and made him live away from the court with his wife. When he was eighteen in 1772 he had asked to join her council. Catherine had responded:

> I told you that your request needs mature consideration. I do not think your entry onto the Council would be desirable. You must be patient till I change my mind.[21]

She did not even allow him to raise his own sons, Alexander and Constantine. As each of the boys was born she took him to train at her own court.[22] As early as the 1760s, only a few years after coming to power, she enunciated the policy of government that characterized her reign:

> The Sovereign is absolute; for there is no other authority but that which centers in his single Person that can act with a Vigour proportionate to the Extent of such a vast Dominion.[23]

Throughout her reign she had time for more than affairs of state. She hunted dressed as a man. She enjoyed dancing. She collected paintings. She patronized architects and craftsmen from Italy and France. She founded a Society for the Translation of Foreign Books and made her courtier Princess Dashkova director of a Russian Academy of Science. She read, she corresponded and met with leading men of letters. She spent time with her grandsons and knitted a blanket for her pet whippet.[24]

As another statement of her power, just like the powerful seventeenth- and eighteenth-century male rulers of Europe, she ignored the usual strictures on sexual behavior. Catherine II took many lovers. Only one man, however, came close to sharing power with her, her lover of the winter of 1773–1774 and subsequently her trusted friend, the Russian nobleman and army commander Grigory Potemkin.[25] After their affair ended in 1776 Catherine continued to take lovers, but they were young men without power, easily found and easily dismissed. She enjoyed them. She called them her "pupils."[26] In her early sixties, she wrote of twenty-two-year-old Plato Zubov, her last favorite: "I have come back to life like a frozen fly. I am gay and well."[27] She amused herself teaching him, translating Plutarch with him.

On her death in 1791 she left the power intact for her son and his descendants. A courtier of her grandson Alexander remembered a conversation with the empress in which she explained how she had managed to rule unquestioned. It was not by fiat, she explained. She consulted, "and when I am already convinced in advance of general approval, then I issue my orders, and have the pleasure of observing what you call blind obedience." She considered customs, opinions, consequences. "And that," she told him, "is the foundation of unlimited power."[28] Catherine the Great ruled with such authority because she studied personalities, and imagined reactions; she weighed alternatives and courted approval and support. In these centuries, royal princesses, daughters destined to inherit kingdoms, learned the same lessons and succeeded when they used the same tactics.

Royal Heir and Monarch

From the sixteenth through the eighteenth centuries women not only gained power as regents for their husbands and sons, but also inherited kingdoms in their own right. In the unusual combination of circumstances that characterized this era royal princesses became queens, crowned with all of the panoply and sanctity accorded a prince at his coronation.

By the seventeenth century, dynasties had established their hegemony, sometimes, as in the case of the Hapsburgs of Austria, over more than a hundred years. The principle of inheritance in the direct male line had become custom. In each instance a female, not a male, inherited because no son had been born or had survived. In each instance the royal family's only heir was a daughter. She embodied the rights of the dynasty and the promise of its continuation. Without female royal inheritance the reigns of families like the Tudors, the Stuarts, the Vasas, and the Hapsburgs would have ended.

Governing councils, parliaments, the leaders of the aristocracy, and the royal bureaucracy accepted designation of a daughter as heir because it seemed the only guarantee of continued order and prosperity for the realm. The governing elites of these countries assumed that to turn to a collateral male branch of the royal line, or to another aristocratic family, would mean the end of the unity brought by the success of this dynasty and a return to the wars of landed families competing for power and hegemony.

Inheritance by a woman meant unity in other circumstances as well. In 1688 the Protestant English Parliament took the throne from the Catholic James II and his infant son and established the royal succession on his daughters Mary and Anne, thus avoiding a return to the religious and political strife that had caused the English civil war of the 1640s. The Austrian Empire might have been torn apart had the nobility of most of the Hapsburg dominions not rallied to the aid of the only heir, Maria Theresa, on the death of her father, Emperor Charles VI, in 1740.

Just as circumstances favored inheritance by women, so did changes in the nature of monarchy. Now a woman could perform the duties of waging war and administering in peacetime as well as a man. The ruler no longer had to lead armies into battle. By the sixteenth century all across Europe mercenaries hired from many lands had replaced the local nobility called to military service by their warrior kings. From the 1670s on, government officials, not civilian entrepreneurs, took on the supplying and provisioning of troops. By the 1700s the wars themselves had been separated from populations. Men no longer fought around castles and walled towns but in open fields far from the centers of trade and government. All this could be arranged and supervised by a queen as effectively as by a king.[29]

The administration of the government too had evolved away from the personal rule of the male warrior and his loyal followers to the administration of hired state functionaries and councilors appointed by the monarch and thus dependent on royal favor. In the scramble for place and for rewards, no courtier protested the gender of his benefactor, no bureaucrat gave his obedience to a king and refused it to a queen.

The coincidence of so many women ruling as regents or in their own right from the fifteenth to the eighteenth centuries inspired a learned debate about women's fitness to rule. Given the traditional prejudices and prevailing views of the female nature and of women's proper function and role, many scholars argued against the idea of a female governing or inheriting a kingdom.

Ancient traditions made such a situation appear a contradiction in terms, and such a person a phenomenon to be feared. In 1558 the Scottish religious leader John Knox, eager to denounce both his own queen regent, Mary of Guise, and Mary Tudor, the ruler of England, described the hierarchy of nature, and the hierarchy of earthly beings. He explained that female rule subverted both and turned the natural order upside down. "For nature hath in all beasts printed a certain mark of domination in the male and a certain subjection in the female which they keep inviolate." The female who rules becomes a monster, the men who accept it worse than "brute beasts."[30]

This coincidence of queens also inspired writings by the learned in favor of women inheriting and ruling just like men. Collections of exemplary lives glorified women who acted the traditionally masculine role. Some writers suggested that a woman's rule could be better than a man's because women were moderate.[31] In his 1583 political treatise, the Englishman Sir Thomas

Smith gave the reasoning of many of his contemporaries living under Queen Elizabeth and reflected the views of other men subject to other female monarchs. The woman was noble by birth, he argued. The blood tie, not her sex, became the determining factor when weighed against the disorder that would result in setting aside the rightful heir.[32] From the fifteenth to the eighteenth centuries such practical realities outweighed the ancient prejudices and gave women the opportunity to show their effectiveness as rulers.

THE QUEENS

At the end of the 1350s the Italian Giovanni Boccaccio published his collection of biographies of exceptional women. He explained that to rule, to write, to excel like a man, a woman had to go beyond and to overcome her natural bodily disadvantages.[33] John Case, an admirer of Queen Elizabeth I of England, believed this to be possible. He wrote in his sixteenth-century treatise:

> Nature often makes woman shrewd, hard work makes her learned, upbringing makes her pious, and experience makes her wise. What therefore, prevents women from playing a full part in public affairs? . . . If one is heiress to a kingdom, why should she not reign?[34]

Many royal princesses who ruled in these centuries had the "hard work" and "upbringing"—an exceptional education—that helped each one play the "full part in public affairs" that Case described, and to which their birth entitled them.[35]

Catherine of Aragon, the recipient of an exceptional education herself, enlisted the Spanish churchman Juan Luis Vives to design the proper course of study for her child, the future Mary I of England (1516–1558). She supervised much of the education herself. Mary Tudor learned Latin, French, and Spanish and could understand Italian. A contemporary described her in her late twenties as "excellent and passing in all kind of learning and language as few have been the like."[36] The education of Mary's sister Elizabeth (1533–1603) was as rigorous as any prince's in this age of Humanism, yet retained bits of the traditional lessons and artistic skills meant for girls. At eleven, already fluent in Latin and reading Greek, she presented her stepmother Katherine Parr with a translation from yet another language, French, of Marguerite of Navarre's *Mirror of the Sinful Soul* encased in a needlepoint canvas cover that she had worked herself. From 1548 to 1550 Roger Ascham, the classical scholar and educator, tutored her. (Even after she became queen, Elizabeth I liked to read Greek and Latin with him every day.) She was sixteen when Ascham wrote to a friend of her "dignity and gentleness" and of the "clarity" of her writing. He gave her the highest praise he could imagine: "Her mind has no womanly weakness, her perseverance is equal to that of a man, and her memory long keeps what it quickly picks up." Lest his friend doubt, he wrote, "I am inventing nothing . . . there is no need."[37]

The future Queen Christina of Sweden (1626–1689) had an even more masculine upbringing. Accepting that his wife could never bear a son, Gustavus Adolphus II, King of Sweden, raised his only child, a daughter, like a prince. She learned to ride, to shoot, to fight with a sword. She studied Latin and Greek. On her own she learned German, French, Spanish, and Italian. When she was nine the Riksdag, the Swedish parliament, supervised her education and emphasized the study of the Lutheran religion and Swedish customs and history. A year later her father's chancellor, Axel Oxenstierna, took charge. In 1641, when she was fifteen, he reported on her progress to the legislature:

> I observe with satisfaction . . . that . . . Her Majesty is not like other members of her sex. She is full of heart and good sense, to such an extent that, if she does not allow herself to be corrupted, she raises the highest hopes.[38]

Even as a reigning monarch Queen Christina continued her studies, corresponding with the scholars of the day, like the philosopher René Descartes, writing plays and librettos for performance at the Queen's Academy.[39]

On accession each of these remarkable young queens had the same problem. Whether in the sixteenth or seventeenth century, in England or in Sweden, each had to present herself as authoritative and dominant, a female who could command the same loyalties and service as a male monarch. To create this image and to keep their independence, neither Elizabeth I of England nor Christina of Sweden married. Each successfully convinced contemporaries that although a woman ruling alone she was a monarch to be respected and obeyed.[40]

Elizabeth I made no pretense of being like the warrior kings of the past; instead she created her own grand image, a magnificent virgin queen who must be flattered and courted. She was pleased to be known as "Gloriana," and she consciously cultivated grandeur in the pageantry of her life. By her sumptuous dress, by her makeup, and by her behavior she used the traditional image of the female to awe and manipulate her courtiers. She was vain, capricious, demanding. She expected praise for her beauty, her delicate hands, for the grace of her movements in a galliard. More importantly, she was intelligent and quick-witted. She enlisted the most adventurous of England's nobility into her service as her suitors and military leaders; inspired the support of able administrators like William Cecil, Lord Burghley, and Francis Walsingham; and maintained a following among the leaders of Parliament. Sir Christopher Hatton, a young favorite among Elizabeth's courtiers, described her methods:

> The Queen did fish for men's souls, and had so sweet a bait that no one could escape her network. . . . I have seen her smile—sooth, with great semblance of good liking to all around—and cause everyone to open his most inward thought to her; when on a sudden, she would ponder in private what had passed, write down all their opinions, and draw them out as occasion required, and sometimes disprove to their faces what had been delivered a month before.[41]

In creating her "network," Queen Elizabeth cultivated other images besides the feminine. To some she was—as John Foxe described her in his *Book of Martyrs*—the "savior" who came to deliver England from the Catholic faith so that her subjects might rest in "the miraculous custody and outscape [escape] of this our sovereign Lady, now Queen," rescued from "the strait time of Queen Mary her sister."[42] When circumstances demanded it she created her own version of the warrior king. At the time of the attempted Spanish invasion in 1588 she rode before her troops assembled at Tilbury in an armored breastplate, speaking of her "weak and feeble" female body, but adding defiantly that she had "the heart and stomach of a king, and of a king of England too." She told the assembled foot soldiers, "I myself will take up arms, I myself will be your general, judge, and rewarder of every one of your virtues in the field."[43]

By choosing not to marry and bear children, both Elizabeth of England and Christina of Sweden kept free of a husband's authority and the hazards of childbirth. But both created another problem: the lack of a direct heir. Even more than a king with his sons, or a woman who came to rule as regent for her children, each of these women had to neutralize or eliminate rivals, to still opposition, to administer the realm in peace and war. Each accomplishment meant stability for their subjects. In the sixteenth and seventeenth centuries, however, subjects also believed that the continuation of that stability was only insured by the birth of an heir to the reigning monarch.

Uncertainty about the succession had always been part of Elizabeth I's life from her own birth in 1533. Many in England had not been reconciled to the divorce and the break from the Catholic church that had enabled her father, Henry VIII, to marry her mother, Anne Boleyn. With the birth of Edward, Elizabeth's half-brother, in 1537 she found herself disinherited by an act of Parliament. Edward's early death at sixteen threw everything into uncertainty again. Elizabeth became the second in line after her Catholic half-sister, Mary Tudor, but Mary wanted her own child, for she knew that Elizabeth would rule as a Protestant, not as a Catholic.

Once safely on the throne in 1558 when she was twenty-five, Queen

Elizabeth combined decisiveness and prevarication to ease the religious conflicts among her subjects and to still the fears excited by prospects of a successor who might change the official religion of the state once again. Throughout her reign, Elizabeth adhered strictly to the 1563 Articles of the Anglican faith passed by Parliament. She gave rival claimants to the throne and potential heirs little chance of gaining a following. Her cousins, Lady Jane Grey's sisters, both married secretly. Queen Elizabeth sent one to the Tower, the other lived in disgrace. Mary Stuart, Queen of Scots, the woman in the direct line of succession and a Catholic, claimed the kingdom on her return from France in 1561, plotted with those who would take it for her in 1569 and again in 1586. When Mary Stuart's own subjects rebelled against her and she fled to England for sanctuary, Queen Elizabeth kept her confined and eventually allowed her execution.[44]

When Elizabeth's Parliament and her councilors advised marriage, the queen appeared to accede to their wishes. She sent her diplomats to gather information about suitors. Just as she enjoyed flirting with her courtiers, so it pleased her to flirt with foreign kings and princes, dangling the prospect of a marriage alliance as she held out the prospect of her favors. Even at forty-six, her principal minister, Lord Burghley, had noted for himself that "by the opinion of women, being most acquainted with her Majesty's body," she could still bear a child.[45] At the same time, however, she prevaricated and portrayed herself to her Parliament as more intent on being a "natural mother . . . unto you all."[46] When, in hopes of forcing a betrothal, parliamentary leaders limited her subsidies (as they did in 1566), she found other ways to supplement the royal revenues: cutting expenses, making loans and investments like that in Sir Francis Drake's circumnavigation of the globe. She made as much from that enterprise as she would have from the usual grant from the House of Commons.[47]

On Elizabeth's death, her kingdom passed quietly into the hands of her cousin's son, James Stuart. Her skills and her intelligent manipulation of personalities and circumstances made the peaceful transition possible. By the end of her reign Queen Elizabeth had succeeded in many ways as a ruler. She had accustomed her court and her subjects to her unusual status. The English took pride in their unique "Virgin Queen," an independent, powerful female monarch who governed in her own right.

Queen Christina of Sweden made different choices and left a different set of memories for her subjects. As a little girl, she, like Elizabeth I of England, accepted the rituals of the monarchy and looked forward to the power she would hold. She remembered in her autobiography that when she was six and her father's death had made her ruler of Sweden, she had not understood what was happening but "was enchanted to see all these people at my feet, kissing my hand."[48] She imagined herself a heroine of the classical age of Greece and Rome: a Queen of the Amazons, or Camilla, Queen of the Volscians, who hunted and acted like a man, the "virile" spirit in a female body. Like one of these women—like the "great men," as she called them— she too might become famous through her own deeds. In a childhood essay on one of the male conquerors of antiquity Christina explained that studying their lives

> instructs us, it corrects us, it raises the soul above itself, inflames it, and makes it understand its capacities. It is the noble sentiments and great actions of extraordinary men that fill the soul with virtue and vigor by a kind of happy contagion.[49]

Where Elizabeth cultivated femininity, Christina cultivated masculinity. As she came to rule, she held to the unorthodox upbringing and the masculine preferences and skills given her by her father's early tutelage. She wore her hair shoulder-length like a man, braided or held with a ribbon. In 1653 she received the English ambassador wearing a simple gray skirt and a man's long gray jacket. Her low voice and her energetic, abrupt movements impressed others as mannish. All accentuated the impression she wished to convey. She made no effort to learn "women's work and occupations," as she called them.[50] Instead she meant to be the first among male equals, able to ride for ten hours, to discuss philosophical questions, and to receive parliamentary delegations, just like a king. She succeeded. All of her behavior convinced contemporaries that "she was the daughter of her great father, accustomed to victory and triumph."[51]

For all of her efforts to behave like a king, and her success with her subjects, Queen Christina did not enjoy the role of monarch. Unlike Elizabeth I, who accepted each regal responsibility as a challenge, the Swedish queen feared that her royal obligations threatened her autonomy. Inflexible and unable to bring accommodation in others, she chose to turn away altogether. At the age of twenty-nine she abdicated in favor of her male cousin.

The conflict between Queen Christina's inclinations and her subjects' expectations appeared soon after her eighteenth birthday in 1644. Contemporaries assumed that she would soon marry and then bow to her husband's authority. This was never Christina's intention. Contemporaries said of the queen that "she freely followed her own genius in all things and car'd not what anybody said."[52] She accepted the tutelage of her principal minister and mentor, Axel Oxenstierna, who presided over an efficient modern bureaucracy, but otherwise she meant to be ruled by no one.

As Queen Elizabeth's subjects wanted her married with a designated heir, so from 1647 Queen Christina's parliament pressured her to marry her cousin

Karl Gustavus. To win his appointment as commander in chief to the army she had to make such a promise in 1648, but she never meant to honor it. Like Queen Elizabeth she did not want to accept the authority of a husband. She feared childbirth. She may have been in love with one of her ladies-in-waiting.[53] Unlike Elizabeth of England, however, Christina did not have the desire or the patience to play out the drama that might have allowed her to rule and to have her own way. She could not prevaricate and manipulate. She could not compromise. Instead, as early as 1649, when she was just twenty-three, abdication seemed the only alternative. As queen she felt that she could neither hold power, nor be "free," as she defined it. Rather than marry and make Karl Gustavus her husband, she had him formally accepted as her heir and in 1651 announced her intention to give up the throne in his favor. The Riksdag, the Swedish legislature, protested and prevailed upon her not to go. She did not change her views, however. She defied them again with her conversion to Roman Catholicism. In 1654 the Riksdag acceded to her second announcement. The ceremony of her abdication and Karl Gustavus's coronation took place on the same day. Christina rode out of Sweden with her hair cut, dressed as a man, believing that—Catholic, unmarried, and no longer a queen—she had at last found "liberty."[54]

After her abdication, Christina traveled for two years, finally settling in Rome. There she became renowned for her receptions, her sponsorship of music, plays, and operas. She built her own theater. She collected books and gave her patronage to artists. A diplomat, Cardinal Azzolino, advised her, helped her with chronic problems of finances, and perhaps became her lover. She played with the possibility of becoming Queen of Naples, and returned once to Sweden on the death of Karl Gustavus. She died in Rome in her sixties, having lived free of marriage, free of children, free of the pressures, dangers, and responsibilities that went with power.

Maria Theresa (1717–1780), Empress of Austria, took a different path from either Queen Elizabeth of England or Queen Christina of Sweden. She embraced monarchy, marriage, and children. She was the powerful and effective ruler of vast, diverse territories, the loving wife of Francis Stephen of Lorraine, and an affectionate mother to their ten surviving daughters and sons. Thus, she projected both traditional and unorthodox images to enhance her power: the dutiful wife and mother, and the successful monarch.

Where subjects accepted Elizabeth's and Christina's accession to their fathers' thrones, Maria Theresa had to fight for her crown. She had clear title to the lands of her father, Charles VI. Even before she was born he had prepared for her reign by stipulating the indivisibility of the empire in the Pragmatic Sanction presented to Europe's monarchs in 1713. Yet Charles VI had given Maria Theresa no special education or training for her future role.

He had merely arranged her marriage. On his death in 1740 his treaty did not prevent Frederick II of Prussia from seizing part of the Hapsburg Empire.

Maria Theresa later attributed her trust in "Divine Providence" with giving her the strength and the judgment to survive this, her first test as a female ruler of the Hapsburg dominions. Just twenty-three at the time of her accession, she later described her predicament:

> I found myself in this situation, without money, without credit, without army, without experience and knowledge of my own and finally also without any counsel, because each one of them first wanted to wait and see what things would develop.[55]

She acted quickly to claim the titles Archduchess of Austria and "King" of Hungary. As she explained to Count Kinsky, Chancellor of Bohemia, in 1741, she decided "to stake everything, win or lose, on saving Bohemia," even if it meant "destruction and desolation." She instructed him: "This, then, is the crisis: do not spare the country, only hold it."[56] She noted that he might think her "cruel," but this was of less consequence than appearing the confident, forceful monarch trying to prevent the dissolution of her empire.

Like Elizabeth of England, Maria Theresa capitalized on other images as well. She used the picture of herself as the ideal wife and mother as yet another proof of her dedication and her fitness as a ruler. As she was mother and protector to her children, so she was "the general and first mother" to her subjects. She explained in her *Political Testament* of 1749–1750:

> And dearly as I love my family and children, so that I spare no effort, trouble, care or labor for their sakes, yet I would always have put the general welfare of my dominions above them had I been convinced that I should do this or that their welfare demanded it.[57]

At the time of her accession she insisted that had she not been "nearly always *enceinte* [pregnant], no one could have stopped me from taking the field personally."[58]

Maria Theresa gradually took the power of government into her own hands, skillfully maneuvering and manipulating the men who served the state. The Prussian envoy wrote in 1747 of her blond hair, blue eyes, and heavy figure, but also that

> Her spirit is lively, masterful, and capable of dealing with affairs of state. She possesses an excellent memory and good judgment. She has such good control of herself that it is very difficult to judge from her appearance and behavior what she really thinks.[59]

Her archenemy, Frederick II of Prussia, admired her effect on the officers of the Austrian army: "above all she excited their devotion, talents, and desire to please her."[60] She gained the same ascendancy over her civilian officials. She inherited a powerful council and patiently waited for better times "to rid myself of their excess ascendancy," as she called it.[61] By the late 1740s no one questioned that she ruled. She reorganized the imperial administration, ending what she called the "paralyzing disharmony between all departments."[62] She made the Austrian nobility accept a tax. She promulgated new civil and criminal codes. With her adviser, Haugwitz, she reformed the organization of the army and created a standing force of over 100,000 professional fighting men to replace the old levy system.[63] She waged war and took territory like the most ruthless of her royal male counterparts.

Ruling such an empire as an absolute monarch took prodigious energy. Contemporaries described her day: she rose at five or six in the morning like Catherine the Great, taking time for mass, for prayers, for reading. She saw her family, consulted with her ministers, gave audiences to courtiers and visitors. With only café au lait to sustain her through the morning, at one o'clock she had lunch, a public ritual at court. (Later in life she ate alone.) The afternoon and early evening were given over to audiences, to reading, to walking, and, later in life, to a brief nap. After a light supper and time with her friends, she retired early. Even when she was young she went to bed by ten o'clock, sometimes giving a last hour to reading or writing.

As Maria Theresa grew older, she worried that her heir, the future Joseph II, would not be as diligent and thoughtful as she had been. "You are a coquette of the mind," she told him.[64] She feared Joseph's interest in and sympathy with the "philosophes" (men of letters who wrote suggestions for more liberal government) would lead him to undermine the power of the monarchy. She saw herself as "saving Europe," and with no one to maintain the struggle after her death. Their disagreements only intensified as Joseph played more of a role in the empire. When in 1776–1777 he urged religious toleration, she warned him that he endangered his own soul and "would drag the Monarchy down with you."[65] The War of the Bavarian Succession which he undertook in 1778 brought her condemnation and ended with her negotiating a settlement without consulting him. She saw catastrophe everywhere, "the destruction of our house and monarchy and even the total disruption of all of Europe."[66]

Maria Theresa was correct in her sense of imminent change and the end of the institutions that had characterized the world of the courts that she had labored to sustain and perpetuate. In addition, her death in 1780 signaled another end as well. The combination of circumstances favoring a daughter or a widow and opportunities for the confident and enterprising that had permitted so many royal women to act like men and to wield all the power of the absolute monarch never occurred again. The unique era of powerful female rulers that had lasted from the fifteenth to the eighteenth centuries passed into history.

These women rulers, whether regents or queens in their own right, played key roles in the histories of their dynasties and their countries. The wives of dukes and princes in fifteenth-century Milan and Ferrara held lands against invading armies. Kings' widows in sixteenth- and seventeenth-century France governed for their young sons. Royal princesses all across Europe from the sixteenth to the eighteenth centuries became their fathers' heirs, ruled over vast lands, waged war like kings, and passed their families' hegemony to the next generation.

Yet for all of their success, at the end of the eighteenth century the traditional denigrating attitudes about women's nature, proper function, and role stood intact. Even the female monarchs accepted the prevailing views. They explained their success by their differences from other women.

In the course of the nineteenth century limited monarchies and representative governments based on exclusively male rights to citizenship, with male control of money and property, evolved in Europe. In the wake of the French Revolution and Napoleon even privileged women had less opportunity than before to exercise any direct power outside of the family. The memory of Russia's Catherine the Great, of Queen Elizabeth of England, of the Austrian Empress Maria Theresa survived and were honored. But these powerful women remained aberrations in the popular imagination, not proof of the equal potential of the female mind and character.

5

NEW OPPORTUNITIES

✿

Performers and Composers

From as early as the fifteenth century the courts of Italy, of Burgundy, and then of the great royal dynasties offered new roles and opportunities to women other than those of ruler and courtier. In the twelfth and thirteenth centuries women and men—the trobairitz and the troubadours—played, sang, and composed, moving from court to court in France and then across the Channel to England. In the same way from the fifteenth to the eighteenth centuries skilled and gifted women enriched the formal entertainments and enlivened the leisure hours of families like the Viscontis, the Medicis, and the Dukes of Burgundy, of courts like those of the Valois, the Tudors, and the Hapsburgs. By the eighteenth century even the smallest courts of Europe's duchies and principalities had their own female performers and composers.

Some of these early women court entertainers played jester for their patrons. In the early fifteenth century the dwarf Madame d'Or (as she was known to the court, perhaps because of her blond hair) acted the fool for the Duke of Burgundy, Philip the Good. At the banquet given to celebrate his marriage to Isabella of Portugal, she and Hans, the court's male giant, wrestled after he had emerged from an enormous pie. For the wedding festivities of Charles the Bold she dressed extravagantly and acted as part of the decorations. Jane Bold held equal status at the English court with Will Somers, Henry VIII's jester. She continued as a member of the Tudor court into the reign of Henry VIII's daughter Mary, entertaining the courtiers by dressing like her betters one day, like a clown with a shaved head the next. At the end of the sixteenth century Mathurine amused her patrons Marie de' Medici and Henry IV of France by dressing as an Amazon, telling political

and religious jokes. The sixteenth-century Spanish court of Philip II organized cruel entertainments. The courtiers laughed at the epileptic fits and
hysterical tantrums of Magdalen Ruiz, brought to live at the palace from an
asylum.[1]

Such crude amusements disappeared as royal and aristocratic taste from
the sixteenth to the eighteenth centuries grew more sophisticated, the demands more elaborate. With these changes the opportunities for the talented
young female entertainer, the musician, composer, singer, dancer, or actress
multiplied. Initially, the women and men of the courts amused themselves
by cultivating their own skills in music, dance, and acting. A woman like
Isabella d'Este, in late-fifteenth-century Ferrara, sang and accompanied herself on the lute. Gradually, however, first music and then dance and acting
became the province of professionals; women and men, trained for formal and
informal performances, lived in the palace and held positions as regular
members of the courtly entourage.

Singers and composers were the first women to become professional
entertainers at the Italian courts. In 1580 Laura Peverara, Anna Guarini, and
Livia d'Arco, ostensibly ladies-in-waiting (each had connections with the
aristocracy or the Mantua merchant elite) were recruited to sing for Duke
Alfonso d'Este. A fourth young woman, Tarquinia Molza, joined them in
1583. The duke's officials found and brought others from all over northern
Italy to perform solos, duets, trios, and madrigals in the evenings between
six-thirty and nine-thirty and for special celebrations. The Florentine ambassador reported in June of 1581, "One can give no greater pleasure to the Duke
than by appreciating and praising his ladies who are constantly studying new
inventions."[2] The duke's *maestro di capella* (choirmaster), the most important musician at the court, made 125 to 135 scudi a year and had the income
of a farm the duke had given him, but these women benefited even more
handsomely. Peverara married in 1583 with a 10,000 scudi dowry and an
income of 300 scudi a year; Molza had the same income and both had rooms
in the palace like the maestro's.[3] Women singing in this way became popular
and the practice spread. By 1600 women traveled to perform music in courts
all over Italy. When Queen Christina of Sweden lived in Rome after her
abdication, she had two singers, Angelina and Barbara Georgini, living in
residence for her own theater productions.

Most sixteenth- and seventeenth-century female singers composed their
own songs; some wrote music for other occasions as well.[4] Two women gained
exceptional renown. Francesca Caccini (1587–1640?), the daughter of a
singer and composer, had her singing debut in 1600 and by 1607 had appeared at the Medici court. By 1623 she was the acknowledged star there,

earning 240 scudi a year. Only the Duke of Medici's secretary made more than she.[5] From the age of eighteen on, in addition to performing, she composed for the court's musical needs: songs, religious music, opera, and light entertainments for the intermissions between the acts of plays. She had her own group of women singers she had trained and then conducted in performance. In 1614 she wrote to her collaborator on librettos about the preparations for a new production and the difficulties of serving the rich and titled:

> Although we have rehearsed my music in the presence of Her Majesty, Madame, and the princesses, the Grand Duke has not yet heard it. However, we daily expect to be commanded, especially as one night we were assembled and ready until three o'clock in the morning, but because of the arrival of an ambassador, our performance was postponed to another night.[6]

Unlike Francesca Caccini, the Venetian singer and composer Barbara Strozzi (b. 1619) entertained outside the court. Her talent, however, drew the aristocratic and powerful to her performances. She was the illegitimate daughter of a poet and playwright, Guilio Strozzi, and a servant woman, Isabella Garzoni. Her father trained her as his protegée, founded an "Academy" for the intellectual elite of Venice, and arranged for her to host the meetings. The assembled courtiers discussed and debated, and Barbara Strozzi sang in the interludes.[7] Eight volumes of her vocal music appeared between 1644 and 1664, making Strozzi the most prolific cantata composer of the seventeenth century. Her father died a pauper in 1652, and the dedications of her compositions suggest that she probably wrote and published to support herself.[8]

Catherine de' Medici, the wife of King Henry II of France, made musical entertainments as integral a part of her life at the seventeenth-century French court as they had been in Italy. From then on women musicians and composers easily found sponsors among the French courtiers and their rulers. The new customs and the opportunities they created for women spread to courts throughout Europe. Female musicians performed on a variety of occasions for their patrons and even became their teachers. In France queens like Anne of Austria, wife of Louis XIII, had their own women musicians. Princesses learned to play an instrument from professionals like Marguerite Antoinette Couperin (1705–1778), who succeeded her father as the royal chamber harpsichordist.[9] Marie Antoinette, wife to Louis XVI, gave her patronage to every variety of music and musician, including the prodigy Angélique-Dorothée-Lucie Grétry (1772–1790), whose first divertissement was performed when she was just thirteen.[10]

With music such an important aspect of court life, with so many elaborate festivals and celebrations requiring formal and informal musical accompaniment, whole families of musicians found employment. Anne de la Barre (1628–1688), daughter of a royal organist, was a favorite entertainer of Cardinal Mazarin and of the court composer Jean Lully. The musician and composer Elizabeth-Claude Jacquet de la Guerre (c. 1664–1727?), daughter of Louis XIV's harpsichord maker, became a favorite of the king's mistress, the Marquise de Montespan. She was probably fifteen when her father brought her to court. From the age of seven she had been able to play at sight and to improvise on the harpsichord. Contemporaries marveled that "sometimes for an entire half hour she followed an improvisation and a fantasy with songs and harmonies extremely varied and in excellent taste."[11] With this talent she composed and had published all kinds of music: for opera, for ballet, for harpsichord, for violin, and three books of cantatas for voices. When Louis XV fell ill in 1721 it was her *Te Deum* that was performed at the chapel of the Louvre.[12] She in turn trained her young son to similar success. At eight, like his mother, he was a prodigy on the harpsichord. The daughters of Anne-Renée Rabel, a court musician herself, sang regularly for the king at mass. Marie-Louise Mangot, wife to the eighteenth-century French court composer Jean Philippe Rameau, worked as an accompanist.[13]

With the arrival of opera from the Italian courts in the seventeenth century, acting and dancing also passed from the domain of the aristocratic amateur into that of the professional, both women and men. The first actresses and actors probably came to the French court with the sixteenth-century Italian Commedia dell'Arte (an improvisational style of theater using a set group of characters). By the 1570s the antics of Columbine, Harlequin, Pantalone, and Scaramouche had become a familiar part of the entertainments improvised and written for Catherine de' Medici's court. Later in the seventeenth century, Louis XIV chartered the Comédie Française, uniting the theatrical groups headed by Madeleine Béjart and Marie Champmeslé.

Dancing as an occupation, and ballet in particular, also evolved as a separate profession under the patronage of Louis XIV. The Académie Royale de Musique that he founded in 1672 had a school of dance as well as a school for musicians. Initially professional women and men danced in the interludes between parts of longer musical or comedy entertainments, or as part of divertissements for operas. At the end of the seventeenth century Françoise Prevost became the first female dancer to achieve renown. No descriptions exist of her performances, but an engraving gives a sense of her artistry. She wears draped chiffon and satin and appears poised to run across the stage. This would have set her apart from other dancers whose costumes and performances were more formal and more stylized, much like the steps and dances done by the courtiers themselves. The ballerina commonly wore stiff court dress and shoes with wooden heels.[14] Ballet remained connected to royal and aristocratic families well into the nineteenth century. These families continued to give their patronage to performances and their support to training. Talented ballerinas, like musicians and actresses, continued to find the royal schools an avenue to success and achievement.

Painters

The role of court painter evolved differently from that of performer, composer, and dancer. Painting was already an established craft in the great merchant towns. Painters, like other craftsmen, were subject to the protection and regulation of their guilds. Within this professional framework women artists had numerous disadvantages to overcome. By the seventeenth century, access to training had become problematic for them. Aspiring female painters and sculptors had to depend on fathers, brothers, and family friends. Even when a woman was an active participant in the profession, the physical constants of women's lives could still prevent success. Marietta Robusti (1560–1590) worked with her brother in the studio of her father, the painter Tintoretto. But when Philip II of Spain and Maximilian of Austria both asked her to come to their courts, her father refused and instead arranged her marriage. She died in childbirth four years later.

Participation in the men's workshops, the guilds and academies—and the training, experience, and commissions that they represented—came rarely to women. The eighteenth-century Paris guild of artists, gilders, and framers, though theoretically open to all practitioners, had already begun to close its ranks to the potential competition of women. The lists show women as only 3 percent (130) of the total members from 1617–1777, when in fact many more must have been qualified.[15]

For women continued to paint. They studied from nature, learned perspective and technique. They created their own specialties. Thus, they circumvented the restrictions of the guild masters. They had earned a living this way in the great towns of Italy and the Netherlands, and so they achieved success in the courts of the sixteenth, seventeenth, and eighteenth centuries. Women specialized in still lifes and found patrons among the wealthy aristocratic families. Margherita Caffi (1662–1700), Maria van Oosterwijk (1630–1693), and Rachel Ruysch (1664–1750) all sold their works to the great royal art collectors. From 1708–1716 Rachel Ruysch enjoyed the title and rewards of court painter to the Elector of the German Palatinate.

As in the towns of these centuries, the other specialty women were allowed to dominate at court was portraiture. In sixteenth- and seventeenth-century England, women worked as illuminators and then turned their skills to the miniature portrait. Esther Inglis (b. 1571), the calligrapher, probably earned her living through the dedication of her manuscripts to members of the government and the royal family. Henry VIII requested that Levina Bening Teerlinc (c. 1520–1576) come from Flanders to paint miniatures. He paid her a life annuity of £40 a year, more than his court portrait artist, Hans Holbein. It was she who set the style made famous by Nicholas Hillyard in the 1570s. The tradition of female miniaturists at the English court continued. Susan Penelope Rosse (c. 1652–1702) completed a small portrait of Nell Gwyn, King Charles II's mistress.

The sixteenth-century King of Spain, Philip II, gave similar commissions to his protegée, the portraitist Sofonisba Anguissola (1527–1625). She was the daughter of an Italian noble family of Cremona, and her father encouraged her and her sister to study painting. Philip II rewarded her lavishly with a pension, a dowry on her marriage, and the position of lady-in-waiting to the court. Women artists found other patrons among the Hapsburg rulers. Caterina van Hemessen (1528–1587?) did commissions for Margaret of Hungary, the Holy Roman Emperor Charles V's regent in the Austrian Netherlands. As well as miniatures and portraits, van Hemessen painted large-scale religious scenes. She returned with Margaret to Spain, living on the pension she had been granted. In 1692 a later Hapsburg, King Carlos II, gave the title of court sculptor to Luisa Roldan (1656–1704). She had been trained by her father, worked in his studio, and came to Madrid and to the court originally with her husband, also a sculptor. Her own workshop produced religious figures in polychromed wood and religious scenes in terra-cotta.

In the eighteenth century at the court of the French Bourbons, women portraitists found commissions and established reputations for themselves. The Venetian Rosalba Carriera (1675–1757) began her career designing patterns for her mother, a lacemaker. She supplemented the family income painting snuffboxes and miniatures which she sold to the visiting aristocrats from other countries making their "grand tour." She went with her sister's family to Paris in 1720 and became the sensation of the court with portraits done in pastels, a particularly delicate medium requiring subtle use of shading and color.[16]

So successful was Carriera that other artists copied her technique. Among these was the father of the eighteenth-century French court's most successful female painter, Marie-Louise-Elizabeth Vigée-Lebrun (1755–1842). Vigée-Lebrun, the future favorite of Queen Marie Antoinette, trained with her father. Her first portrait of significance was of the Duchess of Chartres, and by the time she was twenty she had enough commissions to support her mother and brother with her painting after her father's death.

Her success depended upon her talent, her willingness to work long hours, and her skills as a courtier. Vigée-Lebrun painted in the morning. From 1776 on, she remembered, "It was difficult to get a place on my waiting list. In a word, I was the fashion."[17] She ate a midday meal at home, rested, and then had another sitting. She made it a rule to put off all social engagements while the "precious" daylight lasted. Even the birth of her daughter did not stop her. She continued working on a commission, *Venus Tying the Wings of Love,* through much of the labor.[18]

On an ordinary workday, after the second sitting, she would either go out for the evening or entertain guests herself. Even then she sometimes was preoccupied with her work. One evening, dressed in white to dine with the Princess de Rohan Rochefort, she went to look at the day's portrait and, lost in thought, sat down on her palette.[19] If she had guests herself, after a few hours of conversation, of charades, of reading aloud, she served a meal of fish or fowl, with salad and vegetables, at about ten. If out, she expected to be home by midnight.[20]

It was Queen Marie Antoinette who made Vigée-Lebrun the artist of choice for the French court, and earned her by her own estimate more than a million francs by 1789.[21] She did her first portrait of the queen in 1779. Marie Antoinette ordered two copies, one for herself and one for her brother, Joseph II, Emperor of Austria. In the end Vigée-Lebrun painted everyone in the royal family. She remembered the queen, her patron, admiringly in her memoirs, praised "the splendour of her complexion" and "her grace and kindness." Vigée-Lebrun told of the occasion during her first pregnancy when Marie Antoinette bent to pick up the paint box the artist had dropped, remarking: "Leave them, leave them. You are too far gone with child to stoop."[22]

For artists like Elizabeth Vigée-Lebrun, composers like Francesca Caccini, and dancers like Françoise Prevost, the world of the courts offered the opportunity to use their talents, to gain recognition and rewards. But these women were few, especially when compared to the numbers of men who achieved recognition in the same professions. In countless ways it remained harder for women to succeed, even in the world of the courts. Historians can never know how many gifted young women never gained access to the courts at all. The eighteenth-century composer Wolfgang Amadeus Mozart's father organized his son's first tour not for the five-year-old boy alone, but for both him and his older sister. They continued to play together publicly. In 1775 they appeared before Empress Maria Theresa of Austria.[23] The boy received mounting acclaim and commissions. Marianne Mozart has slipped away from the historical record.[24]

Courtesans

The extravagant new world of the courts offered some women another way to gain status, wealth, and influence irrespective of birth and outside of marriage and family. An enterprising young woman could become a courtesan, the lover and companion of an aristocratic or royal male patron. Prostitutes, women who sold the sexual use of their bodies, had always existed in Europe. Like prostitutes, courtesans initially gained a reputation and rewards because of their appearance and their willingness to have sexual intercourse. Here the similarity ends, for courtesans had to learn to do more for the men they served. They cultivated many talents. Some wrote poetry, others composed music. The most successful learned to create a luxurious environment that charmed and entertained the men who became their patrons.[25] Much like the Greek "hetaerae" who graced the gatherings of powerful men in the classical era in Athens, Italian courtesans of the sixteenth and seventeenth centuries presided over discussions of male Roman and Venetian writers and courtiers. In seventeenth- and eighteenth-century France royal courtesans, with the title of *maîtresse-en-titre* (official mistress), had their own households and apartments in the palace, held rank, and exercised influence. They enjoyed a position at court equal to the most illustrious members of aristocratic families.

The world of the courts made possible the revival of this ancient occupation and its institutionalization. Courtesans and the indulgent environment they created signified for the men of these centuries the luxury, opulence and display available with the new commercial wealth. This was a world that delighted in representations of the human body, which the sixteenth-century Italian sculptor Benvenuto Cellini called "the most perfect of all forms."[26] Nymphs, the goddess Venus, and half-clad shepherdesses decorated the paneled walls of ducal and royal palaces. Carved in marble and limestone, they lined the *allées* of the formal gardens and gathered water in the fountains. Engravers used their images to illustrate erotic and pornographic writings. Craftsmen carved them into the handles of knives and forks. Courtesans personified the pleasurable and the sexual for men.

Although nothing of the religious or secular law about illicit intercourse had changed, the courtiers and the kings of the fifteenth to the eighteenth centuries felt that they could periodically ignore the religious strictures and the traditional values decreeing the sanctity of marriage and the sinfulness of "fornication." In the gatherings over which women like the Italian courtesans presided, men took flirtation, dalliance, and illicit liaisons for granted. They discussed treatises from earlier centuries like *The Art of Courtly Love*, by Andreas the Chaplain (Andreas Cappellanus), and debated the questions posed about the nature of love and how it might be enjoyed outside of marriage.[27] By the eighteenth century such attitudes and sexual infidelity had become commonplace at royal courts all across Europe. Having married and produced a male heir, aristocratic wives and husbands felt free to have affairs with others. In the France of King Louis XV, the courtier Emilie du Châtelet married in 1725 at nineteen and bore three children. By 1726 her husband had returned to taking young peasant women as mistresses. When she took her first two lovers her husband's only comment was that he hoped she would choose better the next time.[28]

The figure of the courtesan had emerged from the luxurious and dissolute world of the papal court at the end of the fifteenth century. Though meant to be celibate, the princes of the church hired the women that the Master of Ceremonies of the Papal Court called our "respectable prostitutes" and kept them as mistresses. Alexander VI, elected Pope in 1492 at the age of sixty-one, had already fathered six children with his lover, Vanozza dei Cattanei. Giulia Farnese, just seventeen, joined him in the Vatican and became the model for the Madonna the artist Raphael frescoed on the walls of the pope's private apartments. In many ways Roman courtesans lived as if they were respectable. They went to church and sat with other women. They established themselves in elegant apartments near the Osteria dell'Orso, Rome's most prestigious inn. They maintained households of numerous family members including their own children and servants.[29] When the spiritual anxiety and reforming zeal of the late-sixteenth-century popes led to periodic bans, the courtesans fled to other towns. In Venice the custom arose of five or six patricians sharing the cost and the services of one woman. During the day she did as she chose; evenings followed a schedule, with a night reserved for each of her patrons.

In the seventeenth century Venice surpassed Rome, and the magnificence of the courtesans drew visitors to both towns. The 1565 catalog of *Most Honored Courtesans* gave the traveler a listing of the 215 women working in Venice, complete with the names of their agents, their addresses, and their prices.[30] Proud of their courtesans' renown, the Venetians called them "our praiseworthy prostitutes," and like the Roman government set special taxes for the women to pay.[31]

Whether in sixteenth-century Rome or eighteenth-century Versailles, courtesans began their careers much like prostitutes. They had their youth, their beauty, and their sexuality. To gain the status, rewards, and potential influence of the aristocratic or royal favorite, however, required the help of others, as well as money and training. At the beginning of the sixteenth century Lucrezia da Clarice, a Roman tavern dancer, caught the attention of Zoppino, a notorious pimp, who taught her how to dress and speak and then introduced her to Cardinal Campeggio, Bishop of Bologna, who became her patron. In the eighteenth century Mme. de Pompadour (1721–1764) and Mme. du Barry (1743–1793), royal mistresses to France's King Louis XV, were specifically trained by aristocratic male mentors and then presented to the royal patron to win the rewards of favor for their sponsors and themselves.

Parents sometimes initiated and schooled their daughters in the arts of the profession. Fiammetta (born c. 1465), the first Roman courtesan whose wealth is documented, came to Rome from Florence with her mother at the age of fourteen. Pope Sixtus IV became her patron, ensuring that she had a house and country property outside the town. The famous Imperia (1481–1512) had been introduced by her parents who hoped that she would make their fortune. Her mother, Diana, had been a prostitute; her stepfather, as part of the Sistine Choir, was a member of the papal household. From the age of seventeen Imperia went from success to success. Her house in Rome with a nude Venus on the exterior wall had been a gift from a nephew of Pope Pius II. The papal secretary to Leo X became a benefactor. Most loyal and generous, however, was the banker Agostino Chigi, the executor of her will, the father of her two daughters, and their protector after her death.

Such success came to a courtesan like Imperia because of the pleasing environment she created for her patron and the many talents she displayed. As mistress to the Roman banker Chigi she wrote sonnets and madrigals and presided over a circle of intellectuals that congregated at "La Farnesina," his elegant small palazzo (palace) outside the town walls.

Most famous of the Venetian courtesans was Veronica Franco (1546–1591). Both she and her mother were listed in the Venetian traveler's guide of 1565, both at the low fee of only two scudi.[32] By the time she was in her twenties, however, she had acquired the attention and patronage of Venice's patrician and merchant families. In July of 1574 Henry III of France stopped in Venice. Given his choice of the town courtesans by the Venetian government (the Signoria) he picked Veronica Franco after seeing an enameled miniature. She pleased men by her beauty and sexual talents. She considered herself like Venus and was proud of her artistry: "So sweet and appetizing do I become when I find myself in bed with he who loves and welcomes me, that our pleasure surpasses all delight."[33]

Most important, however, and the source of her fame, were her other talents and skills. Her poems and letters suggest the special environment she fashioned for the men who sought her out. In her drawing room and her bedroom they could, as she described it, "dine cozily together . . . without pomp or ceremony."[34] She sang and played for musical evenings. Her male guests shared their writings and encouraged her with her own poetry (published in 1575). In one of her verses she challenged a faithless suitor and set her bed as the dueling site:

> Step up and arm yourself with what you will.
> What battlefield do you prefer? the place?
> this secret hideaway where I have sampled—
> unwarily—so many bitter sweets?[35]

A Florentine, Niccolo Martelli, summarized the gifts of such a woman. He wrote to a friend:

> and the royal way in which they treat you, their graceful manners, their courtesy and the luxury with which they surround you—dressed as they are in crimson, and gold, scented, and exquisitely shod—with their compliments they make you feel another being, a great lord, and while you are with them you do not envy even the inhabitants of paradise.[36]

Seventeenth-century Parisian courtesans played similar roles and displayed the same variety of talents. Ninon de Lenclos (1620–1705) hosted gatherings of courtiers and intellectuals that rivaled those arranged by the most respected and sedate members of society. She made no secret of her enjoyment of the life she led and of the pleasures of sex. She suggested that other women felt the same. She explained, "It is more because of men's maladroitness than because of women's virtue that the latter's hearts resist capture."[37]

At the French court of the seventeenth and eighteenth centuries the role of courtesan attained new levels of sophistication and won the women unprecedented rewards and influence. One week after the funeral of his mother, Anne of Austria, King Louis XIV created the title *maîtresse-en-titre* and thus elevated his mistress to the level of an official of the court, acknowledged publicly as his companion, with rank and status. The *maîtresse-en-titre* had her own apartments in the palace and her own household. Children born of the liaison were second only to the king's legitimate offspring. Even Louis XIV's wife, Queen María Teresa, had to accept her. Some became attendants to members of the royal family, even to the queen. By the eighteenth century a gifted woman like Mme. de Pompadour, *maîtresse-en-titre* to Louis XV, functioned almost like a principal minister to the crown. Though not directly involved in government, she devoted herself to the king's interests, was courted for her access to the royal presence, and enjoyed all of the rewards and power of royal favor.

Although French monarchs had established a long tradition of royal concubines, in the seventeenth century King Louis XIV transformed his extramarital attachments into a permanent institution of the court.[38] Each of Louis XIV's principal lovers gave a different character to the office. The ash-blond delicacy and innocence of Louise de la Vallière (1644–1710), daughter of a cavalry officer and the first recipient of the title, held the king's favor until Françoise-Athénaïs Rochechouart Mortemart, Marquise de Montespan (1641–1707), caught his eye and won his bed with her brash vitality. Montespan, the third daughter of an old noble family, was beautiful and vain. Her sister Gabrielle spoke for them both when she said, "My beauty and the perfection of my wit, are due to the difference which birth put between me and the commonality of mankind."[39] With all of her tempers and demands she kept the king's interest and support for twelve years. Her son, the Duke of Maine, remained Louis's favorite child, and both he and his brother became "Princes of the Blood," raising them above the usual stigma of royal bastards. Even in 1679 when the king's attention was wandering from her to the children's governess, Françoise d'Aubigné Scarron (the future Mme. de Maintenon), Montespan won the title of duchess from Louis XIV and the highest appointment open to a woman at court, Superintendent of the Queen's Household.

Poise, accommodation, and the ability to create a quiet, peaceful environment were the qualities that won the forty-one-year-old Louis XIV's favor for the matronly Mme. de Maintenon (1635–1719). A daughter of the minor nobility and widow of a Parisian playwright, she had been chosen at thirty-three by Montespan to take responsibility for the children born from her liaison with the king. As each baby arrived Maintenon supervised their separate households. Intelligent, discreet, and firm, she gained the king's approval by her care of his children. Later in life the king fondly remembered the maternal scenes of his visits: a little girl at her side, a baby asleep in her lap, his favorite—the eldest boy, the Duke of Maine—in her arms.[40] From 1674 to 1680 she gradually supplanted Montespan and began to accumulate the perquisites of royal favor: a generous pension, a royal governorship for her brother, property, and a title. The widow Scarron became the Marquise de Maintenon. In 1680 she became lady-in-waiting to the wife of the dauphin, a new post created for her—a post as attendant to a legitimate member of the royal family.

After Queen María Teresa's death in July of 1683 Maintenon became the king's consort in all but name. Louis XIV had no further liaisons with other women.[41] In August she moved into the queen's apartments at the palace of Fontainebleau, and the pattern of their life together took shape. Ever discreet, ever supportive, she became part of the royal family circle. In public Louis called her "Madame," the title traditionally reserved to the Queen of France. Potentially, Maintenon had indirect power unlike anyone else's at court, but it meant exacting service. She described the events of an afternoon:

> When the King returns from hunting he comes to my room, the door is shut and no one else is admitted. Now I am alone with him. I have to listen to his worries, if he has any, to his melancholy, his vapors. Sometimes he has a fit of weeping, which he cannot control, or else he is ill.

In the evenings his ministers might come as well; sometimes he worked alone while she embroidered. Mme. de Maintenon was always available: "If they want me in their councils they call me, if not I withdraw a little further off."[42]

The couple grew old together, into their seventies. As Louis XIV sustained his absolute power until the end of his life, so she, his consort, maintained hers as his unofficial adviser and confidante. Almost their last activity together was to burn their confidential papers. Illness struck Louis unexpectedly and weakened him quickly in the spring and summer of 1715. He had Maintenon move to the adjacent bedroom. She wrote down his farewell and kept it with her will:

> He said goodbye to me three times, the first time saying that he regretted nothing except having to leave me, but that we should meet again very soon. I begged him to think only of God. The second time he asked my forgiveness for not always having treated me kindly; he had not made me happy, but he had always respected and loved me dearly.[43]

Mme. de Maintenon ended her life in quiet retirement at St. Cyr, the school for the daughters of aristocratic families that she and the king had founded. As Louis XIV's companion she had shown how a thoughtful woman could rise to new heights of prestige and influence in the courtier's world of the absolute monarch.

Unlike Louis XIV's mistresses, the women chosen by the eighteenth-century French king, Louis XV, were not necessarily from noble families. In an odd gesture of propriety he insisted they be married, but otherwise he waited contentedly while enterprising courtiers groomed women for his approval. In Jeanne-Antoinette Poisson, Madame de Pompadour (1721–1764), the French institution of *maîtresse-en-titre* gained its most sophisticated and polished practitioner. In the course of her time with King Louis XV she fulfilled many different roles: sexual partner, architect and designer, hostess and intellectual companion, adviser and orchestrator of delicate diplomatic maneuvers. In return for her service, she enjoyed all of the rewards and privileges her royal patron could bestow.

Mme. de Pompadour had been trained by a royal official, Normant de Tornehem, who befriended her family when her banker father had to flee the country in 1725 because of a grain scandal. An actor from the Comédie Française showed her how to move and speak; she had lessons on the clavichord, in painting, and in a variety of academic subjects. Married to Tornehem's nephew in 1741 at twenty-one, she was presented to the king. In April of 1745, when she made her first public appearance at Versailles, it was evident that the king had chosen her as royal favorite. Within months she had gained the title Marquise de Pompadour and been formally separated from her husband. She moved to apartments at Versailles, dined with the king one evening, and officially assumed the role of *maîtresse-en-titre.* [44]

Mme. de Pompadour appealed to the king initially and held his favor not so much because of her sexuality but because of the delicate elegance that characterized so much of her behavior as well as her appearance. In her portraits satin gowns are fitted at the bodice and flow in yards of subtly colored material from her waist to the floor. She often wears a white ruff at her neck, small flowers as decoration for the dress and for her lightly powdered, softly curled hair. Her favorite painter, François Boucher, showed her in one portrait with the quill, ink, seal, and wax ready on a small inlaid table beside her. In the corner of the room, music and drawings lie about as if already studied and discarded. She holds a book open in her hand as if she had been reading. All suggest her many interests and the variety of ways in which she engaged the attention and affection of the French monarch.

All of her time and energies went to his entertainment. She had a library of over thirty-five hundred volumes of poetry, novels, philosophy, histories, and biographies to read herself or to share with him. [45] She oversaw evenings and dinners with courtiers. She sang, she acted and arranged performances of operas. She gave her patronage to Voltaire for his play writing. She never tired of interesting Louis XV in her building and redecorating plans for new estates they acquired and for the old palaces of Fontainebleau and Versailles. She reorganized the china factory of Vincennes and began the one at Sèvres. During the Seven Years War, one of her rooms became the king's office where he met with his ministers.

A typical day for Mme. de Pompadour began with mass at eight, then visits to the queen, the dauphine (wife to the heir to the throne), and the king's sisters. Then she received courtiers, wrote letters, and worked on the preparations for the king's supper party. Each day Louis XV chose his companions from a list she had prepared. The entertainments, hunting, or riding filled the afternoon, the rituals of the court the rest of the night. Her duties might not end until two or three in the morning. [46] Occasionally she left the court to see to one of her building projects. In theory she was always to be available to the king. A staircase connected his apartments to hers. When, because of her miscarriages and what appeared to her to be the beginnings of aging after 1751, she no longer fulfilled his sexual needs, she set up a villa in the town of Versailles known as the "Parc aux Cerfs," and staffed it with young women for the king to use. As early as 1754, though only thirty-nine, she felt the strain of her life. The reward of the title "lady-in-waiting" to the queen in 1756 only meant more duties. At forty-two in 1763 she spoke of her life as "like that of the early Christians—a perpetual combat."[47] Undoubtedly weakened by her arduous routine in January of 1764 she contracted a respiratory infection. She died quietly with her power intact, but at an awesome price.

For a woman like Mme. de Pompadour, with all of the opportunity, celebrity, influence, and wealth, the life of a courtesan, whether at the papal court or in the royal palaces of kings, remained a difficult and precarious one. The grandeur and the rewards were counterbalanced by the vulnerability and transience of a life based on keeping the favor of powerful men. Italian courtesans had to contend with the periodic harassment of secular and religious authorities: the sumptuary and residency laws governing prostitutes were used against them. In 1580 Veronica Franco was called before the Inquisition, accused of using spells and not keeping fast days. Only a plea of illness forestalled her questioning.

Roman and Venetian courtesans commonly employed bodyguards, for these women who lived just outside the bounds of the socially accepted roles of their sex remained physically vulnerable. A jealous client stabbed the Roman courtesan Beatrice Spagnola (d. 1539); in the early 1530s Giulia del Moro, known in Venice as "La Zafetta," went with a lover for a picnic on one of the islands in the lagoon. She was raped first by him, then by his friends, and left to find her own way back to the town. Just like their poorer sisters, the prostitutes, courtesans might be the victims of popular anger. When in 1566 Pius II, in a burst of reforming enthusiasm, gave them six days to leave Rome, twenty-four gathered their possessions and left the town. Just outside the walls they were robbed and thrown in the Tiber River. Their success also depended upon their health. Syphilis was a hazard for courtesans as it was for prostitutes. It killed the Medici favorite, Beatrice de Ferrara. Intercourse often meant pregnancy. Louis XIV gave the pretty nineteen-year-old Marie-Angélique, Mademoiselle de Fontages, a duchess's title and a generous pension of twenty thousand écus when he chose her to be the Marquise de Montespan's successor in April of 1680.[48] He dropped Fontages just as quickly when hemorrhaging caused by the birth of a stillborn child the following January left her weak. She died two months later.

The mistresses and companions of kings also experienced other kinds of vulnerability. Some kings delighted in humiliating their women. Louis XIV of France insisted that Louise de la Vallière continue to act the part of his official mistress long after she had requested to go to a convent and for a year after he had turned to the Marquise de Montespan. His contemporary, King Charles II of England, treated his women with apparent gallantry but used them thoughtlessly. William Chiffinch, his Page of the Bedchamber, found women for him. His friend the Duke of Buckingham tried out the young actress Nell Gwyn (1650–1687) before recommending her. Once Charles II had decided he enjoyed her lively antics, he brought her to court with another favorite, the French Louise de Kéroualle, Duchess of Portsmouth (1649–1734), and allowed them to compete for his attention. In 1675 Mme. de Sévigné described the scene to her daughter. The duchess "has been disappointed in nothing"; the king came to her "almost every night in the face of the whole Court." She had been rewarded with titles and "treasure." Then "the actress" took "half his care, his time." In addition, Nell Gwyn "insults her, makes faces at her, attacks her," steals the king from her, and boasts of his preference for her.[49]

Worst of all, few of these women, even the most honored of courtesans, lived very long or died with more than the barest necessities. Imperia, the Roman courtesan, poisoned herself at thirty-one. Tullia d'Aragona (1508?–1551) died in a room at an inn owned by her former maidservant. The bulk of her possessions went to a secondhand clothes dealer for 12½ scudi.[50] Nell Gwyn spent much of her time with the English king, Charles II, asking for loans and borrowing money to maintain the house where she entertained him. To others he gave titles and privileges; for her it was an annuity "at his pleasure."[51] Even with favored treatment the well-born duchess Kéroualle died impoverished. Charles II's successors Queen Mary and King William stopped her pension.

In the sixteenth century in Venice Veronica Franco tried to set up refuges for prostitutes and courtesans. Well-born women of patrician families supported the establishment of one in 1580. Franco herself had a hospice built in 1591, the Casa de Soccorso. When a friend planned to train her own daughter for the profession, Franco bitterly wrote:

> I tell you, you can do nothing worse in this life . . . than to force the body into such servitude, . . . to give oneself in prey to so many, to risk being despoiled, robbed or killed. . . . To eat with someone else's mouth, to sleep with the eyes of others, to move as someone else desires, and to risk the shipwreck of your faculties and your life—what fate could be worse?[52]

Franco knew that the life of a courtesan offered rich opportunities and rewards, but like the life of the less fortunate prostitute it depended in the end on the woman's ability to please men. Even the French royal mistresses with their official status and influence were subject to the whim of their royal patron, their lives defined and limited by their relationship to men.

Writers

In orchestrating an evening at Versailles, Mme. de Pompadour, Louis XV's *maîtresse-en-titre*, called on poets and playwrights for verses and for scenes to amuse the king and the group of courtiers invited to share his evening. Calling on writers for entertainment was an old tradition of Europe's aristocratic and royal courts. As early as the twelfth century, the female trobairitz and male troubadours had composed for similar noble gatherings. The fourteenth-, fifteenth-, and sixteenth-century courts of the wealthy and aristocratic Italian and Burgundian families commissioned prayers for religious observances, celebratory odes for special occasions, even translations and commentaries on the works of others.

By the seventeenth and eighteenth centuries the members of the courtly world read histories, memoirs, and frivolous tales as well as poetic and didactic works. All meant that the need for every kind of writing and therefore the opportunities for writers had proliferated with the increasing number and sophistication of the activities of the courts.

Most courtiers, both women and men, wrote as an avocation, an enterprise that gave them pleasure or gained them recognition for their intellectual abilities. In the sixteenth century Marguerite of Angoulême (1492–1549, known as Marguerite of Navarre), wife of Henry d'Albret, King of Navarre, created a court renowned for its learning and scholarly activity. She herself composed religious verse, love poetry, and a collection of tales, *The Heptameron*, modeled on the Italian Giovanni Boccaccio's *Decameron*. María de Zayas y Sotomayor (1590?–1661?), part of the aristocratic world of Madrid in Spain, wrote verses and novellas.

For others, however, writing was more than an amusement or an intellectual endeavor. For some women it became the means to gain the favor of an aristocratic or royal patron, another way for a woman to gain access to the courts and their rewards. Marie de Gournay (1566–1645), with her writings and edition of the works of the French essayist and courtier Michel de Montaigne, was awarded a pension by the seventeenth-century queens of France, Marie de' Medici and Anne of Austria. In 1786 it was because of her success as a novelist that Fanny Burney (1752–1840) was chosen by Queen Charlotte, wife of England's George III, for an office in the royal household.[53]

Christine de Pizan (1365–c. 1430), the most prolific and the most successful female professional writer in the world of the courts, made her living
from her writing. The daughter of an Italian astrologer brought to France
to serve King Charles V, she married one of the king's secretaries and lived
the life of a courtier's wife until her husband died suddenly in 1390 when
she was twenty-six. Left with three children, her mother, and two brothers
to support, she spent the first fourteen years of her widowhood using what
influence she had with her husband's former colleagues and her father's
former associates to collect fees owed her husband, even suing in the law
courts.[54] This life of pleading and disputation gained her little, however,
and she turned to her own resources. Leaving the running of the household
to her mother, she embarked on a program of self-education and began to
write.

From the late 1390s on, Christine de Pizan developed her talents as a
writer and earned her reputation. The variety of her works attests to her
professional skill and to her determination to succeed: love poems and long
verse allegories presented to potential royal patrons; replies in a dispute on
the female nature and on the literary worth of a thirteenth-century romance;
musings on fate and misfortune; histories to glorify a French king and to
honor women of the past and present; didactic pieces for a dauphin and for
the women of the court; books of instruction in arms; speculations on the
state of the kingdom; religious prayers; and musings on the loss of loved
ones.[55]

Initially Christine de Pizan's popularity depended more on the novelty
of her gender than the originality of her verse. Gradually, however, contemporaries acknowledged her talents. She was in her early thirties when her
verses gained the attention of the royal court. She wrote her first longer
piece, *Letter to the God of Love* (1399), for a competition, a common form
of amusement and entertainment at the courts of the most prominent nobles of northern France. With these early poetic efforts and her involvement in a debate among the learned of Paris over the literary merit and
images of women in the thirteenth-century verse allegory *The Romance of
the Rose*, she established her place as a court writer and began to secure its
rewards. England's Earl of Salisbury gave her son a place in his retinue of
1397. Her daughter gained admission to the aristocratic and royal nunnery
of Poissy; later a niece was dowered by another patron, the Duke of Burgundy. Isabelle, the Queen of France, honored Christine de Pizan with the
traditional royal gift of goblets and tankards after receiving an edition of
her collected works.[56]

Although Pizan continued to write poetry throughout her life (her last known work is a poem about Joan of Arc's victory at Orléans in 1429), increasingly she devoted her time and talents to historical, philosophical, and didactic works. These writings brought her recognition, popularity, and relative financial security. *The Book of the Mutation of Fortune*, presented to Philip the Good, Duke of Burgundy, on New Year's Day of 1404 brought her the duke's favor and her first commission, to write the life of his brother, Charles V, King of France. From 1405 until 1416 she dedicated her works to this family, presented her manuscripts at Christmas or New Year's, and they rewarded her, although never as quickly as she expected. In her autobiographical work, *Christine's Vision*, she described "help" given "reluctantly and not very generously, and even when given, the delay in payment and the need to pursue their treasures to receive it."[57]

Pizan saw her works become popular in her own lifetime. Multiple copies of manuscripts have survived, an indication of how many must have been made. Some of her works were printed in England at the end of the fifteenth century. King Henry VII of England is supposed to have requested her *Book on Arms and Chivalry* from the printer William Caxton.[58] Others were printed in France early in the sixteenth century; *The Book of Three Virtues*, on proper activities and attitudes for women of the different estates, had appeared three times by 1536.[59] Christine de Pizan took an active role in the production of her works. She herself commissioned the English translation of *The Book of the City of Ladies*. She employed scribes, supervised and perhaps copied parts of some fifty-five manuscripts.[60]

This unusual popularity came because of her skill in choosing topics and her ability to use all of the resources available to her. A true professional, she learned to write what pleased and interested her contemporaries. She described herself as like the "embroiderers" who "do a variety of designs according to the subtlety of their imaginations." So she selected pieces from others' works and made her "compilation."[61] She synthesized information and presented it in ways that her audience could understand and appreciate.[62]

Although she was popular in her lifetime, within a few generations knowledge of Christine de Pizan's works was lost. Her writings disappeared until rediscovered in the last half of the twentieth century. In contrast, the novel *The Princess of Clèves* (published in 1678), by Marie-Madeleine Pioche de La Vergne, Countess de La Fayette (1634–1692), not only won recognition for the author but also became an acknowledged classic of the French language.

The Countess de La Fayette came from and married into a family of court officials, part of the *noblesse de robe*. She turned to writing after she had left her husband's country estate to live in Paris. Encouraged by her friends, especially François de La Rochefoucauld, Prince of Marcillac, she wrote her novel. It is the originality of the Countess de La Fayette's approach to a popular seventeenth-century theme—chaste love between a married woman and a courtier—that gained *The Princess of Clèves* its contemporary following and its place in French literature.[63]

From the fifteenth to the eighteenth centuries many women and men of the courts wrote about love. The seventeenth-century Spanish courtier María de Zayas (c. 1590–1661) and the learned sixteenth-century queen Marguerite of Navarre (1492–1549) portrayed its ecstasies and its pain.[64] La Fayette's skill lay in her ability to go beyond the traditional metaphors and formalized idealizations. In the delicately painted scenes of meetings in boudoirs, salons, and gardens the three main characters become real. There are the formal rituals of the court, the standard description of heroine and hero, but all is juxtaposed against the emotional interplay surrounding the princess, her husband, and her lover. The heroine finds herself in one impossible situation after another; her loyalty to her husband conflicts with her passion for her lover; her desire clashes with her fear of pain and loss of reputation. The intensity of the emotions and the sincerity of the princess's pronouncements raise the novel above the traditional moral tales of chaste love. Even the death of the princess's husband does not end the conflict. Unable to quiet the fear that her lover's devotion will prove mutable and "grow cold," she rejects his offer of marriage. "Peace of mind," she explains, can only come by never seeing him again.[65]

By writing, and by her success, the Countess de La Fayette challenged the traditional images of the female nature and the roles and functions usually prescribed for women, even in the world of the courts. Like the musicians, the artists, and courtesans, she achieved independence and definition outside of marriage and the family. At the same time, however, all of these talented women suffered from the disadvantages of being female. Only a few women painters like Elizabeth Vigée-Lebrun even gained access to the rich rewards of the court. A successful courtesan had to satisfy the full range of a lover's physical and intellectual needs. A double standard of skill affected the careers of women performers. The career of the talented female musician often began like that of the eighteenth-century harpsichordist Elizabeth Claude Jacquet de la Guerre, who was first heralded as a freakish prodigy and only later gained acceptance as a professional.[66]

Having overcome the difficulties themselves, most gifted women did not question the limiting attitudes and circumstances that continued to constrain them more than their male counterparts. Talented women embraced the opportunities of this courtly world. Only after 1789 when many of Europe's established institutions came under attack did more than a few women perceive their disadvantages. Then in the nineteenth century increasing numbers of women worked to gain equal access to training, patronage, and the commissions that made it possible to earn a living as a composer, performer, artist, or writer.

6

THE LEGACIES OF RENAISSANCE
HUMANISM AND THE SCIENTIFIC
REVOLUTION

✿

Renewed Access to Learning

Just as traditional attitudes about the female's proper function and role restricted women seeking the professional opportunities of the courtly world, so they hampered women seeking its intellectual rewards. Learning was for men, not for daughters, wives, and mothers. At first, however, in the enthusiasm for the new secular learning of the fifteenth and sixteenth centuries, Europe's privileged and educated men allowed and encouraged the women of their families to join in the intellectual movement known as Humanism, to participate in the philosophical and religious speculations that characterized this rediscovery and new appreciation of Greek and Roman authors. Women, as well as men, read, translated, and wrote critical studies of classical and early Christian texts.

Not since the twelfth century had women had such opportunities for serious study, and then it had almost always been as members of religious orders. From the seventh to the twelfth centuries within the cloistered walls of Europe's monasteries and convents, elite women had been allowed to learn Latin and Greek and to study classical and Christian authors. However, with more restrictive attitudes about women's function and role within the Church such education had become less and less a part of their preparation for a holy life or of the activities permitted to them as nuns. Outside of the Church only an exceptional woman like the poet Marie de France in the twelfth century sought out and found writings from earlier centuries.

This revival of scholarly learning for elite women began as early as the late fourteenth and early fifteenth centuries when queens and noblewomen like Anne of Brittany, wife of Charles VIII and then Louis XII of France, had their own libraries and commissioned their own manuscripts. The Portuguese princess Isabella (d. 1471), after her marriage to Philip the Good of Burgundy, brought scholars to her court and did her own translation of the Greek historian Xenophon.[1] The fifteenth- and sixteenth-century rulers of the Italian courts embraced the new learning eagerly. They gave their patronage to scholars and educated their daughters as well as their sons according to the Humanist attitudes. Eleanora of Ferrara (1450–1493), wife of Duke Ercole d'Este, collected a library of religious and classical texts at her fifteenth-century court. Her daughters Isabella and Beatrice d'Este learned Latin and Greek along with their brothers.

Isabella d'Este (1474–1539) surpassed her mother. As wife to Francesco Gonzaga, Margrave of Mantua, she became a model of how a woman might use her learning in the age of Humanism. She too collected manuscripts and studied the classical authors, but she also made her own translations. She corresponded with the learned of the day.[2] Contemporaries admired her energy and her attainments and called her "first lady of the world."[3]

The queens of Europe's royal courts also began to educate their daughters, a tradition which continued well into the seventeenth century. Isabella of Spain (1451–1504) assembled a library of spiritual and secular, classical and contemporary books. She commissioned the first Castilian grammar and brought the scholar Beatriz de Galindo (1473–1535) from Italy to teach her Latin.[4] She supervised the education of her children as they traveled with her throughout the kingdom. Juana, her eldest daughter, as a young girl made the ceremonial Latin response on a family state visit to Flanders; a younger daughter, Catherine of Aragon, wrote fluent Latin, and later provided a similar education for her own daughter, Mary Tudor. The English Stuarts, the seventeenth-century successors to the Tudors, carried on this tradition. Bathsua Makin (1608?–1675?) tutored Elizabeth (1635–1650), King Charles I's daughter, in Latin, Greek, Hebrew, French, and Italian. Elizabeth of Bohemia (1618–1680), Charles I's niece, corresponded with the French philosopher Descartes, and he dedicated his *Principles of Philosophy* to her in 1643.[5]

Louise of Savoy (d. 1531), sister-in-law to King Louis XII of France and mother to the heir, taught her children French and Spanish herself. She amassed a library of about two hundred volumes including religion, historical chronicles, and romances.[6] Her daughter Marguerite of Navarre's (1492–1549) reputation for learning inspired French scholars to cite her as an exemplar and to favor women's education in general.[7] Like other Humanists, Marguerite of Navarre's studies led her into sympathy with those questioning the traditional religious beliefs and rituals of the Catholic church. Her writings, like *Mirror of the Sinful Soul (Le miroir de l'âme pécheresse)* demonstrated her tendency to mysticism and her enchantment with Neoplatonism. She and her circle brought such ideas and the works of Dante and Petrarch to the attention of the French court. At the time of her death she had begun the seventy-two moral tales for her *Heptameron.* [8]

WOMEN SCHOLARS

In the fifteenth and sixteenth centuries among the prosperous and educated families of northern Italy, a few exceptional young women were allowed to devote their early lives to scholarship, much like their privileged male contemporaries.[9] Cassandra Fedele (1465–1558) read her classically inspired orations in public and thus enhanced her father's reputation. Already considered learned by his contemporaries, he became even more exceptional because he had a prodigy for a daughter. Cataruzza Caldiera (d. 1463, also called Caterina) learned of the Latin authors from a text her father, the Venetian Humanist Giovanni Caldiera, wrote for her. He proudly described her accomplishments to his brother: "[our] little daughter exceeds all others in excellence of mind, in depth of character, and in knowledge of the liberal arts, not according to my judgment alone but to that of the wisest men who flourish in this [barbarous] age."[10] She remained exceptional as an adult. When Caldiera was twenty-six the Florentine poet and Humanist Angelo Pliziano compared her to the Italian philosopher at the court of Lorenzo de' Medici, Pico della Mirandola. Olympia Fulvia Morato (1526–1555) studied with the boys her father tutored. (A Humanist, he was tutor to the family of Renée and Ercole, Duchess and Duke of Ferrara.) From her adolescence she wrote poems and orations. Her letters to family and to friends with her thoughts on education and on religion were published after her death, and one edition was dedicated to another woman famous for her erudition, Queen Elizabeth I of England.[11]

As Humanism spread to northern Europe, families there took the same pride in the accomplishments of their female pupils. Margaret More (1504–1544), the eldest daughter of Henry VIII's lord chancellor, Sir Thomas More, was educated at home with her sisters and brothers until she was twelve and her father went into the king's service. Unable to teach his children himself once he had become lord chancellor, More brought learned men into the household to tutor his daughters and his wards (also young girls) in Latin, Greek, logic, philosophy, theology, mathematics, and astronomy.

More's belief in the education of women and the acceptance of that belief in the royal family for its daughters, Mary and Elizabeth Tudor, encouraged others among the English nobility to arrange tutors for their daughters. The parents of Lady Jane Grey (1537–1554) wanted a prodigy to present as a prospective bride for her royal cousins. The learning proved a refuge for her. She described her time with her "so gentle" tutor as "one of the greatest benefits that ever God gave me." In contrast to her parents, who reprimanded and threatened her, she preferred "Mr. Aylmer, who teacheth me so gently, so pleasantly, with such fair allurements to learning, that I think all the time nothing whiles I am with him."[12]

When Elizabeth Tudor became Queen Elizabeth I, English families planned education for their daughters to make them fit members of the learned monarch's court. Fathers like the courtier Sir Anthony Cooke (who had been a tutor to Elizabeth's brother Edward VI) encouraged their daughters to learn Latin and Greek, to do translations, to write poetry, and to engage in the religious dialogues of the day.[13] Three of Cooke's four daughters excelled: Anne became the mother of the court officials Francis and Nicholas Bacon, Elizabeth acquired a reputation as the English "Sappho," Katherine was a Hebrew scholar. The ideal of a Humanist education and a life of scholarship for daughters survived into the seventeenth century. Some English noble households schooled their young women so well that one, Anne Finch, Vicountess Conway (1631–1679), became a philosopher in her own right. The German philosopher and mathematician Leibnitz acknowledged her influence on his ideas about the inseparability of the body and soul.[14]

Two seventeenth-century women, one from the Netherlands, the other from Italy, became famous for their learning. Anna Maria van Schurman (1607–1678) was educated with her two older brothers. After her father's death she became the protegée of the rector of the University of Utrecht. He arranged for her to hear his lectures and to add ancient Middle Eastern languages to the Latin and Greek she had already mastered. She made her own Ethiopian grammar. She had no desire to marry and instead became the model of chaste, female erudition. Scholars like Descartes came to see her when they visited the Netherlands. (He debated with her on the value of studying Hebrew.) The arguments in her treatise *Whether the Study of Letters is Fitting to a Christian Woman*, published in Latin, French, and English by 1659, were used to justify education for women both in the world of the courts and the households of the privileged.[15]

Elena Cornaro Piscopia (1645–1684) not only studied at a university but through the efforts of her ambitious patrician father became in June of 1678 the first woman ever to be granted a doctorate in philosophy.[16] When her exceptional education and intelligence led her to vow chastity and to refuse to marry, her father made a virtue of necessity. He capitalized on her unusual attainments—six languages, astronomy, and mathematics—and installed her in her own house in Padua with tutors and servants. He then pressed for the Doctor of Philosophy degree from the University of Padua, thus gaining an unprecedented honor for his already illustrious family (with four doges and a Queen of Cyprus). Her doctorate awarded, Piscopia retired quietly to her studies, dying six years later.

For all their reputation and honors, learned women like Piscopia and van Schurman remained exceptional. Humanism had opened scholarship to some privileged women. Only in the nineteenth and twentieth centuries, however, long after the passing of the courts, would such education be open to other European women.

WOMEN SCIENTISTS

In the same way that women responded to and participated in Humanism, so they were drawn to the intellectual movement known as the Scientific Revolution. The excitement of the new discoveries of the seventeenth and eighteenth centuries, in particular, inspired a few gifted women scientists to formulate their own theories about the natural world, to perform their own experiments and to publish their findings. In contrast to those educated strictly and formally according to Humanist precepts, these women had little formal training, and chose for themselves what they read and studied. Rather than encouraging them, their families at best left them to their excitement with the wonders of the "Scientific Revolution"; at worst, parents criticized their daughters' absorption in such inappropriate, inelegant, and unfeminine endeavors.

All across Europe from the sixteenth to the eighteenth centuries these women found fascination in the natural sciences. They corresponded and studied with the male scientists of their day. They observed, and they formulated practical applications from their new knowledge of botany, horticulture, and chemistry. The Countess of Chinchon, wife of the Viceroy to Peru, brought quinine bark to Spain from Latin America because it had cured her malaria. Some noblewomen, like the German Anna of Saxony (1532–1582), found medical uses for the plants they studied.[17] The most gifted of these early naturalists is remembered not as a scientist but as an artist. Maria Sibylla Merian (1647–1717) learned drawing and probably acquired her interest in plants and insects from her stepfather, a Flemish still-life artist. As a little girl she went with him into the fields to collect specimens. Though she married, bore two daughters, and ran a household, between 1679 and her death in 1717 she also managed to complete and have published six collections of engravings of European flowers and insects. These were more than artist's renderings. For example, her study of caterpillars was unique for the day. Unlike the still life done by her contemporaries, the drawings show the insect at every stage of development as observed from the specimens that she collected and nursed to maturity. She explained:

From my youth I have been interested in insects, first I started with silkworms in my native Frankfurt-am-Main. After that . . . I started to collect all the caterpillars I could find to observe their changes.[18]

Merian's enthusiasm, patience, and skill brought her to the attention of the director of the Amsterdam Botanical Gardens and other male collectors. When her daughter married and moved to the Dutch colony of Surinam, their support was important when she wanted to raise the money for a new scientific project. In 1699, at the age of fifty-two, Maria Sibylla Merian set off on what became a two-year expedition into the interior of South America. She collected, made notations and sketches. Only yellow fever finally forced her to return to Amsterdam in 1701. The resulting book of sixty engravings established her contemporary reputation as a naturalist.[19]

Mathematics, astronomy, and studies of the universe also interested these self-taught women scientists. In 1566 in Paris Marie de Coste Blanche published *The Nature of the Sun and Earth*. Margaret Cavendish (1617–1673), the seventeenth-century Duchess of Newcastle, though haphazard in her approach to science, produced fourteen books on everything from natural history to atomic physics.

Even more exceptional in the eighteenth century was the French noble-woman and courtier, Emilie du Châtelet (1706–1749). She gained admission to the discussions of the foremost mathematicians and scientists of Paris, earned a reputation as a physicist and as an interpreter of the theories of Leibnitz and Newton.[20] Emilie du Châtelet showed unusual intellectual abilities even as a child. By the age of ten she had read Cicero, studied mathematics and metaphysics. At twelve she could speak English, Italian, Spanish, and German and translated Greek and Latin texts like Aristotle and Vergil. Presentation at court and life as a courtier changed none of her scientific interests and hardly modified her studious habits. She seemed to need no sleep, read incredibly fast, and was said to appear in public with ink stains on her fingers from her note-taking and writing. When she took up the study of Descartes, her father complained to her uncle: "I argued with her in vain; she would not understand that no great lord will marry a woman who is seen reading every day."[21] Her mother despaired of a proper future for such a daughter who "flaunts her mind, and frightens away the suitors her other excesses have not driven off."[22] It was her lover and lifelong friend, the Duke de Richelieu, who encouraged her to continue and to formalize her studies by hiring professors in mathematics and physics from the Sorbonne to tutor her. In 1733 she stormed her way into the Café Gradot, the Parisian coffee-house where the scientists, mathematicians, and philosophers regularly met. Barred because she was a woman, she simply had a suit of men's clothes made for herself and reappeared, her long legs now in breeches and hose, to the delight of cheering colleagues and the consternation of the management.

From the early 1730s until the late 1740s her affair with the *philosophe* Voltaire made possible over ten years of study and writing. He paid for the renovation of her husband's country château in Champagne where they established a life filled with their work and time with each other. They had the windows draped so that shifts from day to night would not distract them. They collected a library of ten thousand volumes, more than the number at most universities.[23] He had his study; she hers.

Emilie du Châtelet usually rose at dawn, breakfasted on fish, bread, stew, and wine, then wrote letters, made the household arrangements for the day, and saw her children. Then she studied. She set up her experiments in the great hall of the château—pipes, rods, and wooden balls hung from the rafters as she set about duplicating the English physicist Newton's experiments. She and Voltaire broke the day with a meal together. Then more study and more writing. When she had trouble staying awake she put her hands in ice water until they were numb, then paced and beat them against her arms to restore the circulation.[24]

Châtelet made her reputation as a scientist with her work on the German mathematician and philosopher Leibnitz, *The Institutions of Physics*, published in 1740. Contemporaries also knew of her from her translation of Newton's *Principles of Mathematics*, her essay on fire, and her collaboration with Voltaire on his treatise about Newton's *Optiks*.

From the fifteenth to the eighteenth centuries privileged women participated in the new intellectual movements. Like the men of their class, they became humanist scholars, naturalists, and scientists. Unfortunately, many of these women found themselves in conflict with their families and their society. A life devoted to scholarship conflicted with the roles that women, however learned, were still expected to fulfill.

CHOICES FOR LEARNED WOMEN

In the eighteenth century Emilie du Châtelet had been able to study and write, to marry and have children. In the sixteenth century, Marguerite of Navarre, as a queen with her own court, could do the same. Most women, however, discovered that they had to choose. A life of scholarship did not combine easily with a life devoted to the needs of a family. Even the most brilliant, even those supported by their families, rarely escaped the traditional expectations about a woman's life, the roles these expectations prescribed, and the practical hazards of being female. When choices had to be made, their own desires and their scholarship suffered.

In most instances the learned women of fifteenth- and sixteenth-century Italy married when their fathers insisted upon it and found it impossible to combine their responsibilities as wives and mothers with those of their scholarship. Both Ginerva Nogarola (c. 1417–1468) and Cataruzza Caldiera (d. 1463) married, bore children, and abandoned their studies and writing altogether. Cecilia Gonzaga (1425–1451) wanted to retire to a convent and the sanctuary it offered, but her father planned a political marriage for her instead. Cassandra Fedele (1465–1558) accepted marriage reluctantly at thirty-three and continued her intellectual studies but did not complete even one of the three major works she planned. Margaret More married at fifteen and became an exemplary wife and mother of five surviving children. For all of his unorthodox views on the education of females, her father, Sir Thomas More, never imagined any other life for her.

Some women scholars defied their parents and the traditions of European society and made a life for themselves outside of marriage. Costanza Barbaro (born c. 1419), daughter of the Venetian Humanist Francesco Barbaro, took holy orders so that she could study. She chose a convent and the Catholic faith as a protective refuge. Some wanted no outside authority—not family, not church—and like the seventeenth-century Venetian Elena Cornaro took vows of chastity on their own behalf and created their own cloistered world. Best-known among these women who chose their own secular kind of cloistering was Isotta Nogarola of Verona (1418–1466), who took a vow of chastity at twenty-three and retired to a room in her mother's house.[25] From this refuge, resembling a convent cell, she corresponded with the Humanist Ludovico Farnesini and out of the interchanges wrote her two major works, *Dialogue on Adam and Eve* and *Oration on the Life of St. Jerome.* [26]

Learned men of these centuries, like Farnesini, idealized women who made this choice and who lived, as he imagined it, not understanding "what desire is," unacquainted "with pleasure," finding comfort distasteful, wanting nothing but the "labors and studious vigils."[27] To him, as to his contemporaries, these women, by choosing the chaste, studious life, had set themselves apart from all others of their sex. Free of the usual demands on their bodies, devoted to the improvement of their minds, they could overcome the disabilities associated with the female nature; they became almost male. This idea of a transformation through learning and chastity is but a secularized version of the rewards offered for similar female efforts by the early male leaders of the Christian church. They too had encouraged the pious women they advised to take vows of chastity, to be zealous in their studies and devotions, and promised that they would then be "like a man." Isotta Nogarola protested to a Veronese Humanist who had not answered her letter. He chided her for her temper and offered her the following advice: be "joyful, gay, radiant, noble and firm," and thus "create a man within the woman."[28] Other learned men made similar assumptions about the transforming effects of education on the female nature. Contemporaries described Cassandra Fedele as a "miracle," a male soul in a female body. They praised Marguerite of Navarre for her learned writings, so excellent "that you would hardly believe they were done by a woman at all."[29]

Some women gloried in overcoming the limits placed on their sex. Olympia Fulvia Morato wrote:

> I, a woman, have dropped the symbols of my sex,
> Yarn, shuttle, basket thread,
> I love but the flowered Parnassus with the choirs of joy
> Other women seek after what they choose
> These only are my pride and my delight.[30]

At the beginning of the fifteenth century the French writer Christine de Pizan believed this kind of transcendence had made it possible for her to survive:

> . . . from the female I became male
> By Fortune, who wanted it thus;
> So [she] changed me, both body and will,
> Into a natural, perfect man;
> Formerly, I was a woman; now
> I am a man, I do not lie,
> My stride demonstrates it well enough.[31]

As a result of the intellectual excitement generated by Renaissance Humanism, women studied, and some women acquired reputations for their learning. Scholarship and a chaste life earned them the respect of their contemporaries. They had become "like men." Such an accolade, however, offered another message as well. Inherent in this idea of the transforming power of learning lay very negative attitudes about women. It assumed disabilities in the female, disabilities that had to be overcome to make her the equal of the male. In this way an idea first put forward by the Church Fathers in the early centuries of Christianity survived intact into the secular world of Humanism, science, and the courts.

The Querelles des Femmes (The Debate over Women)

The traditional premises about the inferiority of the female nature and the corollaries about the need to control and subordinate women formed the basis of the *querelles des femmes*. This is the name given to the more or less constant debate over the value of women that the learned of the courtly world engaged in for almost three hundred years. In early-fifteenth-century Paris, late-sixteenth-century Venice, and in Paris and London at the beginning of the seventeenth century, writers of tracts, treatises, and pamphlets argued, "What is the nature of woman?" "How does she treat men in and out of marriage?" and "Can she be educated?"

For centuries learned men had posed and then answered these questions. But unlike the classical and early Christian eras, when men had argued among themselves, from the early fifteenth century on women participated in the debate. Women joined the *querelles des femmes* because they could not accept the misogynist descriptions of women put forward by learned and privileged men. They did not believe all women to be so evil by nature. They were outraged at the grotesque caricatures of wives, and at the way in which women of virtue and achievement were ignored or discounted as aberrations. Learned themselves, they saw their own experience as proof that women, like men, had rational minds and could benefit from what their contemporaries considered to be a "masculine" education.[32] They argued that most women had the praiseworthy qualities of the ideal female. Most women were chaste, humble, modest, temperate, pious, and faithful. Women could therefore rule themselves and others competently and thus did not need to be constrained or made subject to absolute male authority. Some women could even perform those roles traditionally assigned to men.[33]

In joining the *querelles des femmes,* these women authors had much to refute. In poetry, in prose, in novellas and books of advice for courtiers, in legal treatises for lawyers and judges, learned men of these centuries presented the traditional negative images of the female. Woman, like her predecessor Eve, was "by nature more ready to believe anything," "prone to every sort of evil," and associated with every vice from greed to miserliness to gluttony and pride. In allegories the female personified Idleness, Wealth, Hate, Villainy, Covetousness, Avarice, and Envy. A sixteenth-century French writer left a list that had the ring and authority of a litany after all of the centuries of men's repetitions: women were "vile, inconstant, cowardly, fragile, obstinate, venomous, . . . imprudent, cunning, . . . incorrigible, easily upset, full of hatred, always talking, incapable of keeping a secret, insincere, frivolous and sexually insatiable."[34]

From the fifteenth to the seventeenth centuries these writings by men characterized marriage to such a creature as a kind of living hell. One argument suggested that marriage for a man was the torture devised by God to make up for the Original Sin. A seventeenth-century English pamphleteer summarized the faults of the average wife; she was profligate with money, a nag, jealous, contradictory, and disobedient. Tales and proverbs cautioned men that "if you want a woman to do anything, you can get her to do it by ordering her to do the opposite."[35] Worst of all was the problem of woman's sexuality—her insatiable, lustful nature—more difficult to satisfy "than to dry up the oceans."[36] She played the wanton and the adulteress, trapping and emasculating the unsuspecting male, like Delilah in the Old Testament.[37]

Although women may have attacked these views in other eras, the refutations of Christine de Pizan, the fifteenth-century French writer and courtier, were the first to endure. In her letters and treatises, written from 1399–1402, she asserted that "it is wrong to say that the majority of women are not good"; she "could tell of countless ladies of different social backgrounds, maidens, married women, and widows, in whom God manifested His virtues with amazing force and constancy."[38] She called men's condemnations "arbitrary fabrication": "nonsense," "futile words," "wicked insults," "slanders," and the lies of a lecher.[39] She believed that she had proved that rather than being the exception, virtuous women were the majority. She supported her argument with examples from a wide variety of sources. In *The Book of the City of Ladies* (1405) she takes on each of the traditional charges against women and, speaking through the voices of the allegorical "Reason, Rectitude and Justice," refutes or ridicules them.[40] She presents a positive picture of marriage in which a "woman brings joy to a normal man." She cites numerous examples starting with Eve, the "companion" of Adam, to contemporaries like Mary of Berry, wife of John, Duke of Bourbon: "well mannered in all things, wise, and her virtues appeared like her countenance." In a later treatise, *The Mutation of Fortune,* she praised other female qualities, the "jewels" given to her by her own mother that formed the basis of her ability to educate herself: "discretion," "consideration," the ability to hear and to remember.[41]

In reading and countering adverse images of women's nature, Christine de Pizan found herself in the position of contradicting some of the most revered authorities of the scholarly and courtly world. Puzzling over how these "insults" could have been fabricated by such learned men, Pizan initiated a new line of argument that other women in the sixteenth and seventeenth centuries would take up with equal force. Christine de Pizan argued that men posed these false images out of jealousy, out of rage at pleasures lost to them, out of physical and moral impotence. Other women went further in their criticism of men, accusing them and condemning their nature. The Spanish writer María de Zayas raged against them: "So many martyrs, so many virgins, so many widowed and chaste, so many that have died and suffered by the cruelty of men."[42] The Venetian Arcangela Tarabotti (1604–1652), forced into a convent by her father to avoid paying a dowry, wrote a pamphlet *(Paternal Tyranny)* condemning men as "pimps and procurers who abused their daughters." Marriage in the eyes of these female advocates became a trial for the woman, not the man. Christine de Pizan argued that bad husbands, not bad wives, predominated, that men were notorious for their lack of chastity and fidelity—"inconstant," "fickle," "changeable," and much worse than any woman.[43] The sixteenth-century Venetian Lucrezia Marinella attacked men hoping that she might "tame and humble 'the proud and ungrateful male sex.' "[44] She had entitled her treatise *The Nobility and Excellence of Women together with the Defects and Deficiencies of Men.* (The one she was answering had been titled *The Defects of Women.*) She listed unpleasant male qualities—avarice, envy, pride, ambition, cruelty and tyranny—as others had listed female ones.

Some of these women blamed men for their lack of opportunities and accomplishments. María de Zayas argued it is "men's cruelty or tyranny" that keeps women cloistered and forbids them education. Others speculated on men's motives in denying the female that which was so easily granted to the male. The seventeenth-century Frenchwoman Charlotte de Brachart suggested "these gentlemen would like to see us plain imbeciles so that we could serve as shadows to set off better their fine wits." The anonymous author of a seventeenth-century English pamphlet, *Women's Sharpe Revenge*, believed it all a plot to "make us more weak by our Nurture."[45]

Education was key in the thinking and reasoning of all of the women from the fifteenth to the seventeenth centuries who chose to defend the female in the *querelles des femmes.* They were adamant in valuing both their own knowledge and the education that they hoped to guarantee to other women. They praised the joys of learning and advocated the development of the female as well as the male intellect. Christine de Pizan, the fifteenth-century French writer, remembered that she had "no lovelier occupation, nor one that makes people more complete," and that she had been blessed: "because God and nature had granted to me, beyond the common run of women, the gift of the love of learning." She had simply "closed the doors of her senses to outward things, and turned to beautiful books."[46]

Other women wrote with similar enthusiasm of their learning and advocated it for other women. The French poet of mid-sixteenth-century Lyon, Louise Labé (1525–1566), wrote: "I trained my wits, my body and my mind with a thousand ingenious works." She encouraged other women to "lift their minds a little above their distaffs and spindles . . . and to let the world know that if we are not made to command, we must not for that be disdained as companions, both in domestic and public affairs, of those who govern and are obeyed."[47] The Spanish writer María de Zayas explained her reaction to books: "whenever I see one, old or new, I put down the sewing cushion and do not rest until I've read it all."[48] The seventeenth-century Dutch scholar Anna Maria van Schurman gave the final justification for female education:

> whatever perfects and adorns human understanding is fitting of a Christian woman. Whatever fills the human mind with uncommon and honest delight is fitting for a Christian woman.[49]

Once given access to learning, many of these women writers saw possibilities beyond the "honeyed things" and "sweet savory" of knowledge that Christine de Pizan had described.[50] Pizan believed that lack of education, not innately inferior qualities, held women back. María de Zayas asserted women's equality on their possession of a soul. "The true reason," she explained, "that women are not learned is lack of opportunity, not lack of ability."[51] With education, given "books and teachers instead of . . . linen, embroidery hoops, and pillows, we would be just as apt as the men for government positions and university chairs, and perhaps even more so."[52] Men denied women education so "that they will not usurp their power," and "consequently, women are oppressed and obliged to exercise domestic chores."[53] The sixteenth-century Venetian Lucrezia Marinella hoped "to wake women from their long sleep of oppression." She honored women's activities, describing the proper purpose of the sex: "not to please men, but to understand, to govern, to generate, to bring grace into the world."[54] Women were more virtuous and thus more godlike. Given these special qualities Marinella argued that women were not inferior, but superior to men. She agreed with María de Zayas that once educated, the possibilities of their achievement seemed awesome. They could assume roles usually reserved to men.[55]

Yet, for all of these writers' rhetorical ability, for all of their arguments, the *querelles des femmes* ended with the traditional premises about women as strong as ever. Men mocked these learned, angry women at the end of the seventeenth century as easily as they had in other ages. The remarks of Jean de La Bruyère, the seventeenth-century French writer on morals, were typical. He compared an educated woman to a collector's special firearm, "which one shows to the curious, but which has no use at all, any more than a carousel horse."[56] The vast majority of European men assumed that women would accept the roles and functions assigned them since ancient times. The changes in a few women's lives brought about by their participation in the great intellectual discoveries and debates of the fifteenth to the eighteenth centuries proved transitory. In the end, the new humanism and the new science simply reaffirmed traditional premises and conclusions about women's nature and the ways in which their lives should be limited.

Science Affirms Tradition

In the sixteenth and seventeenth centuries Europe's learned men questioned, altered, and dismissed some of the most hallowed precepts of Europe's inherited wisdom. The intellectual upheaval of the Scientific Revolution caused them to examine and describe anew the nature of the universe and its forces, the nature of the human body and its functions. Men used telescopes and rejected the traditional insistence on the smooth surface of the moon. Galileo, Leibnitz, and Newton studied and charted the movement of the planets, discovered gravity and the true relationship between the earth and the sun. Fallopio dissected the human body, Harvey discovered the circulation of the blood, and Leeuwenhoek found spermatozoa with his microscope.

For women, however, there was no Scientific Revolution. When men studied female anatomy, when they spoke of female physiology, of women's reproductive organs, of the female role in procreation, they ceased to be scientific. They suspended reason and did not accept the evidence of their senses. Tradition, prejudice, and imagination, not scientific observation, governed their conclusions about women. The writings of the classical authors like Aristotle and Galen continued to carry the same authority as they had when first written, long after they had been discarded in other areas. Men spoke in the name of the new "science" but mouthed words and phrases from the old misogyny. In the name of "science" they gave a supposed physiological basis to the traditional views of women's nature, function, and role. Science affirmed what men had always known, what custom, law, and religion had postulated and justified. With the authority of their "objective," "rational" inquiry they restated ancient premises and arrived at the same traditional conclusions: the innate superiority of the male and the justifiable subordination of the female.

In the face of such certainty, the challenges of women like Lucrezia Marinella and María de Zayas had little effect. As Marie de Gournay, the French essayist, had discovered at the beginning of the seventeenth century, those engaged in the scientific study of humanity viewed the female as if she were of a different species—less than human, at best; nature's mistake, fit only to "play the fool and serve [the male]."[57]

The standard medical reference work, *Gynaecea*, reprinted throughout the last decades of the sixteenth century, included the old authorities like Aristotle and Galen, and thus the old premises about women's innate physical inferiority. A seventeenth-century examination for a doctor in Paris asked the rhetorical question "Is woman an imperfect work of nature?"[58] All of the Aristotelian ideals about the different "humors" of the female and male survived in the popular press even after they had been rejected by the medical elite. The colder and moister humors of the female meant that women had a passive nature and thus took longer to develop in the womb. Once grown to maturity, they were better able to withstand the pain of childbirth.[59]

Even without reference to the humors, medical and scientific texts supported the limited domestic role for women. Malebranche, a French seventeenth-century philosopher, noted that the delicate fibers of the woman's brain made her overly sensitive to all that came to it; thus she could not deal with ideas or form abstractions. Her body and mind were so relatively weak that she must stay within the protective confines of the home to be safe.[60]

No amount of anatomical dissection dispelled old bits of misinformation or changed the old misconceptions about women's reproductive organs. Illustrations continued to show the uterus shaped like a flask with two horns, and guides for midwives gave the principal role in labor to the fetus. As in Greek and Roman medical texts these new "scientific" works assumed that women's bodies dictated their principal function, procreation. Yet even this role was devalued. All of the evidence of dissection and deductive reasoning reaffirmed the superiority of the male's role in reproduction. Men discovered the spermatazoon, but not the ovum. They believed that semen was the single active agent. Much as Aristotle had done almost two millennia earlier, seventeenth-century scientific study hypothesized that the female supplied the "matter," while the life and essence of the embryo came from the sperm alone.

These denigrating and erroneous conclusions were reaffirmed by the work of the seventeenth-century English scientist William Harvey. Having discovered the circulation of the blood, Harvey turned his considerable talents to the study of human reproduction and published his conclusions in 1651. He dissected female deer at all stages of their cycle, when pregnant and when not. He studied chickens and roosters. With all of this dissection and all of this observation he hypothesized an explanation for procreation and a rhapsody to male semen far more extreme than anything Aristotle had reasoned. The woman, like the hen with her unfertilized egg, supplies the matter, the man gives it form and life. The semen, he explained, had almost magical power to "elaborate, concoct"; it was "vivifying, . . . endowed with force and spirit and generative influence," coming as it did from "vessels so elaborate, and endowed with such vital energy."[61] So powerful was this fluid that it did not even have to reach the woman's uterus or remain in the vagina. Rather he believed it gave off a "fecundating power," leaving the woman's body to play a passive, or secondary, role. Simple contact with this magical elixir of life worked like lightning, or—drawing on another set of his experiments— "in the same way as iron touched by the magnet is endowed with its powers and can attract other iron to it."[62] The woman was but the receiver and the receptacle.

Anatomy and physiology confirmed the innate inferiority of woman and her limited reproductive function. They also proved as "scientific truth" all of the traditional negative images of the female nature. A sixteenth-century Italian anatomist accepted Galen's view and believed the ovaries to be internal testicles. He explained their strange placement so "as to keep her from perceiving and ascertaining her sufficient perfection," and to humble her "continual desire to dominate."[63] An early-seventeenth-century French book on childbirth instructed the midwife to tie the umbilical cord far from the body to assure a long penis and a well-spoken young man for a male child and close to the body to give the female a straighter form and to ensure that she would talk less.[64]

No one questioned the equally ancient and traditional connection between physiology and nature: the role of the uterus in determining a woman's behavior. This organ's potential influence confirmed the female's irrationality and her need to accept a subordinate role to the male.[65] The sixteenth-century Italian anatomist Fallopio repeated Aristotle's idea that the womb lusted for the male in its desire to procreate. The French sixteenth-century doctor and writer Rabelais took Plato's view of the womb as insatiable, like an animal out of control when denied sexual intercourse, the cause of that singularly female ailment, "hysteria." Other sixteenth- and seventeenth-century writers on women and their health adopted all of the most misogynistic explanations of the traditional Greek and Roman authorities. No menstruation meant a diseased womb, an organ suffocating in a kind of female excrement. Only intercourse with a man could prevent or cure the condition. Left untreated the uterus would put pressure on other organs, cause convulsions, or drive the woman crazy.[66] Thus, the male remained the key agent in the woman's life. She was innately inferior, potentially irrational, and lost to ill-health and madness without his timely intervention.

So much changed from the fifteenth to the eighteenth centuries in the ways in which women and men perceived their world, its institutions and attitudes. The Renaissance offered the exhilaration of a society in which the individual could be freed from traditional limitations. In the spirit of Humanistic and scientific inquiry men questioned and reformulated assumptions about the mind's capabilities and the description of the natural universe. New methods of reasoning and discourse, of observation and experimentation, evolved and led to the reorientation of the natural universe and more accurate descriptions of the physical world, including man's own body. Yet when it came to questions and assumptions about women's function and role and to descriptions of her nature and her body, no new answers were formulated.

Instead, inspired by the intellectual excitement of the times and the increasing confidence in their own perceptions of the spiritual and material world, men argued even more strongly from traditional premises, embellishing and revitalizing the ancient beliefs. Instead of breaking with tradition, descriptions of the female accumulated traditions: the classical, the religious, the literary, the customary, and the legal—all stated afresh in the secular language of the new age. Instead of being freed, women were ringed with yet more binding and seemingly incontrovertible versions of the traditional attitudes about their inferior nature, their proper function and role, and their subordinate relationship to men.

With the advent of printing, men were able to disseminate these negative conclusions about women as they never could before. From the sixteenth century on the printing presses brought the new tracts, pamphlets, treatises, broadsides, and engravings to increasing numbers of Europeans: pictures of the sperm as a tiny, fully formed infant; works by scholars and jurists explaining the female's "natural" physical and legal incapacity; romances and ballads telling of unchaste damsels and vengeful wives set to plague man.[67]

Although these misogynistic attitudes about women flourished and spread, the defense of women had also begun. In her *Book of the City of Ladies* Christine de Pizan, the fifteenth-century writer, asks why no one had spoken on their behalf before, why the "accusations and slanders" had gone uncontradicted for so long? Her allegorical mentor, "Rectitude," replies, "Let me tell you that in the long run, everything comes to a head at the right time."[68]

The world of the courts had widened the perimeters of women's expectations and given some women increased opportunities. However, for the vast majority of women, still not conscious of their disadvantaged and subordinate status, changes in material circumstances had a far greater impact. From the seventeenth to the twentieth centuries more women were able to live the life restricted in previous ages to the few. In Europe's salons and parlors they found increased comfort, greater security, and new ways to value their traditional roles and functions. For these women, "the right time"—the moment for questioning and rejecting the ancient premises of European society—lay in the future.

VII

WOMEN OF THE SALONS
AND PARLORS

●

LADIES, HOUSEWIVES, AND
PROFESSIONALS

1

WOMEN IN THE SALONS

❧

The Creation and Institution of the Salon in France

On a cold winter's night in Paris, 1625, Louis XIII and his queen, Anne of Austria, presided over a court devoted to elaborate rituals and studied responses. The vast tapestried rooms in their royal palace of the Louvre still attracted courtiers eager for privileges and preferment, but even the king's new adviser, the Cardinal de Richelieu, sought out the more informal and entertaining evenings at the Hôtel de Rambouillet. At this elegant new private house, built near the Louvre in 1608, the Marquise de Rambouillet presided nightly over the witty and polite group she attracted to her home and the new social setting she had designed.[1]

The daughter of a French courtier and an Italian lady-in-waiting to Catherine de' Medici, Catherine de Vivonne (1588–1665) was married at twelve to the twenty-three-year-old Marquis de Rambouillet. Disgusted by the coarse sexual innuendoes and flirtations she encountered at court at sixteen, she designed a new home which broke with the classical Renaissance style of the period. Built of brick, slate, and white cut stone, her large private house featured a side entrance leading into numerous relatively small, well-proportioned rooms, instead of the great hall and central staircase—copying the design of royal palaces—typical of earlier hôtels. Passing through a series of ornately furnished rooms, a visitor arrived at last in the famous *chambre bleue,* whose intimate proportions and daring use of color (Renaissance rooms were invariably paneled in dark red or brown leather) established an elegant and fresh tone.[2] Wealthy by both birth and marriage, the marquise spared no expense in furnishing her new home: the *chambre bleue* had waist-high painted and gilded panels lining its walls, which were hung with patterned blue tapestries encrusted with gold and silver. The room had two alcoves, one with a bay window on the house's interior garden. Paintings and Venetian mirrors hung on the walls; embroidered screens shielded guests from drafts and too-direct a heat from the fireplace. The focus of the room was the blue, damask-hung bed in the second alcove where the Marquise de Rambouillet reclined.[3]

In 1623, the marquise had developed a mysterious malady after the birth of her seventh child which made her unable to bear direct heat from a fireplace. So, swathed in furs in her chilly alcove, she directed conversation from her bed. She continued to do so for the next twenty-five years.

By building the *chambre bleue* and attracting the social and artistic elite to it, the Marquise de Rambouillet invented the salon, a space in which talented and learned women could meet with men as intellectual equals, rather than as exceptional prodigies. By insisting on tastefulness, courtesy, and polite behavior, Rambouillet created a genteel environment where aspiring authors of both sexes were encouraged to share their work, comment on each others' productions, and participate in elaborate discussions and conversational games. Before the *chambre bleue* a woman with artistic or intellectual interests had no place to pursue them, except her own home. In the sixteenth century, a few talented Frenchwomen had been able to establish literary circles: Louise Labé (c. 1520–1566), the poet, did so in Lyon; Madeleine and Catherine Des Roches, mother and daughter, did so in Poitiers.[4] But these provincial groups did not survive their hostesses' deaths and never became an institution. In Paris of the 1620s, in the *chambre bleue*, the Marquise de Rambouillet created the salon in both its senses: the room itself where guests could mingle comfortably (a "salon" is a drawing room, distinguished from the more formal great hall), and the institution where women and men of the intellectual, social, and artistic elites could converse freely. Welcomed by Parisian society, the salon rapidly established itself in France; French cultural ascendancy ensured its spread to other European capitals. In the salon, a woman could not only socialize with the famous men of the day; she could also become famous herself, by developing her own talents in a milieu where they would find a polite and encouraging reception.

While she did not write herself, the Marquise de Rambouillet enabled another woman to do so. She sponsored Madeleine de Scudéry (1608–1701), whose literary talent had suffered from her handicaps in being poor, orphaned, unattractive, and from the provinces. Supported by Rambouillet and encouraged to write by the salon, Scudéry produced lengthy popular novels: *Le Grand Cyrus* (1649), which contained portraits of the group, and *Clélie* (1654), dedicated to the marquise. In *Clélie*, Scudéry championed a new kind of love, in which "gallant lovers would be more respectful and more gracious, and consequently, more pleasing" because women "would teach men to become truly courteous and would never allow them to lose the respect that is the woman's due."[5] In these novels, as in these early salons, the "respect that is the woman's due" became transformed into the salonière's rejection of physical love. Wanting the emotions love inspired but not the pregnancies it could cause, Scudéry, Rambouillet, and others of their circle arrived at the solution of refusing sexual intercourse. Rambouillet withdrew into semi-invalidism; Scudéry never married or had affairs. By refusing physical love, by rising "above" their sexual desires, they transformed themselves into "precious women"—rare, difficult to obtain, delicate—whose chastity gave them power and moral prestige over their suitors. By refusing to have sex, these women freed themselves for a role beyond that of wife or courtesan and gained the power of virtue: they were living up to their culture's ideals about correct female sexual behavior and they reaped the reward of moral authority. Known as the *précieuses*—the precious women—they dominated the first decades of the salon in France.

Soon, however, other salonières were using their sexuality rather than denying it. Ninon de Lenclos (1620–1705), the famous courtesan, established a salon in her home. Rambouillet and her friends had occasionally met as an all-female group: a contemporary portrait shows six women conversing in the *chambre bleue* while two young girls and a puppy cavort on the floor.[6] But at Lenclos's, men were the focus. A woman could reap greater rewards by forming a sexual liaison with a talented or titled man than she could by refusing one. The great French salonières of the late seventeenth and eighteenth centuries were as famous for their love affairs and marriages with great men as for the guests whom they attracted to their salons. Social pressure and traditional attitudes about the impossibility of women and men relating other than sexually contributed to many salonières' use of sexual liaisons. Even if a salonière remained chaste, she would be gossiped about unmercifully: it was assumed that relations between women and men, however intellectual or artistic they might appear, could not remain platonic.

Whether chaste or not, Parisian women established the institution of the salon by the last quarter of the seventeenth century. Aspiring hostesses competed to attract the talented, the witty, and the powerful to their homes. Outside the powerful French court and frequently in opposition to it, these new social circles offered women a new possibility: that of being a salonière, who by her graciousness and skill enabled conversation to flourish, artists to find patrons, and aristocrats to be amused. As a salonière, a woman brought the circles of power into her home. In the environment she created, she could help or hinder not only artistic and literary reputations, but political policies as well. The financial remedies for France's economy were debated in the salons; the king's choices of ministers were strongly influenced by the backing of powerful salonières. Salonières were privy to court secrets; salons were frequented by statesmen and ambassadors as well as intellectuals and artists. Salonières could make or break careers and often provided havens for new political philosophies and the new political opposition to the monarchy. "I look upon you as you are now, in full enjoyment of life, as the happiest of all possible creatures," the author Saint-Evremond wrote Ninon de Lenclos in 1669, when the Parisian courtesan was forty-nine. "There are few princesses who do not envy your lot when they realize the hardships of their condition; and who knows but that the convents contain some saintly women who would gladly exchange their peace of mind for the pleasant excitement which must have been experienced by you."[7] With the salon, women created a space and a society where they presided, a world and a life in the capital but separate from the court, where in France of this era male monarchs ruled.

One hundred and fifty years after Rambouillet's creation—in the second half of the eighteenth century—the salon achieved its greatest influence and prestige in Europe. In the leading capital cities, salons flourished, and their existence signaled an active intellectual and cultural life. Appearing in many nations, the salon reached its apogee in eighteenth-century France. There, where women's influence in the courts—as *maîtresse-en-titre*, as queen, as courtier—increased, women's influence also flourished outside the court, in the salons. In the relatively rigid hierarchical society of pre-revolutionary France, where a person had to prove four quarters of nobility to hold many important posts, the salon allowed both women and men a social mobility which existed nowhere else.[8] The salon mixed elements of the nobility, bourgeoisie, and intelligensia and enabled some women to rise through both marriage and influence. After the first few decades, the career of a salonière was not restricted to the aristocracy: Suzanne Necker, the daughter of a pastor, ran an influential Parisian salon in the 1760s and 1770s, as did Marie-Thérèse de Geoffrin (1699–1777), who was the daughter of a valet and the wife of an ice-cream manufacturer. In the salon, a woman could meet and marry a man of superior social rank or wealth. In the salon, a woman of enterprise could make her way by attracting the famous. The salon became the base of influential women who swayed kings, governments, political opinion, and literary and artistic taste. Suzanne Necker helped to make her husband finance minister of France in 1776 through her influential salon; a generation later, her daughter, Germaine de Staël, led the intellectual and political opposition to Napoleon Bonaparte from her salon.[9]

Intellectually, the salon provided shelter for views or projects unwelcome in the courts: when Voltaire was *persona non grata* with Louis XV because of his critical views of monarchy, he was deluged with invitations from Parisian salonières eager to be his hostess. The great Enlightenment project of the *Encyclopédie*, which sought to categorize, define, and criticize all

1. Silhouette of Hannah More (1745–1833), the English reformer.

2. Mary Wollstonecraft (1759–1797), the English feminist. Portrait by John Opie.

3. *The Ages of Woman,* mid-nineteenth-century French engraving.

4. Empress Marie and Czar Alexander III of Russia with their five children, c. 1885.

5. Portrait believed to be the French novelist George Sand (1804–1876) in 1830, dressed as a man.

6. Popular image of the "new woman" of the 1920s. German advertisement for Kaloderma soap.

7. Marie Curie (1867–1934) (*right*) and her daughter Irène Joliot-Curie (1897–1956) (*left*) with colleagues at the Institute of Radium in Paris, c. 1930.

8. Pilot Lily Litvak (*left*) discussing tactics with members of the Soviet Union's 586th Fighter Regiment, early 1940s.

9. Marie Stopes and nurses at the first English birth control clinic. London, 1920s.

existing knowledge, was suppressed by the French court, but completed in secret with Mme. Geoffrin's social and financial assistance. She welcomed the Encyclopedists to her salon, and their presence was sought by other salonières as well. The great French salonières both competed with and helped each other. Rivals for prestigious, usually male, guests, they often bequeathed their salons to younger female protegées. The Marquise de Lambert, author of two treatises recommending the education of girls, gave Tuesday dinners for men of letters from the 1690s; when she died in 1733, her salon passed to Mme. de Tencin. When Mme. de Tencin died in 1749, she bequeathed her salon to Mme. Geoffrin.[10] The interlocking careers of three great French salonières of the third quarter of the eighteenth century—Marie du Deffand, Julie de Lespinasse, and Marie-Thérèse Geoffrin—demonstrate the partnership and rivalries which united and divided these influential women.

The daughter of an aristocratic mother, Julie de Lespinasse (1733–1776) was kept in ignorance of her illegitimacy until her mother's death when the girl was sixteen. Lespinasse remained as a governess to her legitimate half-sister's children because she had no other possibility of supporting herself. Her world opened up in 1752, when Marie de Vichy-Chamrond, Marquise du Deffand, visited the family château. Deffand (1697–1780) had become a leading salonière in the 1720s: abandoning her husband and provincial life, she established herself in Paris, became the mistress of the Regent of France, and attracted leading philosophers and politicians to her ornate drawing room.[11] The Baron de Montesquieu and David Hume visited when they were in Paris, as did Voltaire, who maintained a lifelong correspondence with her. C. J.-F. Hénault, a leading jurist, was her long-term lover as well as a regular guest at her salon. Throughout the third quarter of the eighteenth century, Deffand continued to attract the younger generation of philosophers and politicians as well. Jean d'Alembert, the Encyclopedist and mathematician, attended regularly, as did Loménie de Brienne, the Archbishop of Toulouse, and Turgot, the economist; both, under the sponsorship of Deffand, would become ministers of finance for Louis XVI. "It is sheer boredom which drives me to fill the house with guests," Deffand explained in the cynical tone which characterized her wit.[12] In the early 1750s, the Marquise du Deffand began to go blind, and she decided to take Julie de Lespinasse to Paris as her companion.

For the next ten years, from 1754 to 1764, these women maintained one of the most brilliant salons in Europe. Part of the attraction was the contrast between the two: aunt and illegitimate niece; one in her sixties, the other in her twenties; one known for her love affairs, the other a virgin; blind patroness and sighted protegée; the cynic and the romantic. Their salon became one of the ornaments of Enlightenment culture, and an invitation to the famous yellow drawing room was prized. In 1764, the two women fought, and visitors were forced to choose between remaining with Deffand or following Lespinasse to her new haven at Mme. Geoffrin's. Deffand successfully retained many of her guests and continued to attract newcomers who appreciated her knowing, cynical style. "There is only one misfortune in life," she proclaimed in this period, "that is, being born."[13]

Her pose of world-weary cynicism was disturbed when at age sixty-nine she fell in love with Horace Walpole, the English author. Twenty years younger than she and never romantically attracted to women, Walpole rejected her, and in her seventies her pessimism intensified. She chided Voltaire for his optimism: "The search for truth is for you the universal panacea . . . you think you have found it, but I think it is unattainable." Just before her death at eighty-three, she wrote a bitter quatrain on old age, which she called "a horrible time; / a somber and cold season, / When humans confuse impotence with wisdom / and fears with reason."[14]

Julie de Lespinasse found a new patroness in Marie-Thérèse Geoffrin. Geoffrin sold three paintings to Catherine the Great of Russia and used the money to provide Lespinasse with a pension and to enable her to furnish her own apartment, spiting her rival, Mme. du Deffand. Soon, Lespinasse was advising Geoffrin on whom to invite to the Wednesday dinners and was attending them regularly herself, along with the painters Boucher and La Tour, the authors Marmontel and Walpole, and the Encyclopedists Helvétius, Fontenelle, Diderot, and d'Alembert. A few years later, it was accepted that one dined at Mme. Geoffrin's, where food was plentiful and conversation was limited to noncontroversial topics, but concluded the evening at Mlle. de Lespinasse's, where any topic could be discussed and argued about. In her white-paneled boudoir with its crimson damask curtains, in her drawing room with its busts of Voltaire and d'Alembert, its ormolu clock on the mantle, her guests were able to criticize tradition and hammer out the new moral imperatives of reason, justice, and equality which were to become the ideals of Enlightenment culture. Hume spent time there, as did Turgot and Diderot, d'Alembert and Marmontel, who recalled after Lespinasse's untimely death at forty-three that "nowhere was there a livelier, a more brilliant, or a better-regulated conversation than at her house."[15]

Lespinasse's guests were mostly men, and in this her salon was typical of those of late-eighteenth-century Paris. A contemporary painting of a reading at Mme. Geoffrin's salon shows an audience of thirty-seven men, but only six women.[16] Geoffrin herself had been one of the few women Mme. de Tencin invited to her salon; Lespinasse was the first woman Geoffrin allowed to attend her famous Wednesday night dinners.[17] Some salons were hosted by men. The focus in all the salons was attracting male guests. Men's superior prestige and power gave them precedence, and a salonière made her name by attracting male luminaries to her drawing room.

Despite this male predominance, both French and foreign commentators stressed and even exaggerated the power of "female influence" in France. Traditional male fears about what might happen if women were "out of place" and influenced government combined with the conspicuous role of women in the French court and salons to produce this view. In France, wrote the Scottish philosopher David Hume, "the females enter into all transactions and all management of church and state: and no man can expect success, who takes not care to obtain their good graces."[18]

While foreigners tended to overemphasize the power of French salonières (usually in comparison to the role of women in their own nations), Frenchmen also exaggerated the extent of their influence. "Only here, of all places on earth, do women deserve to wield such influence," wrote Napoleon Bonaparte in 1795. "A woman, in order to know what is due her and what power she has, must live in Paris for six months."[19] This perception of female influence and associated fears about it led to the end of the salonière's influence when the political climate changed at the end of the eighteenth century.

Salons and Salonières in England and Germany

While salonières never achieved the same prestige and influence in other European capitals as they did in Paris, the development of the salon there was duplicated at later dates in London, Vienna, Rome, Copenhagen, Berlin, and other European cities. Just as in France, women in other capitals complained before the appearance of the salons of having no place to develop their intellectual or artistic interests, and just as in France, the creation of the salons was the work of individual women, who as hostesses insisted on conversation rather than dancing or cards. In London, Elizabeth Montagu (1720–1800) initiated the literary salon in the middle of the eighteenth century by forbidding her guests card games and seating them in a large semicircle to foster conversation. Other hostesses were more casual: Mary Monckton took great care "to prevent a circle" and "pulled about the chairs and planted the people in groups with as dextrous a disorder as you would desire to see," reported the English novelist Fanny Burney in 1782.[20] English salonières prided themselves on their graceful informality, which gave rise to the name of their group, the "Bluestockings." (In an era in which men wore white or black stockings to formal events, blue stockings signaled informality.)[21]

In each European capital, the first salonières were women whose self-assurance enabled them to invent new ways for women and men to meet socially. "The women seem to take the lead in polishing the manners every where," wrote Mary Wollstonecraft, the English feminist, after a trip through Scandinavia in 1795, "that being the only way to better their condition."[22] Like the Marquise de Rambouillet before them, other hostesses insisted that women be treated neither as inferiors nor solely as objects of sexual desire, but as intelligent and sensitive equals.

The social background of the salonières varied. In Paris, salonières tended to be aristocratic; in London, from the middle classes. In the Germanic capitals they were often Jewish: nine of the fourteen women who created salons in late-eighteenth-century Berlin were Jewish.[23] Fanny Arnstein and her daughter Rachel Pereira brought the Jewish salon to Vienna in the first half of the nineteenth century. It was only in the second half of the eighteenth century that Jews had first mingled socially with Christians in Berlin, at the "open houses" kept by Nikolai and Moses Mendelssohn. By the 1780s, some daughters of wealthy German Jewish families had transformed these open houses into true salons, which mixed authors and aristocrats, burghers and actors, women and men, as well as Jews and Christians. Henrietta Herz (1764–1847) credited a "complete lack of links with tradition" for her salon's success in Berlin. "The women at the head of it have thrown all the conventions overboard and are indulging in an exuberant freedom of spirit," remembered Herz in her memoirs.[24] For her, this new freedom was intellectual and spiritual; like the *précieuses,* some other hostesses refused sexual love to enable them to function as salonières rather than courtesans or mistresses. Herz, unhappily married at fifteen, maintained a salon for many years without losing her reputation for chastity. Drawing up a "constitution" for her guests to sign, she enlisted them in a "League of Virtue" whose members swore to treat each other platonically and informally:

> The object of our society is the happiness of all our members through love. Consequently, none has any duty towards another. For love knows nothing of duty. . . . Our members dispense among themselves with all the shackles of conventional decorum. They address one another familiarly [using "du" instead of the formal "Sie"] and exchange brotherly and sisterly kisses.[25]

As in Paris, other salonières found the salon a means to make new liaisons with talented, aristocratic, or powerful men they attracted to their drawing rooms. The eldest daughter of Moses Mendelssohn initiated her own salon shortly after her youthful marriage to a Jewish banker. As Brendel Veit, the respectably married mother of four, she fell passionately in love with the young Christian writer Friedrich von Schlegel. She divorced her husband, left her children, and went to live with Schlegel. Later changing her name and religion, she became a respected author herself, and it is as Dorothea von Schlegel (1763–1839) that she is known to posterity. The sexual opportunities provided by the salon proved irresistible to many, and by the end of the eighteenth century the salon was inextricably associated with the looser sexual behavior previously tolerated only in certain court and aristocratic circles. In England, the relatively sedate bluestocking gatherings were succeeded by salons whose hostesses were not acceptable to respectable female society: by the 1830s the only salons left in London were those of Lady Blessington and Lady Holland, both of whom were best known for their marital scandals. Isolated from other women, they presided over salons whose guests were all men.[26]

In eighteenth-century Berlin, however, some salonières exerted themselves to sponsor younger women of talent. Henrietta Herz supported and remained friendly with Rahel Levin (1771–1833), who opened a rival salon in the 1790s. "Rahel Levin more than any other" was the "finest flower" of Berlin salon culture, wrote Herz generously in her memoirs.[27] Rahel Levin, impoverished because her wealthy father left all his money to her brothers, survived on the patronage of her wealthier friends. Believing that "every man is, and is meant to be, an Original, not a manufactured article," she maintained her salon into the 1820s, able to attract the talented by the force of her personality and the charm of her conversation.[28] In 1826, Franz Grillparzer, the Austrian playwright, visited Levin, who was then fifty-five. "This aging woman, deformed by sickness, looking like a witch or rather a sorceress, began to speak," he reported.

> My tiredness left me and gave way to a sort of intoxication. She talked and talked until about midnight. Did I go away of my own free will or did they turn me out of the house? I do not remember. In all my life I never heard anyone talk so interestingly, so facinatingly.[29]

In England also, salonières aided impoverished but talented women, and this aid made a real difference in enabling women to develop in areas previously restricted to men. Englishwomen of intellect and talent in the first half of the eighteenth century found themselves isolated and discouraged from learning, whatever their rank in society. The aristocrat Lady Mary Wortley Montagu (1689–1762) managed to publish her writings because of her wealth, but she felt uncomfortable in England and spent much of her life in Italy. Uneducated herself, she recommended that her granddaughter "conceal whatever learning she attains, with as much solicitude as she would hide crookedness or lameness. The parade of it," she concluded bitterly, "can only serve to draw on her the envy, and consequently, the most inveterate hatred, of all he and she fools, which will certainly be at least three parts in four of all her acquaintance."[30] For women of less wealth and self-assurance, the lack of any institution which would provide support or encouragement for their talents proved crippling. Elizabeth Elstob (1683–1756), a language prodigy, published the first Anglo-Saxon grammar because of her brother's emotional and financial support. After his death in 1715, however, she was forced to support herself by teaching elementary school and working as a governess. She was not able to publish her other translations or pursue her scholarship; wealthy friends found her a post as governess in the Duchess of Portland's household, but the possibility of enabling a woman to be a full-time author or scholar came only with the development of the salon in England.[31]

By the second half of the eighteenth century, Englishwomen of talent were encouraged and financially supported by the salonières. Their support of Elizabeth Carter (1717–1806) enabled her to avoid Elstob's fate. The "learned Miss Carter of Deal," as she came to be known, was introduced to bluestocking circles by Samuel Johnson. Encouraged by her new friends—especially Catherine Talbot and Elizabeth Montagu—she began to translate the Greek philosopher Epictetus into English for the first time. Carter never married—she called marriage "a very right scheme for everybody but my-self"—and by 1758 she had run out of funds and was forced to support herself by sewing. "Whoever that somebody or other is to write the life of Epictetus, seeing that I have a dozen shirts to make, I do opine that it cannot be I," she wrote a friend.[32] When word of her plight reached Elizabeth Montagu, the salonière exerted herself to get friends to subscribe to the publication of Carter's translation. The £1000 netted by the book and a pension of £100 a year from Montagu enabled Carter to continue both her writing and her yearly visits to the bluestocking salons and gatherings.[33] James Boswell, Samuel Johnson's biographer, described such a gathering of 1779: "The company was Miss Hannah More [the writer] . . . Mrs. Boscawen [famous for her letter writing], Mrs. Elizabeth Carter ["Mrs." was given as a courtesy title to older women], Sir Joshua Reynolds, Dr. Burney [Fanny Burney's father], Dr. Johnson, and myself," remembered Boswell. "We were all in very fine spirits and I whispered to Mrs. Boscawan, 'I believe this is as much as can be made of life.' "[34] In the informal English and German gatherings, as well as the more formal French salons, women created a new space in which they could mingle with men as intellectual and cultural equals.

The Decline of the Salon

Rational conversation, sociability between women and men, delight in the pleasures of this world are the hallmarks of Enlightenment culture. The men who mingled with the Bluestockings and frequented the salons were the men who produced the Enlightenment. It is a tragedy for women that these men, who were aided, sponsored, and lionized by the salonières, produced—with very few exceptions—art and writing which either ignored women completely or upheld the most traditional views of womanhood. Just as there was no Renaissance or Scientific Revolution for women, in the sense that the goals and ideals of those movements were perceived as applicable only to men, so there was no Enlightenment for women. Enlightenment thinkers questioned all the traditional limits on men—and indeed challenged the validity of tradition itself. They championed the rights of commoners, the rights of citizens, the rights of slaves, Jews, Indians, and children, but not those of women. Instead, often at great cost to their own logic and rationality, they continued to reaffirm the most ancient inherited traditions about women: that they were inferior to men in the crucial faculties of reason and ethics and so should be subordinated to men. In philosophy and in art, men of the Enlightenment upheld the traditional ideal of woman: silent, obedient, subservient, modest, and chaste. The salonière—witty, independent, powerful, well-read, and sometimes libertine—was condemned and mocked. A few Enlightenment thinkers did question and even reject subordinating traditions about women. But those who argued for a larger role for women—like the Englishwoman Mary Wollstonecraft in her *Vindication of the Rights of Woman* (1791), the French Marquis de Condorcet in his *Admission of Women to Civic Rights* (1790), the German Theodor von Hippel in his *On the Civic Improvement of Women* (1792), the Spaniard Josefa Amar y Borbón in her *Discourse in Defense of Women's Talent and Their Capacity for Government and Other Positions Held by Men* (1786)—prompted outrage and then were forgotten. Instead, most philosophers and writers reiterated the most limiting traditions of European culture regarding women, often in works which condemned traditional behavior for men. John Locke, the English philosopher, had a profound influence on Enlightenment thought when he argued that every man has an equal right "to his natural freedom, without being subjected to the will or authority of any man." But he thought women (and animals) exempt from "natural freedom" and declared they should be subordinate: he upheld "the Subjection that is due from a Wife to her Husband."[35] The Scottish philosopher David Hume delighted in his visits to the Paris salons of Deffand, Lespinasse, and Geoffrin, so much so that in his autobiography he declared he had once thought "of settling there for life" because of "the great number of sensible, knowing, and polite company with which that city abounds above all places in the universe."[36] Yet he simultaneously condemned France, which "gravely exalts those, whom nature has subjected to them, and whose inferiority and infirmities are absolutely incurable. The women, though without virtue, are their masters and sovereigns."[37]

Like Hume, other Enlightenment authors connected the rule of women—and especially of unvirtuous women—to the end of good government.[38] Women who wielded the indirect power and influence of the salonières and the women of the courts were condemned.

The *Encyclopédie* could not have been written without the support of the salonières, yet there is no mention of the salons within its pages. Instead, articles about women concentrated on their physical weakness, their emotional sensitivity, and their role as mothers.[39] While a few articles discussed equality, most accepted the traditional view that women were men's inferiors and often their opposites. By the eighteenth century, the idea that what is a virtue for one sex is a defect in the other had become a cliché: "An effeminate behavior in a man, a rough manner in a woman; these are ugly because unsuitable to each character," wrote Hume.[40] Moreover, there was concern that each sex remain in its proper place. Joseph Addison and Richard Steele's influential journal, the *Spectator*, consistently condemned women who encroached on male territory by being too independent, too forward, or too "impertinent." "I think it absolutely necessary to keep up the Partition between the two Sexes and to take Notice of the smallest Encroachments which the one makes upon the other," wrote Addison in 1712, expressing a sentiment increasingly common in the eighteenth century.[41]

The salonières had entered the male territory of culture, learning, and politics. Traditionally, European men had feared and condemned any semblance of female influence in politics; the eighteenth century, with its numerous examples of such influence, brought forth a deluge of such complaints, often made by men who themselves enjoyed and profited from the salonières' and courtesans' access to the powerful. Montesquieu, the French political philosopher who frequented Deffand's salon when he was in Paris, argued that chastity was women's most important virtue and its loss made her a source of corruption.[42] Women's influence generally was corrupting: it lead to "luxury, revelry, gambling, love, and all the consequences of these passions," as Jacques-Joseph Duguet, a French moralist, charged in 1750.[43] The upheavals of the French Revolution only intensified these sentiments. Napoleon Bonaparte's personal pleasure in the Paris salons did not prevent him from condemning women's participation in politics and advocating more traditional roles: he argued that disorder results when "women leave the state of dependence where they should remain" and that "states are lost when women govern public matters."[44] G.W.F. Hegel, the German philosopher, who was a frequent guest at Rahel Levin's second salon in Berlin in the 1820s, expanded the criticism, citing traditional fears:

If women were to control the government, the state would be in danger, for they do not act according to the dictates of universality, but are influenced by accidental inclinations and opinions. The education of women goes on one hardly knows how.[45]

Hegel, like Bonaparte, did not envision improving the education of women to enable them to learn how to govern, but rather focused on preventing women from wielding any public power.

The contradiction between male artists' and philosophers' personal pleasure in the company of the salonières and their public condemnation of such women is clearest in the consistent criticism directed at women who dared to relate to men as intellectual or cultural equals. Mme. de Rambouillet's *chambre bleue* and the "precious women" who attended it were best known to both their contemporaries and posterity as the butts of Molière's satires on female pretensions to culture and learning, in his plays *Precious Women Ridiculed* (1659) and *The Learned Women* (1672). The combination of "learned" with "woman"—like that of "bourgeois" with "gentleman"—was funny by definition. Fifty years later, little had changed. Voltaire lived for many years with Emilie du Châtelet (1706–1749), one of the most learned women of her age, who was famous for her scientific writings and her commentary on Leibnitz. Justifying her encroachment into the male domain of philosophy, Voltaire explained in his memoirs that "Mme. du Châtelet did not seek to decorate philosophy with ornaments to which philosophy is a stranger; such affectation never was part of her character, which was masculine and just."[46] Despite his pleasure in his educated mistress, Voltaire wrote very little on women's education, function, or role in society—a glaring omission in a body of work which criticized so many other traditional institutions. He also seems to have been at best ambivalent about Châtelet's intellect. "Emilie, in truth, is the divine mistress, endowed with beauty, wit, compassion and all of the other womanly virtues," he wrote a friend. "Yet I frequently wish she were less learned and her mind less sharp."[47] In this, as in little else, Voltaire was in agreement with his arch-rival, the philosopher Jean-Jacques Rousseau. Rousseau, introduced to the Paris salons by Diderot in the 1740s, condemned the salonières in his influential novel *Émile* (1762):

I would a thousand times rather have a homely girl, simply brought up, than a learned lady and a wit who would make a literary circle of my house and install herself as its president. A female wit is a scourge to her husband, her children, her friends, her servants, to everybody. From the lofty height of her genius, she scorns every womanly duty, and she is always trying to make a man of herself, like Mlle. de L'Enclos.[48]

Ninon de Lenclos, the celebrated courtesan, could only be seen as mannish if learning was equated with men, which it was. "A woman who . . . carries on fundamental controversies about mechanics, like the Marquise du Châtelet, might as well even have a beard," wrote the German philosopher Immanuel Kant, who believed that "laborious learning or painful pondering, even if a woman should greatly succeed in it, destroy the merits that are proper to her sex."[49]

Such sentiments became clichés by the end of the eighteenth century. "If you would be happy, don't wed a blue-stocking," went a French rhyme.[50] The English word *Bluestocking* shifted its meaning in this era: originally designating a woman who attended the English salons of midcentury, it came to be a derogative term for a woman who aspired to learning. "Magazines, journals, and reviews abound with sarcastic comments upon the blue-stockings and their productions," wrote the English *Ladies Magazine* in 1825. "Intellectual acquirement, when applied to a woman, is used as a term of reproach."[51] A scholarly woman could exempt herself from such criticism only by asserting her femininity through performing traditionally female activities: breast-feeding, child rearing, cooking, sewing. "A man is in general better pleased when he has a good dinner upon his table than when his wife speaks Greek," declared Samuel Johnson, a frequent guest at bluestocking gatherings. "My old friend Mrs. Carter could make a pudding as well as translate Epictetus from the Greek and work a handkerchief as well as compose a poem."[52]

The men who socialized with the least domestic women of their era idealized a domestic role for all women. They advocated a life for women defined by their relationship to a man and care of his children and household, rather than by the women's own achievements and talents. "The utmost of a Woman's Character is contained in Domestick Life," wrote Richard Steele in the *Spectator*,

> She is Blameable or Praiseworthy according as her Carriage affects the House of her Father or her Husband. All she has to do in this World is contained within the Duties of a Daughter, a Sister, a Wife, and a Mother.[53]

A constant theme in the influential *Spectator* was praise of the domestic woman, who confined herself to home and family, and the consequent condemnation of the woman of society, who "grows Contemptible by being Conspicuous."[54] Subsequent generations of artists and writers glorified female domesticity. The *Encyclopédie* condemned the salonières and praised domestic women, and Diderot, Mercier, Florian, and other French playwrights made idealized family life the center of many of their dramas.[55] Artists did the same, and glorifying motherhood and home life became common in the eighteenth century. Jean Baptiste Grueze's *The Beloved Mother* (1765) was a popular example of the genre. Surrounded, even buried, by her numerous children, one of whom she has just breast-fed, this mother radiates contentment in her clean, well-ordered home.[56] Diderot, who spent many hours with the salonières, was delighted with the painting: "It preaches population," he wrote,

> and portrays with profound feeling the happiness and the inestimable rewards of domestic tranquility. It says to all men of feeling and sensibility: "Keep your family comfortable, give your wife children; give her as many children as you can; give them only to her and be assured of being happy at home.[57]

In France, where women of the elite invariably sent their children away to wet nurses in the country for the first few years of life, breast-feeding was extolled by male authors. "When mothers deign to nurse their own children, there will be a reformation in morals; natural feeling will revive in every heart ... in the cheerful home life the mother finds her sweetest duties," promised Rousseau near the beginning of his highly regarded novel *Émile* (1762). Elsewhere in Europe, women's domesticity was equally praised. Friedrich von Schiller's poem "The Song of the Bell" paid homage to the busy housewife, who "never rests," and his equally popular "Women's Honors" glorified women's traditional role:

> Honor the women! They plait and weave
> Heavenly roses in earthly life.
> ....
> In the mother's humble cottage,
> They remain with chaste morals,
> True daughters of innocent Nature.

The glorification of female domesticity went hand in hand with a condemnation of other choices. If a woman did not breastfeed, argued Rousseau, "every evil follows in the train of this first sin."[58] Philosophers condemned the nondomestic salonière. Johann Gottlieb Fichte, the German nationalist, frequently attended Rahel Levin's salon in Berlin, but saw no contradiction in arguing that the "second sex" could only find fulfillment in marriage.[59] Heinrich von Kleist, the novelist, was also a frequent guest, but declared in 1800 that "the man is not just the husband of his wife, he is also a citizen of the state. The Woman on the other hand is nothing more than the wife of her husband."[60] By the end of the eighteenth century, German men of letters began to organize all-male social groups, like the writer Achim von Arnim's "Christian-German Table-Society," which excluded from membership "Women, Frenchmen, Philistines, and Jews."[61] Women were supposed to remain with their families, not socialize with strange men in the salons.

As well as extolling the domestic virtues for women, the men of the Enlightenment, like so many generations of European men before them, insisted that chastity was woman's highest virtue and left the double standard of sexual behavior intact. "Modesty" and "chastity" are duties which "belong to the fair sex," argued David Hume. The double standard was essential to "the interest of civil society . . . and to prove this," he concluded, "we need only appeal to the practice and sentiments of all nations and ages."[62] Hume expressed the sentiments of his era. Regardless of whether or not they had sexual liaisons with the salonières, the men of the Enlightenment united in condemning the woman who had sex outside of marriage. Johnson thought she "should not have any possibility of being restored to good character"; Rousseau equated her crime with treason:

> The faithless wife is worse [than the faithless husband]; she destroys the family and breaks the bonds of nature; when she gives her husband children who are not his own, she is false both to him and society, thus her crime is not infidelity, but treason. To my mind, it is the source of dissension and of crime of every kind.[63]

In addition, they believed that women's supposed sexual power, so frightening and threatening to men, must be controlled for the good of society. Women who used their sexual power, as salonières or royal mistresses, could only meet with condemnation.

In opposition to the salonières and the values they came to represent, male writers and artists glorified the innocent, uneducated heroine over the more challenging woman of the world. The most popular novel of the eighteenth century was Samuel Richardson's *Pamela, or Virtue Rewarded* (1743), in which Pamela, a fifteen-year-old servant girl, manages not only to resist her master's attempts to seduce her, but even induces him to marry her.[64] Pamela is presented as a paragon of innocence, purity, and virtue, which "make her character worthy of the imitation of her sex," as Richardson wrote in the moral lessons which conclude the novel. Pamela, always subservient to her husband (she finds it difficult to break her habit of calling him her "dearest master"), shows independence only in her strenuous defense of her virginity; otherwise, she is—with the author's approval—totally dependent on Mr. B——, who manages her life, elevates Pamela and her family to his own status, and tries to convince his wife that he genuinely loves her and is not merely being kind to a social inferior. *Pamela* powerfully restates men's traditional womanly ideal: that of a young, virtuous maiden who remains unspoiled and deferential to the husband who deigns to choose her. Pamela's closest literary ancestor is Patient Griselda.

Pamela provided a model for other literary heroines. The earliest female protagonists of the English novel had been the adventurous, enterprising heroines of Defoe: Moll Flanders and Roxanne. While Defoe's hero, Robinson Crusoe, became popular and influential throughout Europe, Moll Flanders and Roxanne were superseded by Pamela.[65] Other European authors acknowledged Pamela's influence and recreated her in their own writings. The beautiful young woman, often of low station, who becomes a heroine because of her virtue was a staple of literature in this period. There was Charlotte, in Goethe's *Werther*, Luise Miller in Schiller's "Love and Intrigue," Emilia Galotti in Lessing's play of the same name. All shared their prototype's rectitude. Their virtue makes the men who try to seduce them evil by contrast, and as virtuous heroines who resist seduction by evil men these women gained the power of moral authority.

Rousseau's novels extolled traditional women who gained immense moral power by their choice of virtue. Julie, or the "new Heloise," in his novel of 1761 redeems herself by renouncing her lover and dedicating herself to the husband chosen for her by her father. Her reward is the approval of all and a paradisical life. "Julie! Incomparable woman!" exclaims the hero,

> in the simplicity of private life you control a despotic empire of wisdom and good deeds. You are a dear and sacred repository for the entire countryside which each would conserve and defend at the price of his blood. You live more securely, more honorably in the middle of a whole people who loves you, than Kings surrounded by all their soldiers.[66]

Sophy, in *Émile* (1762), achieves a similar power through her choice of virtue. Educated only to be a traditional wife, she "can send [men] at her bidding to the end of the world, to war, to glory, and to death at her behest."[67] Rousseau glorified women's traditional role and insisted that it was the chief means of moral and social reform in a corrupt era. These heroines even appealed to women who were themselves salonières, activists, and writers.[68] Their alternative was to accept the eighteenth-century view that women must be treated differently because they were so much weaker than men. "A woman is naturally more helpless than the Other Sex," wrote Richard Steele in 1711, "and a Man of Honour and Sense should have this in his view in all Manner of Commerce with her."[69] This debilitating view rapidly evolved into the supposedly chivalric assumption that women should be treated more gingerly and gently than men, because of their weaker natures. "When we write of women," stated Diderot, "we must dip our pen in the rainbow, and throw upon the paper the dust of butterflies' wings."[70]

By the end of the eighteenth century, the salonière was repudiated in favor of more traditional women. This change occurred very rapidly during the era of the French Revolution and the Napoleonic Wars (1789–1815). The social and political power which the salonières had wielded in pre-revolutionary France became a leading criticism of the old monarchy, and people of differing classes and political philosophies united in condemning this "female influence." "Women ruled [in the eighteenth century]," the French artist and Marie Antoinette's portraitist, Elizabeth Vigée-Lebrun, remarked in her memoirs. "The Revolution dethroned them."[71] Vigée-Lebrun exaggerated women's powers, but accurately perceived their decline in influence. The revolution unleashed a flood of criticism about women's "unnatural" usurpation of the male domain of politics. Female political activity was outlawed in 1793, and male politicians, journalists, and philosophers condemned women's political influence whether it was republican or monarchist, revolutionary or counter-revolutionary.[72] Manon Roland (1754–1793), famous both for her liberal salon and her death cry "Liberty, what crimes are committed in thy name" as she mounted the scaffold to the guillotine, was excoriated in contemporary newspapers. Mme. Roland "was a mother," pontificated the *Journal of Public Safety*, "but she sacrificed nature by desiring to rise above herself; her desire to be learned led her to neglect the virtues of her sex, and this neglect, always dangerous, ended by leading her to perish on the scaffold."[73] Social prominence and aristocratic rank could not protect a woman from such judgments: Germaine de Staël (1766–1817), the eminent author and salonière, was often disparaged for being too "mannish" by participating in politics. "Miserable hermaphrodite that you are," went one typical denunciation, "your sole ambition in uniting the two sexes in your person is to dishonor them both at once."[74] De Staël's exiling by Napoleon Bonaparte signaled the crushing of the power of the salons. After 1815, new institutions like the new law codes, the press, the universities, and a reformed government undercut the political and social influence of the salonières.[75]

They were morally repudiated as well, and often by the very women and men who had frequented the salons. A significant number of aristocratic women who survived the revolution blamed upheavals on the corruption of the salonières and their "unnatural" influence in government and society. These women urged domesticity, even for women of the aristocracy, and wrote guides for mothers advocating that they train their daughters for life in the home.[76] While salonières continued to exist in nineteenth-century France, their power was eclipsed. Salonières often dwindled into hostesses, shunning the limelight themselves and proving their domesticity by bringing children into the salon. The new ground the salonière had claimed between wife and courtesan eroded away, and women of the nineteenth century found it difficult to follow that model. "With a talent like hers," the French journalist Descluze remarked of the salonière and poet Delphine Gay (1804–1850) in 1825, "it is necessary to frankly label her a courtesan. You are freer and embarrass the others much less."[77] The ancient choice for women, of being considered either a respectable wife or a dishonorable whore, was reasserted in nineteenth-century Europe.

Outside of France, the repudiation of the salonière was associated with the new nationalistic repudiation of French values and manners, now seen as especially repugnant because of Bonaparte's conquests. "What has become of the time when we were all together?" lamented Rahel Levin in 1818. "It went under in 1806 [when Bonaparte defeated Prussia]."[78] In Germany, in Russia, in England, the French salonière—cosmopolitan, sophisticated, and libertine—was discarded in favor of a virtuous, chaste, and unspoiled German, Russian, or English maiden.[79] "I came upon the most charming sight imaginable," wrote Werther, in Goethe's influential novel of 1774, describing his first glimpse of his beloved, Lotte:

> Six children . . . were swarming around a very pretty girl of medium height. She had on a simple white dress with pale pink bows on the sleeves and at her breast, and she was holding a loaf of black bread and cutting a slice for every one of her little ones, according to their age and appetites.[80]

The simple domestic scene stands in moral opposition to the world of the courts and the salons. The moral repudiation of the salonière led to the moral empowerment of her more traditional sister. The supposed power of female influence was rejected in favor of the power of female virtue. Just as the *précieuses* had rejected sexual activity in order to gain both time for cultural life and moral authority, so many women of the upper and middle classes "reformed" their behavior. Women acting in the traditionally male arena of government and politics signaled political corruption; women remaining within their traditional arena of home and family would lead to political integrity. Women would have a more positive impact on society, it was continually agreed, if they devoted themselves to providing good homes for their husbands and children rather than meddling in politics. The traditionally virtuous woman was extolled: "A virtuous woman is little lower than the angels," as Rousseau asserted in *Émile.*[81] By the turn of the nineteenth century, when Rousseau's ideas reached their greatest sway, the angelic comparison had become commonplace. "She was a Phantom of Delight . . ." went William Wordsworth's well-known poem of 1804,

> A Spirit, yet a woman too.
> . . . .
> A perfect woman, nobly plann'd
> To warn, to comfort, and command;

> And yet a Spirit still, and bright
> With Something of angelic light.

If a "perfect woman" is an "angel," how could, or should, a man "command" her? Here again, *Pamela* showed the way out of this paradox. Pamela must walk a moral tightrope: as a woman, a servant, and, later, a wife, she is supposed to obey Mr. B——. But as a virtuous human being, she is supposed to disobey him when he threatens her virtue—in both senses of the word. With the author's approval, Pamela resists both Mr. B——'s sexual assaults and his later attempts to make her life too luxurious: she begs to be permitted to give her allowance to the poor, for instance. This was the path which would win women of the elite the most social respect and room to maneuver in an era in which the salonière was repudiated and women were increasingly urged to confine themselves to domesticity. If a woman remained traditionally subservient and deferential to men and did not challenge women's traditional behavior as the salonières had done, she was now able to claim the new moral authority of women's supposedly superior virtue.

By upholding the most traditional female virtues—modesty, chastity, domesticity—women acquired the authority of the virtuous. With that power, they were able to expand women's influence beyond the family and the household. The most successful women of the era followed this route. Adhering scrupulously to the most rigid codes of respectable female behavior, they won society's approval for their actions, even if these took them outside the home and family. But if a woman claimed prerogatives traditionally associated with men—political participation, education, civil rights—she found herself isolated. If she dared to claim the right to sexual freedom (as so many of the salonières had done), she found herself condemned and ostracized. These two paths and their consequences can be clearly seen in the lives of two contemporaneous but very different Englishwomen: Mary Wollstonecraft, the advocate for women's rights, and Hannah More, the moral reformer.

Mary Wollstonecraft and Hannah More

Wollstonecraft (1759–1797) and More (1745–1833) had much in common. Both were born into the growing English middle classes (Wollstonecraft's father, a silk weaver, inherited an income which enabled him to live independently until he squandered it; More's father was a schoolteacher), and each was raised in moderately comfortable circumstances. Both were intelligent and drawn to books and both became schoolteachers in their early twenties—Wollstonecraft after working as a companion, a governess, and a seamstress, More by working at her father's school. Both recoiled from what they perceived as the vapid foolishness and flirtatiousness they observed in many young women of the elite, and both urged a reformation of these women's behavior through the improved education of girls. Both contended that women must become more virtuous: "The main business of our lives is to be virtuous," wrote Wollstonecraft in her *Thoughts on the Education of Daughters* (1786); More's *Strictures on the Modern System of Female Education with a View to the Principles and Conduct of Women of Rank and Fortune* (1799) embraced the same principle. Both urged that women be more serious, more charitable, and more moral. Intelligent and active, each was invited to join a London literary circle and each became accepted for her mental and literary abilities.

But this resemblance did not last. In the course of their lives, Wollstonecraft and More moved in opposite directions: More, welcomed in the 1770s into bluestocking circles, abandoned them in the 1780s for the more congenial "vital religion" group of evangelical Christians known as the "Clapham Saints." Originally a playwright, poet, and wit, she had restricted her reading to religious subjects by the 1790s. Wollstonecraft's first writings were intensely religious: her *Original Stories from Real Life* (1788) preached Christian virtues and portrayed the ideal woman as "Mrs. Truman," a gentle, domestic, and maternal curate's wife. But as the French Revolution unfolded, Wollstonecraft gradually lost her religious faith and turned instead to a faith in reason and reform. She came to believe that women and men were created equal, endowed with the same potential talents and abilities. "The first object of laudable ambition," she wrote in her *Vindication of the Rights of Woman* (1792), "is to obtain a character as a human being, regardless of the distinction of sex.[82] From this premise, Wollstonecraft concluded that girls should be given an equal education with boys. They would then be able to practice as physicians, run a farm, or manage a business.[83] Rejecting the traditional premise that woman's and man's natures were different, Wollstonecraft ended by rejecting all those traditions which circumscribed women's and men's lives. Her political radicalism led her to reject traditional social hierarchies which determined the lives of all before she concentrated on rejecting traditions which confined women. Enthusiastic about the French Revolution, she wrote a *Vindication of the Rights of Man* the year before her book on women. Living in Paris during the height of the revolution, she rejected the tradition of female chastity as well: entering into an affair with the American adventurer Gilbert Imlay, she bore a daughter out of wedlock in 1794.

The violence and poverty of Wollstonecraft's early life predisposed her to rebel against injustice and exploitation. Her parents had a violent marriage and Wollstonecraft remembered that as a child she had tried to protect her mother from her father's blows. Poverty forced her to become a paid ladies' companion at nineteen; a few years later, she damaged her sight by attempting to support herself sewing long hours. Familiar with poverty and the limited options of women who needed to make a living, she became increasingly concerned with the plight of the poor, whom she thought ill-served in the present state of society. When her publisher offered to establish her in London, she seized this novel opportunity. "I am then going to be the first of a new genus," she wrote her sister. "I tremble at the attempt."[84] Refusing to make a home for her two younger sisters, she argued that she could support them better as an independent writer, and in these years she made good her claim by producing the *Vindication*s of the rights of men and women, both of which sold well.

Through the historical and personal upheavals of her life, she remained a professional writer, producing publishable works under difficult circumstances. In France during the Reign of Terror, when she was pregnant with her first daughter, she wrote a critical history of the first six months of the French Revolution. Abandoned by her lover, Gilbert Imlay, she first despaired (even attempting suicide by throwing herself in the Thames), but later recovered and undertook an arduous business journey for him through Scandinavia in 1795. She published her lengthy letters as a book the next year. They are remarkable for their observations and comments about poor working women. Writing of Swedish servants, she argued that "the men stand up for the dignity of man, by oppressing the women."

> The most menial, and even laborious offices, are therefore left to these poor drudges. Much of this I have seen. In the winter, I am told, they take the linen down to the river to wash it in the cold water; and though their hands, cut by the ice, are cracked and bleeding, the men, their fellow servants, will not disgrace their manhood by carrying a tub to lighten their burden.[85]

In the late 1790s, she focused on the plight of poor women, which made her unusual in both her era and her class. She was aware that "the struggles of an eventful life have been occasioned by the oppressed state of my sex."[86] In her last work, *Maria, or the Wrongs of Woman* (1796), she contrasted the oppression of Maria, her middle-class heroine, against the even darker life of Jemima, a working-class woman. *Maria* graphically portrayed the sufferings of "women, the *outlaws* of the world," and in the novel Wollstonecraft depicted rape and sexual violence, poverty and cruelty, men's power and women's relative helplessness.[87] Being "born a woman" meant being "born to suffer."[88] The novel opens with Maria, the heroine, confined in an insane asylum by her husband, who has seized her child and her property. Powerless in the face of laws which gave total control to the husband, Maria pleads futilely for a divorce. In the asylum, she befriends the servant Jemima. Struggling to survive, Jemima has spent her life in drudgery. "On the happiness to be enjoyed over a washing-tub, I need not comment," she tells Maria,

> yet you will allow me to observe, that this was a wretchedness of situation peculiar to my sex. A man with half my industry, and I dare say, abilities, could have procured a decent livelihood . . . whilst I . . . was cast aside as the filth of society. Condemned to labour, like a machine, only to earn bread, and scarcely that, I became melancholy and depressed.[89]

The novel remained unfinished. The year she wrote it, Wollstonecraft became pregnant by her new lover, the radical English philosopher William Godwin. They married, but retained separate apartments in nearby buildings. She gave birth to a daughter (the future Mary Godwin Shelley, author of *Frankenstein*), but died a few days later from infection. She was thirty-eight years old. The following year, Godwin published her unfinished novel, his memoirs of her, and, most inexplicably and disastrously, her letters to Gilbert Imlay. For over a century after her death, she was primarily remembered for these passionate and despairing letters: as a woman who broke the accepted proprieties by having an affair and who bore the punishment of both an illegitimate child and rejection by her lover. Her impropriety tainted her radical, feminist ideas, and to most of her contemporaries, Wollstonecraft's life seemed to prove the wisdom of adhering to tradition: her rebellion against traditional strictures on women's behavior seemed to lead only to disgrace and death. For the overwhelming majority of women of the upper and middle classes, the price of Wollstonecraft's independence seemed too high. Those who wished a role outside the home followed the far more traditional and respectable path extolled and exemplified by Hannah More.

Even during her sojourn among the Bluestockings, More upheld the traditional Christian virtues and criticized her companions when they did not do so. Although she enjoyed a party in 1775 where she, Elizabeth Carter, Elizabeth Montagu, Fanny Boscawan, and Hester Chapone met for dinner, she thought it wrong that they met on Sunday: "though their conversation is edifying, their example is bad."[90] By 1780, she was complaining to her sister that there were few in bluestocking circles "to whom one can venture to recommend sermons," and a few years later, she abandoned the Bluestockings for the more congenial "Clapham Saints."[91]

More was drawn to the Clapham group by their campaign to abolish the slave trade (she wrote a poem on the subject in 1787), and by the late 1780s she had become an acknowledged leader in their subsequent efforts to reform the manners of the rich, revive piety among the poor, make the Sabbath more religious, and sponsor Christian missions abroad. The French Revolution,

which she despised, only confirmed her conviction that tradition and traditional virtues must be revived and strengthened. Although she never met
Wollstonecraft, she loathed all the younger woman stood for. "I have been
much pestered to read the *Rights of Woman,*" she wrote Horace Walpole
in 1793, "but am invincibly resolved not to do it."

> There is something fantastic and absurd in the very title. How many ways there
> are of being ridiculous! I am sure I have as much liberty as I can make use of,
> now I am an old maid; and when I was a young one, I had, I daresay, more than
> was good for me.[92]

More turned her formidable energies to the reformation of English society
along conservative principles, and she argued that women had a vital part to
play in this endeavor. On women, she wrote in 1799, "depend, in no small
degree, the principles of the whole rising generation," and women's "private
exertions" contribute to the "future happiness" or "future ruin" of the
nation.[93]

In her own activities, she set a powerful example of the claim that
women's moral authority should lead to a wider social role. She and her sisters
started over a dozen schools for the poor in rural mining districts, a philanthropy made possible by an annuity More had received as a young woman
from a suitor.[94] The endeavor was controversial—many still believed the poor
should not be taught to read. More believed in teaching the poor reading,
so they could study Scripture, but not writing, which might encourage rebel-
liousness. Discovering that Sunday was the only day poor children had free
from work, she subordinated her belief in the sanctity of the Sabbath to her
belief in the importance of education and taught on Sunday afternoons, often
walking miles across the hills from one school to another. In these same years,
More produced her most influential writings: the *Cheap Repository Tracts*
(1795–1798), designed to keep the English poor from revolting like the
French and priced to be within their reach. The tracts sold two million copies
in their first three years, a publishing feat unparalleled in the period. Prior
to More, women had ventured to write essays, poetry, plays, and novels; More
pioneered female writing for the mass market of the lower classes. Her tracts
were simply written, entertaining parables which preached humility, forbear-
ance, and deference.

Women of the elite had always been encouraged to perform private acts
of charity; More went further and glorified women's duty to reform society.
She was able to do so in part because of her utter respectability; a lifelong
virgin, her name was never linked with a sexual scandal. She advocated
traditional values for both the poor and women. The poor should be respectful of their "betters," women should embrace the traditional female virtues

of chastity, modesty, and piety. Like many women who adhered to society's proscriptions, she urged such adherence for all women. "Propriety is the first, second, and third requisite for women," she argued in 1799.[95] In advocating "proper" behavior, More drew on many of Western culture's most ancient traditions about women. Her first book of essays for young women had as its epigraph Pericles' admonition that the best woman is she who is not talked about. She wrote Horace Walpole that women would only suffer from increased freedom: "To be unstable and capricious, I really think, is but too characteristic of our sex; and there is, perhaps, no animal so indebted to subordination for its good behavior as woman."[96] Walpole (whom Mme. du Deffand had loved in vain) admired More and called her "Saint Hannah." He told her she might "be one of the cleverest of women" if she did "not prefer being one of the best!"

The opposition of cleverness and goodness came naturally to the generation which witnessed the repudiation of the salonière. More's own life embodied that change: originally welcomed as a Bluestocking, she rejected that role for the more respectable one of moral reformer. She and other women of her generation discovered that the claim to virtue enabled them to be more powerful and independent than they would have been as members of the bluestocking and salonière circles. "The woman who derives her principles from the Bible," More wrote in her popular novel *Coelebs in Search of a Wife* (1808), "will not pant for *beholders*. . . . She lives on her own stock. Her resources are within herself. She possesses the truest independence."[97]

In *Coelebs*, More contrasted the fashionable Lady Bab Lawless with her heroine, modest Lucilla Stanley. Lady Bab is criticized as one of "these self-appointed queens [who] maintain an absolute but ephemeral empire over that little fantastic aristocracy which they call the world."[98] Coelebs, the hero, rejects her in favor of Lucilla Stanley, who rises at six and then spends two hours reading Christian literature. At eight, Lucilla consults the housekeeper and cook, plans the meals, distributes provisions, looks over accounts, and makes breakfast for her parents, "as fresh as a rose, as gay as a lark."[99] She reads with her father for an hour after breakfast and assists her mother in teaching her younger sisters. She is modest, tactful, and above all, virtuous: she and her sisters spend their spare time making useful articles to give to the poor, whom Lucilla visits one day and two nights each week.

Coelebs was immensely popular: it went into twelve editions the year it was published and thirty by 1834, the year after More's death. More's talent for writing and sense of humor aside, the novel seems to have been so successful with the largely female novel-reading audience because of its attractive portrait of an ideal woman.[100] "It has always been the practice of the better kind of country ladies to distribute benefactions," wrote Lucy

Aiken, a contemporary, "but Hannah More by representing her pattern young lady as regularly devoting two evenings to making her rounds among the village poor has . . . made it a fashion and a rage."[101]

Throughout the nineteenth century and into the twentieth, privileged women who wished for a life beyond the home would follow the path More had pioneered. They would seize upon their duty to be virtuous and transform it into the right to reform society. But women like this were a minority. Most of the women who moved into the ranks of privilege found maintaining a household and raising a family full-time work, even with servants to help them. The new wealth, which enabled tens of thousands of families to achieve a standard of living above that of the working class, removed women from the labor force and gave them the privilege of staying at home—like a lady. To most of these women, this change was an improvement, and the concern of the majority was not how to leave the parlor, but how to live happily and comfortably within it.

2

WOMEN IN THE PARLORS

❦

The Impact of a Rising Standard of Living

"I wish I could paint my present situation to you," Elizabeth Gaskell (1810–1865), the future novelist, wrote her sister-in-law in the summer of 1836. "Fancy me sitting in an old-fashioned parlour, 'doors and windows open wide,' with casement window opening into a sunny court all filled with flowers." Married for four years, the mother of a two-year-old girl, and pregnant with her next child, Gaskell was visiting the country house where she had been raised:

> Here are Baby, Betsey [the nursemaid], Mama and Bessy Holland [a cousin]. ... We sit and read and dream our time away—except at meals when we don't dream over cream that your spoon stands upright in. ... Baby is at the very tip-top of bliss.[1]

The wife of a Manchester clergyman, Gaskell was not a wealthy woman, but she could take many comforts for granted. A vacation in the country, servants, luxurious foods, some leisure time, and the parlor itself distinguished her life and those of her peers from most women in earlier periods of European history. By the second half of the eighteenth century, more and more women could live in ways previously restricted to the few. For these European women, the economic growth generated first by commercial capitalism and, in the nineteenth century, by industrial capitalism enabled them to afford the servants, homes, food, clothing, child care, and leisure which had previously demarcated the lives of the elite from those of the people. For women, the increase in the wealth of the nation, the rise in the standard of living, and the growth of the middle classes enabled tens of thousands to live in a new way. While very few could build a salon, most could now afford a parlor, itself a symbol of the new, more comfortable middle-class way of life. For these women, massive changes in the economy were manifested in concrete improvements in everyday life. The means to employ a maidservant, to spend more time caring for their children, to live in larger and better-furnished homes, to feed and clothe their families better, to read and write—such were the tangible results of economic growth for these women. For them, the most important single development was the ability to hire a female servant to aid in the work of running the household and helping in the family business, if there was one.

Women ceasing to do heavy housework or labor in a family business were criticized from the early centuries of commercial capitalism, for this signaled a family's attempt to rise in society. A wife who no longer had to labor—either within the home or outside of it—demonstrated her husband's success in providing for his family. "The tradesman is foolishly vain of making his wife a gentlewoman," wrote Daniel Defoe about England in the early eighteenth century. "He will have her sit above in the parlour, receive visits, drink tea, and entertain her neighbors, or take a coach and go abroad; but as to the business, she shall not stoop to touch it; he has apprentices and journeymen, and there is not need of it."[2] Despite contemporary complaints that there was something inappropriate and ludicrous in families' attempts to better their social status, that goal became possible for increasing numbers of people as the economy generated increasing wealth. First in England, later on the Continent, women who could afford to do so began to detach themselves from working in the family's business.[3] In the past, women's work in a family enterprise had to be done in addition to her increasing domestic duties: caring for her husband and children and maintaining and supervising the household. Combining both was arduous and hectic, as can be seen in the description of a typical day which a Frenchwoman, Alexandrine Virnot-Barrois, wrote her husband in 1790. The couple owned a textile business and had three small children; the husband was away on a buying trip. She rose at seven, and while she breakfasted, she nursed one child and amused the two others. She then went to her office to go over the books and keep records, after which she inspected the waterproof cloaks produced the previous day. "I number what there is or I send them to be dyed; I mark them and finally do whatever work is at hand. When I have more time, like today," she explained, "I spend a bit of my morning writing to you." In the afternoon, she visited the dyeworks or the agent; if she had enough time, she paid calls on relatives as well.[4]

A study of bourgeois women in an industrial region of France revealed that Mme. Virnot-Barrois's way of life was vanishing by the third quarter of the nineteenth century, as factories and corporations replaced family enterprises.[5] In England, earlier economic development ensured that such women could cease working in a family business at an earlier date. From the middle of the eighteenth century, Englishwomen were also being advised to detach themselves from the physical labor of housework, if they could afford to do so. "In my opinion a lady of condition should learn just as much of cooking and of washing as to know when she is imposed upon by those she employs," wrote Eliza Haywood (1690–1756) in her magazine *The Female Spectator* in 1746. "[More] may acquire her the reputation of a notable housewife, but not of a woman of fine taste, or in any way qualify her for polite conversation, or of entertaining herself agreeably when alone."[6] For a woman, the first step upward in society was to move out of the kitchen and leave the grosser forms of housework to her servants, to switch from laboring herself to supervising others' labor. If she could not afford to do so herself, she could aspire to raise her daughters to be "ladies." "Men of fortune like their wives to command respect, and not to be like as if they had been ladies maids or housekeepers!" wrote Charlotte Palmer, an English novelist, in the 1790s: "I cannot say I should like to see my daughters pickling and preserving, and putting their hands into a pan of flower [*sic*], and all that; let them keep their tambour, their music, their filligree, and their drawing; that's quite enough."[7]

As increasing numbers of women were able to devote more time to their homes and families, the word "lady" changed its meaning to incorporate them. Originally the female equivalent to "lord," "lady" expanded its meaning in all European languages in the centuries of commercial and industrial capitalism to become the female equivalent of "gentleman."[8] While the masculine words retained the traditional distinction between an aristocracy of birth and an elite based on wealth, "lady" came to mean a woman with pretensions to refinement who did not have to labor with her hands, whether for wages, in a family business, or at heavy housework. "A lady, to be such must be a mere lady," wrote Margaretta Greg, an Englishwoman, in 1853. "She must not work for profit, or engage in any occupation that money can command, lest she invade the rights of the working classes, who live by their labour."[9]

Being a "mere lady," however, clashed with the condemnation of idleness in women. The solution, both on the part of women and those who sought to advise them, was to expand women's traditional domestic role so that it would fill the new time available.[10] Women spent more time on mothering, on housekeeping, on social life, on the ladylike "accomplishments" Charlotte Palmer had wanted for her daughters: music, drawing, embroidery, foreign languages. As economic growth enabled more women to leave the job market, they increased the time and energy spent on their homes and families. As a signal of wealth, families stopped educating their daughters to earn a living and instead expected them to be busy and happy at home. Boys of the middle classes were trained to support themselves, either in business or the professions. In contrast, girls in the same families were raised without such training, whether by an apprenticeship, by being sent out as a maidservant, or by education. "My father, who was certainly quite advanced in his ideas, never for a moment contemplated that any of his daughters [there were six] should learn professional work with a view to their living," remembered the English reformer Edward Carpenter about the 1860s.[11] Both mothers and daughters spent increasing amounts of time on the domestic activities considered appropriate to women. In the course of the nineteenth century, these activities were elaborated to a degree previously impossible to women who also had to earn income for their families. New wealth enabled increasing numbers of women to devote their time to domesticity and to hire other women to lighten domestic burdens for them.

The Nineteenth-Century Lady: Her Life and Tasks

Until the end of the nineteenth century, all the physical labor of maintaining a household still had to be done by hand. To have heat or cooked food, wood or coal had to be hauled up from a cellar or in from a yard; fireplaces and stoves needed to be regularly cleaned and stoked; ashes had to be carried out and soot cleansed from the furniture, rooms, and windows. While it was in the course of the nineteenth century that food and clothing production moved out of the home and into the factory, many European women continued to process food and make clothes themselves. "Bread and cakes were made at home, all the preserves for the winter: fruit, from the simplest dried kind to the most complicated jellies, meat in all its various preparations, butter and eggs—everything was prepared and preserved at home for the household's needs," remembered Louise Otto-Peters (1819–1895), the German feminist, of her girlhood in Meissen in the 1820s.[12] Cold running water was piped into wealthy neighborhoods in European cities in the eighteenth century, but the flush toilet was not standard, even in the larger cities, before the 1880s. Until then, families made do with either a small closet with a chamber pot or an outhouse in back. As cities grew, outhouses disappeared, and the job of emptying and cleaning chamber pots, water closets, and later, toilets, fell to women within the home.

Lighting, in the era before gas or electricity, meant work. Women made candles from saved fat; the new kerosene lamps of the early nineteenth century took skill and experience to prepare. "When you interviewed a maid the first question asked was whether she could make a light," remembered Otto-Peters. The procedure was painstaking and time-consuming. Tinder, made by charring old silk stockings, was measured into a tinderbox; then steel and flint were struck together to produce sparks. Once the tinder had caught, a sulphur wick could be lit and placed in the glass lamp, which had already been washed, polished, and filled. The process involved "huffing and puffing," recalled Otto-Peters, and "since the sulphur fumes went right to your throat, an inadvertent cough or sneeze could sometimes blow the light out and you'd have to start all over again."[13] Laundry was done at home, and doing laundry involved bringing large amounts of water to boil, stirring the garments in, scrubbing them on a washboard, wringing them partially dry by twisting or passing them through a roller, and then hanging them up or ironing them dry. Ironing was done by heating heavy flatirons and pressing the wet clothes dry. The increased availability of cotton clothing—itself a product of industrial technology applied to textiles—escalated the amount of laundry a good century before technology was applied to lightening the burden of washing. Although few wove cloth at home, many middle-class women continued to make their families' garments until the growth of industry in the nineteenth century changed handmade clothing from a necessity to a luxury. Throughout the nineteenth century, women continued to hem and embroider sheets and towels, to make their own and their families' shifts, nightshirts, and undergarments, to knit and sew the woolen and cotton outer garments they could now afford in greater numbers.

Even in simple circumstances, the daily duties of food preparation, housecleaning, and maintenance of the family's clothes and possessions took most of a woman's time. In 1800, Dorothy Wordsworth (1771–1855) kept house for her brother William in a small country cottage, enabling him to write the poetry which made him famous. Despite the services of "Old Molly" from a neighboring family, Dorothy Wordsworth's days were filled with household duties. "Dried linen in the morning, the air still cold," went a typical entry in her journal,

> I pulled a bag of peas for Mrs. Simpson. Miss Simpson drank tea with me and supper on her return from Ambleside. A very fine evening. I sate [sic] on the wall making my shifts till I could see no longer.

The next day, she "ironed until dinner time—sewed till near dark—then pulled a basket of peas, and afterwards boiled and picked goose-berries."[14]

Women first spent their family's new wealth on hiring a maidservant to help them with the most difficult and distasteful chores necessary to maintain a decent standard of living.[15] Carrying heavy loads of wood, coal, laundry, or foodstuffs; emptying and washing chamber pots; boiling clothes, menstrual rags, diapers, and sheets before washing them, cleaning cooking pots, dishes, fireplaces, and stoves; laboring in the home workshop if there was one—these were the chores women first shared with or turned over to their servants. Increased wealth led to a tremendous demand for more female servants. By the nineteenth century, the numbers of female servants expanded greatly, and the most common servant was the essential, "maid-of-all-work."[16] "As her means increase every wife transfers every household duty involving labour to other hands," explained an English household guide of 1870.

> As soon as she is able to afford it she hires a washerwoman occasionally, then a charwoman, then a cook and housemaid, a nurse or two, a governess, a lady's maid, a housekeeper—and no blame attaches to any step of her progress, unless the payment is beyond her means.[17]

Few families could support so many servants. But the ability to afford even a single servant (the large majority of households which employed servants had only one) raised a woman from the lower classes and placed her, however precariously, within the middle ranks of society. For men, the dividing line between middle and working class was usually measured in income; for women, it lay in the difference between being a servant and being able to afford one. For the woman able to employ even a single servant to do the heaviest work, the change was one of status as well as labor. As the employer of a servant, a woman, however lowly, became a "mistress," able to order at least one other woman to do her bidding. She also gained time: having another woman labor in the house gave the employer some extra hours. In the late eighteenth and early nineteenth centuries, women tended to use this extra time for their domestic duties: child rearing and child care, furnishing and maintaining larger houses, serving more elaborate meals, and participating in more social occasions and festivities.

In families where the mother did not have to earn additional income, increasing numbers of women could devote the bulk of their time and attention to raising their children. The few population studies done of privileged groups in this period show an increase in numbers of births as well as family size. Births rose substantially in the English peerage between 1740 and 1815. A study of thirty-odd bourgeois families in the Nord of France showed that the average number of births rose throughout the nineteenth century, from 4.7 per family in the 1780s to 6.8 in the 1880s.[18] Relatively well-off women were the first to benefit from the application of new medical techniques to childbirth. By the end of the eighteenth century, obstetrics had become a separate branch of medicine, and forceps, used to extricate an infant from the birth canal, were widely used by doctors. "Pregnant women should become the object of an active benevolence, a religious respect, a kind of worship," wrote a French physician in the article on pregnancy in the *Dictionary of Medical Sciences* for 1812–1822.[19] As medical men began to specialize in the care of pregnant and birthing women of wealth, so these women responded by increasingly insisting that these new "male midwives" attend their deliveries. Initially most pronounced in England, women's reliance on male doctors grew throughout Europe in the nineteenth century. "Prosperous mothers almost always call a man-midwife to attend them," wrote a German doctor of Berlin around 1840. "The less well-off content themselves with a midwife or a so-called handy-woman [*Wickelfrau*] if they do not deliver in a clinic."[20]

By having a male physician attend her home delivery, a woman gained the advantage of medical expertise without risking the filth and infection prevalent in hospitals. The most important improvements in obstetrics came in the second half of the nineteenth century. Anesthesia was first used for childbirth in 1847 in England; five years later, it was administered to Queen Victoria at the birth of her eighth child. The Queen was delighted: the doctor "gave that blessed Chloroform and the effect was soothing, quieting and delightful beyond measure," she wrote in her journal.[21] Resisted by physicians, the demand for some kind of painkiller during childbirth was insisted upon by privileged women and became widespread by the early twentieth century.[22] Joseph Lister's proof of the efficacy of asepsis (1865) and the invention of absorbable sutures made tremendous differences in obstetrics and gynecology, since the possibility of infection was greatly decreased and new surgical techniques could be used. With anesthesia, antisepsis, and absorbable sutures, the caesarean became an operation a mother could survive, instead of an almost inevitable death sentence. By the last quarter of the nineteenth century, doctors were successfully repairing ruptures, removing fibroids from the uterus, excising diseased ovaries and fallopian tubes. National maternal mortality rates began to decline in the 1880s, and wealthier women were the first beneficiaries of these new developments.[23]

In addition to birthing more children, privileged women succeeded in raising more of them to adulthood in the late eighteenth and nineteenth centuries. One study of French ducal families showed that one out of every five children under fifteen died in the 1750s; fifty years later, the figure was down to one out of every seventeen.[24] Children's mortality, which declined in Europe's population as a whole in the course of the nineteenth century, dropped first among these wealthier families in the second half of the eighteenth century. Evidence suggests that it was the mother's increased care and attention, made possible by economic growth, which contributed most to the greater survival of infants and children.[25]

As women came to have more time for child care, they made child care more time-consuming. As wealth increased, relatively few women adopted the aristocratic model of motherhood, in which child care was left almost completely to servants. While some upper-class women continued to have their children raised by wet nurses and nannies, and educated at boarding schools, far more chose the Enlightenment model of active motherhood. Mary Wollstonecraft, Hannah More, and Jean-Jacques Rousseau agreed on few subjects, but they were united in their conviction that the world would be improved if women of the privileged classes became active, devoted mothers, breast-feeding their children themselves, not swaddling them, and becoming their chief teachers during their early years. For Europe's population as a whole, better distribution of food and the decline of infectious diseases led to the decline of child mortality (the death of children between five and eighteen) in the course of the nineteenth century and of infant mortality (the death of children under five) by the turn of the twentieth century.[26]

Even with the aid of servants, women found that these new standards of maternal care and attention took most of their time and energy. By 1845, Elizabeth Gaskell was thirty-five and had four children: Marianne, eleven; Meta, eight; Florence, three; and Willie, an infant. She also had three servants: a nurse, a nursemaid, and a cook. Her typical day, described in a letter to her sister-in-law, was tiring and complicated. Gaskell had the two youngest children sleep in her room, and she woke their nurse at 6:00 A.M. While the nurse cared for the three older children, Gaskell breast-fed Willie and then breakfasted with the entire family at eight-thirty. She read the Bible to the older children and prayed with her servants and the children. Then, she watched Willie for an hour and ordered the meals for the day. After answering her letters, she dressed Florence and took her to the park from eleven-thirty to one, when she returned to nurse Willie and have dinner with the family at one-thirty. She supervised all the children while the servants had their dinner and then had "two hours to kick my heels in." At five, the family gathered in the drawing room and Gaskell supervised the older girls in feeding the younger children and folding up the laundry, "this by way of feminine and family duties." At six, she breast-fed Willie while the nursemaid bathed Florence. At six-thirty, tea was served to the parents and the older children. "After tea, read to M.A. [Marianne] and Meta till bedtime while they sew, knit or do worsted work," concluded Gaskell. "From 8 till 10 gape."[27]

To privileged women who lived through the decades where conditions for children and mothers slowly improved, continued deaths made more impact than the new statistics which showed more children surviving. Death still threatened where it always had: in childbirth, in infancy, in childhood, and in old age. Nursing the dying remained women's province until the growth of hospital care in the twentieth century, and the death of a child retained its devastating force. A few weeks after Elizabeth Gaskell wrote her description of busy family life, her baby, Willie, died at ten months from scarlet fever. Three years later, she wrote a friend about the effect of his death on her:

> I used to sit up in the room so often in the evenings reading by the fire, and watching my darling *darling* Willie, who now sleeps sounder still in the dull, dreary chapel-yard at Warrington. That wound will never heal on earth, although hardly any one knows how it has changed me. I wish you had seen my little fellow, dearest dear Annie. I can give you no idea what a darling he was—so affectionate and reasonable a baby I never saw.[28]

As well as devoting more time to their children, middle- and upper-class women began to devote more time to their homes. Wealth enabled families to build homes with more rooms whose function became more specialized. By the eighteenth century, even city apartments and small country houses featured a parlor, a smaller and more domestic version of the elite's salon or drawing room. Originally the room in a convent where nuns were permitted to speak with outsiders, the parlor evolved into the family "living room," set aside for sitting and talking and separated from rooms meant for eating or sleeping. New homes had separate kitchens, dining rooms, servants' quarters, nurseries, and bathrooms.

As rooms proliferated, so did furnishings. An 1811 English watercolor of a prosperous middle-class family in their parlor shows a spacious room with an ornate carpet, floor-to-ceiling bookshelves, and a marble fireplace topped with a mirror. Candelabra and small ornaments decorate the mantelpiece, and numerous small paintings hang on the walls. Furniture includes a desk and chair, an upholstered sofa, and an embroidered firescreen—and only a portion of the room is shown in the painting.[29]

New possessions had to be cared for. Women sponged, brushed, and beat the dirt out of carpets. They wound, dusted, and polished clocks. They protected upholstered furniture from stains with crocheted and embroidered antimacassars and dustcovers. Increased wealth meant the housewife could aspire to new standards of decoration and cleanliness. As the numbers of objects in rooms expanded, so did the labor necessary to maintain them. A typical guidebook, *The Housewife,* published in Germany in 1861, contained thirty-nine chapters on subjects ranging from budgets to laundry, from what to feed servants to how to polish jewelry. Chapter 13, "On the Cleaning and Polishing of Rooms and Furniture," contained nineteen subsections, and did not even deal with bedrooms, which had a chapter of their own:

> 1. Cleaning the Oven. 2. Scouring Polished Oven. 3. Dusting Walls and Curtains. 4. Laying and Cleaning Carpets. 5. Sweeping Floors. 6. Beating and Cleaning the Sofa. 7. Cleaning and Freshening Up the Furniture. 8. Cleaning

Mirrors. 9. Cleaning Painted Surfaces. 10. Window-washing. 11–14. Scrubbing, Cleaning, and Wiping Up the Floors. 15. Cleaning Oilcloth Carpets. 16. Waxing Floors. 17. On the Laying and Care of the Tablecloth. 18. Remarks on Choosing Material; Cutting and Hanging Window Curtains. 19. Cleaning Wallpaper.[30]

Increased wealth meant higher standards of living, and higher standards of living meant increased labor on the part of women in the home. In a poem on women's work, Mary Collier (1689–c.1763), an Englishwoman who had both run her own household and worked in other women's, complained in a passage on washing up about the increased labor that increased possessions demanded:

> When night comes on, and we quite weary are,
> We scarce can count what falls unto our share;
> Pots, kettles, sauce-pans, skillets, we may see,
> Skimmers, and ladles, and such trumpery,
> Brought in to complete our slavery.[31]

Improvements in household technology led first to higher standards of housework, rather than increased leisure for the woman who performed the housework. The appearance of the washing machine in late-nineteenth-century England meant that linens were then washed once a week, instead of once every two weeks as they had been in the early years of the century.[32] Privileged women could continue to do housework at an increasingly hectic pace or they could free themselves from a portion of the heaviest household labors by leaving them to their servants.

Women newly freed from the physical labors of the household looked to guidebooks and magazines to help them master this new way of life. In the family's attempt to rise in society, the husband was expected to make enough money to support the new way of life; the wife was expected to ease the social transition by adopting new manners, spending money appropriately, and training her children to the standards of a higher social class. Literature advising women how to do this proliferated in the second half of the eighteenth century. Almost all these guides urged women to expand their duties as mothers: to devote themselves full-time to their babies and small children and to spend the bulk of their time raising their daughters. European women could easily obtain books or journals which advised them on courtesy, household management, cookery, and letter writing, as well as child rearing. This new literature of domesticity, which advised women how to improve themselves as mothers, housekeepers, and "ladies," was itself indicative of another great change in the lives of European women: the tremendous increase in their literacy during the course of the eighteenth century.

By the middle of the eighteenth century, enough European women were literate to constitute an acknowledged portion of the reading public. While women's ability to read and write trailed behind men's (by twenty to twenty-five percentage points in this era), literacy became standard for girls above the working class. By 1750, 40 percent of Englishwomen and 27 percent of Frenchwomen could sign their names—not an adequate test of literacy, but one of the few ways of assessing it before the nineteenth century.[33] More people of both sexes could read and write in the cities and in Protestant countries, but even in Catholic nations enough women were reading by the eighteenth century to call into existence an entire literature of their own. The earliest magazine directed exclusively at women was the English *Ladies Mercury*, which appeared in 1693. In the next half century, magazines for women were published in major European nations. While early publications were often ephemeral (like all magazines in this era), editors succeeded in attracting the new female reading public. J. C. Gottsched's *The Sensible Women Critics (Die vernünftigen Tadlerinnen)* sold two thousand copies a week when it first appeared in Leipzig in 1725. A generation later, women themselves were editing and writing scores of magazines aimed at other women.[34] One of the more successful ones in England was Eliza Haywood's *Female Spectator*, a solid fifty to seventy pages of essays and romances, which appeared monthly from 1744 to 1746; a bound copy of the entire run went into seven editions by 1771. In France, the *Journal des Dames (The Ladies' Journal)* was published from 1759 to 1778; from 1764 on, it was edited by women. By the third quarter of the eighteenth century, the interested female reader could choose from a number of magazines designed specifically to appeal to her. Priced to be within reach of all above the working class (sixpence in England in 1800), those magazines which survived were aimed at a woman busy running a home and family, but with enough education and leisure to read. Composed of numerous short features, these new magazines soon hit upon a successful formula which included romances, short informative essays on natural science or geography, recipes, household instruction, fashion, and some poetry or music.[35]

As magazines for women proliferated, so did "ladies' libraries." These were bound anthologies on various topics, meant for women to read at home. Half educational and half entertaining, they often substituted for school among girls who were taught at home. Religion, literature, history, and the arts were the most popular topics, and some sets also included geography and science. With the ability and leisure to read, women became a major segment of the reading public. By 1800, an estimated three quarters of English novel readers were women, and the new libraries from which one could rent books were created in response to women's growing demand for something to read at an affordable price.[36] Unlike men, women were often criticized for reading. Too much study was condemned because it was unfeminine: "I would particularly recommend to [women] to avoid all abstract learning, all thorny researches, which may . . . change the delicacy in which they excell into pedantic coarseness," urged Charlotte Lennox, editor of the English *Lady's Museum*, in 1760. "Who would wish to see assemblies made up of doctors in petticoats, who will regale us with Greek and the system of Leibniz?"[37] But frivolous reading was also frowned upon. The silly girl who believes life is like novels was made fun of in Charlotte Lennox's *The Female Quixote* (1752) and Jane Austen's *Northanger Abbey* (1817). By the nineteenth century, a female character's novel-reading is often a signal of her laziness, frivolity, or self-indulgence.[38]

Instead of "idly" reading novels, young women of the privileged classes were expected to spend their time in activities like light housework, sewing, drawing, making music or items for the home. To convey a superior social status, such work had to be either nonutilitarian or produce a useful object meant for the poor. Women put away everyday sewing when company arrived and kept a piece of "fancy-work" for show. Just as not having to train a daughter to make a living conveyed a family's superior social position, so her ability to perform nonessential "woman's work" of this sort symbolized her claim to ladylike status, and thus her family's gentility. "The intention of your being taught needlework, knitting and such like is not on account of the intrinsic value of all you can do with your hands, which is trifling," wrote Dr. Gregory in his popular guide for daughters of the English middle classes in 1770, "but to enable you to fill up, in a tolerably agreeable way, some of the many solitary hours you must necessarily pass at home."[39] By the nineteenth century, the word "work" itself had acquired a secondary meaning: the delicate crafts like embroidery which privileged women performed in their homes. In 1862, the English novelist Edward Bulwer-Lytton referred to "that tranquil pastime which women call work" and knew that his audience would understand that he meant one of the "elegant arts" for ladies which were not done for money.[40]

As wealth continued to grow in the nineteenth century, as a result of capitalism and industry, more and more women were able to achieve middle-class status: the way of life made possible with one servant. But in addition, increasing numbers of families were able to move into the upper middle classes: the way of life made possible with three or more servants. In mid-nineteenth-century England, it took an income of £150 a year to afford one servant; £500 a year to afford three.[41] These relatively well-off women devoted their energies to the same activities as their predecessors, but on a larger scale. They hired more servants, lived in bigger houses, ate and dressed more elaborately, bore and raised larger families, and taught their daughters to be "ladies." By the second half of the nineteenth century, each of these activities had achieved a measure of elaboration made possible only because women were devoting their full energies to them. An English magazine article of the 1850s listing "Elegant Arts for Ladies" added "the making of Feather Flowers, Hair Ornaments, Flowers or Fruit in Wax, Shell Work, Porcupine Quill Work, the gilding of Plaster Casts, Bead and Bugle Work, and Seaweed Pictures" to the more traditional music, embroidery, and drawing.[42] The profuse interior decor of the era, in which rooms were crammed with objects, furniture and windows were swathed with materials, and walls, shelves, and tables were covered with bibelots and artwork, was itself a product of all that female energy going into ornamenting the home. Isabella Beeton's (1837–1865) immensely successful *Book of Household Management,* first published in England in 1861, reflects the elaboration of life in this expanded wealthy class.

While Beeton was mindful of the constraints of a life with one servant, she directed much of her advice to the mistress of a large household, who was compared to the commander of an army. Beeton defined eighteen different kinds of household servants, from the ladies' maid to the valet, the scullery maid to the footman. Recipes were given for servants' dinners, family meals, and elaborate entertainments. A May dinner for twelve called for two soups, two fish dishes, seven meats, two kinds of fowl, seven puddings, and fruits, nuts, cakes, and ices.[43] Beeton recommended that company dinners be served in the new "Russian" style, in which servants cleared each course before another was brought out, instead of the old-fashioned and less labor-intensive French manner, in which all the courses were placed on the table together (and often grew cold before they were eaten.)

In the era between 1850 and the First World War, meals reached their greatest size and elaboration. Wealth, servants, and the demands of social status led hostesses routinely to offer lengthy and complicated meals. In the autumn of 1885, Charles Darwin's son and daughter-in-law had the Kelvins to dinner. Both couples were members of England's intellectual aristocracy, well-off, but not immensely wealthy. Dinner, prepared by the household's three servants, consisted of nine courses:

Clear Soup
Brill and Lobster Sauce
Chicken Cutlets and Rice Balls
Oyster Patties
Mutton, Potatoes, Artichokes, Beets
Partridges and Salad

Caramel Pudding and Pears with Whipped Cream
Cheese Ramekins and Cheese Straws
Ice
Grapes, Walnuts, Chocolates and Pears[44]

As with food, so with clothing, furniture, and household accessories of all kinds. Women freed from household labor or work outside the home devoted their energies to elaborating the activities considered appropriately "feminine." By the 1880s, a typical French bourgeois trousseau consisted of 42 chemises, 30 nightgowns, 43 pairs of stockings, 60 handkerchiefs, 72 aprons, 96 dish cloths, 72 towels, and 24 antimacassars, all made and embroidered by the bride-to-be.[45] Although men's clothing was now bought ready-made or from a tailor, women's, girls' and children's clothing continued to be made at home. If it was not, it was often decorated at home with trim and embroidery, which showed off a woman's talent for elaborate handiwork. Women prided themselves on the tininess of their stitches, the complexity of their embroidery, the tastefulness of their wardrobe. Women's clothing became elaborate and profuse and signaled social status as well as age.[46] Girls still put their hair up and let their skirts down when they wanted to be treated as women rather than schoolgirls; older women tended to restrict their wardrobes to darker and more "serious" colors. As a child in the 1890s, Gwen Raverat (1885–1957), an English writer, remembered being astonished by the amount of clothing a "young lady" she shared a room with had on, but the items she recorded were typical of winter wear for the period. Under a pink flannel blouse with a high white collar, a leather belt, and an ankle-length blue skirt, this young woman wore two petticoats, one flannel and one alpaca, black wool stockings held up by garters attached to a boned corset, a set of wool underwear, a set of cotton underwear, and a frilled and embroidered petticoat, bodice, and underdrawers.[47]

As objects were elaborated, so were social events. Wedding ceremonies, funerals, engagement parties, dances, dinners, and afternoon visits became elaborate and intricate rituals, requiring all a hostess's energy and social skill if they were to proceed properly and demonstrate her family's status to the world. An Englishwoman who wrote as "Mrs. Alfred Sidgewick" was invited to a typical bride's "kaffeeklatsch" in Germany in the 1890s. Given to show off her new home to her female friends, the kaffeeklatsch took careful planning and organized execution. "About twenty ladies were invited," remembered Sidgewick, "and they were solemnly conducted through every room of the flat from the drawing room to the spick-and-span kitchen, where every pan was of shining copper and every cloth embroidered with the bride's monogram." The guests had been to school with the bride and "had helped to adorn her home with embroidered chair backs, cushions, cloths, newspaperstands, footstools, dusterbags, and suchlike, all of which they now had the pleasure of seeing in the places suitable to them." After the tour of the apartment, the guests were served tea, coffee, hot chocolate with whipped cream, and a variety of rich cakes. "There is often music as well as gossip, and before you are allowed to depart there are more refreshments, ices, sweetmeats, fruit, little glasses of lemonade . . ."[48] Women had always been told that ideally they should be domestic creatures, devoting themselves to their husbands and families. The increasing wealth of nineteenth-century Europe enabled larger numbers of women to try to make this ideal a reality. In an era which brought so much change to men, many believed that women should remain within their traditional role and functions.

The Nineteenth-Century Lady: Ideals and Restrictions

The increasing wealth of the period enabled Europeans to exaggerate the traditional opposition of the sexes. In the new encyclopedias of the eighteenth century, women and men were defined as opposites.[49] Women were emotional; men were rational; women were passive; men were active. Women were gentle; men were aggressive. A woman's virtues were chastity and obedience; a man's, courage and honor. Women were meant for the home; men for public life. Supported by the economic changes of the period, these ancient ideas held firm in an era in which much else was swept away. Slavery, the divine right of monarchs, distinctions between aristocratic and bourgeois men, and literal views of religion all came under powerful and sustained attack during this period. But traditional views of women endured and even were strengthened in the new developments of the age: in law codes and republican governments, in medical and scientific thought, in images of women, even in the clothing of the period. Men's lives changed radically as new kinds of work expanded: in business, the professions, government bureaucracies, and industry. What was perceived as the relative lack of change in the lives of the women within their families heightened the supposed opposition of the sexes, which consequently seemed all the more natural and eternal.

Changes in the economy, changes in politics, changes in social and moral standards convinced many men that woman should not change. She should remain what they had always wanted her to be: an obedient wife and mother within the home, defined by her relation to a man. The home was increasingly seen as a precious refuge from the cares of the male world of business and industry. Changes in men's lives led many to insist that women remain in their traditional role, maintaining the women's "sphere" of the home as a necessary opposite to the man's "sphere" of the rest of the world. "Without woman present-day life would be unbearable for every sensitive [male] soul," wrote the German literary critic G. G. Gervinus in 1853,

> because the woman of today, like the Greek citizen of ancient times, is removed from the common bustle of life, because she is not concerned with a sense of status, does not suffer the degradation of lowly occupations, the turmoil and heartlessness of work.[50]

As economic changes continued to shift production out of the home and as increasing numbers of women no longer needed to work for wages outside the home, "work" and "home" came to be seen as two opposite worlds associated with the two "opposite" sexes. As always, this duality was seen as both natural and desirable. "Man for the field and woman for the hearth; / Man for the sword and for the needle she . . . All else confusion," the English poet Alfred, Lord Tennyson, wrote in a popular poem of 1847.[51]

This opposition was heightened by education. In the privileged classes, boys were commonly sent to school while girls were kept at home. "School parted us," wrote the English novelist George Eliot of herself and her brother.[52] Boys were trained to be physically strong and courageous, to read Latin, to learn how to support themselves. Girls were given an opposite curriculum: they were to be taught to be good housewives and mothers, to master the elegant accomplishments which signaled social status, to be religious, obedient, and self-effacing. The distinction in education which Rousseau had portrayed so vividly in his influential novel *Émile* (1762), in which Émile was educated to think and work for himself, while his future wife, Sophie, was restricted to the most traditional female activities, was widely imitated in this period. Boys were trained to be "men" and taught that they should support their sisters financially if necessary; girls were trained to be women and taught to provide their brothers with the womanly support services wives gave their husbands. "No woman in the enjoyment of health should allow her brother to prepare his own meals at any time of the day, if it were possible for her to do it for him," advised Sarah Stickney Ellis in her popular guidebook *Women of England* (1839).

> No woman should allow her brother to put on linen in a state of dilapidation, to wear gloves or stockings in want of mending, or to return home without finding a neat parlor, a place to sit down without asking for it, and a cheerful invitation to partake of necessary refreshment.[53]

Girls and boys, women and men were seen as ideally each other's opposites. Clothing came to embody and symbolize this opposition. Before the nineteenth century, women and men of the same social rank dressed far more like each other than like members of their sex in different strata. Female and male aristocrats alike wore similarly shaped neck ruffs, or hats, or sleeves, depending on the fashion of the day. Both wore expensive and fragile laces and silks, velvets and taffetas, and appeared in similar color schemes. Various colors and fabrics were restricted to social orders, not the separate sexes. The sumptuary laws of the fifteenth and sixteenth centuries more often sought to regulate the use of finery on lines of status than on those of sex. Aristocratic women and men in the late eighteenth century powdered their hair or wore wigs, used white facial makeup, rouge, and beauty patches, carried fans and handkerchiefs, wore silk stockings and high heels. Formal dress for both sexes was light-colored and impractical: made of silks and satins, embroidered with gold and silver thread, embellished with lace and ribbons. Both corseted their bodies or padded them where necessary; both wore garments cut so tightly they could not be put on unaided.

Fifty years later, all this had changed. At the Imperial French court of Eugénie and Napoleon III (who reigned from 1852 until 1870), the women dressed as elaborately and impractically as their courtly predecessors of the eighteenth century. Their light-colored dresses, with large sleeves, sloping shoulders, tiny waists, and immense, crinolined skirts would have been recognizable to a court woman from the earlier period. But the men had transformed themselves. Napoleon, Emperor of France, and his courtiers wore tailored woolen business suits. Dressed in dark jackets and trousers, with lighter-colored vests and white shirts, they were remarkable for the somberness and uniformity of their clothing. Instead of a crown, the emperor often wore a top hat. He could wear what he wore at court to an office or stock exchange without attracting attention.

By the mid-nineteenth century, an interest in fashion or clothing had become associated with femininity, and men who persisted in dressing flamboyantly or unusually were criticized as being unmanly and effeminate.[54] In a relatively short period of time, men changed from dress appropriate to their rank to dress appropriate to their sex. Their clothing became increasingly standardized and practical, as the somber costume of the male bourgeois replaced the decorative costume of the male aristocrat. Hair was worn plainly and cut simply. Decoration became restricted to the tie or vest, and too brilliant a display was criticized. Jackets became looser and shorter, making them both more practical and more comfortable. Silk stockings and knee breeches were replaced by the more utilitarian trousers, previously worn only by the lower classes. Men's shoes lost their high heels and became easy to walk in. Makeup, most jewelry, and perfume were abandoned. For men, court dress was often replaced by the business suit or military uniform.

Women's clothing, in contrast, remained decorative, impractical, and cumbersome. As a disregard for fashion came to seem masculine, so femininity was assumed to include a proper "womanly" interest in dress. From the early eighteenth century, women's clothing became softer, more delicate, and light-colored, characteristics which were seen as essentially feminine. From the same era, women began to corset their waists, which had the effect of emphasizing the uniquely female attributes of a woman's torso. Compressing the waist made the breasts and hips seem larger and rounder and thus exaggerated the differences between women's and men's bodies. Emphasizing a small waist persisted into the twentieth century, even though fashions changed from the panniered hoop skirts of the eighteenth century to the crinolines, bustles, and "Gibson Girl" modes of the nineteenth.[55] Into the twentieth century, European women were strikingly differentiated from men by their long hair, their unnaturally small waists, and their ankle-length skirts. Each had become a signal of womanhood and femininity.

It was in the nineteenth century that the state attempted to ensure that the sexes did not cross-dress.[56] The Code Napoléon contained a law forbidding women to dress like men. By the 1820s, the right to wear male clothing, or even women's clothing made more comfortable by shortening the skirt and loosening the waist, was being claimed by women in the radical socialist Saint-Simonian communities. Such women became the butt of vicious caricature, and men of these communities were drawn in women's clothing. Throughout the nineteenth century, women who demanded rights which traditionally belonged to men were routinely drawn wearing trousers. Male clothing symbolized male prerogatives.

Women who wanted the freedom and comfort of male clothing took risks by wearing it. George Sand (1804–1876), the French novelist, became notorious by dressing like a man after she had left her husband and came to Paris in the 1830s. Sand remembered that she had found it difficult to move freely around Paris dressed like a lady:

> I had legs as strong as [my young male friends] . . . yet on the Paris pavement I was like a boat on ice. My delicate shoes cracked open in two days, my pattens sent me spilling, and I always forgot to lift my dress. I was muddy, tired, and runny-nosed, and I watched my shoes and my clothes—not to mention my little velvet hats, which the drainpipes watered—go to rack and ruin with alarming rapidity.[57]

She asked her mother for advice, and her mother told her that when she was young, she and her sister had dressed like men, which enabled them to halve their clothing bills and to "go everywhere." Sand bought herself a man's coat, trousers, vest, hat, tie, and boots. "I can't convey how much my boots delighted me," she remembered.

> I'd gladly have slept in them, as my brother did when he was a lad and had just got his first pair. With those steel-tipped heels I was solid on the sidewalk at last. I dashed back and forth across Paris and felt I was going around the world. My clothes were weatherproof too. I was out and about in all weather, came home at all hours, was in the pit of all the theatres. Nobody heeded me, or suspected my disguise.[58]

Few women had Sand's daring or aplomb, and even if they did, few cities were as tolerant of such behavior as Paris: when the German poet Louise Aston (1814–1871) followed Sand's example in 1846 in Berlin, she was forcibly expelled from the city.[59] In 1857, the painter Rosa Bonheur (1822–1899) thought it wise to apply to the Paris police for permission to dress like a man in order to sketch the slaughterhouses. The prefect of the police issued the permit "for reasons of health" under the authorization of the ordinance of 1800.[60]

The same passionate anger which greeted women's attempts to wear men's clothing also met their attempts to claim the rights of citizens traditionally restricted to men. Both were seen as radical challenges to the traditional roles and opposition of the sexes and both were strenuously resisted. In 1793, the revolutionary French government outlawed all women's political activity, which had been considerable since 1789. As women first began to ask for the right to vote, they were specifically excluded from laws which widened the franchise for men. Before 1832, for instance, English voting laws were based only on property ownership. The Reform Act of 1832, written after a few women had raised the issue of female suffrage, deliberately inserted the word *male* in franchise requirements for the first time. European nations which gave men the vote explicitly excluded women (the one exception was Austria, where from 1848 a few thousand women had a property franchise and could vote through a male proxy).[61] In the second half of the nineteenth century, the French, Prussian, and Austrian governments banned women (and minors) from participating in all political activity and even from attending meetings where politics were discussed. Passed in the wake of women's political participation in the Revolutions of 1848 and nationalist agitation of the 1860s, these laws were sporadically enforced until the twentieth century.[62] In every European nation, women's right to participate in politics was seen as a disturbing and dangerous encroachment on male territory. Just as women's attempts to wear trousers were seen as upsetting the "natural order," so women's attempts to participate politically were viewed as upsetting the natural functions of the sexes. "It is horrible, it is contrary to all the laws of nature for a woman to want to make herself a man," argued a French legislator who convinced the revolutionary Paris government to bar women from its sessions in 1793.

> Since when is it permitted to give up one's sex? Since when is it decent to see women abandoning the pious cares of their households, the cribs of their children, to come to public places, to harangues in the galleries, at the bar of the senate? Is it to men that nature confided domestic cares? Has she given us breasts to breast-feed our children? No, she has said to man: "Be a man: hunting, farming, political concerns, toils of every kind, that is your appanage" [sic]. She has said to woman: "Be a woman. The tender cares owing to infancy, the details of the household, the sweet anxieties of maternity, these are your labors."[63]

The belief that women should remain within their traditional roles and functions was given the force of law in the new codes of the period. These sought to rationalize and centralize, to make the laws uniform and thus extend the power of the nation-state. For men, these changes were often liberating, and many benefited from the democratic gains the laws gave them. Women lost power, both relatively and absolutely. Traditionally excluded from the body politic, some privileged women in earlier eras had benefited from the patchwork of custom and common law which had sometimes allowed them to manipulate a legal system based on male prerogative. But in the new law codes, *all* women were classified as one category and were usually ranked with infants, the insane, and criminals in the eyes of the law. Created in the spirit of reform, the new codes often discarded traditional inequalities among men. Aristocratic privileges were limited or abolished, religious qualifications and tests were often removed, distinctions were based on money rather than birth, and the principle of equality before the law was established. But these gains were for men only. For aristocratic women, especially, the reforms which ended privileges deprived them of rights and powers they had previously been entitled to as members of an elite group. The loss of male aristocratic privileges was balanced by legal gains for all men of property, a category which included male aristocrats. But the loss of female aristocratic privileges was total. Women of the aristocratic and bourgeois classes did not share in the democratic gains won by men, just as they had not shared in the increased economic opportunities for men. "In Russia . . . men and women were equal in not having political rights," wrote a Russian woman after Russian men had been given a limited franchise in 1905.

> Perhaps it was thanks to this that within the Russian educated class there was not such a wall between men's and women's worlds as in Europe. Together we dreamed about freedom; about the rights and responsibilities of citizenship;

together they were to be achieved. Now, suddenly, half way there, could they cut women off from the long list of rights of all?[64]

Russian men were only following the behavior of other European men. Gains for men came first. But the very laws which gave men new rights upheld the most traditional and regressive views of female inferiority and incapacity. Perhaps in compensation for the loss of male hierarchy and differentiated status, all men now gained legal control over "their" women. The new law codes formalized repressive attitudes about women and made them the law of the land.[65]

Now all married women were legal minors, under the guardianship of their husbands. All the wife's wages, inheritance, and property belonged to her husband. Prior to the Danish Law Code of 1683, for instance, married women could own land in their own right as freeholders; under the new code, all property, even the wife's inheritance, belonged to the husband.[66] In some German towns, a woman could transact business in her own name, even against her husband's wishes, and all a wife's "separate earnings" were her own property; under the new Prussian Civil Code of 1794, the husband became the wife's guardian and all her property belonged to him.[67] Under Roman law which prevailed in the south of France before the French Revolution, married women had some legal capacity; under the Code Napoléon of 1804, married women were classified with children, the insane, and criminals as legal incompetents.[68] As in earlier eras, families tried to protect their daughters' property by drawing up marriage contracts. But these contracts often retained provisions which subordinated the wife to the husband: a study of French marriage contracts of the first half of the nineteenth century revealed that all prevented the wife from alienating the property without her husband's consent. Marriage contracts themselves declined in the nineteenth century, as the "romantic" marriage without a contract became popular, but a married woman's only protection for her property remained a marriage contract.[69]

The Code Napoléon, influential in much of Europe, stipulated that one essential basis of the marriage could not be changed, even by a marriage contract: the husband was by law "head" of the household. In Article 231 of the code, "the husband owes protection to his wife; the wife owes obedience to her husband." As "head" of the household, the husband acquired all the traditional legal powers given to married men over their wives; in addition, he alone could decide where they lived, if she could inherit, work, acquire property, give money away, be a witness in a criminal case, or receive official papers. He could read her correspondence and have access to any bank accounts she opened. Until the end of the nineteenth century, she had to have his permission even to open an account. The wife had to obey her husband: living where and how he chose, deferring to him in a conflict, taking his name and giving up her own upon marriage. In the words of the 1836 Code of Russian Laws, the wife's obedience to the husband was supposed to be "unlimited."

> The woman must obey her husband, reside with him in love, respect, and unlimited obedience, and offer him every pleasantness and affection as the ruler of the household.[70]

Even in England, which had no national law code, eighteenth- and nineteenth-century judges upheld the most misogynistic traditions of common law. In 1840, Alexander Cochrane tricked his wife, who had left him to live with her mother, into returning to his house and then kept her confined so she could not leave. The court upheld him on the grounds of legal precedent:

> There can be no doubt of the general dominion which the law of England attributes to the husband over the wife; in Bacon, Abridgement, title "Baron and Femme," it is stated thus: "The husband hath by law power and dominion over his wife and may keep her by force, within the bounds of duty, and may beat her, but not in a violent or cruel manner."[71]

English common law did not differ much from Continental law codes with regard to women: all agreed that the husband could physically punish his wife, but stipulated that such punishment should be moderate.

Laws on adultery institutionalized once again the double standard of sexual behavior. Under the Code Napoléon, a wife found guilty of adultery could be sent to prison for up to two years; her guilty husband went unpunished unless he moved another woman into the house as his concubine, and even then he was only subject to a relatively light fine. Article 324 of the penal code permitted husbands to kill wives whom they caught in adultery; women had no reciprocal right.[72] The fears which had traditionally justified the double standard were maintained in most European law. In 1801, Lord Eldon, the English lord chancellor, argued that "adultery committed by a wife and adultery committed by a husband" differed greatly:

> The adultery of a wife might impose a spurious issue upon the husband, which he might be called upon to dedicate a portion of his fortunes to educate and provide for; whereas no such injustice could result from the adultery of a married man.[73]

As Britain and France gradually instituted civil divorce in the course of the nineteenth century, they retained the double standard there as well.[74] Prior to 1857, each divorce in England required a separate act of Parliament.

Thereafter, the husband could divorce on the basis of his wife's adultery; the wife had to prove his adultery plus another crime: desertion, cruelty, incest, rape, sodomy, or bestiality.

The new law codes favored fathers over mothers, as well as husbands over wives. As legal head of the household, the father had sole authority over the children. Alone, he could decide their education, employment, punishment, and give consent to their marriages. Under the Prussian Civil Code of 1794 (which granted women equal divorce and custodial rights), a "healthy mother" was required to breast-feed her child herself; "how long the child is nourished at the breast is the father's decision."[75] Under English law, the father could take the child from the mother, entrust its upbringing to a third party—who might be his mistress—and refuse the mother the right to visit. While all nations set a legal age at which a young man became an adult, many kept a young woman under her father's guardianship as long as she was unmarried and he was alive. When she married, she remained a legal minor under her husband's protection. In Germany, Scandinavia, and Switzerland, unmarried women were considered to be "perpetual minors"; in Great Britain, France, and Austria, they could manage their own property and sign legal contracts.[76]

Just as the ancient traditions of proper female and male roles shaped the new law codes of the eighteenth and nineteenth centuries, so they shaped the new scientific views of the period. In the new theories of evolution and positivism, in the new biological and medical findings, male scientists and doctors tended to ignore data which did not fit into traditional views of the sexes and to extend theories beyond the limits of data to fit preconceptions of male superiority and female inferiority. "Man is more courageous, pugnacious and energetic than woman, and has a more inventive genius," wrote Charles Darwin in *The Descent of Man* (1871).[77] Scrupulous in backing his theories with biological data when he discussed animals, Darwin argued that among humans, sexual selection and, in modern times, economic pressures had led to male, but not female, competition. Competition had ensured that man would attain "a higher eminence, in whatever he takes up, than can woman." Darwin also asserted that men will pass this eminence "more fully to the male than to the female offspring" and concluded that man had evolved further than woman and "thus man has ultimately become superior to woman." He predicted that "the present inequality of the sexes" would prevail.[78]

Darwin's followers, many of whom sought to apply his theories to human society, were less restrained than he. Herbert Spencer, the influential English social scientist, came to believe that women's bodies ceased their "individual evolution" with the onset of menstruation. As a result, women exhibited "a perceptible falling-short in those two faculties, intellectual and emotional, which are the latest products of human evolution—the power of abstract reasoning and that most abstract of the emotions, the sentiment of justice."[79] Evolutionary theory, so liberating to many of the sciences and social sciences, repeated and reasserted traditional beliefs when it applied to women.

So too did positivism. August Comte, the "father of modern sociology," reiterated ancient beliefs in his influential writings. "The relative inferiority of Woman in this view is incontestable, unfit as she is, in comparison [to man], for the requisite continuousness and intensity of mental labor, either from the intrinsic weakness of her reason or from her more lively moral and physical sensibility, which are hostile to scientific abstraction and concentration," he wrote in 1839.

> This indubitable organic inferiority of feminine genius has been confirmed by decisive experiment, even in the fine arts, and amidst the concurrence of the most favorable circumstances. As for any functions of government, the radical inaptitude of the female sex is there yet more marked, even in regard to the most elementary state, and limited to the guidance of the mere family.[80]

(Comte's later writings, in which he asserted that women would dominate in the future, were far less influential than these earlier books, and were sometimes used as proof that the philosopher had gone insane.)

Nineteenth-century scientific thought tended to bolster ancient traditions about women with new "scientific" knowledge. That men were active and women passive, that men were superior and women inferior, that men could think and reason and women only feel and reproduce were assumptions which not only remained unchallenged by the new biology, but were also supported by it. Writing on women's "innate" passivity in his *Evolution of Sex* (1889), the Scottish biologist Patrick Geddes summed up the "verdict" of evolution in a remark often quoted in late-nineteenth-century European discussions of women's nature, function, and role: "What was decided among the prehistoric Protozoa cannot be annulled by an act of Parliament."[81] Biblical authority for female subordination was confirmed or replaced by biological authority. In medicine, as in biology, preconceived views of the sexes warped knowledge, theory, and practice with regard to women. While the male reproductive system was clearly understood by the late seventeenth century, the female reproductive cycle remained mysterious. The ovum was not discovered until 1827, and fertilization was not understood until 1883. Well into the twentieth century, doctors continued to believe that human fertility was modeled along animal lines: that pregnancy occurred during estrus and not in the middle of the cycle. Accordingly, generations of European women who wished to avoid or achieve pregnancy were advised by scientific and medical authorities that they were most fertile during menstruation and least fertile between their periods—the exact opposite of the truth.

While seventeenth-century anatomists and physicians had seen woman as an inferior and even disabled and deformed variant of man, nineteenth-century medical men stressed the innate weakness and debility of women. The supposedly weakening effects of menarche and menstruation were emphasized, and mothers were warned to take special care in shepherding their daughters through the "brilliant and stormy crisis which is terminated by the appearance of the menses," as an influential French doctor put it in midcentury.[82] Girls—at least in the privileged classes—were to be guarded from physical activity, spicy foods and meat, and, especially, intellectual work during the period itself. Failure to do so would lead to "a disordered pelvic life" and continual ill-health.[83] Menstruation was often portrayed as a tremendous strain on an already weak system. Jules Michelet's influential book *On Love* (1859) argued that because of menstruation "15 or 20 days of 28 (we may say nearly always) a woman is not only an invalid, but a wounded one. She ceaselessly suffers from love's eternal wound."[84]

In addition to speculating about menstruation, European doctors began to make pronouncements about female sexuality. In general, these were more influenced by a doctor's cultural biases than any scientific observation. The Englishman William Acton argued that "modest" women had no sexual desires; his counterparts across the Channel in France assumed they did. Cultural traditions were rarely distinguished from laws of nature, and ancient European traditions were restated in a new scientific guise. In his *Hygiene and Physiology of Marriage* (1849), which went into 172 editions by 1900, the Frenchman Auguste Debay told his readers that "the historical position, that is to say the man lying on the woman, is the natural and instinctive position for the union of the sexes in the human race. . . . The peculiar fancy that some wives occasionally experience to take the husband's place disturbs the natural order." Husbands were advised to arouse their wives before intercourse; wives were advised to fake orgasm if necessary: "this innocent trickery is permitted when it is a question of keeping a husband."[85] Whatever women told each other, public discourse on sexuality was dominated by men.

The same was true of the previously female realm of childbirth. By the nineteenth century, men had appropriated the birthing of privileged women as their own territory. "One sex only is qualified by education and powers of mind to investigate what the other sex has alone to suffer," as the English doctor E. J. Tilt stated in 1853.[86] By the eighteenth century, the male physician had replaced the female midwife in the highest social circles; in the nineteenth century, increasing numbers of middle-class women opted for a male doctor if they could afford one. By midcentury, both French and English physicians were recommending that women give birth lying down rather than in the traditional sitting posture. The physician's ease had superseded some beneficial traditional techniques which used gravity to aid childbirth. Women who could afford to do so gave birth at home: until the 1880s hospitals remained dangerous sources of infection for childbearing women.

The belief that doctoring was for men only was reinforced by all aspects of a culture which exaggerated male scientific aptitude and disparaged female intellect. In the second half of the nineteenth century, a number of popular novels appeared in which the plot turned on the conflict between a young woman's mistaken desire for medical education and her true vocation as a wife and mother (Wilhelmine von Hillern's *Only a Girl*, 1867; Thérèse Blanc's *Emancipated Woman*, 1887; Colette Yver's *The Eggheads*, 1908). In each, the young woman repents, abandons her medical studies, and finds true fulfillment by marrying the hero, usually a doctor himself.[87]

Written by women for women, these novels belong to the genre of popular women's novels, which had become standardized in both plot and message by the nineteenth century. Novels for women were domestic romances: they focused on a young woman's successful passage from girlhood through courtship to marriage and a happy ending as a wife and mother within the home. This formula presented women's traditional function and role in the most attractive possible light. Over and over, these novels justified women's subordination, arguing explicitly that women who followed the traditional path would be rewarded with a good husband, prosperity, and a happy family life. *The Diamond and the Pearl* (1849), by Catherine Gore (1800–1861), one of the most prolific English novelists of the second quarter of the nineteenth century, is typical of this genre. Blanche, the "diamond," is the good sister, who stays home to nurse her aged parents rather than going to parties. Her self-sacrifice refines her into a "lovely and loving woman—pure as a Roman matron, gifted as a muse and feminine as an English gentlewoman."[88] Blanche is rewarded by a happy marriage to a good man.

> Wise in her generation, she knew that it was to the stronger sex she must look for counsel, aid, and protection, and it was a comfort unspeakable to feel that he to whose guidance her future life would be submitted, would soon be there—sustaining her by his strength of intellect—solacing her by the softness of his heart.[89]

"What can mortal woman wish for more?" sighs a cousin in envy, with the author's approval. Like most popular novelists of the period, Gore also portrayed bad women. *Mrs. Armytage; or Female Domination* (1836) demonstrated the dangers of raising a girl too much like a boy. Caroline Armytage "was taught to ride, to run, to settle with the stewards and housekeeper, to parley with farmers, to dispute with tax collectors. A little Latin and a great deal of arithmetic, bounded her accomplishments. . . . She therefore grew to woman's estate and was still positive, still ungovernable." She drives her husband to an early grave and tyrannizes over her children until her son, with the author's approval, drives her out of the house. The second half of the novel shows her "doing penance for her former faults," and learning to be a properly submissive and subordinate woman.

In novels of this sort, in the theater, in magazines and fashion plates, women were presented with idealized images which many sought to fulfill. They trained their daughters to be properly submissive women, knowing that the penalties for nonconformity were severe: social ostracism, failure to find a husband. As the guardian of her family's social life, the woman took responsibility for their social behavior. These women, responsible for raising or maintaining their families' social status, often turned to novels and guide-books to aid them in adopting new manners and customs. For the home-bound women of the privileged classes, images of ideal womanhood were ever-present and were given new force in the nineteenth century by mothers who were now financially able to raise their daughters to fulfill these ideals of the period.

From the mid-eighteenth century on, generations of moralists and writers—many of them women—had followed Rousseau's argument that woman's highest glory was to devote herself to motherhood. *Émile* (1762) opens with a paean to motherhood and breast-feeding; Rousseau's equally popular novel *Julie, or the New Heloise* (1761) ends with the heroine dying "a martyr to maternal love" after she succumbs to the aftereffects of rescuing her son from a freezing river. Well into the twentieth century, European women were continually told that motherhood was women's special role and that mothering was what they should devote themselves to. "Woman's role in society is not at all the same as a man's," went a passage from Augustine Fouillée's *Francinet,* widely used as a school reader in France before World War I.

> A woman's life is entirely interior, and her influence on society occurs in a nearly invisible manner. That is not to say, however, that her role is any less important or her influence smaller; it is only more hidden, that's all. Women exercise their influence first of all on children, and it is remarkable that many illustrious men owed the qualities that made them famous to their mothers' example and precepts.[90]

Repeatedly told that great men had been influenced by their mothers, women were also urged to devote themselves particularly to their daughters' upbringing. "There is no boarding school, however well-run, no grand national establishment, no convent, whatever its pious rule," wrote Henriette Campan in 1822, "which can give an education comparable to that which a girl receives from her mother when the mother is well-informed [*instruite*] and when she finds her sweetest occupation and her true glory in the education of her daughters."[91] (Campan, a former lady-in-waiting to Marie Antoinette, had founded a highly respected girls' school after the French Revolution.) In a culture which emphasized the difference between the sexes, authors stressed the "natural" sympathy and identification between mother and daughter. "As she looks upon this [girl] child, the mother is more and more softened towards it," wrote Albertine Necker de Saussure in her 1838 French guide for mothers.

> A deep sympathy—a sentiment of identity with this delicate being—takes possession of her; extreme pity for so much weakness, a more pressing need of prayer stirs her heart. Whatever sorrows she may have felt, she dreads for the daughter; but she will guide her to become much wiser and much better than herself. And then the gaiety, the frivolity of the young woman have their turn. This little creature is a flower to cultivate, a doll to decorate.[92]

A mother had a new responsibility: to make her daughter "feminine," to cultivate her "womanly" nature, to develop her "heart," that is, her sensitivity, loving kindness, and gentleness. "It will be in the female heart, par excellence, as it always has been," wrote the French novelist George Sand, "that love and devotion, patience and pity, will find their true home. On woman falls the duty in a world of brute passions, of preserving the virtues of charity and the human spirit."[93] Mothers were urged to make their daughters conscious of their different female nature, and the European testimony is that most tried to do so. "By seven or eight years of age," wrote Mme. Gaçon-Dufour in 1787,

> having been taught by our mothers the difference of our sex from that of men— that is, understanding, by being called a little girl or *mademoiselle*, that we were not at all little boys or men, we start to be reserved in our games, controlled in our walk, to lower our eyes, even to blush. . . . This is how our intimate feelings of modesty are created.[94]

Physical punishments were now reserved for boys, and one of the constant criticisms made of leaving the upbringing of children to servants was that servants would resort to physical force. Instead, mothers were urged to use love and psychology. Such methods were extremely effective. "I am going to turn over a new life and am going to be a very good girl and be obedient to Isa Keith" [her cousin and tutor], promised the seven-year-old Marjory Fleming in her diary of 1811. Marjory recorded her misdeeds and Isa's responses. "I confess that I have been more like a little young Devil then a creature," she confessed in 1810,

> for when Isabella went up to the stairs to teach me religion and my multiplication and to be good and all my other lessons I stamped with my feet and threw my new hat which she made on me on the ground and was sulky and was dreadfully passionate.[95]

Marjory added penitently that Isabella "never whipped me but gently said Marjory go into another room and think what a great crime you are committing letting your temper get the better of you."[96] Girls were often told to "sit and think" about what they had done wrong; other women report being made to sit in a corner, or with their back to a group, to sit with a sign around their neck or in a dark room—as was done to George Sand at convent school. "Be seated somewhere," Jane Eyre was told by Mrs. Reed, "and until you can speak pleasantly, remain silent."[97] Control was continually stressed, because a girl had to learn to control the "unwomanly" side of her nature: the sulky, passionate, willful, and independent side which Isabella sought to teach Marjory to repress. "Docility, that internal disposition which naturally leads to the fulfillment of [obedience], may well be the object of peculiar cultivation in young girls," advised Mme. Necker de Saussure.

> Women . . . are called to bear, very often, and perhaps throughout their lives, the yoke of personal obedience. Since such is their fate, it is well to accustom them to it; they must learn to yield without even an internal murmur. Their gaiety, their health, their equality of temper, will all gain by a prompt and cordial docility.[98]

To be obedient, to be self-sacrificing, to put others first, to defer. These were the rules daughters absorbed. "I was mute," recalled Marie LaFarge (1816–1852) of her Paris childhood in the 1820s and 1830s, "because I knew that a young girl ought to be concerned about others without pretending to concern them with herself, and that she ought to use her good sense to listen gracefully and hold her tongue intelligently."[99] Malwida von Meysenbug (1816–1903) recalled that when she began to express her own ideas in Germany in the 1830s, "people deliberately let me hear remarks like this, (speaking of a young girl) 'What a sweet creature! She does not pretend to have opinions of her own.' They wanted to show me how far I had strayed from the proper path."[100] All across Europe, women of the privileged classes wrote of the seemingly endless limits placed on their behavior. "What all was not shocking formerly! What was not unbecoming for a woman, and above all, for a girl!" wrote Fanny Lewald (1811–1889) of her Prussian girlhood,

> A girl could not look at a statue which represented the human form nude, and had to turn away her eyes from a picture with naked figures, and if possible, be shocked and blush; a girl could not undertake the smallest journey alone . . . and to exhibit an independent opinion or interest in general matters—that was absolutely unmaidenly, not even womanly. We were not supposed to have any opinion of our own and it was a recognized rule of womanliness to be sure to begin every sentence with "I believe" or "they say" in order thus to put away from us every semblance of independence, which in itself was regarded as presumptuous.[101]

Modestly brought up by her mother, the ideal girl was sent to a "finishing school" for a year or two to complete her education, develop some "accomplishments," and acquire the social skills to enable her to make a good match. George Sand described her classes at boarding school as amounting, "more or less, to nothing. In practice they consisted of finishing-school lessons."[102] That was the goal. All across Europe, privileged families sought to send their girls to such schools so they could successfully compete in what was increasingly called the marriage "market."

In the all-female worlds in which such girls were raised, men—whether the relatives of friends, schoolteachers, clergymen, or barely glimpsed strangers—became the subjects of intense longings and daydreams. "It is impossible to imagine the importance given . . . to the rather ordinary beings we call men," Marie LaFarge wrote of her days at boarding school:

> I adorned my mind for that being I could not yet see in my imagination but whom I hoped to find in the future, and whom I awaited as the complement to my existence. When I had written thoughts that were lofty, I read them to *him*, when I had conquered a difficult piece of music, I sang him my victory. I was proud to offer *him* a good action, and I dared not think of *him* when I was not satisfied with myself.[103]

After school, the young woman lengthened her skirts and began the search for a husband. By the nineteenth century, there were appreciably more unmarried women than men in the privileged classes, caused both by the better treatment of female babies and the single-sex migrations of propertied men who went abroad. While social reformers concerned themselves with schemes for these "redundant women," as they were called, the women themselves felt a great pressure to marry before they became too old. "In two years I'll be almost an old maid," wrote Stephanie Jullien, a twenty-one-year-old Parisian bourgeoise, in 1833, "and he'll [a suitor of her own age] still be so young."[104] Throughout the letters and diaries of nineteenth-century women run the hopes, negotiations, disappointments, and successes in the marriage market.

Enabling a daughter to marry well was a mother's chief goal, and marriage itself was seen as a woman's main job and the crucial decision on which her future happiness depended. Just as men were expected to compete in their business, so women were expected to compete in their business of marriage. "Married life is a woman's profession; and to this life her training—that of dependence—is modelled," the English magazine *Saturday Review* commented approvingly in 1857. "Of course, by not getting a husband, or losing him, she may find that she is without resources. All that can be said of her is, she has failed in business and no social reform can prevent such failures."[105]

An unmarried woman was seen as a failure. By the nineteenth century, the words for unmarried women—"spinsters," "old maids," "tabbies," etc.— all had pejorative connotations, while "bachelor" remained romantic and attractive. Without a wife, a man remained sought-after no matter what his age; without a husband, a woman "past her first youth" had little status. Supposed to live for others, at best she depended on her parents. Later, she was forced to rely on her married sisters and brothers to give her a place in their households. Untrained to support herself, she could become a governess or paid companion—an equivocal position where she was neither lady nor servant. Poorly paid and insecure, such women were the objects of pity in the nineteenth century. Since a woman attained adult status by marrying (in contrast to men, who achieved it by becoming a certain age), unmarried women remained "old maids," literally, girls who had grown old without maturing in the accepted way. "When being a wife, mother, and housewife is the most important destiny, first and last, for women only," wrote Betty Gleim, a German educator, in 1814,

> when each female being does not have—as each male does—intrinsic value and purpose—then all those unmarried girls are the most useless, most miserable, and most unfortunate of creatures: they have been created for a purpose which they can never fulfill. They have been trained for that and for that alone, and life goes by without them being able to tread the path to which they were so arbitrarily assigned, for which alone they have been prepared for so long.[106]

Mothers made great efforts to ensure that their daughters would make the social contacts which would lead to marriage. Depending on a family's wealth, the *début* into "society" ranged from casual social visits to grand balls in the young woman's honor. It was a way of life which seemed both natural and inevitable. "By just dancing myself dizzy, looking as nice as I could or exploring myself anew through some fresh pair of eyes, I felt I was furthering some momentous, indeed some almost devout purpose," recalled the English-woman Cynthia Asquith of the year she was "presented to society" as a *débutante*. [107]

While marriages were still usually arranged by contract in the privileged classes in the nineteenth century and could be the subject of intense financial negotiations between families, young women increasingly expected to be able to marry someone they could love as well as someone who could support them financially. "Don't think I am talking about a romantic and impossible passion or an ideal love, neither of which I ever hope to know," Stephanie Jullien wrote her father, explaining her views;

> I am talking of a feeling that makes one want to see someone, that makes his absence painful and his return desirable, that makes one interested in what another is doing, that makes one want another's happiness almost in spite of oneself, that makes, finally, the duties of a woman toward her husband pleasures and not efforts.[108]

Jullien and other young Frenchwomen of the bourgeoisie tended to marry men close to them in age; other privileged European women continued to marry older men.[109] Well into the twentieth century, the ideal marriage remained that to an older, wiser man, who would protect and cherish his younger, more inexperienced bride. "I must look up, if not with fear and trembling, at least with deference and a strong sense of inferiority to the husband who is to be obeyed and honored as well as loved," explained one of Catherine Gore's heroines. "I should assuredly degenerate into a mere automaton, a miserable creature of luxury and selfishness, were not my better qualities stirred into activity by the companionship of one far nobler-minded than myself."[110] "Lord, it is you who have given me, / A protector for my weakness," the Comtesse de Flavigny (d. 1883) wrote in a poem published in 1861. As always, the traditional marriage could be a love-match as well. "Now begins a new existence, a beautiful life, a life wrapped up in him whom I love more than myself and everything else," wrote Clara Wieck (1819–1896) at twenty-one when she received permission to marry Robert Schumann.[111] Later, she described her wedding day as "the fairest and most momentous of my life."[112] "O come to my arms, being unspeakably great and beautiful," Caroline von Dachroden (1766–1829) wrote her future husband, Wilhelm von Humboldt, shortly before their wedding,

> Come, taste upon my breast a divine life, encompassed by all the charms of love, then lean to bless your Li [Caroline's nickname]. Her delicious tears, her entire forgetfulness of herself in the adoration she feels for you, her desire to reach at your side a higher level of humanity, will tell you what you have made of her.[113]

By the second half of the eighteenth century, literary convention—especially in England—tended to assume that privileged women were "above" physical passion. "These women have been taught to regard sexual passion as lust and as sin," Dr. Elizabeth Blackwell wrote near the end of her life, "a sin which it would be a shame for a pure woman to feel, and which she would die rather than confess."[114] While there is little doubt that many women of the privileged classes took pleasure in sex, public ideology held that such women should be sexually innocent at marriage and find their greatest pleasure in motherhood. Writing to her brother about his daughter's marriage in 1843, George Sand urged restraint on the husband's part and assumed the bride would be totally innocent, as she herself had been when she married at eighteen:

> Men do not know that what is fun for them is hell for us. Tell him [the groom], therefore to be considerate, and to wait until such time as his wife, under his instruction, may gradually attain to understanding, and be able to respond to his passion. Nothing is more frightful than the terror, the suffering, and the disgust occasioned in a poor young thing as the result of being violated by a brute. *We bring them up as saints, only to dispose of them as fillies.* [115]

Bringing up daughters like "saints," or, as the more usual term went, like "angels," was increasingly possible among the privileged classes of the nineteenth century, most of whom lived in the new cities of the era. In the countryside a girl could hardly avoid knowledge of the facts of life; in the cities utter ignorance was possible. A number of young women recorded in their diaries that they had seen men exposing themselves in the streets, but did not comprehend what was happening.[116] "Seeing and hearing don't take one very far," wrote a Viennese girl in her diary after her older sister had told her about intercourse. "I've always kept my eyes open and I'm not so stupid as all that. One must be told by someone, one *can't* just happen upon it by oneself."

Popular cultural images of the day stressed a young woman's supposed sexual innocence. In drama, in opera, and especially in the ballet, the virginal heroine was worshipped and wept over. Marie Taglioni (1804–1884), the Italian ballerina who was idolized in Paris, London, and St. Petersburg, embodied the new romantic image of women. Dressed like a fairy or a sylph, dancing delicately on her points, Taglioni's tiny hands, feet, and waist seemed to embody the romance of youthful female fragility and innocence. Popular contemporary ballets, like *La Sylphide* and *Giselle,* emphasized the heroine's fragility. Girls of the privileged classes were supposed to be equally fragile and innocent.

Despite this idealized image of a fragile sexual innocent, a real wife was expected to become pregnant as soon as possible after her marriage. Motherhood brought a woman status and prestige. At twenty-one, in 1836, Marie LaFarge walked her three-year-old niece down the Champs-Élysées and pretended the child was her daughter: "I called her my child, and it seemed to me that everyone envied me my beautiful angel, and that I was doubly a woman and doubly worthy of respect."[117] Many privileged women testified to their joy in becoming a mother. It is the "acme" of life, wrote Elizabeth Gaskell, who gave birth five times before she became a novelist, "when all is over and the little first-born darling lies nuzzling and cooing by one's side."[118] Others were more measured about the experience. "Though I quite admire the comfort and blessing good and amiable children are—they are also an awful plague and anxiety. . . . What made me so miserable was—to have the two first years of my married life utterly spoilt by this occupation!" wrote Queen Victoria of England (1819–1901) on the occasion of her daughter's first pregnancy. "What you say of the pride of giving life to an immortal soul is very fine, dear," she wrote in another letter, "but I own I cannot enter into that; I think much more of our being like a cow or a dog at such moments; when our poor nature becomes so very animal and unecstatic . . ."[119]

While motherhood could salvage an unhappy marriage for a woman, an unhappy marriage could also blight motherhood. "Now I am well once again and not pregnant," Sophie Tolstoy (1844–1919), wife of the Russian novelist, wrote in her diary when she was twenty-one and married for three years. "It terrifies me to think how often I have been in that condition."[120] Eventually the mother of thirteen, Tolstoy, like most women of the privileged classes, seems to have been ignorant of any methods of contraception or abortion. Evidence of trying to prevent or end pregnancies is rare among these classes before the third quarter of the nineteenth century, when their birth rate begins to drop. The most famous example is from the Amberley papers, in which Bertrand Russell's grandparents wrote to each other about how Henrietta, Lady Stanley (1807–1895), who had borne nine children in seventeen years, managed to end her tenth pregnancy in 1847 by "a hot bath, a tremendous walk and a great dose," probably of a purgative.[121]

The public ideal that a woman was most fulfilled as a wife and mother within the home proved to be both appealing and powerful. "My father always made fun of so-called *domestic bliss,*" wrote Clara Schumann in her diary in 1841. "How I pity those who do not know it. They are only half alive."[122] Many nineteenth-century novels do not end with the heroine's marriage, but carry on to show her ensconced in domestic bliss with her family. The coziness and security of the vision of the ideal home permeated women's novels of this era. "We pass our evenings *en famille,* sometimes at Albert's [her brother's] home, sometimes at our home, and the hours flow deliciously in company so dear, in the midst of our children," went such a description from the prolific French novelist Mathilde Bourdon's (1817–

1888) popular novel *Real Life* (1858). The women did needlework, the children studied lessons or played,

> our husbands discuss, read, or play chess; at eight o'clock we take tea; Antoinette [the daughter] plays the piano and we leave without regretting the reunions of high society, or the brilliant pleasure parties.[123]

The wife and mother made this cozy, bourgeois life in the parlor possible. She performed or supervised the labor which made this ideal real. In the process, however, she sometimes vanished as a person and almost became one with the home she was so identified with, with the way of life she made possible because of her care and attention. "Mother was—mother. Present like the sun or the lamp or the four walls," wrote Louise Otto-Peters in her memoirs.

> She was there as a matter of course; when I arose and washed, for the breakfast milk, the clean apron, the dry stockings, for everything intimate and superficial, from morning to evening. That was understood, that was a piece of one's own life—altogether different from father.[124]

In the nineteenth century, economic growth and cultural ideology combined to make the domestic ideal for women stronger and more possible than it had ever been before.

The Domestication of the Queen

The domestic ideal was powerful enough to invade royal circles, and beginning in the late eighteenth century, some royal and privileged women championed domesticity even for queens. The powerful and conspicuous eighteenth-century female rulers—Catherine the Great of Russia, Maria Theresa of Austria, her daughter Marie Antoinette of France—were disparaged and feared. Just as the salonières were repudiated in the aftermath of the French Revolution, so were queens who functioned more as monarchs than as women. The year after Catherine the Great's death, the Russian law of succession was changed to give preference to men over women. In Russia, as in the rest of Europe, Catherine and other independent royal women came to stand for the "reprehensible past." In Russia, as in the rest of Europe, the new domestic ideal became important for queens as well as other women, and monarchs were increasingly judged as wives, mothers, and hostesses rather than as rulers. After Catherine, Russian empresses retreated to domesticity. They were praised for being domestic and criticized when they "interfered" in nondomestic concerns.[125] This shift is embodied in the differences between Napoleon Bonaparte's two wives, Josephine and Marie-Louise. Josephine (1763–1814) came to Paris at fifteen and was married the next year.[126] Separated from her husband six years later, she hosted a Paris salon which became a center of political intrigue during the Revolution. In 1795, she had an affair with a director of the new French government; the following year, she married the handsome young general Napoleon Bonaparte. Six years her junior, Bonaparte seems to have been attracted both by Josephine's charm and her political connections. The latter helped in his coup of 1799 and the establishment of his rule. Despite Josephine's infidelities and age (she was past childbearing), the couple remarried in a full religious ceremony in 1804, just before they were jointly crowned Emperor and Empress of France. Five years later, Napoleon divorced her to marry the eighteen-year-old Hapsburg princess, Marie-Louise of Austria (1791–1847).

Marie-Louise, the eldest daughter of the Austrian emperor, had been brought up ignorant of sex: male animals were kept out of her presence, and all references to the sexual act were cut from books she read.[127] Napoleon boasted she was the first virgin he had deflowered, and she fulfilled her maternal duty by bearing a son two years after the wedding. In discarding Josephine, an intelligent companion, for Marie-Louise, Bonaparte, in his marriages as in so much else, reasserted traditional values of European culture. In the last resort, a youthful, fertile wife who could bear a male heir took precedence over one who could not.

The new domestic queenly ideal was most fully embodied by Queen Luise of Prussia (1776–1810), whom Napoleon called "the most admirable Queen and at the same time the most interesting woman I had ever met."[128] A German princess, she had been raised like a girl of the privileged classes: she was taught languages, music, and dancing at home, and she and her sisters visited the poor with gifts, as Hannah More had recommended. Married at seventeen, she was beautiful, graceful, and domestic. "When we were at home of an evening, drinking tea in our small circle, reading now and again, and rejoicing in our little darling," she wrote, "I am so happy that I never want to be anything else in all my life."[129] The mother of eight children, she was called the "mother of the nation" [Landesmutter], and she became a national heroine in 1807, when she pleaded with Napoleon for Prussia, which he had just conquered. Her domesticity and patriotism made her a popular favorite, and her early death in 1810 enshrined the legend. At the end of the nineteenth century, she was still the most admired German woman in a poll on that subject.

As Queen of Prussia, Luise had no political power, but even where queens ruled in their own right, domesticity superseded politics. The reign of Queen Victoria of England (1819–1901; queen from 1837) marked not only the decline of royal power, but also its replacement by domestic values. Politically active in the first years of her reign, Victoria became convinced of the necessity of remaining within the constitutional limits of royal power by her first prime minister, Lord Melbourne, and her husband, Prince Albert. She soon devoted her formidable energies to domestic life, which she came to prize and to use to symbolize her authority, which she based on her femininity as well as her royalty. "Really when one is so happy and blessed in one's home life, as I am, Politics (provided my Country is safe) must take only a 2nd. place," she wrote in her journal in 1846. By then, she had been married for six years and was the mother of five (she would eventually have nine children). Her happiest times were with her family, and her descriptions of those moments echo those of humbler homes:

> The children again with us, and such a pleasure and interest! Bertie and Alice are the greatest friends and always playing together.—Later we both read to each other. When I read, I sit on a sofa, in the middle of the room, with a small table before it, on which stand a lamp and candlestick, Albert sitting in a low armchair, on the opposite side of the table with another small table in front of him on which he usually stands his book. Oh! if I could only exactly describe our dear happy life together![130]

Initially resistant to deferring to a husband, Victoria came to cherish Albert's strength and support. He became her political adviser, and his premature death in 1861 left her bereft and virtually unable to function. Drawn out of retirement by her prime minister, Disraeli, she achieved her greatest popularity in her later years, when the "Widow of Windsor" became both the embodiment and the "mother" of her nation and era. Meticulous about her royal prerogatives, she also cultivated her domestic image: at her Golden Jubilee in 1887, celebrating the fiftieth anniversary of her accession to the throne, she insisted on wearing a lace bonnet rather than a crown.[131] Her Diamond Jubilee ten years later and her funeral in 1901 were occasions of national celebration of the "little woman" who had come to symbolize an age. As mother and grandmother to many of Europe's royal families, Victoria set a style for queens which remains intact to the present. The queen became an embodiment of the domestic ideal: a perfect wife, a perfect mother, a perfect lady. The arbiter of etiquette and ceremony, her function has become that of providing an attractive image rather than actually ruling. Domesticity and patriotism continue to be expected of queens in the twentieth century, when their function has dwindled to that of figurehead. It was the queens of the late eighteenth and nineteenth centuries who established this style, by embracing the new domestic ideals of the period.

The domestic ideals of the nineteenth century provided privileged women—whether royal or middle class—with a structured, traditional and enclosed way of life. Comfortable and secure, privileged and protected, these women were expected to be content within the parlor, within the "sphere" considered appropriate for them. While many were, it was from those same parlors that some women sought both to enlarge their sphere of activities beyond the parlor and to assert domestic values outside of the home—to claim that women had a role "housekeeping" society at large.

The ideal Victorian lady did not venture outside the parlor, did not become anything other than a devoted family member, did not write for a public audience. Ironically, the women whose memoirs and testimonies provided the illustrative material for this section themselves violated the conventional ideals of the period, which did not include that of a woman publishing her views. Marie LaFarge wrote her memoirs after her conviction for her husband's murder. Clara Schumann was one of the most famous concert pianists of her day. Louise Otto-Peters wrote novels, published a women's newspaper in the revolutionary period after 1848, and became a founder of the German women's movement. Mathilde Bourdon, Elizabeth Gaskell, Catherine Gore, Fanny Lewald, and George Sand all were well-known novelists. First by writing, later by actively participating in charitable endeavors and by fighting for better education, some women of the privileged classes created new paths which led to a life outside the parlor.

3

LEAVING THE PARLORS

❦

Writers, Artists, and Musicians

"What am I that their [other women's] life is not good enough for me? Oh God what am I? . . . why, oh my God cannot I be satisfied with the life that satisfies so many people? I am told that the conversation of all these good clever men ought to be enough for me. Why am I starving, desperate, diseased on it?" wrote Florence Nightingale (1820–1910) in an anguished private note when she was thirty-one years old.[1] The beloved elder daughter of a privileged English family, Nightingale moved in the course of the next year from blaming herself for her feelings to blaming the culture and society which so limited women like her. The following year, 1852, she won the bitter, decade-long struggle with her family to study nursing and also wrote *Cassandra*, an impassioned indictment of the confinement of privileged women. "Why have women passion, intellect, moral activity—these three— and a place in society where no one of the three can be exercised?" she asked on the first page of the essay. Chafing at the limits of a life within the parlor, she struggled to create an independent existence for herself, based on her work rather than her family relationships.

As the founder of modern English nursing, Nightingale later provided a model for other women who wished a life outside the home. But in Nightingale's youth, a woman who wished to move beyond the proper womanly roles had to look to male examples for inspiration. By writing, by studying, by working outside the home, privileged women called their sexual identity into question. Acting in ways defined by their culture as unwomanly, they and others concluded they must be acting like men. "You must now consider me . . . a son," Nightingale thought of telling her mother when she left for nursing training, for although nursing had traditionally been performed by women, becoming educated professionally was a male prerogative. In later years, Nightingale often referred to herself as a "man of action" or a "man of business."[2]

The limbo between the sexes where Nightingale found herself was a space familiar to women authors of the day, especially those who challenged contemporary values and attitudes about women. For by writing for the public, and especially by writing critically about women's role in society, women in the eighteenth and nineteenth centuries challenged accepted views of female behavior. Many early women writers used anonymity or male pseudonyms to make their writings more acceptable. Jane Austen published all her novels anonymously. Books often appeared "by a Lady" or "by the author of [the writer's previous book]." "George Eliot" was the pseudonym of Marian Evans; "George Sand" of Aurore Dudevant. The Brontë sisters published as "Ellis, Currer, and Acton Bell." Delphine Gay de Girardin, the salonière, wrote as the "Viscount Charles de Launay." Louise Otto-Peters published initially as "Otto Stern"; her French contemporary, Marie d'Agoult, used "Daniel Stern" throughout her long literary career.

"She reminds you continually of a man," the American author Bret Harte wrote of the English novelist George Eliot, "a bright, gentle, lovable, philosophical man—without being a bit *masculine.* "[3] Eliot had moved beyond the accepted woman's sphere in two ways: by writing novels which dealt seriously with difficult moral issues and by living with a married man. Her male pseudonym initially fooled her readers; when she discovered Eliot was a woman, Elizabeth Gaskell wrote a friend that "I should have been more 'comfortable' for some indefinable reason, if a *man* had written them instead of a *woman.* "[4] Though protected from the charge of aping a man by her marriage and motherhood, Gaskell herself had been socially ostracized and condemned in the press a few years earlier, when she published *Ruth,* a novel sympathetic to an unwed mother. Copies of *Ruth* were burned in Gaskell's home city of Manchester, and this made her question her temerity in writing it: "I think I must be an improper woman without knowing it, I do so manage to shock people," she wrote a friend.[5] Gaskell, too, had published her early writings under the male pseudonym of Cotton Mather Mills; her first novel, *Mary Barton,* was published anonymously.

These women were all aware that they were not being good women according to the accepted ideals of their era and class. For women of the privileged classes, being a bad woman did not mean becoming a prostitute, as it did for their working-class counterparts. Privileged women were not even supposed to know, or at least acknowledge that they knew, that prostitution existed. Rather, for these women, being "bad" meant acting in an unwomanly manner: being independent, going beyond the accepted limits of female behavior, relying on their own opinions or judgments rather than others'.

"There is something unfeminine in independence," wrote Mrs. Sandford in her 1831 English guide for women. "It is contrary to nature and thus it offends."[6] Given such limitations, women who wished for more than a conventional life as a wife and mother within the home were forced to look beyond women's roles.

For privileged women, the only model of independence and achievement was a male one. Freedom was possible only if one acted as if one were a man: by doing, by writing, by relying on one's own moral authority, by breaking the conventions of femininity. The price, except in very rare and isolated instances, was that one had to reject not only the traditional woman's role, but also to face rejection from those women who followed it. Florence Nightingale received her greatest opposition from her mother and sister. Mary Wollstonecraft found support in London not from the bluestocking women, who preferred the traditional values of Hannah More, but from the largely male circle which gathered around her publisher, Joseph Johnson. "The entire social order . . . is arrayed against a woman who wants to rise to a man's reputation," wrote Germaine de Staël, the French salonière, in 1800, after her independence in speech and writing had alienated most of her female contemporaries.[7] A woman who wanted independence had to reject the traditional woman's role in order to be herself. In the process, she was often forced to reject traditional women as well. "I don't like young ladies. I can't abide women," Mary Clarke, later Mme. Mohl the Parisian salonière, wrote to Florence Nightingale in the 1830s. "Why don't they talk about interesting things? . . . If your friend is a man, bring him without thinking twice about it, but if she is a woman, think well."[8] Nightingale herself consistently claimed that only men had helped her in her struggles; most women had been unsympathetic, if not hostile.

The result was that an exceptional woman tended to identify with like-minded men rather than other women and to see herself as an "honorary man"—a unique exception to the constraints and rules which limited the lives of the other women. This lasted well into the twentieth century. "My upbringing had convinced me of my sex's intellectual inferiority," wrote the French author Simone de Beauvoir (1908–1986) of her student days in Paris in the 1920s.

> This handicap gave my successes a prestige far in excess of that accorded to successful male students. . . . Far from envying them, I felt that my own position, from the very fact that it was an unusual one, was one of privilege. . . . I flattered myself that I combined "a woman's heart and a man's brain." Again I considered myself to be unique—the One and Only.[9]

In adopting this attitude, de Beauvoir was following the example of her illustrious countrywoman George Sand (1804–1876). In the nineteenth century, no woman author captured public attention like Sand, and her career extended the limits of acceptable behavior for privileged women. The child of a dancer and an officer, Aurore Dupin was brought up by her aristocratic paternal grandmother at the Château de Nohant, which became her future home. Married at eighteen to the Baron Casimir Dudevant, she was pregnant one month later. At first, she lived conventionally in the country with her husband and child, but by 1824, she had fallen in love with another man. She renounced romance, remained with her husband, and bore him another child in 1828. Two years later, however, she was in love again and this time decided to leave her husband and go to Paris. She bullied him into giving her an allowance and permitting her to live in Paris six months of the year, arranged for a tutor for her son, and took her young daughter with her. "I am going to prove that my idea of married life is not that I should be endured as a burden, but sought and claimed as a companion," she wrote a friend as she prepared to leave.

> To live! how wonderful, how lovely that will be, in spite of husbands, and troubles, and embarrassments, and debts, and relations, and gossip, in spite of deep anguish and wearisome vexations! To live is to be in a state of constant intoxication! To live is happiness! To live is Heaven![10]

Quickly adopting the man's wardrobe which enabled her to roam the city comfortably and cheaply, she was welcomed by the young male Romantic artists and writers of the day who found her a delightful and brilliant novelty. Notorious because of her male clothing, cigar-smoking, and affairs, she soon became famous through her success as an author. In her first major novel, *Indiana* (1832), Sand passionately asserted a woman's right to follow her feelings of love, even if this led outside marriage, or to new marriages. Her years as a wife had convinced Sand that women's greatest oppression lay in not being able to love freely, and she thought that divorce (which was illegal in France in her own day) would be vastly preferable to the "sacrifice" and "despair" of a bad marriage. "One must love, say I, with all of one's self," she wrote in her memoirs, "or live a life of utter chastity."[11] Following her feelings, she left her lover, Jules Sandeau (whose name had inspired her pseudonym) and embarked on a series of romantic liaisons: with the writer Prosper Mérimée, the poet Alfred de Musset, the composer Frédéric Chopin.

A prolific author, George Sand usually produced two novels a year, many of which turned on the romantic dilemmas and erotic longings of her heroines. She was friends with the great men of her day—the writers Balzac and Flaubert, the painter Delacroix, the critic Sainte-Beuve—and she herself retained her hold on the public's imagination by the quality of her writings and the unconventionality of her life. She seemed to touch all the living currents of her day. A romantic Catholic, she continued to believe that women had a special mission to preserve "the virtues of charity and the Christian spirit" in a materialistic world. A romantic socialist, she wrote for and supported the short-lived French republic of 1848. Always an individualist, she believed that "as things are, [women] are ill-used," but when a group of radical women asked her to run as a candidate for the legislature in 1848, she angrily and haughtily repudiated them, arguing that the day when women could take part in politics was far away.[12] Throughout her long and productive life, she remained a source of inspiration for other women writers and a symbol of the dangers into which writing could lead a woman.

By 1840, the pejorative noun "George-Sandism" had appeared in English, French, German, and Russian to denounce the way of life embodied by Sand. Many were struck by the difficulties she was faced with: "There are many dark and thorny crowns / Which God gives to children in this world," wrote Louise Otto-Peters in a poem addressed to Sand, "but the most painful that God in his anger has crowned a woman's head with is—genius."[13] Even the most sympathetic found it difficult to assimilate Sand into conventional concepts of womanhood: "Thou large-brained woman and large-hearted man, / Self-called George Sand!" began one of the two poems Elizabeth Barrett Browning wrote about Sand in 1844. In each, Barrett reconciled her confusion about a woman who acted like a man by portraying Sand as an incorporeal angel of no sex.[14] But Sand forced other women to question their own lives.

> If you are not like us and will not suffer
> And smile, it is not your fault,
> You cannot do it. That is why you fight, where we suffer
> And you break the fetters that we still bear,

wrote the German poet Ida von Reinsberg-Düringsfeld.[15] Many privileged girls and women testified that they read her novels in secret and dreamed of acting like her or her heroines.

By insisting on being both a woman and a genius, by claiming the masculine prerogative of loving freely outside marriage, by succeeding as a novelist despite her condemnation by society, Sand broke new ground for some women of the privileged classes. Her success was based on her talent and her consistent claim to her own moral authority, both present in her first great success, the novel *Indiana* (1832). Married to a man she does not love, Indiana insists that her feelings should take precedence over her husband's legal and conjugal rights. "I know that I am a slave, and you are my lord," she proclaims defiantly.

The law of the country has made you my master. You can bind my body, tie my hands, govern my actions; you are the strongest and society adds to your power; but with my will, sir, you can do nothing. God alone can restrain and curb it.[16]

Sand and other women writers claimed the right to assert their convictions against the world's conventions. In so doing they appropriated rights contemporary male Romantic writers were asserting: that being true to one's innermost feelings took precedence. This was new and different from most earlier female writings. By the nineteenth century, thousands of European women were writing for a living, primarily novels which were read by other women. Most of them, like Catherine Gore, upheld conventional values in their works.[17] But it was also in the nineteenth century that women writers began to assert women's moral authority against the conventional limits on women's behavior. They portrayed independent morality as superior to a meek acceptance of conventional behavior. When the heroine is publicly crowned a poetic genius and wins the love of the hero as well in Germaine de Staël's *Corinne* (1808); when Elizabeth Bennett repudiates Lady Catherine de Bourgh's snobbish values in Jane Austen's *Pride and Prejudice* (1813); when an unwed mother is portrayed as the moral superior of her seducer in Elizabeth Gaskell's *Ruth* (1853); when Dorothea Brooke realizes that she is wiser and more humane than her older, clergyman husband in George Eliot's *Middlemarch* (1871), women writers were insisting on a woman's right to make her own moral judgments. Authors who lived far more conventional lives than George Sand found common ground with her in their basic claim that woman, no less than man, was a moral being who must decide her own fate and work out her own salvation. The women who read these works, taught from childhood to defer and subordinate their views, found in novels not only confirmation of their intimate feelings, but also a strengthened claim to moral authority and individual rights. The first great expansion of woman's domain in the eighteenth and nineteenth centuries occurred in the pages of books. Those who dared not flout society in their personal behavior, as Sand had done, were able to claim the new authority of their own feelings, even if these contradicted society's prescriptions for women. "If I were a hunter in the open fields / Or just a bit of a soldier / If I were a man, no matter how small / Then Heaven would have blessed me," wrote Annette von Droste-Hülshoff (1797–1848), the German poet who published her collection of poems under her initials so as not to offend her family.

> Instead I must sit so delicate and pure
> Like a well-behaved little girl

And only secretly may I loosen my hair
And let it blow in the wind![18]

Of all the arts, writing was the one women found easiest to combine with the domestic duties expected of them. Harriet Martineau (1802–1876), the English writer, remembered that she had first written secretly, "before breakfast, or in some private way."[19] Other women writers did the same. But painting and music could not be hidden so easily. In addition, painters and musicians in this period labored under a handicap women writers did not have. Their fields were considered part of the polite "accomplishments" a young woman of privilege was expected to master. A little drawing, watercoloring, and painting, some ability to sing or perform on an acceptable instrument, like the piano or the violin, were considered appropriate adornments to a young woman. Thus serious women painters and musicians, unlike women writers, faced the problem of being dismissed as amateurs, unworthy of serious consideration. In addition, as artists whose works needed to be publicly displayed or performed, they encountered more barriers than their counterparts in literature, who could publish under male pseudonyms or keep their writing secret from their families. The result was that while the nineteenth century saw the emergence of great female novelists and poets, female artists and musicians had a far more difficult time and, in the case of painting, actually lost ground achieved in the eighteenth century.[20]

Traditionally, European women became artists if they were born into artistic families and were trained by their fathers. By the late eighteenth century, some of these women were able to reach the pinnacle of their professions. Both Angelica Kauffmann (1741–1807) and Elizabeth Vigée-Lebrun (1755–1842) achieved an international success which few if any women have achieved since.[21] Both were able to establish themselves as major artists, Vigée-Lebrun as a portraitist, Kauffmann as a historical painter. Both were admitted to their nation's academies, the chief means for an artist to succeed. Despite the barring circumstance of her husband's being an art dealer, Vigée-Lebrun became a member of the French Royal Academy because of Marie Antoinette's intervention. During the French Revolution, she lived in many European cities, returning eventually to Paris. Immensely prolific, she produced close to a thousand paintings. Angelica Kauffmann insisted on entering the prestigious and previously male territory of historical painting and achieved great success. She and Mary Moser, who specialized in flower paintings, were admitted to the British Royal Academy when it was formed in 1768.

But Kauffmann and Moser were the only women who gained entrance to the Royal Academy until the 1920s.[22] German women were also excluded in the nineteenth century from academies which had admitted them in the eighteenth.[23] While Frenchwomen continued to be allowed to exhibit in the salons, few women artists achieved any prominence in the first three quarters of the nineteenth century. Rosa Bonheur (1822–1899), the French animal painter, was an exception: in 1865, she became the first woman to be awarded the prestigious Cross of the French Legion of Honor, which Empress Eugénie conferred on her in part to show that "genius has no sex."[24] But with the exception of Bonheur, whose example loomed over other women painters, most women found themselves in the position portrayed by the English artist Emily Mary Osborn (1834–after 1893) in her 1857 painting entitled *Nameless and Friendless*. [25] Accompanied by her younger brother, whose presence was necessary for chaperonage, the young female artist stands dejected, with a drooping head and downcast eyes, as a male art dealer skeptically examines her painting. She is the only woman in the shop.

For much of the nineteenth century, women artists labored to be admitted to art schools and exhibits, salons and museums. It was not until the third quarter of the nineteenth century that some women felt sure enough of themselves to challenge the established artistic circles and seek recognition as outsiders, in dissident or avant-garde movements. One of the first was the French artist Berthe Morisot (1841–1895), who participated in the genesis of the new Impressionist movement. She had paintings at the first Impressionist exhibition of 1874 and took part in each subsequent exhibition except that of 1879, when the care of her infant daughter prevented her participation.[26] While her work was often called "feminine," both she and her fellow Impressionists considered her a significant artist, equal in talent to men of the movement. She had her first one-woman show in Paris in 1892.

By the 1890s, other European women were also able to follow their own artistic visions. In Germany, Käthe Kollwitz (1867–1945) and Paula Modersohn-Becker (1876–1907) won contemporary recognition as important artists, Kollwitz as a political graphic artist, Modersohn-Becker as a pioneer of Postimpressionism. Both portrayed women whom artists had previously ignored. Kollwitz sketched poor working women, showed women as revolutionaries in her series on the Peasants' War, and used forceful images of mothers and children to illustrate her socialist and pacifist concerns. Her powerful self-portraits, portraits, and statues became familiar icons of the left in the first three decades of the twentieth century.[27] Modersohn-Becker also chose women and children and the poor as her subjects. Part of the artist's colony at Worpswede, she sketched in the local poorhouse and in peasants' homes. In 1900, she made the first of her four trips to Paris, Europe's artistic capital, where she was drawn to the work of new artists.

Married to the German painter Otto Modersohn in 1901, she experienced conflict between her marriage and her art in the few years of life which remained to her. Her most powerful and enigmatic self-portraits (some of which challenged convention by painting the artist naked and pregnant) date from 1906, when she stayed in Paris for over a year. From there she wrote her husband, "Let me go, Otto. I do not want you as my husband." Three days later, she repented and wrote, "My harsh letter was written during a time when I was terribly upset. . . . Also my wish not to have a child by you was only for the moment. . . . I am sorry now for having written it."[28] In February 1907 she became pregnant for the first time in her life; she died of a heart attack a few days after the birth of a daughter in November. She was thirty-one years old.[29]

While some women artists had won their place as equals by the turn of the twentieth century, women musicians found acceptance only as performers. As composers, they disparaged their own work and still have not been completely accepted as equals. Fanny Mendelssohn Hensel (1805–1847), the daughter of an illustrious Berlin Jewish family and sister to the more famous Felix Mendelssohn, found her musical talent consistently disparaged by both her father and her brother. "Music will perhaps become his [Felix's] profession, whilst for *you* it can and must only be an ornament, never the root of your being and doing," her father wrote her in 1820. "Remain true to these sentiments and to this line of conduct, they are feminine, and only what is truly feminine is an ornament to your sex."[30] Her brother discouraged her from publishing her work, believing, like her father, that her home and family should come first. In 1836, she wrote of her discouragement to a friend:

> My songs lie unheeded and unknown. If nobody ever offers an opinion, or takes the slightest interest in one's productions, one loses in time not only all pleasure in them, but all power of judging their value. . . . I cannot help considering it a sign of talent that I do not give up, though I can get nobody to take an interest in my efforts.[31]

Although she rarely performed her own works in public (she became a concert pianist in 1838), Mendelssohn was able to publish two volumes of her music just before her death in 1847.

Her experience as a composer was shared by her countrywoman Clara Wieck Schumann (1819–1896), who became internationally famous as a pianist and teacher, but not as a composer. A musical prodigy, Schumann began performing publicly at nine. Encouraged to compose by both her father and her husband, Robert Schumann, she consistently disparaged and undervalued her own work. "I once thought I possessed creative talent, but I have given up this idea," she wrote in her diary when she was twenty;

a woman must not desire to compose—not one has been able to do it, and why should I expect to? It would be arrogance, though indeed my father led me into it in earlier days.

Although she thought "there is nothing greater than the joy of composing something oneself, and then listening to it," she dismissed her composition as "only a woman's work, which is always lacking in force, and here and there in invention."[32] Her marriage in 1840 to Robert Schumann, "whom I love more than myself and everything else," convinced her to subordinate her career to his own. She believed that the care of her family (she had eight children) and household should come first. "But children and a husband who is always living in the realms of imagination, do not go well with composition," Robert Schumann wrote in their joint diary in 1843. "She cannot work at it regularly."[33] Robert Schumann's mental breakdown forced her to support the family, which she did by concertizing. After 1853, she stopped composing altogether, and in the second half of the century she devoted herself to performing the works of male composers: Chopin, Robert Schumann, and her protegé Johannes Brahms. Like many women of talent, she tended to associate with men whom she thought even more talented than herself, and she considered it her mission to promote Schumann's and Brahms's works rather than her own.

Still, with all these limitations, women artists and musicians, like women writers, claimed new territory and widened the options of privileged women in the course of the nineteenth century. By 1900, a woman's desire to write, paint, or perform music in public was capable of fulfillment, and the initial battles for study and recognition had been won.

Charity Workers and Social Reformers

Success in the arts demanded talent and so was restricted to those women who possessed it. Many more privileged women found their way to a life outside the parlor by following the path pioneered by Hannah More. More claimed that a privileged woman had a moral duty to be benevolent: "The care of the poor is her profession."[34] It was no longer enough for a woman to be a good wife and mother within the home; she should also venture beyond it to aid those less fortunate than she, as More herself had done by teaching poor children to read in the Sunday schools she established. She and women like her insisted that a moral woman should practice her morality in society at large as well as in her home. "The family is . . . too narrow a field for the development of an immortal spirit, be that spirit male or female," wrote Florence Nightingale in *Cassandra*.[35] Nightingale's hard-won expertise in nursing led her to public health work and sanitation reform for the London poor, the British army, and the Indian Sanitation Commission. By accepting the traditional premise that women and men had different natures and talents, women like More and Nightingale were able to argue that women's domain should be expanded beyond the home. By adhering in the main to traditional values, increasing numbers of privileged women were able to venture into areas untraditional for women of their class: the prisons, the schools, the hospitals, the homes of the poor, the streets of the new cities. To women expected to confine their activities to domesticity, the easiest way to a life outside of the parlor lay not in overt rebellion, but in the virtuous path of charity work.

The right to work with the poor and disadvantaged had to be fought for. Traditionally, privileged women had given alms and patronized religious charities, but such patronage usually did not mean the active work of raising money or visiting the recipients. Most often, it simply involved the gift of one's name and money. Some charities endowed their patronesses with social prestige, and being accepted into such a charity signaled social acceptance for bourgeois women, much as being invited to join a club did for men. The Queen Luise League and the German Red Cross functioned this way in Imperial Germany, for example. But from the late eighteenth century, some privileged women defied convention by taking a more active philanthropic role. For years, Catherine Cappe, an English Unitarian, worked unsuccessfully to have the York County Hospital open its female wards to visits from charitable women. Finally, in 1813, she attached her demand to a large donation, which the hospital reluctantly accepted.[36] In 1817, such women in Lille founded the Society for Maternal Charity to help poor women who had recently given birth. To be a member, the *dame patronesse* had to contribute not only money and goods, but also her time: she had to visit poor women's homes herself or pay a fine.[37] Such activities were unconventional. "It is only within a short time that the prejudice against women appearing alone in public has begun to pass away," explained the Dutch writer Elise Haighton in 1884.

> How was this change brought about? By women simply doing, while strictly adhering to propriety and decorum, what society had been pleased to call improper. . . . The definition of the word *womanly* broadens every day, and when society hesitates to give us what it is our right to have, our women associate and establish for themselves organizations similar to those from which they are excluded.[38]

The fruits of such efforts could be very impressive. Prior to the work of Elizabeth Fry (1780–1845), women prisoners, many of whom had their children in jail with them, were confined in squalid conditions. At Newgate, the women's prison in London, three hundred women, some awaiting trial, some convicted murderesses, were kept in two wards. "Here they saw their friends, and kept their multitudes of children; and they had no other place for cooking, washing, eating and sleeping," a friend of Fry's wrote in 1818.

> They all slept on the floor; at times one hundred and twenty in one ward, without so much as a mat for bedding; and many of them were nearly naked. . . . Every thing was filthy to excess, and the smell was quite disgusting. Every one, even the Governor, was reluctant to go among them.[39]

Fry, a Quaker minister, the wife of a banker, and the eventual mother of ten, first visited Newgate in 1813; three years later she decided to work with women prisoners on a sustained basis. Everything had to be invented. Fry wanted both to improve the women's daily living conditions and to enable them to earn a decent livelihood on their release. She recruited other privileged women to help her, and they began to visit the prison daily. She first organized a prison school, where children and younger women were taught to read, to knit, and to sew. Religious tracts and pamphlets were distributed once a week, and Fry persuaded wealthy women to donate supplies for the prisoners to work on. The garments they manufactured were sold in a prison shop. One volunteer ran the shop, others recruited privileged female "guests" to come buy every Friday. The prisoners received the profits, half when the goods were sold, half on their release. On Saturdays, Fry instituted a day of cleaning and washing and arranged for prisoners to be supplied with soap, clean water, rags, and mops—all previously lacking. By 1818, Newgate, formerly " 'this hell upon earth,' already exhibited the appearance of an industrious manufactory, or a well-regulated family," wrote an admiring contemporary.[40]

Initially Fry's success impressed the men of her day, and because of her acknowledged expertise in 1818 she became the first woman to be called to testify before a British parliamentary committee. In 1821, Fry founded the British Society of Ladies for Promoting the Reformation of Female Prisoners, and in 1827, she published *A Sketch of the Origin and Results of Ladies' Prison Associations, with Hints for the Formation of Local Associations*, to enable other women to benefit from her experience. When British prison reform eliminated many of Fry's innovations in the 1830s in favor of stricter discipline and the "silent system," she found a new audience for her work in Europe, where she was invited to tour and advise on prisons in France, Germany, Switzerland, Belgium, the Netherlands, and Denmark. Inspired by her work at Newgate, the German pastor Theodor Fliedner, with the help of his wife, established a refuge for women prisoners at Kaisarswerth in 1833. By midcentury, Kaisarswerth had grown to include a lunatic asylum, an orphanage, an infirmary, and a training hospital for nurses, teachers, and visitors to the poor. It was Kaisarswerth which Florence Nightingale fought so hard to enter for her nursing training in 1852; by 1870, forty-two similar communities with over twenty-one hundred "sisters" had been established. Fry "inspired troops of like-minded exalted Protestant Christian women to follow her example in visiting prisoners," Fliedner wrote in a tribute to her.[41]

Fry urged privileged women to be active in charity work, whatever field they chose to reform. "I rejoice to see the day in which so many women of every rank, instead of spending their time in trifling and unprofitable pursuits, are engaged in works of usefulness and charity," she wrote in 1827.

> Earnestly is it to be desired that the number of these valuable laborers in the cause of virtue and humanity may be increased, and that all of us may be made sensible of the infinite importance of redeeming the time, of turning our talents to account, and of becoming the faithful, humble, and devoted followers of a crucified Lord who went about DOING GOOD [sic].[42]

Women's claim to moral authority enabled them to move beyond the protected sphere of the parlor or the convent. From the thirteenth century on, the papacy had only grudgingly approved uncloistered orders for women. This policy changed in the 1840s, as the need for more uncloistered women to staff the Catholic charities was recognized. The population growth of the eighteenth and nineteenth centuries and the immense new cities of the period had overwhelmed the existing charities, whether Catholic or Protestant, and the state had not yet assumed the work of welfare. Without the efforts of charitable women, the destitute often had nothing. "William Davis and his wife, living in Blackbird-Alley, St. John's Street, Bethnal Green, when first visited were found in a most wretched condition," went a typical report from a London charitable society in 1803,

> both of them ill of a raging fever, lying upon a bedstead with only a gardener's mat to cover them—no linen to wear,—their three helpless children almost naked, and nearly starved,—no one to assist them until visited by the Strangers' Friends Society.[43]

In Catholic nations, young women of privilege entered the new nursing and service orders in large numbers. (Convents were restricted to the relatively wealthy: most orders still required that a postulant bring a dowry for her support and refused to accept anyone who had previously worked as a servant.) The Sisters of Mercy, the Franciscan Sisters of the Poor, the Sisters of St. Charles, the Daughters of Divine Providence, the Sisters of Christian Instruction were but a few of the new, uncloistered nineteenth-century Catholic women's orders. These women nursed, taught, founded orphanages and refuges, and, in large numbers, provided the service personnel for the numerous new Catholic and Protestant missions in Asia and Africa. There were hundreds of thousands of such women in Europe, and in the nineteenth century their numbers far outstripped those of men in religious orders. By 1876, for instance, there were 113,750 women in Catholic religious orders in France; the male equivalent was 22,843.[44] Working in schools and hospitals, in venereal wards and leprosaria, in insane asylums and old-age homes, these women provided a multitude of social services which would not have existed without them. Catholic lay women's groups also expanded during the same period: prayer groups, service organizations, and charitable societies all testify to the philanthropic energy of privileged women in the nineteenth century. By 1900, a woman of privilege could participate easily in charity work without making it a full-time vocation or career. "Doing good" had become an accepted part of the womanly role, even if it involved visiting prisons or slums.

Women invented new ways to raise money for their charities. The practices of charity "visiting," of making clothes and other useful objects for the poor, of organizing collection and distribution drives were all new. The thrift shop, the charity bazaar, the ladies' sale, the fundraising dinner and dance were all improvised by women of privilege who rarely had ample money of their own to donate to their charitable concerns. The extent of such charitable work was considerable. By 1885, for example, Swedish women had established refuges for destitute women, old-age homes, numerous societies for the promotion of female industry, and a fund for the destitute. They had founded a Society for the Relief of the Poor, a home for out-of-work maidservants, a children's hospital, a home for children with incurable diseases, an asylum for "idiots," and a school for the deaf and dumb.[45]

Women tended to concentrate on charities which aided other women and children.[46] The conditions they observed in the homes of the poor and existing charitable institutions often led them to demand that women—with their special feminine character and abilities—should have a greater role in society. To their eyes, nineteenth-century society needed exactly those qualities traditionally associated with women: emotional warmth, tenderness, and nurturing love. This conviction led a number of privileged women to demand a wider sphere of action and better education for themselves and their daughters, so that they could properly carry out "woman's mission," defined in a guide popular in both France and England in the 1830s as "the establishment of peace, and love, and unselfishness, to be achieved by any means, and at any cost to themselves."[47] This attitude flourished in the second half of the nineteenth century: it accepted the traditional belief that women were (or ought to be) self-sacrificing and used it to enable some of them to work actively for causes they advocated. "I believe that nothing whatever will avail but the large infusion of Home elements into workhouses, hospitals, schools, orphanages, lunatic asylums, reformatories, and even prisons," wrote the English reformer Josephine Butler, in 1869,

> and in order to attain this there must be a setting free of feminine powers and influence from the constraint of a bad education, and narrow aims, and listless homes where they are at present a superfluity.[48]

The daughter of an influential English family of liberal reformers, Josephine Grey Butler (1828–1906) had been brought up to form her own opinions on moral issues. Her family had a long tradition of championing unpopular causes, like the abolition of slavery. Married to a like-minded clergyman, she remembered being repelled by the hypocrisy they encountered at Oxford University in the 1850s. Elizabeth Gaskell's novel *Ruth* was under discussion, and "this led to expressions of judgment which seemed to me false—fatally false," Butler wrote in her memoirs.

> A moral lapse in a woman was spoken of as an immensely worse thing than in a man; there was no comparison to be formed between them. A pure woman, it was reiterated, should be absolutely ignorant of a certain class of evils in the world, albeit those evils bore with murderous cruelty on other women. One young man seriously declared that he would not allow his own mother to read such a book as that under discussion—a book which to me seemed to have a very wholesome tendency though dealing with a painful subject. Silence was thought to be the great duty of all on such subjects.[49]

In Catholic nations, the care of prostitutes had been traditionally left to nuns: the Order of the Good Shepherd was renewed in 1828 to provide houses of refuge for them, and 116 such houses existed on the Continent by 1906. In Protestant nations, few such systems existed, and women of privilege found, like Butler, that they were not even supposed to acknowledge the existence of either prostitutes or prostitution. When Elizabeth Barrett Browning referred obliquely to the practice in her epic poem, *Aurora Leigh* (1856), she wrote a friend,

> What had given most offense in the book . . . has been the reference to the condition of women in our cities, which a woman oughtn't to refer to, by any manner of means, says the conventional tradition. Now I have thought deeply otherwise. If a woman ignores these wrongs, then may women as a sex continue to suffer them; there is no help for any of us.[50]

Motivated by a similar sense of solidarity with all women, Butler befriended an unwed mother in prison for child murder and nursed her in the Butler home. This personal charity was continued for many decades, but in 1864, a family tragedy motivated Butler to greater action. Her only daughter, age six, fell to her death while running to greet her parents. Thereafter, Butler found she had lost all interest in life. "I only knew that my heart ached night and day and that the only solace possible would seem to be to find other hearts which ached night and day, and with more reason than mine."[51] Characteristically, she went not to some respectable, established charity, but to pick oakum in sheds near the Liverpool docks, where homeless women, many of them prostitutes, sought to earn a few pennies. (Oakum is the untwisted fiber of old ropes, used to caulk ships. "Picking" it was tedious, painful, and unpleasant and so was assigned to prisoners and the poor in nineteenth-century England.) The women "laughed at me, and told me my fingers were of no use for that work," remembered Butler, "but as we laughed, we became friends." Curious about their lives and sympathetic to their plight, Butler spoke to the women as fellow Christians and equals. "Did I speak to them of their sins? Did I preach that the wages of sin is death? Never! What am I—a sinner—that I should presume to tell them that they are sinners? That would have stirred an antagonism in their hearts, a mental protest."[52] Instead, she preached forgiveness and offered her own personal charity. She soon realized that personal charity was not enough: after befriending numerous young women and establishing a Home of Rest for them, Butler concluded that prostitution was a deep-rooted social problem whose basic cause was economic. It "is not caused by female depravity or male licentiousness, but simply by underpaying, undervaluing and overworking women so shamefully that the poorest of them are forced to resort to prostitution to keep body and soul together."[53] Her work among prostitutes rapidly convinced her that they were but the most miserable of an oppressed group which included all women. Women received neither education nor training to prepare them for decent jobs; no wonder, she reasoned, that many found prostitution their sole source of livelihood. In 1866, she became active in the North of England Council for the Higher Education of Women. In 1867, she became one of the 1,498 English women to sign a petition for the vote. In 1868, she wrote for and edited an influential book, *Women's Work and Women's Culture*. In 1869, she dedicated her energies to the cause which was to bring her fame and move her to political activities far beyond what was considered acceptable for privileged women: the abolition of state-regulated prostitution.

Since the early nineteenth century, European nations had increasingly adopted the Napoleonic system of *maisons tolerées:* making prostitution legal in brothels controlled by the police where prostitutes were examined for disease. Women who refused examination or were found diseased were sent to locked prison-hospitals. In the growing cities, in the ancient ports and garrisons, the power of the state enabled police to try to control and regulate prostitution. Medical knowledge of the period focused concern on the rapid spread of venereal disease; the sexual double standard ensured that medical inspection was applied only to women, which made any attempt to control the disease futile. Nevertheless, in 1867 the International Medical Congress in Paris proposed making such inspections uniform throughout Europe. In 1864, the first English attempt at state regulation, the Contagious Diseases Acts, gave special police the right to arrest and examine any woman suspected of being a prostitute in any of eleven ports and garrison towns.

To Butler, prostitution epitomized man's inhumanity to woman; she found the institutionalization of such a system insupportable. "I feel as if I must go out into the streets and *cry aloud*, or my heart will break," she told her husband in 1869. "Go and God be with you!" he replied. Butler organized on two fronts. She began speaking publicly to working-class audiences, an extraordinary act in an era in which "ladies" did not speak from public platforms. Second, and equally extraordinary, she claimed that women possessed civil rights and liberties, and she organized women of her own class into the Ladies' Association for the Repeal of the Contagious Diseases Acts. These actions lost her the support of a number of previous allies who thought she should continue working for the less controversial cause of women's education. In her first year of political activism, Butler spoke at nearly one hundred meetings, established, edited, and wrote for *The Shield*, which defended women's rights, swayed the outcome of a by-election, and submitted a protest to Parliament signed by 251 prominent women, among them Florence Nightingale. In 1871, Parliament bowed to "the revolt of the women" and their political allies in Parliament, established a royal commission on the subject, and invited Butler to testify. When reform failed in England, she went on a tour of Europe in 1874. Revolted and appalled at what she saw in the state brothels and locked prison-hospitals where diseased prostitutes were confined, she founded the British and Continental Federation for the Abolition of Government Regulation of Prostitution, modeled on the antislavery organizations of her youth. Continually active, Butler wrote and spoke on conditions in Europe, on the trade in young girls, on the insanity of setting the legal age of consent for women at twelve.

Finally in the 1880s her efforts bore fruit in England. The victory of the Liberal party and the constant pressure applied by Butler and her Ladies' Association—often in the form of prayer meetings—led to the suspension and then the abolition of the Contagious Diseases Acts and the raising of the age of consent for women from twelve to sixteen. "By this terrible blow which fell upon us [the passage of the Contagious Diseases Acts], forcing us to leave our privacy and bind ourselves together for our less fortunate sisters," concluded Butler, "we have passed through an education—a noble education.

God has prepared in us, in the women of the world, a force for all future causes which are great and just."[54] To women like Butler, philanthropy and social work led to the conviction that women of privilege must have a larger share in the running of the world. She was but one of the many English-women active in charity work who came to believe that women should have the vote in order to influence society.

Josephine Butler was exceptional in her sense of solidarity with all other women. Her empathy, lack of snobbishness, and sense of equality enabled her to discard contemporary class biases. But most other privileged women, even those who worked actively in charity, shared the views of their classes. Poverty was a sign of inferiority; misfortune signaled moral lapses and divine displeasure. Hedged about by propriety themselves, such women saw no reason to relax morality when dealing with the less fortunate. To aid an unwed mother, to provide for an illegitimate or unbaptized child, to countenance civil marriage conflicted with the moral standards these women had been taught to embody. Such women often resisted pleas to relax their standards and instead insisted on the strictest behavior from the recipients of their charity. The 1890 regulations of the Protestant Parisian "Temporary Day and Night Shelter for Unemployed Women Workers or Servants without a Position," for example, stated that

> order and silence are required in the workroom [where residents were required to sew twelve hours a day] as well as in the dormitories. Work must be done carefully and regularly, and with entire obedience to the orders of the directress, who has complete authority. . . . Any person who causes disorder or trouble or who is guilty of rudeness or any infraction harmful to the good order of the house, will be dismissed immediately by the directress, whenever this occurs.[55]

Even worse, many privileged women saw poor women's only respectable way to earn a living as domestic service. "It is the one occupation they can follow in life," the English reformer Louisa Twining testified to Parliament in 1861, and many charities run by Englishwomen trained poor girls only for domestic service.[56] While Twining was sympathetic to poor women and believed any other option was unrealistic, other privileged women strenuously resisted any reforms which would deprive them of their servants' labor. In 1900, for instance, a French feminist congress refused to endorse a provision allowing maidservants one day off a week on the grounds that, given an unsupervised holiday, the domestics would only earn extra income working as prostitutes. The German Housewives Union strongly opposed the limitation of servants' hours to thirteen per day in the 1920s, and campaigned for a "service year" in which the state would compel young women to work as servants.[57] Charity work rarely eradicated class barriers.

The Battle for Education and Professional Training

Charity work, however, often did convince privileged women that better education was vital, both for themselves and the women they sought to aid. Charitable women in the Netherlands "became conscious of their imperfect training," wrote Elise van Calcar in 1884, "often discovering that even for this work, they had not received the proper preparation."[58] In charity work, a privileged woman might find herself called upon to function as a secretary, a treasurer, a head of an organization. Better education was needed to enable her to balance the books, to take shorthand, to prepare a report, to speak in public. Better education could also train poor women to earn a living other than as prostitutes or domestics. When Jessie Boucherette helped to found the Society for Promoting the Employment of Women in London in 1859, she rapidly decided that lack of education was the greatest obstacle preventing women from earning more than a subsistence wage. She soon established a school to give "a solid English education to young girls and teach older women to write a letter grammatically, to calculate rapidly without a slate, and to keep accounts by single and double-entry."[59]

By the last quarter of the nineteenth century, most European nations were providing some free primary education for girls as well as boys: by then it was acknowledged that the new nation-states and industrialization both required that future mothers and women wage earners be able to read, write, and do simple arithmetic.[60] Providing education beyond this for girls was controversial and went counter to all the traditional beliefs in separate roles and functions for women and men. Pressure for girls' higher education came especially from women of the professional middle classes, who often found they had to battle for what their brothers took for granted: the right to go to school, to become teachers, and finally, to attend universities and medical schools. "You must fight your own battles still," Josephine Butler's husband, George, urged the North of England Council for the Higher Education of Women around 1870. "At all times, reforms in the social position of women have been brought about by efforts of their own, for their own sex, supplemented by men, but always coming in the first instance from themselves."[61] From the second half of the nineteenth century, increasing numbers of privileged women entered the campaigns to obtain better education for themselves and their daughters.

The struggle was complicated by two factors. First, women were often deeply divided as to whether girls should be educated differently from or the same as boys. Generally, European women pushed first for a differential education which stressed special training for "womanhood," primarily in the form of less science and mathematics and more religion and needlework. But by the third quarter of the century, it was apparent that with a differentiated education, a young woman would never be admitted to university, and thus an automatic limit to women's development had been created. Many educators who began by advocating a specialized education for girls came to champion an equal education for at least the most talented among them. Second, in Catholic nations, girls' education was traditionally in the hands of nuns. In France, and to some degree in Italy, the battle for higher education for girls was connected to anticlerical and liberal reform movements which sought to separate women from the Church. "France is not a convent, women are not in this world to become nuns," argued the French legislator Camille Sée, who fought through the law which established public secondary schools for girls in France in 1879, "they are born to be mothers and wives."[62] Like privileged women, most anticlerical or republican men initially believed that girls' education should be easier and less rigorous than that for boys: the French public secondary schools gave girls no Latin and Greek (essential for university admission), less science and philosophy than boys, and more literature, home economics, and hygiene. In each European nation, however, whether Catholic or Protestant, whether liberal or conservative, higher education for girls was a hard-fought battle which took many years and was not won until the twentieth century.

In Imperial Germany, for instance, the academic secondary schools and universities for men were admired throughout the world; facilities for privileged girls consisted in 1871 of a hodgepodge of about two hundred finishing schools of widely varying quality. The Revolutions of 1848 had led to the creation of a short-lived women's college in Hamburg and some demands for equal education for girls, but the collapse of both the revolutions and German liberalism in the 1850s and 1860s put an end to such efforts. The college only lasted for two years and inspired no imitators.[63] The General Association of German Women stopped asking for equal education and limited its demands to better education for future mothers. A meeting of teachers in girls' secondary schools, held in Weimar in 1872, graphically demonstrated the obstacles and prejudices faced by women who wanted more education for girls. The 54 women teachers present were consistently outvoted by the 110 male delegates, who also excluded them from important sessions. At the formal luncheon which concluded the conference, the sexes ate separately: caviar, eels, and cutlets were served to the men, bread and butter to the women.[64] The male majority resolved that men should take the important positions in girls' schools and that the curriculum should focus on producing good wives, "so that the German man will not become bored in his home through the shortsightedness and narrow-mindedness of his wife and thus be crippled in his devotion to higher interests."[65] Infuriated, the women delegates vowed to press for more positions for women teachers, although the demand for equal education was not made in force until the 1890s.

Opponents to girls' higher education used all the traditional arguments: it would weaken them as future mothers; their frail physiques would "degenerate" if they used their brains too much; women were intended to be subordinate.[66] "The Girls' School can only fulfill its task in a satisfactory manner," went one statement from a conference of male teachers,

> if men and women teachers work together in the school, just as father and mother work together in the family, with the male influence predominant at the higher level, the female at the lower level. *The leadership of the State Higher Girls' School belongs to men.*[67]

Such arrogance angered even women who were conservative on every other issue. Helene Lange (1848–1930), a teacher who believed in a differentiated education to enable girls to "be the good spirit of the family," was roused in 1887 to publish a pamphlet claiming a larger role for women in the education of girls.[68] This pamphlet made her famous, and in 1888 she founded the General Association of German Women Teachers. The next year, 1889, she established college preparatory courses for girls in Berlin. The German academic establishment resisted her demands, especially in Prussia, the largest and most influential of the Imperial German states. The Prussian minister of education ruled in 1891 that secondary girls' schools should remain controlled by men and, three years later, decreed that girls' schools' curricula should be nine hours to the boys' twelve, that Household Arts be compulsory, and that time be given each week to needlework. Needlework was important because it taught "the cultivation of feminine precision, neatness and patience, circumspect diligence in the production of modest works."[69] In 1893, Lange changed her courses to the exact equivalents of those taught boys; three years later, her students were passing the exams in Greek necessary for admission to German universities in that era.

Lange had traveled in England in the 1880s and made contact with the English educational reformer Emily Davies (1830–1921). In England, Davies had succeeded in getting women's colleges established at Cambridge by first training girls to pass the exams required of university men. Lange adopted the same strategy in Berlin, and it began to succeed near the end of the century. Baden established the first state academic high school for girls in 1893 and other German states followed in the next decade. Academic high schools were the key to university admissions. their female graduates were the first to attend college. The University of Baden became the first in Germany to allow women to matriculate in 1900. By 1914, there were 4,126 German women attending German universities.[70]

The prestigious University of Prussia in Berlin was the last major European University to admit female students, which it did in 1908. But other nations than Germany had also made higher education for women difficult to obtain. The first university in Europe to admit women was the University of Zurich in 1865. Although other Swiss universities began to accept women in the late 1870s, the large majority of female students there throughout the nineteenth century were Russian women studying for medical degrees, which they were forbidden at home. Admitted to Russian universities in 1876, women were barred again from 1881 to 1905, following the assassination of Czar Alexander II by a woman, Sophia Perovskaya (1854–1881). The University of London first granted college degrees to Englishwomen in 1878; the more prestigious English universities of Oxford and Cambridge allowed women to attend college, but did not grant them degrees until after the First World War. (Cambridge did not grant women full university voting rights, which normally accompanied a university degree, until 1948.) In France, Julie Daubié, the first woman to pass the baccalaureate examination, was granted the degree in 1861 only because of personal pressure applied by Empress Eugénie. Until 1924, public secondary education for girls was differentiated from that for boys and did not prepare female students for the baccalaureate degree. Only a few private schools prepared girls for university, and as a result the numbers of women in French universities remained low: 4,254 in 1913 compared to 37,783 men.[71]

The battle to win entrance to medical schools was as difficult as the battle for higher education in general. By seeking to become doctors, European women were challenging what was perceived by the nineteenth century to be a male prerogative.[72] The opposition from men, both students and doctors, was extreme. Even men sympathetic to women's desire for higher education drew the line at medicine. F. D. Maurice, the liberal English clergyman who had helped to found one of the first colleges for women in the 1840s, argued in 1855 that he did not believe in educating women

> for the kind of tasks which belong to OUR [men's] professions. In America some are maintaining that they should take degrees and practice as physicians. I not only do not see my way to such a result; I not only should not think that any college I was concerned in should be leading to it; but I should think there could be no better reason for founding a college than to remove the slightest craving for such a state of things, by giving a more healthful direction to minds which might entertain it. The more pains we take to call forth and employ the faculties which belong characteristically to each sex, the less it will be intruding upon the province which, not the convention of the world, but the will of God assigned to the other.[73]

When the first American woman doctor, Elizabeth Blackwell (1821–1910), was placed on the British Medical Register in 1859 because of her American and French diplomas, the British Medical Association ruled the following year that holders of foreign degrees not be allowed to practice medicine in England. Blackwell had a French midwifery degree; a few women were allowed to study at the Faculty of Medicine in Paris, thanks to the intervention of Empress Eugénie. They were not allowed to become interns, however, and so were prevented from completing their studies.[74] When Elizabeth Garrett Anderson (1836–1917) became qualified to practice medicine in England by passing the apothecaries' examination (whose old guild regulations did not explicitly exclude women), members changed the regulations. When Sophia Jex-Blake (1840–1912) applied to the medical school at the University of Edinburgh in 1869, she was told that a single woman could not with propriety attend classes. Organizing a group of seven women, she completed her first year without trouble. The second year, the women seeking to enter the required course in anatomy found their way blocked by male students, who barricaded the doors to the hall, threw mud at the women, and shouted obscenities. When the women made their way into the hall, they found their male classmates had placed a sheep in the room, explaining that they understood that now "inferior animals" were no longer to be excluded from the classroom.[75] Four years later, the university won a lawsuit which allowed them to refuse the women their degrees. When Jex-Blake sought to have women placed on the medical register through the midwifery license in 1875, the entire board of midwifery examiners resigned in protest.

In France, the first woman to demand to be allowed to be an intern was burned in effigy by her classmates. In Spain, Pilar Tauregui, one of the first women to attend medical school, had stones thrown at her in class in 1881. The following year, she and three other Spanish women who had passed their medical examinations were refused degrees and instead were issued certificates which did not allow them to practice as doctors.[76] In Russia, the government opened medical-school classes to women in 1872, in part because eighty-five Russian women had already studied medicine at Zurich where they were exposed to radical political ideas, in part because the shortage of doctors was so great. Conditions for the early female classes were rigorous in the extreme: of the first group of ninety, only twenty-five graduated. Twelve had died, two from suicide and seven from consumption. By the eve of the First World War, Russia had the greatest number of women doctors in Europe: fifteen hundred, about 10 percent of the medical profession.[77] Germany had five hundred, Great Britain slightly less than five hundred, France just under six hundred.[78]

Access to higher education, however limited and difficult to obtain, enabled some young women of talent not only to move into previously male professions, but also to revolutionize their new fields of study. Female genius found it possible to develop, and the pioneering women scholars of the late nineteenth and early twentieth centuries, no less than George Sand seventy-five years earlier, claimed new territory and domains for other women. Being an intellectual, a woman, and a scholar, became possible, even if still difficult and demanding. "Ah! how harshly the youth of the student passes. . . . And yet in solitude / She lives obscure and blessed, / For in her cell she finds the ardor / That makes her heart immense," wrote Manya Sklodowska (1867–1934)—later famous as the scientist Marie Curie—in 1894, when she was studying physics at the Sorbonne in Paris. Born in Warsaw, Sklodowska had been sent to secondary school, but after she graduated, her father's income declined. She gave up her own formal education and worked to enable her older sister to study medicine in France. From 1885 to 1891, Sklodowska was a governess for various wealthy families. "I shouldn't like my worst enemy to live in such a hell," she wrote her sister about her first position. "I learned that the characters described in novels really do exist."[79]

Despite the demands on her time from her job as a governess, she continued her own education in the sciences, which interested her especially. "At nine in the evening I take my books and go to work," she wrote in 1886. "I have even acquired the habit of getting up at six so that I work more—but I can't always do it."[80] She was then reading a physics text, a series of lectures on anatomy and physiology, and Herbert Spencer's *Sociology*. Her sister's marriage to a doctor enabled Sklodowska to stop working as a governess and enter the Sorbonne in 1891. The next year, she moved to cheap lodgings near the university—"in a quarter of an hour I can be in the chemistry laboratory, in twenty minutes at the Sorbonne," she wrote her brother. At night she studied at the Library of Sainte-Geneviève, where heat and light were free. Her dedication and talent brought success: in 1893, she placed first in the physics examinations; in 1894, first in mathematics. That year, she met Pierre Curie, an eminent French physicist whom she married in 1895. As Marie Curie she expanded the boundaries of modern physics, giving the lie to all those generations who had argued that women could not succeed in the "masculine" field of hard science. Working with her husband, she explored the unstudied phenomenon of radioactivity, for which the couple shared a Nobel prize with Antoine Becquerel in 1903. Curie bore two daughters during these scientifically creative years: Irene in 1897, Eve in 1904. "I have a great deal of work, what with the housekeeping, the children, the teaching and the laboratory," she wrote her brother in 1905, "and I don't know how I shall manage it all."[81] The next year Pierre Curie died in an automobile accident; although heartbroken, Marie Curie kept on with their work. "On the Sunday morning after your death, Pierre, I went to the laboratory with Jacques for the first time," she wrote in a diary she kept during this period.[82] A few weeks later, she was given her late husband's post as professor of physics on the Sorbonne's Faculty of Science, becoming the first woman ever to be appointed to a university position in France.

She continued her work on radiation and in 1911 became the first person to win a second Nobel prize, this time for chemistry. During the First World War, she trained 150 female x-ray technicians and herself drove an ambulance equipped with x-ray machines to field hospitals near the front. "X-ray had only a limited usefulness up to the time of the war," she wrote later. "The great catastrophe which was let loose upon humanity, accumulating its victims in terrifying numbers, brought up by reaction the ardent desire to save everything that could be saved and to exploit every means of sparing and protecting human life."[83] After the war, she became director of the Radium Institute in Paris and began to receive the many medals, prizes, and honorary degrees which paid tribute to her unique achievements. "We salute in you a great scientist, a great-hearted woman who has lived only through devotion to work and scientific abnegation," went the tribute from the Academy of Medicine in 1922, when she became the first woman to be admitted,

> a patriot who, in war as in peace, has always done more than her duty. Your presence here brings us the moral benefit of your example and the glory of your name. We thank you. We are proud of your presence among us. You are the first woman in France to enter an academy, but what other woman could have been so worthy?[84]

Curie also saw herself as a pioneering exception. "It isn't necessary to lead such an antinatural existence as mine," she remarked to admirers in the 1920s. "What I want for women and young girls is a simple family life and some work that will interest them."[85] Her matter-of-fact addition of "work that will interest them" to the traditional set of expectations for young women of privilege is a measure of how greatly their world had widened. By pioneering, Curie and others like her enlarged the world of all privileged women. Whether they became professors or not, the old barriers which had seemed so eternal had fallen, and the next generation of talented women found the path to the professions easier than Curie's arduous route from governess to Nobel laureate.

"Oh how happy I am! And thankful!!" wrote Karen Danielsen, a fifteen-year-old German girl in 1901, when she had finally won her father's permission to attend the new academic high school for girls in Hamburg. Her father, a Norwegian steamship captain, had opposed her going for two years but had finally given in to the pressure applied by her, her mother, her brother, and assorted female friends:

> Three of Mother's friends came one after the other yesterday, to work on Mother to send me to the Gymnasium [academic high school]. Mother spoke with Father afterward. . . . Beside that Berndt [her brother] heard a lecture on "the woman question" by a gentleman who greatly praised the Gymnasium for girls. Berndt had to tell Father the whole lecture. When he goes to Tante [aunt] Clara she will work on him too. Today Berndt is going to a lady who can inform us about admissions, age, courses, etc. I believe more and more that I "must" get there.[86]

Later famous as Karen Horney, the psychoanalyst, at high school Karen Danielsen loved her studies, but felt torn between what seemed to be the rival claims of scholarship and romance. "Two questions are agitating me all the time," she wrote when she was twenty and about to graduate. "The one is: Graduation? Berlin? Study? And the second is: Ernst? The second is the more exciting because, as yet, there is no answer but the uncertain promise of a growing relationship. And then of course, I am first of all a woman."[87] Like Curie, she chose study and was accepted as the only woman student in her year at the medical school at the University of Freiburg. At first, she felt isolated: "The first semester—never will I forget those first days here, days full of desperate loneliness, of the disconsolate feeling of being forsaken." There is a photograph of her from this era, the only woman in the center of a line of six men. They stand smiling, dressed in suits and ties with canes held over their shoulders like swords; she wears a white blouse and long black skirt, looks warily into the camera and holds a skull under her arm.[88]

A few months later, she had made friends and was enjoying her studies: "my working power is borne on the wings of joyousness in work," she wrote in 1907. By then she had met Oscar Horney, a lawyer whom she married and had three children with before they separated in 1926. Like Marie Curie, Karen Horney (1885–1952) was able to combine marriage, motherhood, and a career in which she challenged the basic premises of her profession. Psychoanalyzed herself in 1910, she became a respected professor at the Berlin Psychoanalytic Institute from 1918 to 1932. In the 1920s, she produced a series of papers which challenged Freud's theories about women, arguing that men's envy of women's ability to give birth was at least as important as women's supposed envy of men's penises.[89] Leaving Germany in 1932, she established herself and her three daughters in the United States, where she founded the Association for the Advancement of Psychoanalysis, the American Institute of Psychoanalysis, and the Karen Horney Institute. Widely published and internationally famous, she, like Marie Curie, proved that once freed from artificial restrictions on intellectual development, there was no limit to women's abilities to achieve in fields previously restricted to men.

Endowed with extraordinary talent, energy and determination, Curie and Horney rivaled or surpassed the most talented men of their day. In so doing, they raised other women's aspirations and pushed back the boundaries which had previously limited women's life of the mind. They also proved that a woman need not deprive herself of a husband and children in order to succeed as a professional. But Curie and Horney were truly exceptional, and few possessed the energy or ability to follow their paths. While the opening of the universities enabled some women to reach the summit, it was the expansion of primary and secondary education which most affected the lives of privileged European women between 1875 and the Second World War. For not only did this expansion lead to a higher level of schooling for many young women, and thus the possibility of new ways to make a living, it also opened a new respectable profession to them. If a nation's children were to be educated, someone would have to teach them. Both tradition and women's settling for far lower salaries than men ensured that they would predominate as teachers of young children and girls.

As a teacher, a single woman (most European school systems required that a female teacher resign her position if she married) gained a status and independence which her counterpart within the home—the governess—sorely lacked. In the nineteenth century, becoming a governess was the last resort of an impoverished woman of the privileged classes, and governesses were the objects of pity and scorn. Edward Carpenter, the English reformer, remembered that in the 1860s "there was only one profession possible for a middle-class woman—to be a governess—and that was to become a pariah."[90] Throughout the memoirs of women who worked as governesses runs a constant strain of frustration and dissatisfaction: the possible love of the children she taught could not compensate for the insecurities and snubs which her indeterminate position—somewhere between a servant and a lady—called forth from both her employers and her fellow workers. "For the first time in my life I feel sorry that I am not a man," the Dane Mathilde Fibiger wrote when she became a governess at eighteen.[91] "Such a strong wish for wings—wings such as wealth can furnish—such an urgent thirst to see—to know—to learn," wrote Charlotte Brontë in 1836, when her job as a governess prevented her from joining a friend on a trip abroad. "I was tantalized with the consciousness of faculties unexercised—then all collapsed and I despaired."[92] Brontë's novel with a governess as heroine, *Jane Eyre*, was severely criticized when it appeared in 1847 because of Eyre's insufficient gratitude to her employers, but Eyre's bitter experiences could easily happen to a woman forced to earn her living teaching others' children in a home that was not her own.

Forced by poverty into the role of governess, most women found that a lifetime of such work led to an impoverished old age. The following case, from the records of the English Governesses' Benevolent Institution, is typical:

> Miss Margaret ———, aged seventy-one. Fifty years a governess having been left an orphan at three years old, and the uncle who meant to provide for her being lost at sea. Assisted her relations as far as possible from her salaries. She is now very feeble, and her health failing fast. Her entire support is an annuity of fourteen guineas.[93]

Founded in 1843 to provide both a job register and pensions for governesses, the Governesses' Benevolent Institution helped to found the first college for women in England. Queen's College, established in 1848, was a teachers' training school which sought to enable governesses to earn more money. It "began the Woman's Education Movement," remembered Frances Mary Buss (1827–1894), later the founder of an important London secondary school for girls. It "opened a new life to me and to most of the women who were fortunate enough to become students."[94]

As a teacher, a woman became an accepted and generally respected public figure. Teaching gave her a position in which she depended on her own merits, rather than the benevolence of the family who employed her. Although teachers were very poorly paid and often worked under daunting conditions, many recalled their happiness in the job. "The first time I cashed my check, I thought the shining sovereign passed across the counter was magical, for I should have been glad to do the work for nothing," remembered the Englishwoman Sara Burstall of 1882. "There is no profession in which people have less money," wrote the Frenchwoman Louise Michel (1830–1905) of her school-teaching days in Paris in the 1850s, when she had to work all but one week of the year to support herself. "But except for the struggle for existence," she added, "I have never been unhappy as a teacher."[95] The great reward of the job was the ability to influence others, and numerous memoirs of the period testify to the affection and respect which could grow between a schoolteacher and her students. In the early days of teacher training, teacher and student might find themselves trading places. Molly Hughes (1866–1956), a pupil at Miss Buss's school in the 1880s (where she was shocked to find that one of the school's requirements was that no girl could enter who could not sew a buttonhole), was a member of the first class of the new College of Teacher's Training at Cambridge in 1885. In 1892, Hughes taught her first class in teacher's training and found "sitting meekly among my students, none other than Miss Armstead, the classics mistress of the North London Collegiate School, who was the finest teacher I had ever sat under."

She smiled happily at me now and again during that first hour, but as soon as possible I approached her . . . to ask why on earth she had come to me for help in a business at which she herself was so brilliant. "Oh, I know absolutely nothing about psychology" said she, "nor logic nor hygiene nor all those other things you mention. It's all so exciting."[96]

In France and Italy, where governments sought to secularize education by replacing nuns with lay instructors, teachers carried the extra responsibility of representing the state in its conflict with the Church. "My classroom is long and narrow, very low, with a worm-eaten, shaky door. . . . The desks are old and shaky too, but my enthusiasm is great and my aspirations are boundless," wrote a young Frenchwoman in 1899. Describing her first day, she recalled the "crowd of little people" (she was expected to teach eighty-five students) and the interference of the Catholic principal and the nun who had previously taught there:

I have passed the day organizing my classes and getting things underway. The principal came into my classroom this morning, made the sign of the cross, and all the pupils chanted prayers for a half-hour. What could one do? And this morning, Sister Mélânie, coming in stealthily, took a seat at the farther end of my classroom, gave me a friendly little nod, and started the little ones on syllable exercises. Now I understand the words of the academy inspector when he gave me this "position of responsibility." "You will need patience and tact, Mademoiselle."[97]

Teaching provided the greatest number of new jobs for women above the working-class level in the years before World War I. In England, for instance, in 1911, 183,298 women were employed as teachers and comprised 72.8 percent of all teachers employed, while 77,000 women worked as nurses.[98] As extensions of women's traditional duties into the public sphere, teaching and nursing were seen as "naturally" suited to women. But additional new areas of employment opened for women in the period before the First World War. The developing industrial and state economies created a new "tertiary" sector, neither agricultural (primary) nor industrial (secondary), which slowly came to be seen as appropriate for women. As postal clerks and telegraph operators, women made their first entries into government work, where they were prized for their attention to detail and their willingness to accept lower salaries than men. By 1906, Frenchwomen comprised 22 percent of such employees; by 1911, Englishwomen were 51 percent of their nation's post office clerks and 35.9 percent of the telegraph and telephone operators.[99] Office work expanded tremendously in this era, and the male clerk began to give way to the female clerical worker and clerk-typist, who again accepted lower wages than her male equivalent. In the new department stores, women filled the sales positions; in the new banks, they became tellers; in the new commercial enterprises, they were the shop assistants and clerks. While these jobs were neither prestigious nor well-paid, they did provide a new option for women above the working-class level. Previously restricted to the home, these women could now make a respectable living outside it. As a teacher, a nurse, a typist, a clerk, a woman could support herself without resorting to manual labor or prostitution. The opening of new work and educational opportunities to women of privilege was seen at the time as a tremendous widening of women's previously restricted roles and functions. "It was from the educated women of the middle classes that issued the demand for better instruction and better paid employment for their sex," wrote Kirstine Frederiksen, a Danish teacher, in 1884,

> and it has been thought that the movement would keep within those limits. But women are human beings: give them an education and a competency, and they must have all the rest.[100]

Winning "all the rest": entry into the professions, the right to vote, civil equality before the law, would occur in the twentieth century. It was then that European women achieved political and legal equality with men. It was also then that they perceived both the opportunities and limits of that hard-won equality.

4

OPPORTUNITIES AND LIMITS:
CHANGE AND TRADITION
IN THE TWENTIETH CENTURY

✿

The Impact of World War I

Possibilities for a life outside the home and family multiplied for privileged women during World War I (1914–1918). "The demand for educated women has risen phenomenally during the six months since the war began," wrote Magda Trott, a German journalist, in 1915. "Women have been employed in banks, in large commercial businesses, in urban offices —everywhere, in fact, where up to now only men had been employed."[1] With so many men drafted, the numbers of German, English, and French women engaged in white-collar work—the professions, commercial and clerical jobs, government and public services—roughly doubled during the war.[2] While most of these jobs were on the clerk-typist level and would increasingly be filled by young working-class women aspiring to a job more pleasant and prestigious than domestic service or factory work, many also represented real gains for privileged women who needed to earn money. In addition, privileged women flocked to nursing and wartime service charities. In England and Germany, the large women's suffrage associations instantly transformed themselves into wartime service organizations for the duration. "In the midst of this time of terrible anxiety and grief," Millicent Fawcett, head of the largest English women's suffrage organization, wrote in the first week of the war,

> it is some little comfort to think that our large organization, which has been completely built up during past years to promote women's suffrage, can be used now to help our country through the period of strain and sorrow.[3]

The English National Union of Women's Suffrage Societies became the Women's Active Service Corps; the League of German Women's Associations *(Bund Deutscher Frauenverein)* formed the National Women's Service.

In belligerent nations without large suffrage organizations, privileged women took an equally active war role, organizing canteens and charities for soldiers and their families, joining the Red Cross and similar groups, writing to soldiers, rolling bandages, and raising money for relief organizations.[4] Although women had traditionally provided these services in wartime, never had they done so on such a massive scale. Many accepted the traditional division of roles between women and men, seen in the anthem of the German National Women's Service:

> Women's hands work busily in service to our dear Fatherland;
> We must cook and sew and nurse, because wartime makes many wounds.
> This mobilizes us girls too, and if we can't accomplish so much
> We knit socks for soldiers; we are here for a labor of love.[5]

But World War I, with its new call to patriotism and its unprecedented impact on society, inspired some privileged women to go beyond traditional tasks. Many young women of the upper and middle classes, whose male relatives and friends were joining the military forces in massive numbers, felt that rolling bandages and knitting socks was not enough. "I do not agree that my place is at home doing nothing or practically nothing," Vera Brittain (1893–1970) wrote her parents in 1916, after she had left a scholarship at Oxford University to train as a front-line nurse. "I wanted to prove I could more or less keep myself by working, and . . . not being a man and able to go the Front, I wanted to do the next best thing."[6] In earlier struggles, whether wars, revolutions, or nationalist movements, a handful of privileged women had fought and suffered with men. What was new in the First World War was the relatively widespread feeling that women's traditional support service was not enough in a struggle which demanded that men risk their lives. "It all would not be so difficult / If only there were not this burning shame," as the German Andrea Frahm wrote in a poem she entitled "At Home."

> They go amid bullets for you
> You read about it evenings by lamplight.
>
> They sleep stretched out in the wet grass—
> Your nice warm bed is in front of you.
>
> You can hug your loved ones close in your arms—
> They see in dying a strange face.[7]

Privileged women now insisted on active participation, often over objections of their families or men who rebuffed their efforts. At the outbreak, Dr. Elsie Ingliss (1867–1917) offered to provide the British army with fully staffed female medical units, but was rejected by the War Office with the words,

"My good lady, go home and sit still." Undaunted, she offered her services to Britain's allies, who accepted. By 1917, she had organized and administered fourteen medical units for the Belgian, French, Russian, and Serbian armies. She herself directed the Serbian unit in the field until her death from exhaustion in 1917.[8] Some French women doctors replaced their male colleagues in nonmilitary hospitals; others joined the Health Service as nurses in order to aid the war effort.

Nursing in the First World War could be arduous and dangerous, for women served near the front lines, close to the battlefield slaughter. "The hospital is very heavy now—as heavy as when I came," Vera Brittain wrote her parents in December 1917;

> The fighting is continuing very long this year, and the convoys keep coming down, two or three a night. . . . Sometimes in the middle of the night we have to turn people out of bed and make them sleep on the floor. . . . We have heaps of gassed cases at present . . . the poor things burnt and blistered all over with great mustard-coloured suppurating blisters with blind eyes.[9]

Near the front, a woman might drive an ambulance, like Marie Curie, or organize a hospital, like Elsie Ingliss. She might be wounded, like the French nurse Charlotte Maître, who, in the words of a citation for bravery,

> Posted as a volunteer at the front line, has borne the dangers and the fatigue of the life of the Front in the underground shelters, has shown in the face of repeated bombardment an exemplary courage and decision. Wounded by shell bursts while carrying out her duty, refused to allow herself to be evacuated. Has contracted two serious infections in the course of her service while caring for men with contagious diseases.[10]

She might even be executed by the enemy, like Edith Cavell (1865–1915), an English nurse in Belgium. Running a Red Cross hospital in Brussels, Cavell helped to smuggle allied soldiers out of German-occupied Belgium. Her execution for espionage by the Germans caused a great outcry in Britain, but all nations executed female as well as male spies during the First World War.

The rough equality of the battlefield eroded traditional ideals of chivalry and protectionism. In Russia, privileged women organized combat battalions after the February Revolution of 1917; the Winter Palace was unsuccessfully defended by a women's battalion against the Bolsheviks in October 1917. The French Secret Service enlisted Marthe Ricard (1889–1982), an early woman pilot, to spy for them during the war; afterward, in tribute, they decorated her with the previously masculine award of the Legion of Honor. Although women in active service did not ignore the horrors of the war, they often spoke of its liberating effects on them. "I spread my wings and tested my strength," wrote Russian Grand Duchess Maria Pavlovna, who served as a nurse at the front. "The walls which for so long had fenced me off from reality were now finally pierced."[11]

On a very basic level, in combat and service, in the munitions factories and dockyards, behind the wheels of trucks and the dashboards of airplanes, women demonstrated their competence at jobs traditionally held by men and thus proved their equality with men. In some nations, this led to their winning the vote. "There are thousands of such women, but a year ago we did not know it," stated British Prime Minister Asquith after Cavell's execution in 1915. A year later, Asquith gave up his long-standing opposition to female suffrage, since "women have aided in a most effective way, in the prosecution of the war." Public opinion in England had changed, largely as a result of women's war work, and this was a major factor in gaining Englishwomen over thirty the vote in 1919. (Men voted at age twenty-one.)[12]

While historians are currently in disagreement about the exact importance of the war in winning women the vote in Protestant European nations, contemporaries associated such changes with the war. German suffrage groups, some of which had been bitter rivals for years, joined together in their first common plea for the vote in October 1918, when the old government was in collapse.[13] The new German Weimar Republic enfranchised women in 1919; the new Russian government had given women the vote in 1917. In the Scandinavian nations, except for Sweden, women had won the vote before or during the war, because of events unrelated to the military struggle; but in other cultures, the vote and the war seemed connected. Belgium enfranchised only war widows who had not remarried, widowed mothers of sons who had been killed in the war, and women who had been imprisoned by the Germans. Sweden, the Netherlands, Austria, and Luxemburg joined Great Britain and Germany in enfranchising women in 1918 and 1919. At the time, the end of the war and the new right to vote seemed to usher in a new era. "The War was in the past and the downtroddenness of women was in the past," remembered Peggy Wood, who was a young woman in Northern Ireland at the time. "Now we were all wise and sensible and had the League of Nations and women were, you know, free. We had education and we could do anything."[14]

This impression of major change being brought to women's lives by the war was reinforced by two factors. First, any change gained immediate publicity. The city of Glasgow's decision to hire two female streetcar conductors in March of 1915 was widely reported in the British press as an important breakthrough.[15] One woman conductor, where there had previously been none, convinced thousands of passengers that the war had revolutionized women's work. Second, there were striking and dramatic changes in women's dress and manners during the war, especially for privileged women. At the time, such changes seemed revolutionary. "In manners, customs, and social conditions the results have been equally, if not more astonishing [as the changes in law]," wrote the Englishwoman Ray Strachey in 1928.

> A hundred years ago a [privileged] girl could go nowhere unprotected; today there is nowhere she cannot go. The chaperon has vanished with the crinoline, and freedom and companionship between the sexes have taken the place of the old uneasy restraint. In work as much as in play, in study as much as in games, there is now little divergence, and scholarship, athletics, travel, tobacco, and even latchkeys and cheque books are no longer the sole prerogatives of man.[16]

The changes in clothing and behavior were the most shocking. Until 1914, women's skirts were ankle-length, as they had been since the thirteenth century. This tradition ended during the war: skirts began to rise as early as December 1914; by the winter of 1915–1916, they were 10 inches off the ground. Despite subsequent fluctuations in length, daytime clothes for women never returned to ankle-length again. To people for whom the glimpse of an ankle had been the subject for music-hall ditties and rowdy remarks, the revelation of the female leg—made all the bolder by the new, transparent silk stockings of the day—was shocking. England's mass-circulation newspaper, the *Daily Mail*, railed against the "extraordinarily short skirt" of 1915, "revealing, as it does, the feet and ankles and even more of the stocking."[17] In addition, underclothing—and the ideal shape of the female figure—changed from the "hour-glass" silhouette produced by a tight corset to a thinner and more "boyish" shape. The omnipresent corset, which by the twentieth century was being worn even by working-class women, was discarded and replaced by the loose chemise or the brassière. (In the mid-1920s the bra itself was temporarily discarded for the flattening breast band, but this style only lasted about five years.) Slimness, rather than just the tiny waist, became prized, and fashionable women began to diet and watch their weight.

In addition to raising their skirts and leaving off corsets, women began to "bob" their hair—not only cutting it, but cutting it short. Like the long skirt, long hair was a traditional insignia of womanhood, and became all the more so in the early twentieth century, when men wore their hair short and were generally clean-shaven. Hair was "woman's crowning glory," and most women never cut their hair in their lives. The resulting mass was heavy, hot, difficult to wash and dry, and potentially dangerous if a woman worked with machinery. During the war years, perhaps for ease, perhaps for fashion, women began to cut their hair short, and this too was a change which did not reverse itself. The "Eton crop," the "Bubikopf," the "gamine" haircut all became standard during the 1920s.

By shortening their skirts and their hair, women gained freedom of movement. Corsets, long skirts, and long hair were cumbersome and restrictive and did not suit such modern activities as riding a bicycle or participating in sports. A long skirt was "a nuisance and even a danger," remembered the Englishwoman Helena Swanwick (1864–1939) of her first time on a bicycle in the 1890s. "It is an unpleasant experience to be hurled on to stone setts [paving stones] and find that one's skirt has been so tightly wound around the pedal that one cannot even get up enough to unwind it."[18] Swanwick's solution was to wear trousers, but in the 1890s she only dared wear them at night. By the 1920s, fashionable women could wear trousers for sports activity without being singular. The 1920s also saw the appearance of the woman athlete who became idolized in the mass media much like her male counterpart. A few of the most famous in the 1920s included Gertrude Ederle (1906–   ), the American who became the first woman to swim the English Channel in 1926 (and swam it faster than any man); Suzanne Lenglen (1899–1938), the French tennis champion who pioneered short sleeves and bare legs for women on the court; and Sonja Henie (1910–1969), the Norwegian Olympic skating champion and subsequently a Hollywood movie star. They gave girls new models of activity to aspire to.

In demanding freedom of action in clothing and sport, women were appropriating two areas previously off-bounds to them. On the one hand, they claimed masculine privileges: short hair, trousers, unhampered movement. But they also claimed the physical freedom of their girlhood, when their skirts were short and their hair was worn loose. Both were threatening to contemporaries, who were further disturbed by women's new use of cosmetics, and freedom in sexual matters.

In the nineteenth and early twentieth centuries, makeup was considered a woman's signal that she was sexually available. Associated with prostitutes, cosmetics were not widely used by "respectable" young women of either the privileged or the working classes. But during World War I, these boundaries broke down. All sorts of women began to wear cosmetics, and now cosmetics were meant to show: bright red lipstick, white face powder, pink rouge. Women mascaraed and shadowed their eyes and painted their fingers and toenails bright red. The time and attention such cosmetics demanded tended to fill up the time gained by wearing simpler or looser clothing, and by the mid-1920s, lipstick, rouge, powder, eye makeup, and nail polish all had become standard among younger urban women. Prodded by the influence of

Hollywood, popular magazines, and mass media advertising, women began to speak of "putting on their face" before they went out in public or on a date.

In addition, sexual mores changed for privileged young women. Previously chaperoned and kept sexually innocent, young women began to be able to go to public dances, to date, to smoke and drink in public, and to be more knowledgeable about the "facts of life." By the end of the nineteenth century, demographic studies show that the upper and middle classes had reduced their birth rates through the use of contraceptive devices. The cheap production of rubber in the 1880s made the condom available to men down through the upper strata of the working class; the simultaneous development of the diaphragm gave an effective means of contraception to women with the money to afford it. (The diaphragm had to be fitted; contraceptive jelly was essential and expensive; and the device had to be washed easily—all of which put it out of the reach of working-class women in this period.) Women's magazines began to advertise contraceptive devices and methods to terminate pregnancy by the 1890s, and by then, well-brought-up young women were mulling over "free love," sexual desire, and even contraception and abortion. "The dawn of a new time is breaking," Karen Horney wrote in her diary in 1903 when she was seventeen.

> A girl who gives herself to a man in free love stands morally way above the woman who, for pecuniary reasons or out of a desire for a home, marries a man she does not love. . . . Perhaps more of the next generation will become mothers, true mothers, whose children are children of love. For how difficult it is today for a young girl to admit that she is having a child [out of wedlock]. The immorality of abortion will cease in that time—which perhaps will never come.[19]

Horney's reference to abortion would not have occurred a generation earlier; a generation later, contraception and abortion had become subjects of public debate. In 1918, the Englishwoman Marie Stopes (1880–1958) published *Married Love* and *Wise Parenthood,* advocating birth control; both became best-sellers and were widely translated. Stopes successfully opened the first English birth-control clinic in 1921; two years later, she received a letter of support from Lady Constance Lytton, who wrote that birth control "has been practiced in our family and by their numerous friends for generations."[20] In Scandinavia and Germany as well as Great Britain, contraception became available through doctors and clinics in the 1920s. The Church of England approved its use in certain cases in 1931, and in 1933 a "Malthusian Ball" to raise funds for birth-control clinics was a success of the London social season and attracted many aristocratic patrons.[21] Catholic nations were far stricter in this regard, but there is evidence that privileged

Catholic women were also starting to use contraception. "I am a practicing Roman Catholic, but am perfectly convinced from a very practical experience—that some sort of Birth Control is *absolutely essential*—and *is* practiced by all *thinking* Catholics," an Englishwoman identified as "Mrs. L.D." wrote Stopes in 1922.[22] By the 1950s, birth rates in Catholic nations were as low as those in Protestant ones, indicating the widespread use of contraception.

The heroine of Victor Margueritte's 1922 novel *La Garçonne* [*The Bachelor Girl*] embodied the "new" woman of the 1920s for many. Monique Lerbier has bobbed her hair, wears makeup and short skirts, plays sports, dances, takes courses at the Sorbonne, and holds an interesting job. Disgusted by her fiancé's decision to continue keeping a mistress, she retaliates in Paris by having love affairs with men and women, drinking, and taking drugs. Throughout, she is presented as a decent person whose own morality has been outraged by the hypocrisy of the sexual double standard. She finds happiness at last in a good marriage to a broad-minded engineer. Shocking at the time, the book was immensely popular: it sold 300,000 copies its first year in print; by 1929 it had sold 1 million in France alone and had been translated into twelve languages.[23] *La Garçonne* epitomized all the changes in the lives of some privileged young women which contemporaries found so disturbing: new clothes, new manners, and new sexual mores. My father "was obsessed by the 'well-bred' young lady idea: it was a fixation," remembered Simone de Beauvoir of this era.

> My cousin Jeanne was the incarnation of this ideal: she still believed that babies were found under cabbages. My father had attempted to keep me in a state of blissful ignorance; . . . he now accepted the fact that I read what ever I liked: but he couldn't see much difference between a girl who "knew what's what" and the *Bachelor Girl* whose portrait Victor Margueritte had just drawn in a notorious book of that name.[24]

De Beauvoir was fourteen when *La Garçonne* appeared in 1922 and was beginning to ask her family for the new freedoms available to privileged women. In her long and varied life as an intellectual, writer, and activist, she both enjoyed women's new opportunities and perceived the limits of those freedoms.

Born in 1908 into a traditionally minded bourgeois French family, de Beauvoir was educated at a Catholic girls' school. Both family and school inculcated traditional ideals of womanhood. "I had the habit of obedience, and I believed that, on the whole, God expected me to be dutiful," she wrote in her memoirs. "My mother's whole education and upbringing had convinced her that for a woman the greatest thing was to become the mother of a family; she could not play this part unless I played the dutiful daughter."[25] Her classmates and the older girls at school aimed at marriage rather than a career:

> Once they had passed their school-leaving certificate they would follow a few lecture-courses on history and literature, they would attend classes at the École du Louvre or the Red Cross where they would learn how to decorate china, make batik prints and fancy bindings, and occupy themselves with good works. From time to time they would be taken out to a performance of *Carmen* or for a walk round the tomb of Napoleon in order to make the acquaintance of some suitable young man; with a little luck, they would marry him.[26]

In families with money, "you either had to get married or become a nun," but the de Beauvoirs, like many other bourgeois families, found their standard of living in decline after the war. They sought to educate their daughter, rather than provide her with a dowry, which they could not afford. "My parents, by helping to push me, not into marriage, but into a career, were breaking with tradition," remembered de Beauvoir; "nevertheless they still made me conform to it; there never was any question of letting me go anywhere without them nor of releasing me from family duties."[27]

Between 1924, when she graduated from school at sixteen, to 1929, when she began her long relationship with Jean-Paul Sartre, de Beauvoir battled her parents over the new freedoms and fashions. When she was eighteen, she joined a social-work group because "it allowed me to spend an evening away from home"; the next year, she bobbed her hair with the encouragement of her best friend. "My mother was furious . . . and refused to allow me the luxury of having my hair set."[28] With no income of her own, de Beauvoir was under her parents' control, but she and her younger sister managed to assert their privacy and freedom: in 1927, when they were nineteen and seventeen, they begged their mother to stop censoring their correspondence. "She replied that it was her duty to watch over the safety of our souls, but in the end she gave in. It was an important victory."[29] The next year, when she was twenty, de Beauvoir tried makeup. "Just before we left home, Madeleine [a twenty-three-year-old cousin] put a little rouge on my cheeks as a joke," she recalled.

> I thought it looked very pretty, and when my mother ordered me to wash it off, I protested. She probably thought it was Satan's cloven hoof-mark she saw on my cheeks; she exorcised me by boxing my ears. I gave in, with very bad grace. However, she let me go out and my cousin and I wended our way to Montmartre.[30]

It was the end of chaperonage which made the greatest difference in de Beauvoir's life. She had first traveled alone when she was seventeen: her mother put her on a train and her hostess met her at the other end. "I was proud of my independence, but slightly worried," she recalled; "at every station I was on the lookout: I should not have liked to find myself alone in a compartment with a strange man." Five years later, she was meeting fellow university students on her own, had gone out on dates, and was walking around Paris alone, even at night. Her overwhelming sensation was one of freedom:

> In the Boulevard de la Chapelle, under the steel girders of the overhead railway, women would be waiting for customers; men would come staggering out of brightly lit bistros; the fronts of the cinemas would be ablaze with posters. I could feel life around me, an enormous, ever-present confusion. I would stride along, feeling its thick breath blow in my face. And I would say to myself that after all life is worth living.[31]

De Beauvoir's privileged status, intelligence, and strong will enabled her to live as free a life as possible for a European woman in the mid-twentieth century. She graduated from the Sorbonne and supported herself thereafter, first as a teacher in an academic high school, later as a novelist and writer. She was able to be sexually active without marrying or being ostracized: her lifelong but unexclusive commitment to the philosopher Jean-Paul Sartre provided a model of an adult relationship between woman and man without the religious, legal, or social framework of wedlock. Contraception and abortion—which although illegal in France were accessible to some women with money and connections—enabled her to have sex without motherhood. Her active participation in the leading intellectual movement of her day—existentialism—led her to political involvement, and her opinions on the important social and political issues of the day were sought out and respected. Like George Sand a hundred years earlier, Simone de Beauvoir became publicly known as a major female writer and thinker, and like Sand she insisted on her personal and sexual freedoms no less than her intellectual and political rights. In many ways, Simone de Beauvoir's life embodied the new options and possibilities gained by privileged European women in the post–World War I era.

But her life also embodied the limits these women faced. De Beauvoir came to believe that all women, even the most privileged, were oppressed by the attitudes of their culture and society toward womanhood itself. In her most famous work, *The Second Sex* (1949), de Beauvoir critically analyzed Western culture as male-dominant and oppressive to women. Rejecting female biological inferiority, she argued that "women are not born but are made," made to see themselves as deficient, as inferior, as secondary. "What particularly signalizes the situation of woman is that she—a free autonomous being like all human creatures—nevertheless finds herself in a world where men compel her to assume the status of Other."[32] De Beauvoir's growing feminist awareness of the limits on the lives of even the most privileged European women points to the other side of the interwar period: the reassertion of traditional values and limits on women's lives.

Traditions Reasserted

The right to vote, equal education, the new sexual freedom, changes in fashion and mores all gave many European women real gains toward equality and freedom. But the period after the First World War also witnessed a reassertion of the traditional limits on women's lives. The powerful nation-states of the interwar years controlled and to some degree reversed women's gains. In both high and popular culture, among supporters of regimes and their opponents, in representative democracies, fascist states, and socialist societies, women's traditional roles and functions were reasserted and often given the force of law. The opposition to women working for pay outside the home, to women making sexual and reproductive choices, and to women achieving equal education and participation in politics was powerful and effective. As in previous eras, traditional attitudes continued to limit the lives of even the most privileged European women.

After World War I (except in Soviet Russia), women were removed from the "male" jobs they had taken up in wartime. Governments which previously had urged women into such employment now used a mixture of penalties and rewards to stop women working for pay outside the home and to open these jobs for the returning men. Starting in 1918, the French government offered women a bonus payment to induce them to leave traditionally male positions.[33] Weimar Germany's new socialist government implemented Demobilization Decrees from 1919 to 1923 which called for dismissals in the following order of priority:

> First: Women whose husbands had a job
> Second: Single women and girls
> Third: Women and girls who had only 1–2 people to look after
> Fourth: All other women and girls[34]

During the inflation of 1923, the Weimar government ordered that all married women in government employ could be fired on the grounds of economic emergency; the following year, the economy improved, but the order remained in effect until 1928. That year, the Berlin Labor Court ruled that the marriage of a female government employee could be automatic grounds for dismissal, a ruling upheld by the German Supreme Court. In 1930, all German parties except the Communist voted to fire married women in government service if their husbands were employed.[35]

Opposition to women working for pay outside the home cut across other political divisions in Europe. In Great Britain, coalition, Labour, and Conservative governments successively barred most working women and virtually all married working women from unemployment benefits which were then being widely extended to men.[36] Margaret Bondfield, the first female English cabinet minister, implemented the most stringent legislation of this sort, in the hard times of 1931. But in the 1920s, Great Britain, like other nations, pressured privileged women to leave the job market. In the early 1920s, Charing Cross Hospital, where Edith Summerskill (1901–1980) had been a medical student during the war, "closed its doors to women [medical students] and later like other hospitals agreed to accept only a small quota. . . . So powerful are all the forces of prejudice and custom," Summerskill added bitterly, "that they have prevailed despite the shortage of doctors." From 1923 to 1935 the London County Council required that all female teachers who married must resign—a requirement common in much of the rest of England as well.[37]

Governments also resisted paying women equally to men, even when they performed identical jobs. Although the Versailles Peace Treaty of 1919 contained articles which promised women equal pay, only the Scandinavian governments even paid lip service to them. In Denmark, an act of 1919 mandated that women employed by the government be paid equally to men, instead of at the old rate of two thirds. Local Danish authorities sabotaged the act by reclassifying jobs, hiring men instead of women in previously female sectors (like the teaching of girls), and trying to have women excluded from "breadwinner" bonuses. A German court decided in 1922 that women teachers deserved to be paid less than men because "men teachers were contributing to the material restoration of Germany by training workmen, whereas the women were only making housewives."[38] Although the International Labor Organization called upon all governments in 1936 to honor their commitment to equal pay made in the peace treaty of 1919, nothing was done until after the Second World War.

Instead, powerful voices demanded that women return to their traditional roles. In 1931, Pope Pius XI called for an end to married women's employment on the grounds that it debased "the womanly character and the dignity of motherhood, and indeed of the whole family, as a result of which the husband suffers the loss of his wife, the children of their mother, and the whole family of an ever watchful guardian." The Pontiff went on to warn that this false liberty and unnatural equality with the husband is to the detriment of the woman herself, for if the woman descends from her truly regal throne to which she has been raised within the walls of the home by means of the Gospel, she will soon be reduced to the old state of slavery (if not in appearance, certainly in reality) and become as among the pagans the mere instrument of man.[39]

As European economies worsened under the impact of the Great Depression, women's work outside the home was increasingly blamed as the cause of men's unemployment. "All things considered, machines and women are the two major causes of unemployment," argued the Italian dictator Benito Mussolini in 1934. "It is necessary to convince ourselves that the same work that causes woman to lose her reproductive attributes furnishes man with an extremely powerful physical and moral virility."[40] More outspoken in their statements about traditional female and male roles, fascist governments sometimes followed the precedent of democracies of the 1920s in removing women from the labor force. In Germany, Weimar labor legislation anticipated some Nazi measures of the 1930s.

Closely connected to attempts to force women from the labor market were attempts to force them to return to their traditional role of mother. In this area as well, fascist and socialist governments followed the democracies in making methods of avoiding pregnancy illegal. France took the lead, outlawing information about and the sale of contraceptive devices as well as abortions in 1920. Subsequent laws passed in 1923 and 1939 strengthened the penalties for abortionists and women who received abortions.

The other side of coercion was encouragement. As modern nation-states outlawed contraception and strengthened penalties against abortion, they also instituted a system of rewards for women who bore large numbers of legitimate children. In 1920, France instituted "Medals of the French Family," in order "to raise the birth rate, which our country must undertake in order to retain the rank in which victory [in the First World War] has placed us and to enable us to enjoy all its fruits."[41] Drawing on a division of awards as ancient as Sparta and Rome, modern European states decided to honor a woman for feats of motherhood as a male warrior was honored for feats of bravery. Like him, she would be able to wear a medal; like his, her medal was not just given to any mother. A Frenchwoman had to have five legitimate children, healthy and over one year old, and she had to be married or a widow to be eligible. (Divorcées, no matter how prolific, were automatically disqualified, as were all mothers of illegitimate children.) Five to seven children earned the mother a bronze medal, eight to ten, a silver one, and over ten, vermeil. A Mother's Day holiday was proclaimed.[42]

Other nations followed France's lead. In 1933, Mussolini equated raising the birth rate with fighting for one's country in his call to "Win the battle of motherhood!" for Italy. Christmas Day, 1933, was proclaimed the "Day of the Mother and Child," and mothers who had borne fourteen to nineteen children could meet Mussolini and be given both a medal and a copy of Pope Pius XI's encyclical, in which he condemned contraception, abortion, and sex in marriage without the intent to procreate, as well as the employment of married women. The same year, the Nazis came to power in Germany: some of their first decrees were concerned with the "birth war," as a fascist book of 1936 called it. New laws made Mother's Day a national holiday, established government loans for married couples where the wife did not work outside the home, and outlawed the sale of contraceptives. Prosecutions for abortion rose to seven thousand a year. Abortions remained illegal, except for "racially undesirable" women, like Jews and gypsies. Eugenic sterilization was mandated for women and men classified as "feeble-minded," "schizophrenic," or physically or psychologically "invalid." "Racially desirable" married couples could erase their marriage loan if they had four children, and in 1939 the government instituted the "Honor Cross of the German Mother," which was awarded to mothers of four or more children and bore the slogan "The child ennobles the mother."[43] Mothers of nine children or seven sons—whichever occurred first—received a special award. Members of the Hitler Youth and the armed forces were instructed to salute a woman wearing the motherhood medal.[44] Hitler himself strenuously reasserted traditional views of women's role. "The use of contraceptives [by Aryan women] means a violation of nature, a degradation of womanhood, motherhood, and love," he told his close associates.

> Nazi ideals demand that the practice of abortion . . . shall be exterminated with a strong hand. Women inflamed by Marxist propaganda, claim the right to bear children only when they desire. First furs, radio, new furniture, then perhaps one child.[45]

Abortion was associated with Marxism because the Soviet Union had been the first European nation to legalize abortion, in 1920. But in 1936, Stalin outlawed all abortion, which had become Soviet women's chief method of avoiding childbirth, since industrial priorities ensured that contraceptive devices were either unavailable or of poor and unreliable quality during these years. In 1944, divorce was made almost unattainable, the distinction between legitimate and illegitimate children was reintroduced, and rewards were instituted for prolific mothers.[46] Five or six children earned a woman a bronze or silver medal; seven to nine gained her the Medal of Honor. Mothers of ten or more children received the title of Heroine Mother, as they still do in the Soviet Union today.[47] Further incentives included bonuses of pay and time off for children over a certain number and earlier retirement for the mother of numerous children. In the late 1930s, Sweden's socialist government began to provide state loans to married couples, free maternity care, a motherhood bonus, tax exemptions, reduced rents, and public housing for large families, all in an effort to raise its birth rate. Other nations followed Sweden's policies. In general, the European maternity legislation of the turn of the twentieth century, instituted to make it easier for poor women to have more children, was added to in the twentieth century. Nations provided and continue to provide rewards and incentives to parents to maintain or raise the birth rate.[48]

Just as the legislation of the interwar years limited the freedoms women won in the period, so the practice of political parties limited the equality women had won. A number of important European nations—France, Italy, Switzerland, Bulgaria—refused to give women the vote until after the Second World War. Others hedged the vote with so many restrictions that the number of women who could actually vote was tiny. In 1925, for instance, Hungary gave the vote only to those women who were over thirty, or were the mother of three children, or had a secondary-school diploma. Belgium and Portugal had equally restrictive franchises for women. In still other nations, women were deprived of the vote when a conservative or fascist government came to power. (Or elections ceased to be held entirely, which deprived both women and men of the vote.) German women voted from 1919 to 1933, Polish women from 1921 to 1932, Greek women from 1929 to 1936, Spanish women from 1931 to 1939.[49] But even in those nations in which women won the vote, traditional values and attitudes ensured that women's participation in politics continued to be limited. Both women and men were under the illusion that the vote alone was sufficient to reverse millennia of subordination and that nothing more needed to be done to achieve women's equality with men. "It will never again be possible to blame the Sovereign State for any position of inequality," claimed Stanley Baldwin, the British prime minister, in 1928, when the government extended the vote to women between twenty-one and thirty (these younger women had previously been considered too irresponsible to vote). "Women will have, with us, the fullest rights. The ground and justification for the old agitation is gone, and gone forever."[50] Women who had spent their lives fighting for the vote tended to agree. "The 'Woman Question' in Germany no longer exists in the old sense of the term; it has been solved," declared Marie Juchacz (1880–1956) in 1919, in the first speech made in the Reichstag by a female deputy. Juchacz, a socialist, asserted, "We women now have the opportunity to allow our influence to be exerted within the context of party groupings on the basis of ideology." "Women's righters" must now become "party politicians," declared Marie Stritt, the former head of the League of German Women's

Associations.[51] The problem with this strategy was twofold: women did not flock to political life, and few welcomed them when they did. Newly enfranchised women voted much the way men did; parties' attempts to win the women's vote focused on traditional "women's" issues: social services, housing, education, and strengthening the family. A few women's parties were formed, but they failed: a Woman's List in a Stockholm election in 1927 received only 0.6 percent of the ballots cast.[52]

Left-wing parties were the most hospitable to women, but even they made little effort to recruit women or to give them important roles in political life. By 1922, 8 percent of the total Communist party membership was female in the Soviet Union; seventeen years later, in 1939, the figure had risen only to 14.5 percent.[53] Most socialist parties merged their women's groups into the larger party organization and expected that the women would follow men's lead and confine themselves to traditional women's issues. (The German and British socialist women's organizations were ended in 1928; the Russian, in 1930.) A study of female deputies to the Weimar parliament showed that they spoke most often on social welfare issues and least often on foreign policy. Greatly outnumbered by male deputies, these women spoke only one quarter as frequently as their male counterparts.[54] The number of women in the Reichstag declined from forty-one to thirty-eight between 1919 and 1930; the Nazi victory in 1933 eliminated all women from the government.

Where socialist women's groups were not merged into the central party structure, they found themselves powerless and isolated. Writing of the French Group of Socialist Women in the 1920s, Germaine Picard-Moch complained bitterly that

> when for example Louise Saumoneau speaks of the Charter of the Party, she has for an audience: Suzanne Buisson, Marthe Louise-Lévy, Citizen Osmin, Germaine Picard-Moch; when Suzanne Buisson speaks about the CGT, she has for an audience: Louise Saumoneau, Marthe Louis-Lévy, Citizen Osmin, etc. . . . and that's the way it always is.[55]

Women who wanted to run for office found themselves up against traditional prejudices. In 1934, Edith Summerskill, by then a physician and Labour party activist, was offered a "hopeless seat" to run for Parliament "because no man wanted it." She accepted the challenge and won the election, becoming one of a handful of Englishwomen members of Parliament for many years. "At this stage of our social progress," she later wrote, "I had to accept the fact that my reproductive organs were what decided my political future, not my experience, knowledge, intelligence, or indeed any other qualification which should commend a potential candidate."[56]

By the 1920s, left-wing governments in power began to appoint a woman to minor cabinet posts, usually ones having to do with areas traditionally considered female. "I have yet to find any Government of any political colour," wrote Summerskill, "which does not feel that it has satisfied its obligations to women by appointing *one* woman only of the highest rank."[57] Early female cabinet ministers were orthodox party members who consistently put their party's needs above those of women when a conflict occurred. Token women in government, they were as powerless as token women in the professions or any other institutions. "The fact that there is a woman here and there has not altered the status quo in the slightest," wrote the German psychologist Alice Rühle-Gerstel in 1932.[58]

In the face of such minor gains, older women who had spent years fighting for women's rights often became disillusioned, and such disillusionment became widespread among activist women in the interwar years and was intensified by the rise of fascism. The triumph of fascism, first in individual nations, later on the battlefields of Europe, ensured that millions of women and men would have to confront both fascist tyranny in general and the fascists' particular view of women's roles. For women, fascism in theory reasserted the most traditional limits on their lives; in practice, it often demanded new sacrifices and public service. Privileged women especially found their lives circumscribed by fascist values and prescriptions, but their status and wealth (unless they were in "racially undesirable" groups) protected them from the government's attempts to make them enter the labor force or military service.

> We want to have wives again,
> Not playthings decorated with trifles.
> No woman in a foreign land has
> The German wife and mother's gifts.[60]

In both their early propaganda and later pronouncements once in power, the German Nazis paid lip service to the ideal of the traditional woman, which in practice they restricted to women of the relatively privileged classes. While the Nazis initially claimed they would remove all married women from the job market, in practice their efforts were directed at professional women—some unmarried—in the higher echelons of the economy. The Nazis immediately dismissed all female school administrators and some other female civil servants from government employ: Gertrud Bäumer, the conservative, unmarried head of the League of German Women's Associations, was the first. This firing gave a signal to other active women to retreat from public life. A quota of 10 percent for female students was set in the universities. Professional women were also stigmatized in a variety of degrading ways: in

1937, for instance, it was ordered that women with advanced degrees could no longer be addressed with the honorific "Frau Doktor"—that title was now reserved for the wives of male doctors or professors.[61] When the Nazis appealed to women's supposed propensity for self-sacrifice, women, especially privileged women, did not respond as expected. Women in general supported the Nazis in lower numbers than men; middle- and upper-class women in particular were able to avoid Nazi attempts to mobilize their services as the war intensified.[62]

In World War II, middle- and upper-class women remained relatively sheltered from the worst effects of the war as long as their money and social position could free them from hardship. Treated relatively mildly unless they were Jews, middle- and upper-class women were usually spared the harshest impact of the war and occupation. Some took active roles in fascist movements, although their efforts were discouraged and they were eliminated from positions of power once fascist rule was established. Others collaborated with fascist governments or armies of occupation. Still others joined resistance networks and aided the struggles against the Germans and Italians.[63]

As nurses at home or the front, as soldiers in home defense units or partisan groups, or as combat-support staff, some women of the upper and middle classes played active parts in the war. A few resumed roles they had left after the First World War: a spy in 1916, Marthe Ricard also worked in the French resistance in the Second World War when she was in her fifties. In all the European resistance networks, women played important subsidiary roles: carrying messages and goods, helping to print underground journals, smuggling children, Allied soldiers, and Jews to safer locations. They were usually punished as severely as men by Nazi occupation forces, as in the case of Edwige de Saint-Wexel (1923–). Seventeen when the Germans occupied northern France in 1940, Saint-Wexel joined a student protest in November at the Arc de Triomphe in Paris. Arrested by the Gestapo, she was beaten, burned, and jailed for three months in solitary confinement with a dislocated ankle. Starved and not allowed to wash, she was released, crippled and filthy, to show her family and others the folly of resistance. Returning to school, she joined a resistance group—risking recapture—and worked throughout the war smuggling Jews and British pilots out of the country. She escaped rearrest and today is a bank executive in Paris.[64]

Saint-Wexel and other women like her functioned as equals in resistance organizations. One of the most active female leaders of a resistance network, Marie Madeleine Fourcade (1909–) found that her sex enabled her to escape suspicion in a world which still considered women passive and subordinate. Fourcade was a thirty-year-old former wife of an officer and mother of two when she was asked by a French specialist in military intelligence to "organize the underground side" of an intelligence network in France in 1940. "I'm only a woman; who will obey and follow me?" she remembered asking. "That's a good reason to use you," he replied. "Who would suspect a woman?"[65] Between 1940 and 1944, Fourcade successfully organized and directed one of the largest and most successful resistance organizations in Western Europe. Forced to rebuild a number of times because of spies and betrayals, Fourcade passed invaluable information, as well as hundreds of British and American pilots downed in France, across the Channel to England. The men who met her as secret agents or co-conspirators were shocked that "Hedgehog"—her underground nickname—was a woman. Fourcade herself used this prejudice against women to her advantage when she was finally captured by the Germans in 1944. Placed in a cell in a dilapidated local prison in the south of France, she was able to squeeze out through the bars in the middle of the night, carrying her dress in her teeth. Making her way through back roads and villages, she struggled to reach a château to warn her second-in-command. "I found myself in a field in which some old peasant women were busy gleaning," recalled Fourcade.

> I began to do the same, picking up ears of corn and bits of dandelion. Out of the corner of my eye I could see the German soldiers setting up road blocks, striding up and down the bridge, stopping all the women who went over it and checking their papers. But they paid not the slightest attention to the gleaners below. I moved forward, bending double.[66]

Fourcade crossed successfully and was able to alert her male subordinate.

For most European women, even privileged women, the war created a climate of its own, in which normal life was suspended for the duration. "It was long, very long," remembered Françoise Giroud, then a French journalist, of the German occupation of France,

> and sometimes unspeakable. Any situation which offers a great number of people the opportunity of behaving badly is one to be avoided at all costs. When cowardice, venality, and betrayal become not only useful, but are also esteemed and rewarded, they come into their own, it's only to be expected. The same people who under normal or unpressured circumstances would never show this side of themselves are triggered into action.[67]

For women in Western Europe especially, largely spared the worst horrors of Nazi brutality, the paramount memory of the war years was of shortages—shortages in food, in clothes, in consumer goods generally. "Anyone who wasn't in France in those days cannot understand what it means to be hungry for consumer goods," recalled Giroud. "From nylon stockings to refrigerators, from records to automobiles . . . in 1946 in France there was literally nothing. If I remember correctly, we even had to have ration tickets for bread." Many remembered their overwhelming desire to return to normal life. "As the last guns rumbled and the last all clear sounded all the squalor and discomfort and roughness that had seemed fitting for so long began to feel old fashioned," wrote the Englishwoman Anna Scott James in 1952. "I wanted to throw the dried egg out of the window, burn my shabby curtains and wear a Paris hat again. The Amazons, the women in trousers had had their glorious day. But it was over. Gracious living beckoned once again."[68] Just as after the First World War, there was a brief "baby boom," temporarily reversing the overall decline in the birth rate. In "the post-war era, I lived egocentrically, concentrating on my private life," recalled Giroud.[69] In this, she and tens of thousands of other European women continued to fulfill traditional expectations about women's role and function.

The defeat of fascism and the revelation of genocide discredited fascist policies and ideology in much of Europe after the Second World War. But until the mid-1940s, fascism looked like the winning side. As one European nation after another came under fascist domination, material reality seemed to support traditional ideals of women's role. Allied victory did little to dispel these ancient beliefs which had been so forcefully rearticulated by the fascists. Throughout the interwar years and beyond, wider conceptions of womanhood were on the defensive, if not defeated, and both political and military battles seemed to point to a return to traditional roles and values for women. Unfortunately for women, so too did some of the most progressive elements of European culture.

The Mixed Blessings of Psychology

In the 1920s and 1930s, European women who wanted wider roles came under attack not only from the most regressive elements of their culture, but also from some of the most progressive and potentially liberating cultural movements. The role of psychology and the new psychoanalytic movement was crucial. On the one hand, it acted as a liberating force, rationally examining previously "unmentionable" topics. Masturbation, sexual molestation, homosexuality, fetishism, masochism and sadism, and the inability to attain orgasm became subjects of public discourse, and the gain to those who had previously suffered in isolation was immense. The exploration of the unconscious; the delineation of the mind's logic and dynamic; the invention of the psychoanalytic method; the new awareness of the importance of the past in general and sexuality in particular to the development of the human personality changed the way Europeans thought about both the world and themselves. On a personal level, analysis and therapy enabled thousands of women and men to lead more fulfilling and productive lives. The "talking cure," which

she and her therapist invented, enabled the hysteric "Anna O." of Josef
Breuer's and Sigmund Freud's early studies to develop into Bertha Pappenheim (1859–1936), the head of the German Jewish Women's Organization
and a leading feminist in twentieth-century Germany. "Analysis has helped
me unspeakably," wrote Karen Horney to her therapist, Dr. Karl Abraham,
in 1911.[70]

On the other hand, the "new" values and behavior prescribed for women
by psychology were often indistinguishable from traditional, limited conceptions of women's nature and function. For women, psychology, like so many
other progressive intellectual movements before it, reiterated in new language
the oldest and most limiting strictures about their lives. It bolstered the
ancient male-dominant traditions anew with the authority of the youngest of
the sciences. Moreover, psychology, claiming to be the science of the human
psyche and its motives, confronted its critics with a closed system. Those who
disagreed with some of its axioms were attacked for being unaware of their
own unconscious motives. "We shall not be so very greatly surprised if a
woman analyst who has not been sufficiently convinced of the intensity of her
own desire for a penis also fails to assign adequate importance to that factor
in her patients," wrote Sigmund Freud in one of his last works, *An Outline
of Psychoanalysis* (1938).[71] Undermining to professional therapists, this attitude also restrained those women who, whether directly exposed to psychoanalytic teachings or not, found their lives newly circumscribed by the
authority of psychological findings.

The restrictive nature of psychology with regard to women came from two
main sources. First, psychology, like all European intellectual movements
before it, took man as the standard and saw woman as an inferior variant of
the male. Karen Horney protested this view: "Till quite recently the minds
of boys and men only were taken as objects of investigation," she wrote in
1926.

> Psychoanalysis is the creation of a male genius, and almost all those who have
> developed his ideas have been men. It is only right and reasonable that they
> should evolve more easily a masculine psychology and understand more of the
> development of men than of women.[72]

Second, while Freud occasionally warned against confusing natural instincts
and the teachings of a particular culture, both he and his followers ignored
these warnings often enough to create a psychology which upheld the most
patriarchal traditions of European culture, rather than critically analyzing
them. Freud and most of his followers drew on European culture to confirm
their theories of the mind. Most of this cultural heritage was accepted
uncritically, and in the process some very ancient ideas about women were

revived and given new life.[73] That women were "incomplete" men, that civilization was a male creation to which women were "hostile," that women's moral authority and sense of justice were less than men's, that women's "biological destiny" was motherhood, that women were most fulfilled when they gave birth to a boy—all of these theories were given new authority in the work and writings of Freud and many of his fellow psychoanalysts. By rearticulating traditional views which subordinated women and by giving them modern medical and scientific authority, psychiatrists and psychoanalysts helped to buttress and strengthen the traditional view of woman: that her existence and being should be defined by and dependent upon a stronger, wiser, and, ultimately, healthier man.

The work of Sigmund Freud was of central importance, both because of his virtual invention of psychoanalysis and his continuing influence as a thinker and culture hero. Freud consistently argued that "penis envy" was the central concept in the development of female personality and that it determined the girl's "peculiar" (when compared to boys) formation of the Oedipus complex. Arguing that both boys and girls saw women as "castrated men," Freud theorized that the boy resolved his fear of being castrated by identifying with his father and breaking his earlier identification with his mother. The girl, in this scenario, has a far more difficult passage: she must accept that she and her mother have already been "castrated"—they do not have and never will have a penis. Girls fall "victim to 'envy for the penis,' which will leave ineradicable traces on their development and the formation of their character and which will not be surmounted in even the most favorable cases without a severe expenditure of psychical energy," he wrote in his essay "Femininity" (1933), which summarized his thoughts on women. This penis envy must be successfully transformed into, first, sexual desire for a male penis in the vagina, and second, maternal desire for a child—which Freud viewed as a penis substitute—in the womb. For Freud, female development was also more complex and difficult than male development for two additional reasons: the necessity of coming to terms with masochism and the need for a girl to transfer her sexuality from her clitoris to her vagina.

Although Freud occasionally warned against a crude association of passivity with feminity and activity with masculinity, he often ignored his own warnings and tended to use these ancient stereotypes himself. "One uses masculine and feminine at times in the sense of *activity* and *passivity*, again, in the *biological* sense, and then also in the *sociological* sense," he wrote in an essay on puberty. "The first of these three meanings is the most essential and the only one utilizable in psychoanalysis." Building on this, Freud tended to associate sadism with masculinity and masochism with femininity. Believing that femininity has "some secret relationship with masochism," Freud identified masochism as "feminine" when it appeared in women or men.[74] Following the same logic of stereotypes, Freud and a number of his followers considered clitoral stimulation a "masculine" activity and argued that the healthy and mature woman will "transfer" the source of her sexual pleasure from an "active" stimulation of her own clitoris (itself considered a "masculine" organ because it was seen as the analogue of the penis) to the relatively "passive"—and therefore more "feminine"—reception of the penis into her uniquely female organ, the vagina. "Along with the abandonment of clitoral masturbation a certain amount of activity is renounced," Freud wrote approvingly in "Femininity." "Passivity now has the upper hand, and the girl's turning to her father is accomplished principally with the help of passive instinctual impulses."[75] A woman's failure to develop in this way was seen as immature or deviant development. Freud suggested that such a failure might be the result of too much clitoral masturbation when she was a child. Reiterating another European tradition (and one particularly strong in the Jewish culture in which he was raised), Freud asserted that the high point of a normal woman's life came when she gave birth to a son; when, in terms of his theory, the woman created a penis substitute of her own and solved her problem of penis envy. But even this "solution" left women inferior to men in many crucial areas. "The fact that women must be regarded as having little sense of justice is no doubt related to the predominance of [penis] envy in their mental life," he wrote in his 1933 essay summarizing psychology's findings about women. "We [psychologists] also regard women as weaker in their social interests and as having less capacity for sublimating their instincts than men."[76] Such beliefs were as ancient as European culture itself.

While Freud resisted making such assertions into dogma, many of his followers and popularizers—some of them women—did not. Instead, they expatiated on the most repressive and limiting of his ideas about women. Both Princess Marie Bonaparte in Paris and Helene Deutsch (1884–1982) in Berlin argued that since biology necessitated that women feel pain at various points in their lives (menstruation, the breaking of the hymen, childbirth), the ability to derive pleasure from pain was a mature female adjustment to reality. Both analysts stressed the innateness of female masochism and the "immaturity" of clitoral sexuality. Others formulated new theories, but continued to rely on traditional stereotypes. Carl Jung's investigation of the mythic images common to all cultures emphasized the importance of female and male archetypes in understanding the human psyche; again, the age-old distinction between male action and female passivity was retained in a new conceptual model of human nature, and traditional judgments were reasserted. "No one can evade the fact, that in taking up a masculine calling, studying, and working in a man's way, woman is doing something not wholly in agreement with, if not directly injurious to, her feminine nature," Jung wrote in a 1927 essay, "Woman in Europe." "Those women who can achieve something important for the love of a thing are most exceptional, because this does not really agree with their nature. The love of a thing is a man's prerogative."[77]

Psychology also undermined female authority in the areas of child rearing and sexuality. Nineteenth-century "experts" on motherhood were usually other women, who claimed authority from their own womanhood or experience as mothers. Twentieth-century "experts" tended to be male psychiatrists and doctors. Concerned first with the physical health of children, their interest expanded to include mental health as well, and in most instances the wisdom and authority of the mother herself were criticized and undermined. "About 30 years ago the maternal instinct was considered an infallible guide in the upbringing of children," wrote Karen Horney in 1933. "When this proved inadequate it was followed by an equally overstressed belief in the theoretical knowledge about education."[78] By prescribing authoritatively in areas previously left to mothers and nurses, psychiatry completed the takeover of motherhood by male "experts" which had begun in the nineteenth century.

Not only did mother no longer "know best" how to raise children, now her practices were ridiculed or condemned as ignorant and harmful. By identifying women primarily as mothers and simultaneously diminishing the mother's ability and authority, doctors and psychologists undercut woman's traditional territory without offering her more room in what was still seen as a "man's world." From advocating rigid feeding schedules to regulating the timing and method of toilet training, experts sought to prescribe how best to mother. Women who did otherwise were criticized for hindering their child's development, if not worse. Just like the eighteenth-century philosophers who attacked mothers for failing to breast-feed their own children, twentieth-century experts criticized women for warping their children and so contributing to society's decline if they did not nurture correctly. Thus, for women, the genuine health gains in some new child-rearing practices were offset by the erosion of traditional maternal authority.

The same was true with regard to sexuality. From the third quarter of the nineteenth century, "sexologists" (all of them male) sought to delineate and study what was seen as "healthy" and what was seen as "deviant" sexuality. By the beginning of the twentieth century, sexologists and doctors began to be concerned about privileged women's relative lack of sexual pleasure. Their solution was twofold. First, women should cultivate their "femininity" and abandon sexual practices identified as "masculine" (like masturbation or clitoral stimulation). Second, men should take responsibility for their wives sexual pleasure. "There is no such thing as a frigid woman, there are only incompetent and insensitive men," went a slogan of the German Sex Reform League in the 1920s.[79] While such an attitude was an improvement on earlier views which often ignored women's sexual pleasure, it gave the male ultimate authority in the sexual act and reinforced women's supposed sexual passivity.

Increased sexual sophistication could lead to new limits on women's behavior as easily as it led to new freedoms. New knowledge about masturbation might be accompanied by new and authoritative warnings about allowing girls to practice it. New knowledge about the female orgasm might lead to new guilt about being "frigid" or "immature." The demand that privileged women now be sexually knowledgeable could be experienced as being just as coercive as the previous insistence that they be sexually ignorant. The double-edged nature of increased sexual knowledge for women is most clearly seen in the change of attitudes about female homosexuality.

Traditionally, European culture had condemned male homosexuality and ignored female homosexuality. Lesbians were rarely persecuted unless they passed as men. By the eighteenth century, the northern European assumption that privileged women were less highly sexed than men fostered the view that sex between women could not exist, and women thus enjoyed a far wider range of behavior in their relations with women than men did with other men. Women regularly slept in beds together, spent months in each other's company, and expressed both love and passion for each other without being criticized. When Lady Eleanor Butler and Sarah Posonby, both from respected Irish families, eloped in 1778 to live together, they were considered a romantic curiosity rather than perverted degenerates.[80] When in 1811 two female Scottish schoolteachers were accused of improper sexual contact with each other, they were acquitted on the grounds that "the crime here alleged has no existence."[81]

This ignorance had ended by the beginning of the twentieth century. New words appeared to describe sexual love between women, and "lesbianism" "Sapphism," "tribady," "Uranism," and "inversion" were all coined to describe what was increasingly attacked as an "unnatural" way of life. Sexology, popular stereotypes, and the new sophistication brought by psychoanalysis led to a new wariness about lesbianism and had the effect of curtailing the range of behavior allowed to women of privilege. Novels like the Englishwoman Clemance Dane's *Regiment of Women* (1917) and films like the German *Madchen in Uniform* (1931) stressed the dangers of lesbian teachers in all-girl schools. Mothers were now warned to guard against such tendencies in their daughters and to protect them from the sexual aggression of women as well as men. By the 1920s, crushes on other women, a too-boyish or

-mannish style of dress and behavior, even an "unfeminine" ambition to succeed in a male profession were interpreted as warning signals of sexual "inversion." While later generations romanticized the life of lesbians in the 1920s in capitals like Paris and Berlin, those who moved in those circles stressed the limits. "In spite of the freedom, or illusion of freedom, in the Weimar Republic," wrote the lesbian doctor Charlotte Wolff of her life in Berlin in the twenties, "lesbians were watched by the police, and from time to time lesbian clubs were raided. . . . It was an ambiguous situation all around."[82]

From Domesticity to Gracious Living: Privileged Women Since 1945

Even after the Second World War, privileged women found many traditional attitudes still firmly in place. Nothing in the postwar era diminished the prestige of motherhood for women nor its appropriateness as an ideal for all women. Writing in France in the 1960s, Béatrice Marbeau-Cleirens complained that "since woman is able to be a mother, it has been deduced not only that she should be a mother, but that she should be nothing but a mother and can find happiness in nothing but motherhood."[83] What changed in Europe since 1945 was not the traditional attitudes about women's proper role and function, but rather a rise in material wealth which only increased responsibilities and obligations for privileged women. The rise in the standards expected in child rearing and housekeeping was accompanied by the disappearance of the female domestic servant who traditionally helped to perform such work. Unless they move in the wealthiest circles, privileged European women have had to provide child care and do housework alone in the home with some new help from technology. At the same time, both the economy and social expectations have placed pressure on them to work outside the home as well. One result has been a sense of strain among those women who considered themselves and were considered by their society to be relatively privileged and well-off. "There is growing anxiety and depression among ordinary women which results from the impossibility of living up to newspaper and magazine propaganda about women who effortlessly organize a career (not a "job" you note), husband and children while still finding time to keep a 24-hour-perfect hairstyle," wrote the Englishwoman Shirley Conran in 1978.[84]

Conran's solution was that arrived at by other European women: put family and home first and job second. "I don't believe women have (or should have) the job-pride which men have and can more easily shift 'downwards' in employment, especially if it's going to improve the quality of home life for the children," she reasoned.[85] Sociological studies of European women confirm this decision. "We know from surveys that most young girls of today orientate themselves to a future with the family as their main occupation and employment as a supplementary job," concluded a study of West Germany in the 1970s.

> With very few exceptions, they were also trained for an existence in which the family is the focal point. Basically, hardly anything has changed since grandma's times. For this reason, the mass rejection of the role of housewife cannot be anticipated.[86]

A study of women in postwar France reached similar conclusions, finding that Frenchwomen were still often forced to choose between a family and a career and were penalized in their careers if they attempted to be working mothers in a society which provided few support services for them. A postwar English survey found that girls continued to follow very different employment strategies from boys of the same social class, from their own brothers.

> Sons of middle-class families will as a rule adopt middle-class occupations. This is not generally so among the daughters of the upper and middle classes, who can frequently be found doing the same type of work as girls of humbler background . . . the result of a wide-spread attitude among girls who regard their gainful employment as a temporary phase.[87]

The mass of privileged European women have accepted the traditional view that a woman's chief duty is to her husband and children. If women work outside the home, the work should be subordinated to the needs of their families. This "double burden" of family duties and labor is the primary reason why privileged women, no less than their working-class counterparts, have failed to reach job or pay equity with men. Professional women, whether doctors, lawyers, scientists, or politicians, cluster in the lower and middle ranks of their fields.[88]

Exceptional women, like Marie Curie, have been able to perform both jobs successfully. But even the most successful have testified to the strains involved, strains their male counterparts never felt. Edith Summerskill, a physician, mother, and member of Parliament for much of her life, wrote her daughter Shirley in 1955 of the pressures she experienced:

> I remember well, when you and Michael had certain childish complaints and I was compelled to stay in the House of Commons for a long sitting, how my mind wandered from the speaker and I would creep out to phone Nana for the latest. Women have their physical and mental reserves of strength tapped every day by

these demands on their very finest feelings—their unselfish love and devotion to their families.[89]

Shirley Summerskill Samuel (1931–) followed in her mother's footsteps and also became both a physician and a member of Parliament. But women of the most exceptional talent and energy have often been able to triumph over adverse social circumstances and negative attitudes about women. "A Jenny Lind cannot be stopped in her singing, nor a Siddons [Sarah Siddons, a famous English actress] in her dramatic career, nor a Currer Bell [Charlotte Brontë's pseudonym] in her authorship, by any opposition of fortune," declared the English reformer Harriet Martineau (1802–1876), in 1859. "But none of us can tell how many women of less force and lower genius may have been kept useless and rendered unhappy, to our misfortune as much as their own."[90]

In the last quarter of the twentieth century, as in Martineau's day, most privileged European women continue to choose traditional roles and embrace traditional values. "I don't think women were made to be completely equal," an English housewife stated in the early 1970s. "Most of us like to be feminine, and feel that someone is going to give up a seat on the bus for us!"[91] A study of West German textbooks used in public schools in the 1970s showed little girls and boys still reading such pronouncements as, "A girl is almost as good as a boy."[92] The most advantaged European women continue to define their lives in terms of their relationships with men. Their husbands' success is their success; their husbands' goals, their goals. They remain primarily wives and mothers. No longer restrained by laws, education, or the economy, women recreate traditional attitudes to justify their lives. "Men are our legislators, our employers, our husbands, our lovers, our lives would make no sense without them," the French women's magazine Elle told its readers in 1970. "To persecute them, to hurt them would do nothing but make them run away."[93] Today, as in ages past, the most successful strategy for a privileged European woman is to ally herself with a successful man, given a society and culture which still value boys and men more than girls and women.

VIII

WOMEN OF THE CITIES

•

MOTHERS, WORKERS, AND REVOLUTIONARIES

1

FAMILY LIFE

❦

Two Lives

Lucy Luck (1848–1922) and May Hobbs (1938–), born ninety years apart, were both poor Englishwomen who spent most of their lives in London. Living in a large modern city, their experiences differed from those of women in earlier periods: the metropolises of the late eighteenth and nineteenth centuries provided a new environment as well as new opportunities and dangers for the increasing numbers of European women who came to live in them. Their two lives exemplify many of the changes and constants in the lives of urban, European working-class women during the nineteenth and twentieth centuries.[1]

Lucy M. (she concealed her maiden name in her memoirs) was born in Tring, a large town about thirty miles northwest of London. Her father, a bricklayer, deserted the family when Lucy was three; she described him as "a drunkard and a brute." Lucy's mother, already ailing with the cancer that would kill her ten years later, was left with four children: Lucy's crippled nine-year-old sister, her six-year-old brother, and an infant boy. There was no way Lucy's mother could support the children herself, so she applied to the parish for relief. By 1848, England had ended the earlier system of "outdoor relief" in which the needy were given charity in their own homes; instead, Lucy's mother was told she had to enter the local poorhouse to receive any assistance. She did so as a last resort—throughout working-class memoirs of the period runs the dread of going on public assistance and living in the prison-like conditions of the wards—and Lucy remembered that when her mother reached the poorhouse (carrying her baby and walking the other children the five miles from her home), she "sat down on the steps with one of us on each side of her, and one in her arms, crying bitterly over us before she took us in."[2]

Separated from her mother, Lucy M. grew up in the poorhouse. (Children were routinely separated from parents to discourage people from entering.) Her most vivid memories were of the horrible food:

> We used to have tin mugs for our gruel or milk, whichever we had, and wooden spoons. One morning, my mug was half full of dry crumbs and the half-cold thick gruel did not wet it. I leave you to guess what it was like when I stirred it up. I have thought of it many times, particularly when I see paperhanger's paste. You may think this is a lot of foolish talk, but it is the truth.[3]

Lucy M. learned to read and write in the poorhouse, but when she reached age eight, she, like the other poorhouse children, was considered to be old enough to start earning her living. She was sent to work in the Tring silk mill and boarded with a family paid by the parish to keep her. "The first day I went to work I was so frightened at the noise of the work and so many wheels flying round, that I dared not pass the rooms where men only were working, but stood still and cried," she remembered. "But, however, I had to go. . . . I was too little to reach my work, and so had to have what was called a wooden horse to stand on."[4] Regulations called for children under eleven to work only half a day, but Lucy found she was expected to work with her foster family at home in the afternoons and evenings. In one household, she sewed; in another, she plaited straw used to make hats. At thirteen, she was considered able to support herself, and the parish payments to her foster families ended.

For the next five years, until she was eighteen, Lucy M. lived a wandering, rootless life. She first worked as a domestic servant in various homes. In the first, her mistress dismissed her because she was too young; in the second, she was "treated like a dog"—forbidden to sit down for her meals and called "Ann" because her employer thought it a more suitable name for a servant than "Lucy." In her third post, the father of the family "did all he could, time after time, to try and ruin me, a poor orphan only fifteen years old."[5]

She left domestic service and became an apprentice straw-plaiter, working for her room and board. But straw-plaiting was seasonal work, done only from January to June. That summer of 1864, Lucy M.

> wandered about, trying for work, sometimes getting just enough to get some food; but I could not get any more and my boots were almost worn from my feet, often causing me to go wet-footed. How often I was tempted to lead a bad life, but there always seemed to be a hand to hold me back.[6]

Women's wages were so low that a "bad life"—crime, prostitution, or being kept by a man—was a real temptation. Like most unmarried women who had to make money, Lucy M. found she needed a position which provided her with shelter and food as well as wages: there was no noncriminal way she could support herself alone.

At eighteen, she found such a job, sewing in a workroom and boarding with the family who ran it. She began to be courted by Will Luck, a twenty-one-year-old plowman. Her new job threatened her eyesight: "I had sat night after night at work until 11 or 12 o'clock, using a rushlight candle, and my eyes had begun to get so bad that I could hardly see."[7] Her master advised her to marry Will, telling her, "he will never hit you." Alone, Lucy M. would have had to arrange another post for herself; married, she had the security of her husband's income as well as her own. In these circumstances, she consented to marry. Will "was a steady, saving chap," she wrote in her memoirs—her only description of her husband.

Married in December 1867, Lucy Luck was pregnant by June 1868. "Things went on in a general way," she wrote, "living happily together but working very hard [she continued her sewing and straw-plaiting], and we found it difficult to keep things straight, and as time went on I had three little girls."[8] The third daughter died at eight months; the second daughter caught the croup when she was four. As the girl lay dying, her anguished mother remembered that

> six weeks before this she had done something wrong. I beat her and told her the bad man would have her, and when she was dying she said, "You beat me and told me the bad man would have me. Is he going to have me now?" Just imagine my feelings then! I said to her, "No, he will not have you," and she said, "No, I am going to heaven with Sally, ain't I, Mother?" [Sally was her friend who had just died]. . . . At 11 o'clock on Monday morning, she called for her father. He was fetched from the plough-field and she clung around his neck.[9]

She died that afternoon. When the couple's fourth child was born, the family needed more than Will Luck's thirteen shillings a week to live on. When he asked for a raise (his first, and his wages were low by contemporary standards), he was told to leave in a month if he was not satisfied with his job. Leaving the job meant leaving the cottage where they lived, but Will Luck "had thoroughly made up his mind to go to London, and try to get work there."

Cities grew in the eighteenth and nineteenth centuries because of decisions like this one: more jobs could be found there than in the countryside. Will Luck went to London where he soon found work as a horsekeeper for a railway; Lucy Luck followed, bringing the children to lodgings in Paddington Green. "I shall never forget my first two or three months in London," she wrote.

I think I cried most of the time, for my husband was on night work, and I amongst strangers and thinking of my poor child I had so recently buried. I would have given anything to go back to the country.[10]

But she made the transition and soon arranged to do straw-plaiting in London for the same firm which had employed her in the country. She summarized the rest of her life briefly: "I have been at work for forty-seven years, and have never missed one season, although I have a large family [she eventually had seven surviving children]. . . . In my busy seasons I have worked almost day and night."[11] In the months between July and January, when there was no straw-plaiting work, Lucy Luck continued to help support her family: "I have been out charring and washing, and I have looked after a gentleman's house a few times, and I have taken in needlework."[12] She bought fabric on sale and made her children's clothes. She prided herself that she had done her "best to bring them up respectable," and when she was in her seventies, it was her children who persuaded her to write down "her life's story."[13] Her daughter remembered that "Mother used to sit and write at night when she couldn't sleep."[14]

In her ability to write, her success in raising seven children to adulthood, and her own longevity, Lucy Luck's life embodied advances in the standard of living which had reached even Europe's poorest women by the nineteenth century. A hundred years later, May Hobbs benefited from further advances.

Like Lucy Luck, May Hobbs was born to poor parents: her mother worked as a presser and machinist in a men's suit factory; her father was a foreman in a brewery. Unlike Lucy Luck, however, May Hobbs was the first child in her family and was born in a hospital rather than at home. In Luck's day, a woman ran a 34 in 1,000 chance of dying in childbirth or shortly afterward in a hospital; by 1938, when Hobbs was born, hospital births were as safe as home births for normal deliveries and safer if there were complications. Like virtually all Europeans of her generation, Hobbs's life was affected by World War II: at age two she was evacuated to the country, where she lived until she was six. When she returned, she found that her parents "could not have me or did not want me. Even after all these years I have never found out the exact reason."[15] Like Luck, she was raised by foster families; unlike Luck, she grew to love one of her foster mothers "as a mother." By Hobbs's day, local parish charity and the poorhouse had given way to the "welfare state," which subsidized and supervised the needy at home.

In addition to benefiting from a century of social reform, Hobbs was also the beneficiary of a century of educational reform. In Lucy Luck's day, free public education did not exist in England; by May Hobbs's time, education was free, public, and compulsory for all children fifteen and younger. Child

1. An English servant, c. 1865.

2. A weaver and loom overseer in an English textile factory, 1909.

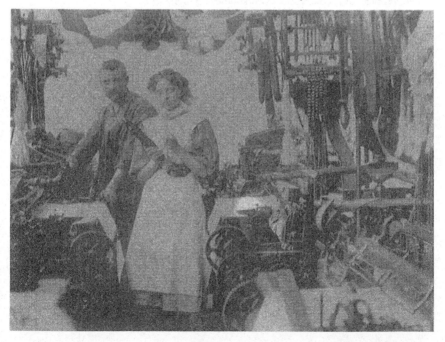

3. Revolutionary Parisian market women marching to Versailles in October 1789.

4. Women in the St. Pancras workhouse, London, 1901.

5. Temporary crèche established by the Soviet Union, early 1920s.

6. Nazi card, Germany, 1930s. The text reads: (*left*)
"Honor card for German mothers with many children. Protecting the mother is the noble duty of each
member of the German [racial] community." (*right*)
"Mother is the most beautiful name in the world."

7. Mother and child leave their bombed-out home. England, early 1940s.

8. Captured Jewish resistance fighters from the Warsaw ghetto. Poland, 1943.

9. Schoolgirls in France, 1957.

labor had been ended. Hobbs had six more years' schooling than Luck, and although she found school boring and alienating, she learned to write fluently. Her schooling differentiated between girls and boys: "When I asked to do metalwork and woodwork at school," Hobbs remembered, "I was told that only boys do those lessons. Maybe I only asked because I was so bored with cookery or maybe I was having a bit of rebellion."[16] She remained in the "domestic science" program and fantasized about being a secretary, but when she left school, she lacked the qualifications for secretarial work and ended up in a series of factory jobs. Factory work enabled single women to avoid domestic service, but Hobbs found that even in the 1950s, when she was a teenager, factories fired women when they married. She worked in food and light goods manufacturing—areas traditionally female and traditionally paid less than jobs in heavy industry.

Luck prided herself on not having resorted to crime, even when hungry and ragged; Hobbs made no clear distinction between poverty and crime and at twenty spent a month in Holloway prison for receiving stolen goods. She admired women like her foster mother—or like Lucy Luck—who strove to keep their families together no matter what the labor involved. "It was always the mother that bore the brunt of all the worry," she wrote,

> that saved and plotted to keep children fed, that found new lodgings when the landlord turfed them out. It did not matter whether your mother was a brass [a prostitute], a roller [a woman who enticed men so their wallets could be stolen], a hoister [a shoplifter], or a charlady. The fact was she loved you and looked after you. She held the family together. That was how important the mother was in that society. It was strange the way a family would break up when the mother died—brothers and sisters drifting apart in all directions.[17]

Like Lucy Luck, May Hobbs married at eighteen. "Like a lot of teenagers, I just thought marriage would mean independence and a life of roses and stars," she remembered, "not a rude awakening."[18] She complained that she had been kept ignorant about sex—adults told her the stork brought babies, and when she menstruated for the first time, she was told she had cut her leg: "the older people's attitude was terrible in that area—really Victorian."[19] Lucy Luck, a true Victorian, never mentioned menstruation, sex, sexuality, or childbirth in her memoirs.

When Lucy Luck married, marriage was for life: divorce was virtually unobtainable for women of her class, and she needed her husband's income to survive. Luck mentioned nothing about her feelings for her husband of fifty-seven years, either because of her reticence or because she saw her marriage more as an economic partnership than a romantic relationship. Hobbs expected "roses and stars"; when they did not materialize, she left her husband after two years and one child and moved in with another man who became her common-law husband.

Both women began having children as soon as they married, although Hobbs could have obtained contraception. Two of Luck's children died young, reflecting the high mortality rates of her day; Hobbs raised four delicate and premature babies, thanks to hospital care and free medical service through the National Health system. She hated the hospital:

> I could not bear to be pulled about by doctors who treated you as a moron and herded their patients from place to place as if they were a load of cattle in a market place. After every child I had I was told I ought not to have any more, and when I asked to be sterilized, they told me I was too young. Makes you laugh, don't it.[20]

Both Luck and Hobbs had to find work compatible with child care: each of their families needed the mother's income as well as the father's to survive. Luck did piecework at home and went out charring; Hobbs tried selling secondhand clothes at a market stall, but also ended up cleaning—in her case, in offices at night, when her husband could watch the children. She benefited from the Family Allowance: by her day, England, like most European nations, paid women who had children in an effort to raise the birth rate.

In her memoirs, Lucy Luck recounted exploitation: going hungry in the poorhouse, slaving for low wages, being laid off when her work was "out of season." But she neither complained nor expressed any hope for change: although her life saw the creation of the Labour party and the extension of political rights to women and men of the working class, they are not mentioned in her account. Luck summarized the bulk of her life in her memoirs and she may have done this because she believed the bulk of her life contained no important changes. When she and her husband were turned out of their cottage by the landlord, they had no option but to move in with family or go to a poorhouse. In contrast, when Hobbs was evicted for nonpayment of rent in the 1960s, she and her children were housed by the state in decent hostels and then provided with a council flat at a subsidized rent.

Generations of working women before Hobbs had been conscious of their disadvantaged circumstances, but there was little possibility for them even to hope to change what seemed like eternal traditions. Hobbs aspired to more. Shortly after she began to do night cleaning, she joined the local branch of the Labour party. She began to do political work, canvassing for candidates and lining up votes. Involved in strikes on her jobs, she found herself blacklisted by cleaning companies in the early 1970s. She organized women night workers into the Cleaners Action Group and led a number of successful strikes for higher wages and better working conditions in these years. "May had Whitehall [the British government offices] on its knees," wrote the *Daily Mail*, a popular newspaper. "At the Admiralty, engineers stopped servicing the phones, dustmen left the bins overflowing, mail deliveries were halted and no bread, milk, or beer got through to the canteen."[21] Hobbs brought change to women like herself, and she aptly titled her memoirs *Born To Struggle*. At the end of her account, she summed up her philosophy:

> I think of my Jenny [her foster mother] and thousands like her who slaved their guts out in return for a raw deal. So that is my work for the moment: Cleaners Action, working for the homeless, campaigning for the rights of young mothers and their children, and anything else where justice needs to be fought for in the face of reactionary governments, big business, bureaucracy and the parts of society which say they couldn't care less. If I am a militant it is what I see going on makes me into one.[22]

It was not just what she had seen. Lucy Luck witnessed far more misery and saw little hope for change. Although much in Hobbs's life was similar to conditions a hundred years earlier, change had occurred. In her memoirs, Luck continually apologized for her temerity in writing about her life at all: "I am a poor scholar and afraid I shall make a failure of it," was how she began.[23] Hobbs possessed a self-confidence brought in part by progress: one reason that she wrote was to correct the erroneous picture of lives like hers she thought "middle-class authors" had drawn.[24] The lives of European women in the new modern cities, like Lucy Luck's and May Hobbs's, show both constants and changes in the period between the eighteenth century and the present.

Moving to the Cities

The new cities of modern Europe were not only larger and more populous than earlier towns, they differed in what they provided the women and men who lived in them. Cities had more complex and varied economies, more public institutions, more cultural life, more specialized neighborhoods and districts. By the late seventeenth century, several European towns had grown large enough to be considered cities. By 1700, there were 700,000 people living in London; 360,000 in Paris; over 100,000 each in Madrid, Amsterdam, Venice, Vienna, Naples, Copenhagen, and St. Petersburg.[25] Decisions like Lucy and Will Luck's—to move to the city where there was work—swelled urban centers, and cities and towns grew continuously in both size and population. As population grew in the countryside, increasing numbers moved into the cities until, by 1940, the growth of the cities halted the absolute growth of Europe's rural population.

By 1750, about 10 percent of the population of Europe lived in cities of over 10,000 people.[26] The expansion of government and industry accelerated this growth and the opportunities for employment. Capital cities became enormous: London passed the 1 million mark by 1800; by 1890, over 4 million lived there. In 1890, Paris had over 2.5 million inhabitants; Berlin, Vienna, and St. Petersburg each over 1 million.[27] Other cities grew as economies industrialized. By the mid-nineteenth century, large cities other than the capitals tended to be centers of industry: Manchester, Liverpool, and Glasgow had over 300,000 inhabitants; Marseilles, Lyon, and Hamburg over 150,000.[28] Movement to the cities paralleled a nation's industrialization. By 1851, about one third of the inhabitants of England and Wales lived in cities, but only about one tenth of the populations of Germany and France.[29] In 1800, seventeen European cities had 100,000 inhabitants; by 1890, there were 101 such cities. By the end of the nineteenth century, about one third of the *entire* European population lived in cities of over 10,000.[30]

Although cities offered jobs and opportunities to earn income, living conditions remained abysmal. Contemporaries recorded the filth and crowding, especially in the poor neighborhoods which were the fastest-growing and most populous. In 1839, Flora Tristan (1803–1844), the French socialist, made her fourth trip to London. This time, she insisted on seeing the slum area of St. Giles, known as "Little Dublin." Located in the center of the city, off Tottenham Court Road, the neighborhood housed about 150,000 people. "Hardly have you gone ten paces when you are almost suffocated by the poisonous smell," wrote Tristan, "with evil-smelling soapy water and other household slops even more fetid lying everywhere in stagnant pools."[31] Tristan found that "the wretched inhabitants wash their tattered garments themselves and hang them on poles across the street, shutting out all pure air and sunshine," but what shocked her most was the human degradation she witnessed:

> Picture, if you can, barefoot men, women and children picking their way through the foul morass; some huddled against the wall for want of anywhere to sit, others squatting on the ground, children wallowing in the mud like pigs. . . . I saw children *without a stitch of clothing*, barefoot girls and women with babies at their breast, wearing nothing but a torn shirt that revealed almost the whole of their bodies; I saw old men cowering on dung hills, young men covered in rags.[32]

Such squalor led to disease and death: throughout the nineteenth century, the death rate in cities was higher than in the countryside and often higher than the urban birth rate. The growth of cities came from internal migration, from families moving in. Single men tended to migrate long distances, to industrial or capital cities; single women went to nearby towns. "Woman is a greater migrant than man, only she travels shorter distances," commented the nineteenth-century sociologist Adna Weber.[33] Women outnumbered men both in very large cities and in the smaller towns, where many found work as domestic servants.

Cities remained crowded, both inside and outside homes. Throughout the nineteenth century, the one-room dwelling was standard for working-class families; by the end of the century, only the most successful expected to live in two rooms.[34] In this era of large families, these rooms were jammed. In a typical alley in Belfast of 1837, 147 people lived in eighteen rooms, about eight to a room.[35] Families with any spare space took in lodgers, and beds were often let by the day or night, shared by workers on different shifts.

Crowding, coupled with no sanitation, contributed to disease. Poor sections had no paving, streetlights, sidewalks, or, most importantly, safe supplies of water. A study of Manchester done in the 1840s found only thirty-three privies for 7,000 people, an average of 215 men, women, and children to each.[36] Cholera, caused by drinking water contaminated with sewage, infested European cities until the third quarter of the nineteenth century. With no adequate water supply, keeping minimally clean became an immense labor. In his first medical report to the city of London in 1849, Dr. John Simon reported that in order to obtain water—"this first necessary of social life"—thousands of persons "wholly depend on their power of attending at some fixed hour of the day, pail in hand, beside the nearest standcock; where, with their neighbours, they wait their turn—sometimes not without a struggle, during the tedious dribbling of a single, small pipe."[37]

Under such conditions, raising a family became heroic labor. Frédéric Le Play, the French social scientist, described such a life in Paris of 1856. A family of two parents and two children lived in a small room 15 feet by 30 feet, at the top of a tenement building. The wife had borne four children in five years; two of them had died. The woman did piecework for her husband, a tailor, until eleven every night. "In these conditions," wrote Le Play,

> the total number of hours she works must be estimated at 12 hours a day, 365 days a year. During these hours of labor, in addition to the work done with the tailor, the woman does the household tasks. She cleans the room, makes the beds, dresses the children, and prepares food for the meals. Every week she washes the major household linens, the children's clothes, and even her dress when they need it. She soaps these things at home in a glazed earthen dish, then she goes down to the courtyard near the pump . . . to rinse them.[38]

It was women who washed the clothes and hung them out their windows to dry, causing the shadowing of the alleys which bothered Flora Tristan. It was women who waited in line for water and hauled it back home. Life in one room called for labor and logistics rather than domesticity. Most rooms lacked the most minimal cooking facilities; all lacked running water. Furniture was sparse: a few beds or pallets, a table, some chairs, perhaps a wardrobe. Furniture and clothing—and, *in extremis,* bedding—could be pawned for food money. "Thousands of us have *scarce any domestic affairs to look after,*" a woman who signed herself "A London Mechanic's Wife" wrote angrily in 1834, "when the want of employment on the one hand, or ill-requited toil on the other, have left our habitations almost destitute."[39]

As well as noticing poverty and squalor, contemporaries remarked on women's heroic labors to keep their families together and to raise their children. Henry Mayhew, who investigated the London poor in the 1850s, reported many cases of mothers sacrificing for their children. The poorest people Mayhew encountered were Irish refugees from the potato famine, who clustered in slums like the one which so shocked Flora Tristan. But even in direst poverty, women tried to maintain family life. A "pinched and wretched" Irishwoman told Mayhew she wished she were in prison, except for her "childer." She sold oranges in the street "to keep a bit of life in us." Her husband did less well than she—when he sold, people assumed he was lazy. "He goes sometimes whin I'm harrud tired. One of us must stay with the childer, for the youngist is not three and the ildest not five. We don't live, we starruve."[40] Labors like this supported life in the cities and contributed to the population growth which made large cities possible.

Partnership with a Man

Lucy Luck found it difficult to survive on the income she earned alone. Women's wages remained one half to two thirds that of men's. This was usually enough to provide food for one person, but not shelter, heat, light, or clothing. "If there is one opinion that is widespread among the popular classes, it is that a single woman cannot earn a living in Paris," wrote a French social observer in 1900. "A good half of young workers, if not the majority, find themselves with this alternative: to live in privation or to marry."[41] Remembering the same era in the East End London slum where she grew up, Grace Foakes (1901–) thought married women had the hardest time of it, "what with child-bearing, poor housing, unemployment and the constant struggle to make both ends meet." But marriage was still the best option for a poor city girl. "It was the aim of every girl to get married, and those who did not were looked upon with pity and were said to be 'on the shelf.' "[42]

Working-class girls were continually exhorted to ignore feelings of romantic passion and to look instead for a good provider. Marriage without property had to be a working partnership. "You cannot expect to marry in such a manner as neither of you shall have occasion to work," went an eighteenth-century manual for female servants, and both women and men expected their partners to work when they were able.[43] Economic necessity did not preclude love, any more than it did in the more privileged classes, where marriages involved dowry negotiations in this era. "My young man's that good ter me I feel as if somethink nice 'ad 'appened every time 'e comes in," a poor married Lambeth woman told a middle-class visitor around 1910.[44] Nineteenth-century social observers often remarked on (and deplored) the propensity of the very poor to form unions young and have many children.

In the cities and towns, poor couples often did not marry. Marriage fees remained high, and working-class people saw little wrong with the *union libre:* a marriage in all but law, where a couple lived together, contributed to their joint household, and often raised a family together. Whether a couple had a formal marriage license or not, strict rules governed the behavior of poor married women who wished to remain "respectable." "Each woman kept to her own man," remembered Grace Foakes, "and would not have dreamed of doing otherwise."[45] Working-class wives were also expected to be moderate in their drinking and cursing. They had to provide food and clothing for their families, as well as enforce "respectable" behavior. This meant not letting the girls of the family "run wild" and keeping the girls and boys separated from each other. "The damned houses were those where, the neighbours knew, an incestuous relationship had borne a fruit which walked the streets before their eyes," recalled Robert Roberts of his Salford slum community in the years before World War I.[46] Observers of working-class life continually attest to the efforts made, while living in one or two rooms, to preserve the sexual decencies among family members.[47]

To be respectable in her community, a poor woman also had to defer to her man, at least in public. Peasant customs were imported to the cities and towns. "It would not be unusual to find houses where the women always ate standing up by the fireplace, seeing that nothing was lacking on the table where the men were seated," wrote a French observer about village life in the second half of the nineteenth century,

> and everywhere it appeared normal and reasonable that women should be ordered about in a tone of voice which brooked no answer; if they did not obey, they would at least give the appearance of obedience in public; husbands who lacked authority were the laughing-stock of the country.[48]

In Salford, men who helped their women work around the house were called "mop rags" or "diddy men."[49]

A woman expected to defer and, if asked, justified her deference because of her husband's higher wages and earning capacity. In marriage, "the man should be the head, the woman the crown. . . . The wife must have her right, but the husband has more rights," stated a sixty-nine-year-old East Prussian woman to an investigator in 1909. "I also think it's right that men are better paid than women."[50] A good woman was expected to put her husband and children first. "The women seldom get new clothes; boots they are entirely without," concluded a report on a poor working-class London district in 1913. "The men go to work and must be supplied, the children must be decent at school, but the mother has no need to appear in the light of day."[51] "Sunday all day he's in bed," went a rueful English ballad of the eighteenth century,

> While his shirts and his stockings I'm washing. . . .
> I drink water and he drinks strong beer,
> O, I wish that I'd never got married.[52]

The man's diet was consistently better than his woman's: a study of Lambeth from 1909–1913 showed that more than twice as much was spent to feed the father as to feed the mother and children.[53] Mothers were expected to economize by eating less. "Now the children are getting bigger I find they eat all the dinner and there is none left for me," a London woman told a social worker in 1913. "Looking back, I remember that we children never expected mother to have any dinner. She always took a bit of bread."[54] As men's wages rose in the early twentieth century, they tended to pocket the increase and left their wives to manage on a fixed amount.[55] Managing was women's business. "Economy was the keynote of every woman in this community," wrote Grace Foakes of the East End of London in the early twentieth century.

> These women, if they thought up some fresh idea for economising, would run to tell their neighbours so that they too could practise it. Here are some of the things thought up by my mother. When times were very hard she invented a meal which I doubt had ever been tried before. She cut a slice of bread for each of us, put each slice in a separate mug, then covered each with boiling water. When the water was absorbed, she would pour off the surplus, add a knob of margarine and some salt and pepper and mix it all together. She called it "Pepper and Salt Slosh" and it was surprising how good it tasted, especially if we were cold and hungry.[56]

Women who failed to manage or spent too much could expect to be hit. A man's casual violence toward "his" woman was normal in working-class communities: some wives expected a slap if they had "asked for it" by being fresh or disobedient.[57] Others resented male violence: "What a bitter mixture for a poor woman to take!" wrote an elderly Englishwoman about a bad marriage in 1834, and if she offers the least resistance, it is thrust down her throat with his fist, possibly with the loss of a tooth or the spilling of a little of that blood which he thinks so inferior to his own. As he is lord of the castle, he is master and must be obeyed.[58]

Neighbors intervened only if a man beat another man's woman or if the woman's murder seemed likely.[59] In his investigations of the London poor in the 1850s, the social scientist Henry Mayhew found women famous for the beatings they received: one costermonger's woman was nicknamed Cast-Iron Poll, "her head having been struck with a pot without injury to her"; another was called Black-Wall Poll because she usually had two black eyes.[60] Such women stayed with their men and were extremely reluctant to prosecute: they could survive with some violence, but not without the man's income. "Sometimes the men were very cruel to their women, especially when in drink," remembered Grace Foakes.

I have heard many a woman screaming and shouting as a drunken man gave her a good hiding. The following day she would emerge with black eyes and swollen face, yet would not utter a word against her husband—and woebetide anyone who did! Not a word would she have against him.[61]

For poor women, as well as the more privileged, life with a man, however unsatisfactory, was better than life without one. Consistently the poorest women were those who had to support themselves without a man.[62]

Raising Larger Families

Lucy Luck was one of two surviving children of a family of four; she herself was able to raise seven children to maturity. Achievements like this underlay the tremendous growth of European population in the eighteenth and nineteenth centuries. In previous eras, slow population growth had periodically been interrupted by famines and disease. What was new in Europe of the eighteenth century was the immense and *sustained* increase in population. In 1750, the population of Europe (including European Russia) numbered about 145 million people, a figure which showed no great increase over previous eras. By 1800, this figure had reached 187 million, by 1850, 265 million. Between 1850 and 1914, Europe's population doubled again; by 1914, there were 468 million Europeans, even though during that period an additional 26 million had emigrated.[63]

The dynamics of this population "explosion" are still imperfectly understood. Two factors are crucial in determining population: the birth rate and the death rate. A rising live-birth rate combined with a stable or decreasing death rate produces growth, and this is the situation which had evolved in

Europe by the middle of the eighteenth century.[64] By then, the first trustworthy national population statistics appear, as nation-states began to require the registration of all births. (Sweden, which then included Norway, did so first in 1749.) These figures show an excess of births over deaths.[65] Birth rates probably rose because of improved economic conditions, in which food production and distribution increased and more opportunities of earning income appeared. Better economic conditions allowed couples to start having children at a younger age. More food enabled more mothers to birth healthier children who survived in greater numbers. Throughout European history, women and men had regulated the size of their families in response to local economic conditions; this process continued in the seventeenth and eighteenth centuries. Areas in which families could earn more show more population growth than areas where they could not.[66] The age-old tactics of deciding to marry or not, regulating the age of marriage, and practicing contraception within marriage (in this era, abstinence and withdrawal—*coitus interruptus*—remained the most common techniques) allowed some control over population growth. The decimation of Germany caused by the Thirty Years War (1618–1648), in which the population was almost halved, was reversed in less than a century by a very high German birth rate—much higher than in the rest of Europe during that period.[67]

Behind the rising population figures of eighteenth- and nineteenth-century Europe lie the successful efforts of women, most of them poor and uneducated, to keep increasing numbers of children alive. While the *long-term* result of industrialization favored a decrease in family size, the *short-term* effect favored high rates of fertility, as food production and distribution and earning opportunities expanded. In this period also, death rates declined among infants and young children, most probably because improved diets and sanitation led to the decline of infectious diseases.[68] Child mortality rates dropped in Europe after 1865; infant mortality (children under one year), after 1900.[69] What increased was not only the number of births per woman, but the number of children who survived, and thus, family size. In past eras, it had been necessary to bear two children to raise one adult: infant and child mortality claimed one out of two. By the mid-eighteenth century, circumstances had changed and family size increased.

The timing of changes in rates was different in different European nations. For reasons which are still not understood, France led this growth. There, family size peaked at 4.9 children per family in the first quarter of the eighteenth century, making France the most populous nation in Europe. In the rest of Europe, family size peaked in the late nineteenth century at slightly under five children per family—and eight births were still necessary to produce five survivors. Great Britain had the highest birth rate in Europe between 1800 and 1850; family size there reached 6.16 children in the 1860s. From 1850 to 1900, Russia, with a far larger population base, matched the British rate of growth.[70]

In looking at the human lives which are the background to this growth, the sense of responsibility with which most women faced motherhood is striking. "God grant me seven [children]," went a nineteenth-century Russian lullaby,

> Don't take from the five [I have],
> But increase them to ten.[71]

Mothers sacrificed for their children in a variety of ways—eating less, working harder, making do—and believed that such sacrifices justified their own existence. "I am glad to think now that we did our best for our children," wrote the Englishwoman "Mrs. Wrigley," a plate-layer's wife, of her life in the 1890s when she raised five boys. "I have gone without my dinner for their sakes, and just had a cup of tea and bread and butter."[72] For poor parents, children might be a burden when they were little, but remained the only security for the future in the absence of any government provision for the elderly.

Women took responsibility for their children's survival. Both population growth and contemporary sources attest to the success of the average woman in keeping her family alive and together. Even under conditions of extreme poverty, cases of mothers abandoning their families are virtually nonexistent. If a parent abandoned the family, it was almost always the father—like Lucy Luck's father. "If the father is dead, the family suffers," went a nineteenth-century Sicilian proverb. "If the mother dies, the family cannot exist."[73] In addition, there is evidence that the improved economic conditions of the nineteenth century led to better survival rates for infant girls: in industrialized nations, female life expectancy surpassed that of men, while in more agrarian economies, like Ireland and Italy, life expectancy remained the same for both sexes as late as 1900.[74]

Increased life expectancy for the female sex in the nineteenth century, however, did not mean that adult women outlived men in any greater numbers than in previous centuries. (From as early as the fourteenth century, European urban women outlived men if they survived past age forty.) Between the ages of twelve and forty, European men outlived women well into the twentieth century.[75] First, maternal mortality remained high: until the 1880s, motherhood was fatal for one woman out of every twenty.[76] Poor women faced increased risks with childbearing due to poor diet and health, the absence of any prenatal care, the need to keep working to the day of delivery, and the dangers of hospitals. Well into the twentieth century, women who could afford to do so gave birth at home. In 1866, a European woman faced a 4.7 chance out of 1,000 of dying during or shortly after childbirth if she delivered at home; in a hospital, her chances of death soared to 34 per 1,000.[77] In the hospitals, childbed (puerperal) fever was the great killer, caused largely by doctors ignorant of asepsis.[78]

Second, even if the risks of maternity are taken into account, "excess female mortality" prevailed in many nations into the twentieth century. Women dying in greater numbers than expected during what are normally their healthiest years was inextricably connected to conditions of poverty in which women were devalued, in which they ate less and had to work harder. The growing economies of the eighteenth and nineteenth centuries rested on increased agricultural production and piecework in the home—work done primarily by women. A study of nineteenth-century Bavaria revealed that a wife's work load was greater than her husband's until the early twentieth century.[79] If women and men receive equal treatment, more men than women die of tuberculosis. But in Victorian England, women's deaths from tuberculosis outnumbered men's until the 1880s, and half of the young women who died, died of tuberculosis.[80] Throughout the nineteenth century, because of factors which favored the survival of children without improving the health or working conditions of women, being a mother became harder than in earlier eras, if only because caring for more surviving children took more time and work than a smaller family. Excess female mortality in the child-rearing years bears witness to the price this toil exacted, and for much of the nineteenth century, young women died in greater numbers than they should have.

By the last quarter of the nineteenth century, both maternal mortality and family size began to decline. The use of asepsis, the development of anesthesia, and the growth of germ theory all contributed to making hospitals safer for women who gave birth there. For reasons which are still obscure, family size declined first in France, just as it had first risen there. By the 1880s, French family size was down to three children, a figure the rest of Europe did not reach until the early twentieth century.[81] Throughout Europe, family size declined from the top of the social pyramid down: upper- and middle-class families declined in size first, and by the end of the nineteenth century, families in the upper levels of the working class became smaller as well. The decline in family size resulted from two factors: in an industrialized society, children became an economic liability rather than an asset, and contraception became more easily available.[82]

Illegitimacy and Unwanted Children

The movement into cities seems to be connected to the dramatic rise in European illegitimacy rates, which occurred in the eighteenth and early nineteenth centuries. Children born to unmarried women traditionally comprised 2–5 percent of all births; at the end of the eighteenth century, the rate increased to 15–20 percent.[83] In preindustrial Europe, peasant and working-class couples often had premarital sex in the literal meaning of the words: intercourse occurred after a promise of marriage had been given and accepted. Marriage often took place when the bride was pregnant or after children had been born, and no social stigma attached to this.

But as male mobility increased, this traditional system broke down in rural areas. Couples had intercourse, but faced with fatherhood in a countryside in which work was difficult to find, increasing numbers of men moved into the cities, sometimes abandoning the pregnant women. Responsibility for the child increasingly fell on the woman alone. Eighteenth- and nineteenth-century laws reinforced the man's ability to decline the father's responsibility of providing for his child. Under the Prussian Civil Code of 1794, the natural father owed up to one quarter of his income for the maintenance of his child, even if he and the mother were not married and she had had intercourse with other men. In 1854, this law was changed: now the concept of natural fatherhood disappeared, and the conditions under which an unwed mother was entitled to child payments narrowed greatly.[84] The Italian Civil Code of 1866 allowed for *no* paternity suits except in cases of rape or abduction; Italian women did not gain the right to sue for paternity until 1975.[85] The prevailing European attitude was stated by an Irish legislator in 1837: "Irish females should be . . . guardians of their own honor, and be responsible in their own persons for all deviations from virtue."[86]

The migration of young, single women into city households as servants also contributed to illegitimacy rates: many of the women who left their babies in foundling hospitals were unwed servants pregnant by fellow servants or by their masters.[87] A study of the "pregnancy statements" required by law from unwed mothers in eighteenth-century Aix-en-Provence in France revealed that between one half and two thirds expected their sexual partner to marry them. Wilhelmine Kähler, a German textile worker, recorded her experience in the 1860s. She sewed overcoats at home with her mother, "day in, day out, hard drudgery for a mean existence." She met a "higher employee of the firm" who courted her with "flowers and presents." She fell in love with him and for a few months lived a "blissful" existence. "But oh! Happiness lasted only a short time," she remembered.

Soon I knew that the happiness of being a mother awaited me. I didn't hesitate to share this discovery with my heart's dearest, because I imagined it would also fill him with the greatest happiness. "But kid, I never meant I would marry you—no, I can't marry you." His answer hit me like a death sentence. I nearly went crazy. Rejected by my parents and sisters, overcome with despair, I was disgusted with everyone. Embittered and angry with the world, I moved to another place.[88]

Kähler's testimony is typical; in the cities as in the countryside women expected marriage to follow intercourse, especially if they became pregnant, and it was men's refusal to marry which led to the high illegitimacy rates of the period.[89] Women attempted to control their fertility through traditional remedies, like nursing their children for long periods. Grace Foakes remembered that in the East End of London in the early twentieth century,

mothers would breast-feed their babies until they were two years old in the hope of keeping themselves from having another baby, for it was a common belief that they couldn't conceive while breast-feeding. (As far as I know there were no contraceptives at that time.) I do not think people knew how to prevent more babies arriving.[90]

Kähler's baby died shortly after it was born. She did not have to face the problem of the mother, wed or unwed, with unwanted children. Faced with an unwanted pregnancy, the vast majority of women chose, like Kähler, to have the child. This decision often led to the enormous families of the era and a state of constant pregnancy and child rearing for the mother, whose health might deteriorate under such a regime. "I was born in Bethnal Green, April 9th, 1855. . . . I was my mother's seventh child," recalled "Mrs. Layton," who wrote her memoirs when she was in her seventies, like Lucy Luck.

Seven more were born after me—fourteen in all—which made my mother a perfect slave. Generally speaking, she was either expecting a baby to be born or had one at the breast. At the time there were eight of us, the oldest was not big enough to get ready to go to school without help.[91]

Women who sought to end unwanted pregnancies consulted other women or midwives—the European male medical establishment resisted performing abortions, even therapeutic abortions, on poor women well into the twentieth century. Women tried traditional remedies to end pregnancy: the herbs savin and rue, the dangerous ergot fungus which caused convulsions, and large doses of alcoholic spirits were widely used in the nineteenth century. If these remedies failed, women tried combinations of violent exercise, purgatives, laxatives, bleedings, and hot baths. The final and most dangerous remedy was some form of curettage: introducing a foreign object, usually metal, into the uterus and trying to scrape off the fetus and placenta. Before asepsis, these attempts risked the pregnant woman's life through infection and hemorrhage and were usually resorted to only by the most desperate pregnant women: those who were not married.[92] By the end of the nineteenth century, when abortions were safer, tens of thousands of European women sought them from doctors and quacks, from midwives (whose function in many communities became that of an abortionist as doctors took over child delivery), or through "medications" touted in magazine advertisements to restore menstruation.[93]

Once the child was a living reality, the mother who could not raise it herself had three options. She could pay another woman to nurse and raise it, she could abandon it, preferably to a foundling hospital, or she could let it die or kill it. Hundreds of thousands of European women resorted to these methods, usually to ensure the survival of themselves or their older children.

In earlier centuries, many women had hired others to wet-nurse their babies. "Women of fortune here," wrote the English feminist Mary Wollstonecraft from Sweden in 1796, "as well as everywhere else, have nurses to suckle their children."[94] In the course of the nineteenth century, privileged women generally abandoned the use of wet nurses, but poorer urban women who worked outside their homes began to use them in greater numbers.[95] A study of nineteenth-century Parisian wet nurses revealed that the majority of both wet nurses and the women who used them were poor.[96] In a survey of five hundred wet nurses in the Moscow area in 1888, most replied that their motive for taking in babies was poverty.[97] Wet-nursing usually meant depriving one's own child of breast milk, weaning it early or feeding it less. Most wet nurses boarded the babies in their own homes; the mother paid a fee for its care. The fees were low and conditions were often very poor. "Baby-farming" (as this practice was called) often led to the child's death, and to some degree this system worked as an indirect method of eliminating an unwanted child.[98] Mothers who used this system usually had other children at home and could not afford to stay home from their jobs to care for the new child.

Baby-farmers themselves left few records. One Englishwoman, Margaret Waters, wrote an account of her life as a baby-farmer five days before her execution for child murder. She started caring for children in her home because of poverty; after her husband's death, she was unable to support herself by sewing and began to take in lodgers. One, an unmarried mother, asked Waters to care for her child; others followed suit. Paid £10 per baby, Waters could not afford funeral expenses if the baby died. She left the bodies in the street, which led to her eventual arrest. That year, 1870, there were 276 dead babies found in the streets of London.[99]

The abandonment of babies was so widespread that by the eighteenth century, most European cities had established foundling hospitals to receive them. These new hospitals institutionalized care for unwanted children, but they could not keep the abandoned babies alive in anything like the rates achieved by individual mothers. Outside the hospitals, infant and child mortality was about 50 percent; inside, it ranged from 65 to 75 percent.[100] The numbers of children abandoned were tremendous: forty thousand per year in France of the 1780s; nearly fifteen thousand admissions to the London Hospital between 1756 and 1760; five thousand admissions a year to the St. Petersburg Foundling Hospital in the 1830s.[101] Although most of these babies were the children of unwed mothers, between a third and a half were abandoned by both parents.[102] Poverty was the chief motive, and rates of abandonment fluctuated with economic conditions in a given region.[103] By the last quarter of the nineteenth century, governments turned to trying to aid individual mothers in raising their own children, either by providing cash payments or daytime crèches, and attempted to extricate themselves from the murderous business of baby care.

Given the option of abandonment, only women whose own survival depended on their not having a child resorted to infanticide. Of the eighty-seven women executed in Nuremberg between 1513 and 1777 for infanticide, all but four were unwed mothers.[104] Domestic servants were especially vulnerable: pregnancy was almost automatic grounds for dismissal, and most of the women tried for infanticide in the eighteenth and nineteenth centuries were servants. The case of the French servant called "la petite Loisy" was typical. Dismissed by her mistress for being pregnant in 1847, she gave birth to a boy, crushed his skull, and buried him in a field.[105] By the nineteenth century, statistics show that infanticide was practiced equally on girl and boy babies, in contrast to earlier periods when girls tended to be killed or let die in greater numbers.[106] This sort of infanticide was almost certainly caused by the mother's desperation.

Neglect and exploitation could also cause the death of a child. A German working-class woman who identified herself as "Martha" wrote to Louise Otto-Peters's women's newspaper on the subject of "angel-making" in 1849. "In order to earn money by means of them and then 'get rid' of them," she stated, "children are mutilated intentionally and made sick, so that by their misery they will inspire pity on the arm of the female beggar and finally die." Martha believed poverty caused parents to do this: "people just want to make some money, and because they know no other way of earning a living, they make angels."[107] Even under conditions of extreme poverty, however, women rarely resorted to infanticide or harming their children. Such practices died out in the third quarter of the nineteenth century, when the rising standard of living and the increased availability of contraception made such desperate measures unnecessary.[108] While many European women in the late nineteenth and twentieth centuries continued to bear children they had not wished for, increasingly women gained some measure of control over their reproductive capacity. With that, infanticide disappeared for the first time in European history and the chances of a child's survival generally improved.

2

EARNING INCOME

❦

UNTIL THE LAST QUARTER of the nineteenth century, few women of the laboring classes, whether married or single, could afford not to earn money. A woman's income was essential to her own or her family's survival. The most remarkable aspect of women's paid labor is how little it changed through the centuries. Women traditionally performed a variety of jobs throughout a lifetime; they continued to do so. Women traditionally worked in areas associated with women's traditional tasks of housekeeping and family maintenance—food provision, textile manufacture, cleaning, and child care; they continued to do so. And women traditionally accepted one half to two thirds the money paid to men for the same job; they continued to do so. In comparison to male workers, women worked for the lowest wages in the least prestigious sectors of the economy. The major occupation of women who earned income was domestic service, and this remained true until 1940.[1] The availability of servants' jobs in the cities, which led women to migrate from the countryside, had more impact on women's lives in the nineteenth century than industrialization. This can be seen in the Danish economy, which was moderately industrialized by 1890. Of 232,588 women who worked for income outside the home (about one third of the female population over fifteen), 121,184—over half—were servants. Workers in industry numbered 40,729, while 30,878 labored in agriculture. There were 16,045 women classified as "diverse contract workers," a category which included some pieceworkers, but these Danish statistics, like those of other nations, almost certainly undercounted women's efforts to earn money.[2]

The entry of women into a trade was a signal that the trade was being downgraded, and this did not change with the growth of cities or industrialization. "In manufacturing towns, look at the value that is set on women's labour," wrote Frances Morrison, the wife of a radical English editor, in 1834, "whether it be skillful, whether it be laborious, [just] so that a woman can do it. The contemptible expression is, it is made by woman and therefore cheap. Why, I ask, should women's labour be thus undervalued?"[3] The answer was that it always had been, and neither new economic forms nor the development of new industries changed this tradition.[4]

If a woman had to earn income, either to support herself or her family, she had no choice but to accept the traditionally low rates of pay for women. In 1848, the Statistical Society surveyed a poor parish in East London. They discovered that while men's incomes ranged widely and averaged 20 shillings, 2 pence a week, women's incomes were uniformly low and averaged 6 shillings, 10 pence. There were 229 women living without men in the parish (and only 125 men living without women) and two thirds of the women had at least one other person, usually a child, dependent upon them for support. Some of the women classified as "widows" were "widowed only by the abandonment of husbands," the society reported. Given the disparity between female and male income, the society labeled the men without women as "single," but the women without men "unprotected."[5] These conditions did not change over time. In 1903, the women's committee of the Manchester and Salford Trade Unions Council reported that

> a man thinks himself badly off if he cannot earn more than seventeen shillings a week. It is no exaggeration to say that there are thousands of girls in Manchester who think themselves lucky if they bring home seven shillings at the weekend, and more older and skilled women who can never hope to earn more than twelve shillings to the end of their lives. . . . These are surely the wages of the poorest poor.[6]

But women alone supporting themselves or children were the "poorest poor." Women's supposed "willingness" to work for extremely low wages (at or even lower than the subsistence level) has always been caused by their desperation to support themselves and their children, if they had them. Low wages for single women were rationalized on the grounds that they had only themselves to support. Low wages for married women were rationalized on the grounds that they had a husband to support them. Both rationalizations ignored reality and rested on the traditional devaluation of women and their labor in European culture.

Employers exploited women's need to earn income. In 1844 an English factory inspector testified before Parliament that "a vast majority of the persons employed at night and for long periods during the day are females; their labour is cheaper and they are more easily induced to undergo severe bodily fatigue than men."[7] The married woman with children was even cheaper and more tractable. A manufacturer testified at the same hearings that he employed only women at his power looms and "gives a decided preference to married females, especially those who have families at home dependent on them for support; they are attentive, docile, more so than unmarried females, and are compelled to use their utmost exertions to procure the necessities of life."[8] As the London Ladies' Committee for Promoting the Education and Employment of the Female Poor put it in 1804, women were "confined, most frequently, to a few scanty and unproductive kinds of labour."[9] The *average* woman worker resembled the poorest and most exploited male worker.

Child Labor

Children worked as soon as they could; both girls and boys helped not only with chores, but often with the paid labor which supported their families. Until the second half of the nineteenth century, when nation-states began to provide and enforce free public primary-school education, child labor was expected in poor families. Little girls had special responsibility for caring for younger children, and they began to do so when they were very young. A German girl born in 1860 and identified only as "Kathrin" told an interviewer that her parents, both of whom worked in a nearby factory, left her in charge of her three-year-old sister and six-month-old brother when she was four. "My mother told and showed me how I must do things until she came home again. I always had to stand up on a bench in order to reach in the little wagon where my brother lay, sometimes things were a mess." The next year, when she was five, she and her sister "had to go into the forest with a small wagon [to gather wood], we also always had to take our brother with us, two or three times each day. Sometimes we had to be out the whole day alone, trying to find a big pile [of wood]."[10] Her father beat them if they had not collected enough.

Parental harshness seems to have been caused primarily by the strains of struggling to survive. Families labored together, especially where they earned income by doing piecework in the home. Home industry was ill-paid, and the low income forced parents to have their children help out. "Just to earn the necessities, mother and we children had to work too," remembered Luise Zietz (1865–1922), later a German socialist, of her childhood in the early 1870s.

> We had to pluck apart and oil the raw wool, to run it through the "wolf," which compressed it further, then it had to be put through the carding machine two times. A pair of dogs, who switched off, drove this machine by a large treadwheel, and when one of the large dogs died on us, we ourselves had to get down in the wheel. . . . Spinning was a terrible torture for us children. We crouched hour after hour on the low stool behind the spinning wheel at the horrible monotonous and exhausting work, just spinning, spinning, spinning.[11]

Poverty forced parents to send children out to labor as well. The English social investigator Henry Mayhew interviewed an eight-year-old "watercress girl" who had been selling in the streets since she was seven. She lived with her mother and stepfather, who were "very poor," and she had helped earn income from the time she was five, sewing at home with her mother. Mayhew found "something cruelly pathetic in hearing this infant, so young that her features had scarcely formed themselves, talking of the bitterest struggles of life, with the calm earnestness of one who had endured them all," but the watercress girl herself was proud of her ability to earn income. "I'm a capital hand at bargaining. . . . They can't take me in," she bragged; "I ain't a child and I shan't be a woman till I'm twenty, but I'm past eight, I am."[12]

This sort of child labor was traditional and unexceptional. But the new industrial labor of children in factories and mines began to cause concern. There, children usually worked under the supervision of a foreman rather than their parents and were expected to keep up with adult schedules and work loads. In 1832, Elizabeth Bentley, then twenty-three, testified before an English parliamentary committee about her childhood in a linen factory. She had started when she was six, working from six in the morning until seven at night in the slow season, from five in the morning until nine at night in the six months of the year when the mill was busy. She had forty minutes off at noon; that was the only break in the day. She worked as a "little doffer," taking the full bobbins off frames and replacing them. If she fell behind, she was "strapped," and she testified that the last doffer was always strapped. At ten, she transferred to the carding room, where the foreman used straps and chains to beat the children "up to their labour." She was asked, "Were the girls so struck as to leave marks upon their skin?" and she replied, "Yes; they have had black marks many times, and their parents dare not come to him [the foreman] about it, they were afraid of losing their work." The work in the carding room pulled her arm bones out of their sockets and she became "considerably deformed . . . in consequence of this labour."[13] Children in mines worked equally hard. "I hurry [push carts loaded with coal] with my brother," eleven-year-old Eliza Coats told an English parliamentary commission in 1842.

> It tires me a great deal, and tires my back and arms. . . . I can't read; I have never been to school. I do nought on Sundays. I have had no shoes to go in to school. I don't know where I shall go if I am a bad girl when I die. I think God made the world, but I don't know where God is. I never heard of Jesus Christ.[14]

Testimony like this shocked public opinion, and nations began to outlaw the labor of children under ten in mines and factories. Efforts were begun to provide free, compulsory primary-school education to children, but educating girls was usually seen as less important than educating boys.[15] Public primary schools for all boys were mandated in France in 1833; equivalent schools for girls did not become law until 1881. "Girls do not have to learn much," Emma Goldman (1869–1940) remembered her father telling her in Russian Poland in the 1870s. "All a Jewish daughter needs to know is how to prepare gefilte fish, cut noodles fine, and give the man plenty of children!"[16] Mayhew found that London street-sellers were reluctant to educate their girls: "They say, 'What's the use of it? *That* won't yarn a gal a living.' "[17] "No learning ever was bestow'd on me," the English washerwoman/poet Mary Collier (1689–c. 1763) wrote in 1739,

> My life was always spent in drudgery:
> And not alone; alas! with grief I find,
> It is the portion of poor woman-kind.[18]

Conditions improved in the third quarter of the nineteenth century, when primary schooling for girls was provided and enforced. But schooling caused problems in poor families, especially where the daughter's income was crucial to survival. As a child in Austria, Adelheid Popp (1869–1939) sewed and knitted at home with her widowed mother. She missed school so often that the authorities imprisoned her mother for twelve hours on Easter Sunday in punishment. Popp and her mother were too poor to give up the girl's income. "What use was it to go to school when I had neither clothes nor food?" Popp wrote later. Mother and daughter moved into town to avoid the authorities, and the girl entered a workshop at ten where she crocheted shawls on a piecework basis. The work paid so little she crocheted late into the evenings at home and before she went to work in the mornings at six;

> after my mother had wakened me, she gave me a chair in the bed, so that I might keep my feet warm, and I crocheted on from where I had left off the previous evening. In later years a feeling of unmeasured bitterness has overwhelmed me, because I knew nothing, really nothing, of childish joys and youthful happiness.[19]

Poor mothers, like Popp's, suffered as a result of their daughters' schooling, since they lost both their help at home and the income they had earned. When an English member of Parliament voted in 1899 to raise the school-leaving age from ten, angry mothers in the industrial Lancashire district spat at him. "I will never forget the first time my daughter brought in her first week's wages," stated a Lancashire mother. "The two shillings four pence seemed to go further than any sovereign [21 shillings] had done."[20] Conflicts like this continued until the late-nineteenth-century rise in adult incomes made children's wages no longer essential to a family's survival.

Single Women

By her early teens, a young woman was considered to be able to support herself. From the late eighteenth century on, there were four chief ways for a young city woman without children to earn money: domestic service, factory work, street-selling and manual labor, and prostitution. Piecework in the home paid less and was largely the province of married women with children who needed to earn income in their own homes. Single women most often went to live in other people's homes as domestic servants. From the second half of the eighteenth century on, domestic service jobs for women expanded tremendously, and between one third and one half of all women who earned income outside the home did so as domestic servants.[21] These positions increased for two reasons: as wealth grew, more families could afford to hire a servant, and as new urban and industrial jobs opened up for men, they left domestic service to women. In the first half of the eighteenth century, between one third and one half of all town servants in France were men; by the 1790s, eight to nine tenths were women.[22] By 1901, 91.5 percent of all English servants and 82.9 percent of all French servants were women.[23] Both women and men left domestic service if they could: in factory towns, young women worked in the mills rather than other people's homes, and families who wanted to employ servants had to import them from other districts.[24]

Workers abandoned domestic service because of bad working conditions and degrading treatment by employers. Domestic service retained its paternal treatment of workers longer than any other sector of the economy, and servants often came under different law codes than the rest of the population. In a dispute, the master's word took precedence; in French law, the master's affirmation that he had paid a servant's wages was all that was needed. A separate law code for German servants was in force from 1810 to 1918; until 1850, employers were allowed to punish their servants with "a leather whip whereby a moderate number of strokes can be given on the back over the clothing."[25] Until 1914, French servants were not covered by other laws written for workers' protection: there were no restrictions on child labor, no limitation of hours, no guarantee of a weekly day of rest, no retirement or insurance benefits, and servants could be fired for virtually any reason.[26] "Maids had no liberty, status, or privileges," remembered an Englishwoman who had worked as a servant before World War I.

> As I see it, the system was a continuation of slavery, except that you were able to hand in your notice and leave instead of having to stay for life. One did not answer back when lashed with the tongue or suffering humiliation. I always had

[to ask] to be allowed to post a letter at the letter box less than 400 yards away. If one became ill, permission had to be obtained to attend a doctor's surgery. To my way of thinking, those were cruel, unjust times.[27]

Male servants could find jobs in other fields; if they remained in domestic service, they commanded better wages and working conditions than female servants. Article 37 of the German Servants Code, for instance, required that male servants be provided with their uniforms, while women supplied their own working clothes.[28] In 1798, an Englishwoman wrote that a fashionable footman,

> whose most laborious task is to wait at table, gains, including clothes, vails [tips], and other perquisites, at least £50 per annum, while a cook-maid, who is mistress of her profession, does not obtain £20, though her office is laborious, unwholesome, and requires a much greater degree of skill.[29]

As the number of male servants decreased, they became prized as status symbols and emblems of wealth, and they often moved into the more prestigious and supervisory servants' roles, like butler or majordomo. In 1863, the satiric English magazine *Punch* published a cartoon showing the relative status of male and female servants: a haughty butler carries a letter upstairs on a silver tray; he is followed by a female drudge hauling a full coal scuttle. A filled coal scuttle or large bath jug weighed about 30 pounds.[30]

Young women could not find other jobs as easily as men, and from the second half of the eighteenth century, a majority went into domestic service. Poverty determined their choice: female servants consistently came from the poorest families.[31] Lucy Luck became a servant when she was thirteen, and European poorhouses and orphanages routinely sent girls into service when they were twelve.[32] Domestic service was the province of young, unmarried women: a study of Bayeux in 1796 showed that 67 percent of the female servants were between fifteen and thirty.[33] Women tended to work as servants for about ten years, saving up their wages to make it possible for them to marry.[34] Domestic service as a job for young women was often justified on the grounds that it would prepare them to be good wives and mothers; some servants testified that both the job and their youthful entry to domestic service only prepared them to be subservient and obedient. "I went out to service too soon," wrote the Englishwoman Hannah Cullwick, who began in 1841 when she was eight;

> at the Charity School i [sic] was taught to curtsey to the ladies and gentlemen and it seem'd to come natural to me to think them *entirely* over the lower class as if it was our place to bow and be at their bidding and I've never got out o' that feeling somehow.[35]

Employers insisted on deference. "I was once in the nursery bedroom when Anna came in panting with a can of water," wrote M. Motherly, the female author of *The Servant's Behavior Book,* published in London in 1859.

> As I spoke to her she sank down on a chair saying "Excuse me, ma'am, I am so tired," but I could not excuse her. She acted very rudely. It would have been but a small effort to stand for a few moments, however tired she might be and girls who are not capable of such an effort are not fit for service.[36]

A similar French book of 1844 warned against allowing the servant to dress too well: "she will no longer give all the respect which she should; she will be less submissive taking orders, less exact in the accomplishment of her duties."[37] Until the First World War, when large numbers of young women began to find jobs other than domestic service, employers could command a great deal of work from their servants. The following is a typical description of the duties expected of a female servant, from a French manual of 1896.

> The maid of all work should get up at six, fix her hair, get herself ready, and not come down to the kitchen without being ready to go out to the market. From 6 to 9 o'clock, she has the time to do many things. She will light the furnace and the fires or get the stove going. She will prepare the breakfasts, do the dining room, brush the clothes and clean the shoes. When the masters arise, she will do their rooms, will put water in the water closets, carry up wood or coal, and carry down the excrement. For all these tasks, she will put on oversleeves and a white apron and take care to wash her hands. Then, when the dining room is restored to order, the tableware washed and put away, the cooking utensils cleaned, she can, before the preparations for dinner, do a special chore for each day of the week. For example, on Saturday, the thorough cleaning of the kitchen and all its accessories; on Monday, the living and dining rooms; on Tuesday, the brass and copper; on Wednesday, washing; on Thursday, ironing . . .[38]

From the eighteenth century on, some women servants were able to articulate what such labor was like. "Alas! our labours never know no end," wrote Mary Collier in 1739,

> On brass and irons we our strength must spend;
> Our tender hands and fingers scratch and tear:
> All this, and more, with patience we must bear.[39]

A hundred thirty years later, nothing had changed. "I have been so driven at work since the fires begun [since she had to prepare and clean up the fireplaces which heated the house] I have had ardly [*sic*] time for anything for myself," Harriet Brown, a young English servant, wrote her mother in 1870.

> I am up at half past five and six every morning and do not go to bed till nearly twelve at night and I feel so tired sometimes I am obliged to have a good cry.[40]

Dorothy Hill, another English servant, estimated she had shifted three tons of coal in one exceptionally cold week, keeping fires roaring in all the rooms.[41] Female servants routinely carried heavy loads of coal, wood, laundry, produce, and water.

Like the mistress of the house, female servants were expected to be "on call" at all hours and on all days of the week. Employers granted time off reluctantly and expected servants to give up their free time if the family needed it.[42] A survey of German domestic servants in 1917 revealed that over half worked sixteen hours a day; their employers successfully resisted efforts to reduce the working day to thirteen hours.[43]

In addition to working long hours, female servants usually lived in squalid conditions. They generally were housed in the basement or attic. They slept in kitchens, hallways, alcoves, on mattresses or pallets on the floor. When bathrooms existed, servants were often forbidden to use them. In 1906, a French doctor surveyed servants' quarters in Paris. He found that

> maids' rooms hardly ever have fireplaces. Windows are rare and most often air is let in only through a sky-light or vent. . . . Sometimes the window looks out on the street. Otherwise, she gets her air from a small courtyard from which rise the insipid and fatty odors of unwashed sinks, of dung heaps and waterclosets.[44]

Such conditions remained typical well into the twentieth century, and employers took comfort in the thought that the living conditions the young woman had left behind were probably worse.

The advantages of domestic service were the food and wages, which were competitive with other jobs available to women.[45] The chief disadvantage was the servant's vulnerability. If she got sick, she was usually fired.[46] Around 1885, Josefine Joksch, a Viennese children's nurse, caught a sore throat from her employer's child. "My employer appeared and ordered me to leave the children's room and to pack my bags, since I couldn't stay in the house one minute longer with an 'infectious' disease," she remembered. "What did they have in mind for me? Did they intend to put me in a hospital?"[47] Instead, she was taken to the old "servant's alley" in Vienna, where she was left to fend for herself as best she could. If the servant was not dismissed for illness, she was often left to take care of herself. "The [servant] girl slept in a windowless attic, up a ladder out of the kitchen," recalled an Englishwoman of a visit to Hanover in 1901.

> While we were there, everyone had influenza rather badly; when the girl fell ill she just went up into her dog-kennel, and stayed there, alone in the dark, for

several days. Nobody went near her, for the presumption of supposing that she was not able to do her usual work.[48]

Older servants were routinely dismissed; only well-off or very benevolent families could afford to keep on an aging servant not up to the labor of the job.[49] Pensions usually were nonexistent; servants were expected to save for their old age. "The number of old servants who are paupers in workhouses is immense," Prince Albert told the newly formed Servants' Provident Society in 1849.[50] But most young women worked as servants only as long as it took to amass the dowry which enabled them to marry. They did not expect to remain in service all their lives, and some of their toleration of bad conditions may have been based on their perception of their situation as temporary.

By becoming a domestic servant, the young working-class woman crossed into a more privileged world. She was expected to adapt quickly to a new way of life often very different from what she was used to. "My duties seemed easy to do, although I couldn't understand the reason for many orders," wrote Josefine Joksch about the Viennese job she had lost because of her sore throat.

> Why, for example, did they order me to take care that the children not touch the floor in bare feet? What did Frau Preuss mean by this? Was this possibly injurious to their health? At home children splashed about in water puddles even late in the fall.[51]

If well-treated, the young woman could easily come to identify with her "new family." Helene Demuth (1820–1890), given to Jenny Marx as a present by her mother in 1845, remained loyal to the Marx family all her life, refusing offers to marry, going without wages, and even putting her clothes in pawn for them when necessary.[52] She "reserved to herself the right of 'speaking her mind' even to the august doctor," remembered a family visitor of the late 1870s. "Her mind was respectfully, even meekly accepted by all the family."[53] Demuth, Jenny and Karl Marx, and one of the Marx's grandsons were all buried in the same grave.

In 1851, when her mistress was pregnant with her fifth child, Helene Demuth gave birth to a baby also fathered by Karl Marx. She chose to have her son raised by foster parents in order that she could remain to care for the Marx children.[54] Young female servants remained sexually vulnerable to the men of the household, both fellow servants and their employers. Boys of privileged families often began their sexual experiences with the family's maid, whose youth and dependence made it difficult for her to resist. Anna Sachse Mosegaard (1881–1954) first worked as a servant when she was fourteen. An orphan, she was touched that her employer told her that he would not "be a distant master, but rather a good father" to her. However, "before

I could count sixteen springs, my master stood lusting in front of my garret's door. He, who had wanted to be a father to me!"[55] Little testimony remains of servants' experience of such affairs. Sexual relations between master and maid furnished a major theme in both pornography and literature in the eighteenth and nineteenth centuries, showing the attraction of such fantasies for men.

Unlike Helene Demuth and the Marx family, almost all female servants who became pregnant were fired. Most unwed mothers had been servants, as were most women who left their children in foundling hospitals. The servant who had had a child would find it difficult to get another position. If she could find no other sort of work, she would have to turn to prostitution. A study of Parisian prostitutes in 1830 revealed that almost 28 percent—the largest group—had previously been domestic servants; a study of the same city in 1900 asserted that over half had been servants.[56] "How did I come to take to this sort of life?" an English prostitute called "Swindling Sal" told Henry Mayhew in the 1860s.

> I was a servant gal away down in Birmingham. I got tired of workin' and slavin' to make a livin', and getting a _____ bad one at that; what o' five pun' a year and yer grub, I'd sooner starve, I would.[57]

Sal told Mayhew she normally earned between £3 and £4 a week as a prostitute. The only field other than prostitution which could offer a young woman anything like a living wage was factory work, and the pay was nowhere near as high. But factory work had the advantage of being socially respectable, while prostitution was not.

In industrial cities, young women chose to work in the mills, rather than become servants. Seventy-six percent of all the fourteen-year-old girls in the Manchester/Salford region in 1852 worked in factories.[58] The majority of female factory workers were young, unmarried women, and like servants, they expected their employment to last only until they married.[59] "The single women often look upon their work as merely a temporary necessity," reported the Women's Committee of the Manchester and Salford Trade Unions Council in 1895.[60] Some women were frightened by their first weeks in the factory, much as Lucy Luck had been. "At first I was highly terrified by the noise and the proximity of clashing machinery," remembered Alice Foley, who worked in a Lancashire textile mill in the 1890s. "It was a vast unexplored region, stifling, deafening and incredibly dirty."[61] Anna Boschek, who started working in a Viennese spinning mill in 1891, when she was seventeen, was terrified she would receive no pay at all for her first week's work: her machine kept jamming as the yarn broke. Undone, she stood by her machine and cried until a fellow worker showed her the tricks of the trade.[62] Others remembered that camaraderie had made the job pleasant. "We were happy," stated a Belfast woman who started at the linen mill in 1898, when she was twelve. "You stood all day at your work and sung them songs." "Wonderful times then in the mill," remembered another Belfast woman who worked in the same mill from the time she was eleven. "You got a wee drink, got a join [pooled money with others to buy food], done your work and you had your company."[63] Factory women had each other's support and were not as isolated as domestic servants. Some prided themselves on their "ladylike" behavior; others enjoyed the freedom being in groups provided. "The usual tricks played on a learner were played on me," remembered Harry Pollitt of his entry into a Lancashire mill in the 1890s,

> but the [female] weavers, being a cut above the cardroom operatives (as they thought) played only polite, lady-like tricks on me. It was left to the buxom girls and women in the cardroom to break me in by taking my trousers down and daubing my unmentionable parts with oil and packing me up with cotton waste.[64]

In factories, however, the traditional devaluation and subordination of women prevailed. Although women earned more in factories than they could in other jobs, their wages remained between two thirds to one half of men's wages, even for identical work. Men who ran the cutting machines in French shoe factories in the 1890s, for instance, earned 6 francs a day; women who did the same, 3 francs, 25 centimes. Men who worked at the machines which cut threads into screws made 4 francs a day; women who did the same, 2 francs, 50.[65] A study of wages paid in the Lancashire cotton mills in 1833 showed that while girls and boys could earn roughly the same, male factory workers started to pull ahead in their late teens. By age thirty-five, the height of their earning power, men averaged 22 shillings, 8 pence a week; women, 8 shillings, 9 pence. In old age, the gap began to close as male earnings dropped, but at age sixty-five the average man earned more than twice as much as the average woman.[66]

Lower wages for women caused part of this earnings gap; relegation of women to less prestigious and less well paid factory positions also contributed to it. Women entering factories were often excluded from the best-paid jobs, which required special skill or supervisory duties. When they tried to move into such positions, male workers often objected, since the entry of women usually meant a downgrading of the work and the lowering of wages.[67] Most factory women worked in textiles: by 1900, a quarter of a million Lancashire women labored in the cotton mills. Studies of these mills in the 1830s and 1890s show that women there worked in the least skilled and prestigious jobs.[68] While industrialization opened factory jobs to women, it also mechanized activities done by women—like spinning—and moved them to the factories, where they were often performed by men for more pay. Never equal in factory jobs, women even lost access to skilled positions in the factory as the century progressed.

In a typical Lancashire textile mill, work began in the carding room, where machines cleaned, fluffed, and carded the cotton into a long roll. Men supervised the machinery and the other workers; ten times as many women worked as "tenters," tending the machines. (By the 1890s, "tenter" had come to mean an unskilled attendant to a skilled workman.) Next, the cotton roll was fed to a "mule-spinner" which spun it into yarn. In 1834, some women had worked as mule-spinners in the Lancashire cotton mills; fifty years later, when a few women applied for the job, male workers struck in protest. They succeeded in keeping women away from the prestigious work of mule-spinning.[69] In the spinning rooms, in both the 1830s and the 1890s, men worked as overseers and spinners, women and children as their assistants: the "piecers" who pieced the thread together when it broke, supplied new rolls of cotton, and removed the finished skein of thread or yarn from the machine. The cotton was then wound and twisted for weaving. Men worked as overseers; women as winders. By the 1890s, the skilled jobs of placing warp threads on a beam and dipping the yarn in sizing was done exclusively by men paid double what the women winders earned.[70]

Most women in textile manufacturing worked as weavers: there were over 10,000 women weavers in the Lancashire cotton mills in 1834; by the 1890s, they numbered well over 100,000. In England, some weaving had been performed out of the home in workshops as early as the fifteenth century, but generally industrialization changed weaving from a male occupation performed at home with the aid of family members to a female occupation performed in the factories under the supervision of male overseers. Men tended to monopolize both the wider looms and the larger six-loom machines. Such factors contributed to men's higher earnings at weaving, even when both women and men were paid on a piecework basis.

Women factory workers also were subjected to more fines for petty infractions than men—a common employer's practice in the nineteenth century. In 1913, a woman doctor, M. I. Pokzovskaia, reported on factory conditions for Russian women and discovered that fines were widespread:

> Fines are imposed for: late arrival, work which is not found to be up to standard, for laughter, even for indisposition. At a certain well-known calendar factory in St. Petersburg the women workers receive 0.45 rubles a day, and the fines have been known to amount to 0.50 rubles a day.[71]

While Russian factories were notoriously bad, fines for women factory workers were standard elsewhere in Europe. When she started work at twelve in a Zurich silk factory in 1874, Verena Knecht found that her foreman regularly fined her 1 franc a week so that her wages would be lower than those of older women.[72] The foreman of an English patent screw factory, which employed hundreds of girls and young women, testified proudly in 1864 that he had ended accidents caused by girls' getting their dresses caught in the machinery by fines:

> Finding that some check was necessary, I put on a fine for it, and no one was caught [in the machinery] for three months after. It might be thought, as was indeed said to me, that the fright and the loss of the dress were punishment enough, but it is clear that the fine was wanted.[73]

Employers could treat factory women poorly because women both protested and organized less than men. "Their estimate of their own position is a low one, and they seem to think . . . that any display of independence on their part would oust them from the labour market entirely," asserted the Manchester and Salford Trade Unions Women's Committee in 1895.[74] But there was another factor as well. In 1834 an Englishwoman wrote in a radical newspaper, "There is a jealousy in the men against Female Unions. . . . What can be the cause of this but the tyrannical spirit of the male?"[75] Until the last decades of the nineteenth century most male trade unionists opposed both women's factory work and women's unions. In 1866, both the French and German delegates to the International Working Man's Association (the "First International") called for higher wages for men and demanded women's return to their "rightful work" of home and family.[76] Attitudes like this, and factory women's own view of their work as temporary, contributed to women's low wages and poor working conditions.

For many young women, the factory was a place to meet a future husband—as ever, a woman's chief economic security. Factory women prided themselves on their ability to find a good husband: the following nineteenth-century folk song—here in its "weaver" version—was sung throughout the British Isles, and other jobs and tools were substituted to fit local conditions.

> You'd easy know a weaver
> When she goes down to town,
> With her long yellow hair,
> And her apron hanging down.
> With her scissors tied before her
> Or her scissors in her hand,
> You'll easy know a weaver,
> For she'll always get her man![77]

Some married women continued working in the factories and in the mid-nineteenth century brought their babies to work with them. Some Russian factories allowed mothers to breast-feed their children, but docked them for the time it took. In 1875, Aurelia Roth, then fourteen, started working at a German glass factory. "I saw children of all ages in this dust-crammed place. Even new-born babies were there," she remembered. "Most of the little ones lay in a cradle near the polishing boxes and screamed unbearably. . . . Wherever I looked, I saw dirt—fruit scraps, tobacco ashes, etc. lay on the floor. . . . The children played in this morass, even food was given to them here."[78] Factory owners began to forbid the bringing of children, and by the twentieth century some insisted that women resign if they married.

Contemporaries portrayed women in factories as "girls": unmarried young women whose numbers were exaggerated because of the novelty of their work. "Where are the girls? I'll tell you plain," went an English folk song of the early nineteenth century,

> The girls have gone to weave by steam,
> And if you'd find 'em you must rise at dawn
> And trudge to the factory in the early morn.[79]

Although single women were far more likely to be working as servants, both at the time and afterward, the factory girl symbolized working-class women and, often, the horrors of working-class life in general. The following verse is from a lengthy English ballad of the early 1840s, written to enlist support for the passage of the act limiting factory work to ten hours a day:

> Ye! who alone on Gold are bent,
> Blush! at the Murder'd Innocent,
> Let not Old England's glorious pride
> Be stain'd by black Infanticide!!
> But let Humanity's bright Ray
> Protect from greedy Tyrant's sway
> The poor defenceless Factory Girl![80]

Reformers of all political persuasions stressed the sexual vulnerability of women in factories, prey to the advances of their bosses, but there is little direct or reliable evidence on this matter.

Factory work and domestic service were the best-paid and most secure respectable work a young woman could find in the cities. Other jobs were more arduous; as in earlier centuries, women performed the heaviest and most disagreeable tasks for low wages in order to survive. In the English census of 1841, the fourth-largest group of women who earned income were laundresses, a job infamous for its long hours, hard labor, unhealthy conditions, and poor pay.[81] A study of Parisian washerwomen in the late nineteenth century revealed that the women arrived at the Seine River at six in the morning to begin their work and rarely were done before seven at night. Seventy-five percent of them had hernias from carrying the heavy loads of wet clothing.[82] Laundering in an employer's home was no easier. People hired laundresses to do the "heavy washing": boiling sheets, woolens, and cotton clothing. The laundress was expected to work hard and long: in England, they were often hired to appear a few hours after midnight and work through until the evening.[83] By the end of the nineteenth century, most laundresses worked in small laundries rather than homes, but the hours remained lengthy: a study of Scottish laundresses in the 1890s showed them occasionally working a 37.5-hour nonstop shift; French laundresses in the 1880s still worked a 15- to 18-hour day.[84] "Our Toil and Labour's daily so extreme," wrote Mary Collier, the English washerwoman/poet who worked as a laundress until she was sixty-three, "That we have hardly even time to dream."[85]

The English census of 1841 revealed that the numbers of young women in other occupations declined as they increasingly entered domestic service and factories. There were 3,157 women over twenty who made boots and shoes, but only 560 under twenty; 22 female coopers over twenty, none under twenty.[86] The census of 1841 listed 11,394 females under twenty in London earning a living by needlework and "slop-work"—the coarsest and cheapest sewing—but most often this work was done by married women with children who needed to earn income in their homes. Many women's efforts to earn income were irregular and went unreported: Mayhew estimated there were over ten thousand female street-sellers in London alone. Women sold fruits and vegetables, milk, shoelaces, and bootblacking. They predominated in the sale of "the most disagreeable and offensive commodity," as an 1804 description of London put it—the spoiled offal fed to cats and dogs.[87] They sold old clothes, bits of furniture, flowers, and food and drinks they had made at home. In the 1850s, London street-food included

> hot-eels, pickled whelks, oysters, sheep's-trotters, pea-soup, fried fish, ham-sandwiches, hot green peas, kidney puddings, boiled meat puddings, beef, mutton, kidney and eel pies, baked potatoes, all kinds of tarts and fruit pies, hard candies, ice cream, cakes, tea, coffee, cocoa, ginger-beer, lemonade.[88]

As in past eras, the most lucrative item a woman could sell on the streets was the sexual use of her body. Well into the twentieth century, prostitution remained a woman's most profitable means of earning income, and a prostitute could earn in a day what other working-class women made in a week.[89] Traditionally, women were more likely to become prostitutes in two circumstances: in times of extreme want and in port cities where there were large numbers of single men with money. The first condition continued in the nineteenth century: from 1846 to 1848, when there were famines in Germany, more women became prostitutes, but left the trade rapidly when the economy improved.[90] As cities mushroomed in the nineteenth century, increasing numbers of women earned income as prostitutes. Prostitutes were difficult to count: many hid their business and moved rapidly in and out of it. There were estimated to be about 30,000 prostitutes in England and Wales in the 1850s and 1860s; between 100,000 and 200,000 in Germany by the turn of the twentieth century.[91] Women's low wages and men's higher income led many to become prostitutes. "The inadequate pay of the urban working woman sometimes drives her . . . to complete her budget by the sale of her body," wrote the Frenchwoman Julie Daubié (1824–1874) in 1866. "This is called the fifth quarter of the day."[92] Young prostitutes almost always came from the poorest groups and often had one or both parents dead. Little else distinguished them from other working-class women.[93] Young women moved easily in and out of prostitution and many regarded it as a temporary stage in their lives.[94] French studies revealed that most had first had sexual intercourse when they were sixteen, usually with a man of their own background.[95] In her memoirs, Celeste Vénard (1824–1909) recounted how she came to register as a prostitute in Paris in 1840. An illegitimate child, she claimed her mother's lover had tried to rape her when she was fourteen. She ran away and spent four days on the streets, when she was taken in by Thérèse, a registered prostitute. Two days later, she and Thérèse were picked up by the police and taken to the prison-hospital of Saint-Lazare, where she met Denise, an older girl. "All the things she told me just danced about in my mind," Vénard wrote. "I saw myself rich, and covered with lace and jewels. I looked at myself in my little bit of mirror; I was really pretty."[96] Denise gave her the address of a brothel, and when she was sixteen, Vénard registered as a prostitute.

Registration was the chief method by which nineteenth-century men sought to regulate the lives of prostitutes. France instituted registration during the Napoleonic era, and most other European nations followed suit. In Paris, the young woman went with her birth certificate (she had to prove she was at least sixteen) to the *deuxième bureau* of the prefecture of police. An officer interrogated her to discover the occupations of her parents; whether or not she still lived with them, and if not, why she had left them; how long she had lived in Paris; whether or not she was married; whether she had children living with her; and her reasons for registering. If she was married, her husband's permission had to be obtained. The mayor of the commune of her birth was supposed to verify her testimony. She was then given a medical examination in the police dispensary, and if she was free of symptoms then associated with venereal disease, she was allowed to ply her trade.[97] She had to submit to a police-supervised medical examination once a week, and if she contracted venereal disease, she was sentenced to a locked prison-hospital. Police had the power to require women they suspected of being prostitutes to submit to these medical examinations. In the prison-hospitals, the women were required to sew eleven hours a day and were punished by being confined in a straitjacket.[98] By 1857, this French system had been adopted by Prussia, the Netherlands, Belgium, and Norway on a national level; Italy and Denmark used it in parts of their nations, and almost every major European city outside of Great Britain and Spain had followed suit on the local level. England instituted such regulations in various port and garrison cities under the Contagious Diseases Acts, in force from 1864–1886.[99]

Nineteenth-century nation-states gave police such power over women because of fears of the spread of venereal disease and the hope of protecting male clients from an infected prostitute. Prejudice and ignorance combined to prevent any inspection of men. When the English reformer Josephine Butler suggested the latter in 1871, she received the following reply from the royal commission in charge:

> We may at once dispose of this recommendation, so far as it is founded on the principle of putting both parties to the sin of fornication on the same footing by the obvious but no less conclusive reply that there is no comparison to be made between prostitutes and the men who consort with them. With the one sex the offence is committed as a matter of gain; with the other it is an irregular indulgence of a natural impulse.[100]

Medical ignorance almost certainly worsened the health of infected women. Nineteenth-century doctors spread infection during the medical examinations. In the prison-hospitals, hygiene was minimal: at Saint-Lazare in Paris, itself better than many provincial French hospitals, inmates used common sinks and bidets until the 1890s. Most hospitals treated women by irrigating their vaginas with some douche thought to alleviate venereal symptoms; usually the douche was administered to many women with the same unwashed syringe—an ideal method of spreading venereal disease.[101] French prostitutes called the doctor's speculum the "government's penis."[102] "It is *men,* only *men,* from the first to the last, that we have to do with!" an English prostitute complained to Josephine Butler.

> To please a man I did wrong at first, then I was flung about from man to man. Men police lay hands on us. By men we are examined, handled, doctored, and messed around with. . . . We are up before magistrates who are men, and we never get out of the hands of men.[103]

The chief effect of such regulation was to make the prostitute even more of an outcast than she had already been. Prostitutes had always functioned on the margins of European communities, never fully legal, always vulnerable to the authorities. Registration and regulation tended to increase the numbers of prostitutes who remained in the business for life. Registered women found it more difficult to move into other types of work, and some argued this was the goal of regulation: to make prostitutes into a separate, marginalized, and controllable group.[104]

In the nineteenth century, as in other eras, a few prostitutes were able to achieve great wealth and notoriety. The *grandes horizontales* of the great European capitals became famous through plays, operas, and the popular press. Theirs were the most lurid "rags-to-riches" stories, and some of their careers provided material for the melodramas of the day.[105] "La Païva" (1819–1884), for instance, was born Thérèse Lachmann, the daughter of weavers in the Moscow ghetto.[106] Married at seventeen to a tailor, she had a son. Two years later, she abandoned her family and moved to Paris, where she worked as a prostitute in the slums. By 1841, she had amassed enough fashionable clothing to enable her to appear at the luxurious German spa of Bad Ems, where she formed a liaison with the well-known pianist Henri Herz. He established her in Paris, where she maintained a musical salon, visited by Wagner, Gautier, and others. Turned out by Herz a few years later, she went to London and Baden, another spa, and in 1851 married the wealthy Portuguese Marquis de Païva. From 1856 to 1866, she built herself a showy and sumptuous private house in Paris, funded in part by her Prussian lover, Henckel von Donnersmark. Her Hôtel Païva became a center of Prussian espionage, and when the Prussian army marched into Paris in 1871, all the grand private houses remained shuttered and closed except for La Païva's. The next year, the marquis committed suicide, and La Païva, then fifty-two, married von Donnersmark. He gave her the former Empress Eugénie's diamond necklace as a wedding present. People hissed when she appeared in public, and in 1878 the government forced her to leave France. She and von Donnersmark moved to his castle at Neudeck, in Germany, where she had a stroke. Near the end, she smashed the Venetian mirror in her bedroom so as not to see her physical decline; after her death at sixty-five, von Donnersmark had her corpse preserved in alcohol at Neudeck.

Just as prostitution enabled some young women to achieve fame and fortune, so the theater, opera, and ballet provided poor young women with beauty or talent a route to improve their condition. Beginning in the late sixteenth century, French and Italian women, usually from families of acting players, began to take stage roles. In France Madeleine Béjart (1618–1672) assumed leadership of her family's company of touring players and first made her reputation as an actress. In 1643, she met the young playwright Molière and began a lifelong collaboration with him. Louis XIV brought the company to Paris, where Béjart continued to perform in Molière's comedies; Armande Béjart, raised by Madeleine and perhaps her daughter, married Molière. In 1680, the Béjart company merged with Marie Champmeslé's theatrical group: together they formed the Comédie Française, the repertory company which became a central institution of the French theater. Elsewhere in Europe, family connections also first provided women with access to the theater: the English actress Anne Bracegirdle's (1673?–1748) own parents died when she was young, but a theatrical family raised her; Caroline Neuber (1697–1760), the German actress-manager, was introduced to the stage by her husband.

But once women had been accepted on stage, they could rise either by their talent or by forming liaisons with wealthy or important men, or both. Nell Gwyn (1650–1687) began as an orange-seller at a London theater, where she first appeared on stage at fifteen. She caught the eye of the English king, Charles II, and remained one of his mistresses until his death. Peg Woffington (1714–1760) sang as a child in the streets of Dublin to earn income before she became a celebrated actress in both Ireland and England. Fanny Elssler (1810–1884) rose from obscurity—she was one of a large poor Viennese family; her father was Haydn's valet—to become one of the most famous ballerinas in Europe, whose performance was commanded by Queen Victoria of England.

By the nineteenth century, the state-supported theatrical, operatic, and ballet schools of the European capitals offered a surer avenue to success. Jenny Lind (1820–1887), the "Swedish nightingale," came from a Stockholm slum, but was admitted as an "actress-pupil" to the Swedish Royal Opera House when she was ten. After a very successful international singing career, she ended as a professor of singing at the Royal College of Music in London. Rachel (born Élisa Félix, 1821–1858), sang with her sister in the streets of Paris: their parents were poor Jewish peddlers. Her talent was recognized and she entered a theater school where she first performed on stage at sixteen. Anna Pavlova's (1882–1931) father died when she was two; her mother worked as a laundress in St. Petersburg. She was able to become a ballerina by entering the Imperial Ballet School when she was ten.

As a famous actress, singer, or ballerina, a woman could achieve influence and adulation. Rachel saved the Comédie Française from being burned down by a revolutionary Paris mob in 1848 by her stirring rendition of the *Marseillaise*. [107] The rivalry between Sarah Bernhardt (1844–1923), the illegitimate daughter of a Dutch courtesan, and Eleonora Duse (1859–1924), from an Italian theatrical family, caught the public's attention. The actresses competed for roles and audiences and captured headlines not only because of their performances on stage but also because of their love affairs. The "divine Sarah" mastered all types of theatrical roles, from classical heroines to "breeches parts," like Hamlet or Cherubino. (From their reappearance on the stage in the seventeenth century, actresses often performed men's roles. This died out in the early twentieth century.) Bernhardt lived in London for many years and toured successfully abroad. She continued to act into her seventies, after one leg had been amputated, thrilling and impressing audiences to the end. She was also famous for her corps of lovers and bore an illegitimate son to a French prince in 1864. Her rival, Eleonora Duse, excelled as a tragedienne, and moved easily from the melodramas of her youth to the plays of Ibsen and Chekov with which she ended her career. Although married to an actor, she founded a theater company with one lover and from 1897 to 1902 would only act in plays written by another lover, the poet Gabriele d'Annunzio.

Many actresses, singers, and ballerinas used the sexual freedom traditionally associated with the stage. The dividing line between respectability and prostitution blurred for these women: they lived in a unique milieu where they could function much as men, choosing and discarding lovers at will. Some took lovers for pleasure; others wanted gifts and money. Some of the famous nineteenth-century courtesans started as actresses: Léonide Leblanc entered the theater at fourteen; later she became known as "Mademoiselle Maximum," the most expensive of the Parisian courtesans. Rachel's reply to a prince who sought to buy her favors became famous. He asked "Où? Quand? Combien?" She answered, "Chez toi. Ce soir. Pour rien."[108]

A number of women of the stage chose never to marry. Fanny Elssler, the ballerina, formed a liaison at sixteen with the Prince of Salerno; she lived later with the Baron von Gentz, but decided not to marry in order to maintain her freedom and independence.[109] Other performing women found the stage a path to brilliant marriages. Trained by her parents, also opera singers, the opera star Adelina Patti (1843–1919) made her debut at seven. Her first marriage was to a French marquis; she divorced him and then married a famous tenor. Her third and final husband was a Swedish baron. Mathilde Felixovna Kshessinska (1872–1971) rose to become *prima ballerina assoluta* at the Russian Imperial Ballet, where she met and married Grand Duke Andrei, a cousin of the czar.

These were the success stories which fed the fantasies of working-class girls. While the theater, opera, and ballet did provide young women with a way to rise in society, for most the glamour and gains were temporary. The young women who trained in the state schools, who staffed the choruses and corps de ballets, were notorious for trading their sexual favors. Like prostitutes, they were vulnerable to pregnancy, disease, and male violence. They were viewed as semi-prostitutes, and middle- and upper-class men thronged about the stage doors and dressing rooms. The unsuccessful left no records, and only glimpses can be gained into their lives. The mother of Emma Givry, for instance, was a sixteen-year-old dancer at the Paris Opera who became pregnant by a member of the fashionable Jockey Club and bore an illegitimate daughter in 1842. The daughter had great talent, but burned to death when her skirt caught fire during a rehearsal—a not uncommon accident in this era of gaslit stages.[110] But with all its uncertainties, the stage remained one way a poor young woman could achieve. Marie Lloyd (1870–1922), born Matilda Alice Victoria Wood, was the eldest of eleven children of a poor East End London family; her father was a waiter. She began performing at fourteen, as part of the Fairy Bells minstrel troop; by sixteen, she had changed her name and become a star of the music hall stage; by eighteen, she was married and owned her own house. She once bought boots for eighty poor East End children, and she became a beloved music hall star, both in England and abroad. Her second husband was a coster "king" who became wealthy selling merchandise; her third husband was a jockey. Even successful working-class women found in marriage a security they did not possess if they remained single. Through her talents, Marie Lloyd married comfortably, gained financial security, and achieved international fame. Few working-class women could do so well.

Women with Children

It was the mother's job to contrive her family's survival by taking odd jobs, making do, and managing the home economy. Women who had children usually needed to keep earning: until the last quarter of the nineteenth century male wages alone were rarely sufficient to support an entire family. In contrast to male workers, who tended to perform the same work for a lifetime, women often changed jobs so they could look after their children. Women without children worked in other people's homes or in factories; if a mother continued factory work, she had to arrange to have her children looked after. (Married women and mothers rarely remained in domestic service.) In the nineteenth century, some mothers brought children to the factories with them, but this practice died out and most arranged for another woman to care for small children. "My sister and I were carried out of our bed at 5:30 every working day to be left in the care of Granny Ford," remembered Harry Pollitt of his mother Louisa's care in Lancashire in the 1890s. "Mother came rushing home from Benson's [cotton mill] in the breakfast half hour to give us our breakfast."[111] "My children stays with my sister," a shrimp-seller with two young girls told Mayhew, the English social investigator. "They don't go to school, but Jane [the sister] learns them to sew."[112] In 1894, the Lancashire *Labour Gazette* did a survey of 165 children to find what arrangements their mothers who worked outside the home had made for their care. One fourth were left with grandparents, one fourth with other relatives, and close to one half with neighbors. Only nine children were left with no adult, either alone or to look after each other.[113]

Given the difficulties of arranging child care, most women with children preferred to earn income in their own homes. These women either took in lodgers or did piecework, or both. Both types of work were grossly under-counted in censuses and surveys: the work was "hidden," done at home out of public scrutiny, and many did not want to declare the extra income they made. In the 1851 census of Colchester, England, for instance, Ann Dun-ningham, a sixty-three-year-old widow, was classified as having "no occupa-tion," although she had four "tailoresses" lodging in her home, as well as her daughter and two grandchildren.[114] Providing lodging, food, or domestic services meant that the wife and mother supplied services normally provided for her husband and children to others, usually men. Letting lodgings ranged from running a boardinghouse or small hotel to unrolling an extra pallet in a one-room dwelling. In the growing industrial cities especially, space was at a premium and dwellings were often shared. In 1900, one quarter of the households in Hamburg kept lodgers.[115] Services could range from simply renting space to providing the "attendance, light, and firing" that were standard in English lodging houses: cleaning a lodger's room and supplying clean bed linen, emptying chamber pots and filling basins with clean water, cleaning fireplaces and lamps, and carrying up coal or wood for fires.[116] Women who ran boardinghouses supplied meals as well, usually breakfast and dinner.

The other chief way for a woman to earn money in her own home was to do piecework. Piecework in the home gave the mother work which was compatible with child care: it enabled her to remain at home to supervise her children and to enlist their help. The growth of industry tremendously ex-panded domestic piecework.[117] The uneven pace of mechanization caused booms in various home industries: machine-spun thread led to more weaving at home; mass-produced iron rods to more nail and chain making. These "cottage industries" originated in the countryside; as cities grew, rural women tended both to concentrate on piecework in the home and to "import" this work to the cities, which soon became major centers of home production as well. When Lucy Luck and her family moved into London in the 1870s, she "kept up with her straw work"—plaiting split straws for bonnets and hats—

by transferring to the London branch of the firm which had employed her in the country; she got her eldest daughter work from there as well.[118]

Piecework in the home, especially work which could be done by women, was traditionally ill-paid. Work done by women at home was considered "dishonorable," and earnings below the subsistence level were standard. Men could look for other jobs; women with young children could not. "The masters prefer giving out the work to those women who keep young children," testified a witness at an 1833 English inquiry into lace-making, "because it can be done at the lowest price."[119] The assumption that most women who worked for income at home were married was used to justify the low rates of pay. One might think, argued the Berlin Chamber of Commerce in 1907, "that in home industry most only earn starvation wages." But since most of these workers were married,

> the low earnings of the women home-workers hence become not proof of a low standard of living, but indeed . . . can be used to conclude *that the income of the family to which the woman belongs allows her to dispense with higher earnings for herself.*[120]

The chamber of commerce saw no reason to raise women's rates of pay.

Men's response to the abysmal conditions of women's piecework in the home was usually to try to prevent women from working at all in fields traditionally reserved to men. "It is the ease with which women and children can be set to work, that keeps these weavers in poverty and rags and filth and ignorance," argued Francis Place, who tried to organize London tailors to keep women out, in 1824.

> In a trade, say that of mill-wright, the man alone can work at . . . his wages must be sufficient to keep himself, his wife and a couple of children, and it is so. . . . If a man, in the ordinary run of his trade, has a wife and two children, who can, and by the custom of the trade do, work with him, they will altogether earn no more than he alone would earn if only men were employed.[121]

"Married women must be maintained by their husbands, know housekeeping, and care for and educate their children," argued Berlin tailors in 1803, trying to exclude women from making clothes at home. "The unmarried may work as domestics (for which there is no lack of opportunity locally) or engage in other feminine occupations outside regular manufacture."[122]

The fear that women's entry into a field would lower wages and depress the status of a craft was based on reality. In the eighteenth and nineteenth centuries, skilled women workers found that subcontracting and home production reduced even trades that required training to sweated industries.

Skilled dressmakers, lace-makers, milliners, artifical-flower makers, glove-resses, and cobblers all found their trades becoming "sweated"—character-ized in the nineteenth century by "exacting hard work from employees for low wages."[123] When this happened to men, they could find work elsewhere; women with children could not. By the early twentieth century in Germany, for instance, five times as many women as men did piecework in the home.[124]

This decline of women's skilled work was especially apparent in the largest of the home industries, garment-making. Prior to the nineteenth century, girls were fortunate to be apprenticed to dressmakers and seamstresses: sew-ing was one of the few ways in which a skilled woman could earn a good income. In 1789, for instance, a petition of Frenchwomen in the "needle trades" asked not for an improvement in working conditions, but only that men be forbidden to compete with them:

> we ask that men not be allowed, under any pretext, to exercise trades that are the prerogative of women—such as seamstress, embroiderer, *marchande de mode*, etc., etc.; if we are left at least with the needle and the spindle, we promise never to handle the compass or the square.[125]

Industrialization transformed this. The spindle became mechanized first, and spinning moved into the factories. The needle, its use no longer protected by guilds, whether female or male, became the basis of sweated garment industries in the home, which developed in England by the 1840s, in France by the 1870s, and in Germany by the 1890s. And the seamstress, once considered a skilled artisan, became the most common pieceworker and the most exploited of laborers. As a famous English ballad of 1843 put it:

> With fingers weary and worn,
> With eyelids heavy and red,
> A woman sat in unwomanly rags,
> Plying her needle and thread:
> Stitch! stitch! stitch!
> In poverty, hunger, and dirt;
> And still, with a voice of dolorous pitch,
> She sang the "Song of the Shirt"!
> . . . .
> Work—work—work!
> My labour never flags;
> And what are its wages? A bed of straw,
> A crust of bread and rags.[126]

With an unlimited supply of women who needed to earn income at home, employers controlled a surplus labor force which could be hired or fired at will. The garment industry was seasonal: Lucy Luck found no work in hats for half the year, and Mayhew discovered that most needleworkers could get no work for two to three months a year.[127] French seamstresses called these periods of unemployment or underemployment "la morte."[128] On the other hand, during busy periods, employers demanded inhuman amounts of labor both from women who worked at home and those who sewed in small workshops. "Take this [velvet skirt] to your mother, and tell her if she doesn't get it done by nine o'clock tomorrow morning she will get no work from this house," Mayhew heard a porter of a "first-rate" dressmaking establishment in the fashionable West End of London tell a seven-year-old boy in 1850. Mayhew went home with the boy to a "dirty, narrow street":

> The garret he enters is a sort of triangular-shaped room, about twelve feet square. . . . In the middle of the room is the deal table and around this are seated seven women, dirty, thinly clad, with pale and hollow countenances, weak red-looking eyes, and lean emaciated frames. The one working at the head of the table is Mrs. _____, the boy's mother. Her husband is dead; she is about middle age. She rises and takes the skirt from the boy and demands of him what is wanted. He answers that it is to go in at nine. "Nine in the morning!," she exclaims. "Why I have got six from the City to go home at eight!"[129]

A 1907 survey of Parisian seamstresses who worked at home revealed that 56 percent had to work more than ten hours a day to survive; 13 percent worked more than twelve hours a day.[130]

Pieceworkers in the home also had to pay for supplies and "trimmings," which severely reduced their income (Mayhew estimated such costs equaled one sixth of their gross earnings);[131] later in the century, they had to pay rent on their sewing machines. Throughout the nineteenth century, scandals periodically erupted over the ill-health or deaths of women forced to sew endless hours in workshops which produced fashionable clothing, but the pieceworker in the home was even less protected because working conditions were not inspected. Like female factory workers, pieceworkers were fined and paid nothing for work considered defective. An investigation into the English glove-making trade in 1864 revealed that gloveresses were paid 2 shillings, 6 pence to 3 shillings for a dozen pairs of gloves. "If one of the gloveresses spoil a pair of gloves in any way, she is charged the full retail price, three shillings, sixpence per pair, and the spoiled gloves are thrown on her hands."[132]

Employers also forced payments down in an era of generally rising prices and wages. An elderly woman who had sewn waistcoats for the same company for twenty-six years told Mayhew that when she started she earned 1 shilling, 9 pence for each vest; now she made only 1 shilling, 1 pence. "Work's falling very much," she complained. "The work has not riz, no! never since I worked at it. It's lower'd but it's not riz."[133] Fifty years later, little had changed. In the early twentieth century, a survey of female French home garment workers revealed that 60 percent earned less than 400 francs a year; forty years earlier, the minimum budget of a Parisian working woman reached 500 francs a year.[134] In Berlin, wages which were only at the subsistence level in the 1880s fell as much as 60 percent in the 1890s, leading the Berlin Arbitration Office to state in 1896 that "wages paid in many cases have sunk so low that even with the most strenuous and diligent effort, no existence worthy of a human being is possible for the worker."[135] Most female pieceworkers in Berlin earned less than 10 marks a week at a time when rent and the meagerest diet alone took 8.5 marks. In 1900, Emma Döltz (1866–1950), herself a home seamstress and the daughter of one, published a poem entitled "The Woman Home Worker." Reminiscent of Thomas Hood's "Song of the Shirt" of 1843, it spoke of the unchanged exploitation of women who sewed at home:

> Quickly wash out the eyes,
> Dear God, it's already struck five,
> How short the night, how tired I am,
> I feel as if I had been beaten.
> Quickly make a fire in the cold room,
> Quickly take care of breakfast,
> So that one can just start sewing
> Since I have to deliver, deliver tomorrow.[136]

Well into the twentieth century, governments which came to regulate both prostitution and industries which employed large numbers of women refused to interfere with women's piecework in the home. Work in the home, however exploitative and dangerous, seemed "natural" for women.

Old Women

One unchanging tradition of European history was that the poorest members of society were women alone, especially old women alone. "And after all our toil and labour's past," wrote Mary Collier about domestic service in 1739,

> Six-pence or eight-pence pays us off at last;
> For all our pains no prospect can we see
> Attends us, but old age and poverty.[137]

Parish registers and government reports consistently show more women than men in need of charity: in the 1860s, for instance, the Paris Pauper's Office (Bureau des Indigents) found 50 percent more unemployed women than men.[138] Older women tended to be alone more often than men: throughout the eighteenth and nineteenth centuries, as in earlier eras, widowers remarried at a far higher rate than widows. Poor widows remarried at an especially low rate.[139]

Old women, especially if they were alone, still needed to earn income although the jobs available to them were even fewer and poorer than those for women in general. In his investigations of London's poor, Henry Mayhew found a seventy-year-old woman living in a small back room where she slept on a hay mattress on the floor. She survived by sewing convicts' suits, the worst-paid sewing. She did not qualify for parish assistance and had done this work since her husband died. "I very often want," she told Mayhew. "I wanted all last Sunday, for I had nothing at all then. I was a-bed till twelve o'clock—lay a-bed 'cause I hadn't nothing to eat."[140] This seamstress was by no means the worst off: she still had a home and could support herself, however meagerly. In a "wretched, tumbledown" lodging house used by prostitutes, Mayhew found an old Irishwoman who cleaned the water closets in exchange for a place to sleep:

> This woman was lying on the floor, with not even a bundle of straw beneath her, wrapped up in what might have been taken for the dress of a scarecrow feloniously abstracted from a corn-field. . . . Her face was shrivelled and famine-stricken, her eyes bloodshot and glaring, her features disfigured slightly with disease, and her hair dishevelled, tangled and matted.[141]

Elderly women lived with others, often other widows, so they could stay out of the poorhouse. "We all live in this one room together," a widowed shoe-binder told Mayhew;

> there are five of us, four sleep in one bed; that is the man and the wife and the two children, and I lie on the floor. If it wasn't for them I must go to the workhouse; out of what little I earn I couldn't possibly pay rent.[142]

"I would rather starve than go into the workhouse," declared a woman who sold bootlaces in the street, and women begged rather than give up their independent lives.[143] As beggars, women tended to receive more sympathy than men, and even in the poorest neighborhoods found support and charity. "There's always some neighbor to give a poor woman a jug of boiling water," an elderly woman who could not afford a fire explained.[144] Mayhew encountered an old beggarwoman on the Mall in London and accused her of lying after she asked for charity both for a crippled husband and for herself as a widow. "I's up to so many dodges I gets what you may call confounded," she explained;

> sometimes I's a widder and wants me 'art rejoiced with a copper, and then I's a hindustrious needlewoman thrown out of work and going to be druv into the streets if I don't get summat to do. Sometimes I makes a lot of money by being

a poor old cripple as broke her arm in a factory, by being blowed hup when a steam-engine blowed herself hup, and I bandage my arm and swell it out hawful big, and when I gets home, we gets in some lush and 'as some frens, and goes in for a regular blow-hout, and now as I have told yer honour hall about it, won't yer give us an 'apny as I observe before.[145]

Women seem to have begged rather than turned to crimes like pickpocketing or theft. While urban women committed crimes in higher rates than peasants, they usually worked as assistants and subordinates to men, as decoys, swindlers, or fences.[146] Women's chief "crime" throughout this era was prostitution.

With luck, hard work, and some support from others, an elderly working-class woman could achieve a comfortable old age. Mayhew found a coster-monger's widow living in a "large, airy room . . . laying out the savoury smelling dinner looking temptingly clean." She attributed her ease to the fact that her daughters and sons helped her and none of them drank. She had a "turn-up bedstead thrown back, and covered with a many-coloured patch-work quilt," and "the coke fire was bright and warm, making the lid of the tin saucepan on it rattle up and down." Several warm shawls hung on the walls, which had been papered with " 'hangings' of four different patterns and colours." Her mantle was decorated with crockery, glasses, and a picture of Prince Albert; the wall above the fireplace "was patched up to the ceiling with little square pictures of saints."[147]

Women and Religion

The old woman's pictures of saints signified an important nineteenth-century change in religion in Europe. Working-class men abandoned the Church—especially the Catholic church—in large numbers, and women came to be, as the bishop of Chartres declared in the late 1820s, "the sole consolation of the church."[148] Although church attendance for both sexes was lower in industrial regions, women came to predominate as churchgoers. A study of the percentages of people who took Easter communion in Orléans in 1852 showed that 31.4 percent of the women and 88.9 percent of the adolescent girls participated, but only 6.4 percent of the men and 28.6 percent of the adolescent boys. In the industrial town of Pithviers, ten times as many women participated as men.[149]

For women, religion remained what it had once been for men as well: a chief consolation in a harsh life and the major institution which affirmed their dignity. For poor women especially, churches continued to be an important center of their lives, providing ritual and hope, charity and solace. The first crèche for infants of mothers who worked outside the home was established by the Soeurs de la Sagesse in Paris in 1844.[150] In Catholic nations, primary education for girls remained available only through the teaching orders of nuns, even after school systems began to be secularized in the nineteenth century. In France, which made the greatest effort among Catholic nations to educate children in state schools, nuns still taught nearly half the girls in primary schools in 1901.[151]

What social and charitable organizations existed for poor women—and hundreds were established in the course of the nineteenth century—were usually under the aegis of a church. For poor urban women, as well as for peasant women, the churches remained important: the widow Mayhew visited kept a special bonnet in a bandbox just for churchgoing on Sundays. In Lyons in the late 1850s, a sixty-year-old widow who worked as a silk-reeler was indignantly interrogated by a republican male neighbor. He wanted to know how she could belong to the Compagnie de Jésus, a lay Catholic women's order, when her late husband had been an ardent republican. The *compagnie*, affiliated with a local convent, provided her with vouchers for bread and coal, and she replied "that she was old and poor and that in going to church she found a distraction. Further, when she fell sick the church helped her out a little, whereas 'from the republicans one never gets anything but misery and contempt for riches.' "[152] Such attitudes underlay poor women's continued support of religion and the churches.

The costermonger's widow with the pictures of saints on her wall lived in a comfortable, clean room; she had heat and enough to eat. She dressed warmly and even had her special bonnet for Sundays. In addition to her bed and fireplace, she had a chest of drawers and "a long dresser with mugs and cups dangling from the hooks, and the clean blue plates and dishes ranged in order at the back."[153] After a lifetime of scrimping and coping, of laboring and saving, she did not have to earn additional income in her old age. Until the late-nineteenth-century rise in the standard of living, this was the best a working-class woman could attain.

3

REVOLUTIONS AND REFORMS

❦

LABORING WOMEN had a long tradition of violent protest, in the countryside and the towns, in peasant revolts and bread riots. Individual names passed down: Black Anna of the German Peasants' War of the sixteenth century, Jeanne "Hachette" of fifteenth-century France, "Long Meg" and "Mad Meg" of English folklore.[1] All were portrayed as enraged women of the people whose recourse to violence signified the breakdown of law and order. In the large new cities and the new nation-states, women continued to participate in food riots: when the price of bread or other essentials rose too high, they intervened collectively, going to the markets and stores and distributing the merchandise at what they considered a fair price.[2] In addition, women continued to protest economic conditions. In 1689, for instance, female and male London silk-workers rioted to protest the government's support of woolen cloth.[3] Women both participated in and initiated protests against the introduction of machinery to trades they traditionally performed. In the French city of Troyes in 1791, women spinners gathered into a mob to protest the use of spinning jennys and prevented their installation in the city. In 1819 in Vienne, another French city, they led the protests against a shearing machine, throwing stones and shouting, "Down with the shearer! . . . Let's break it, smash it! Go to it!"[4] In the riots and demonstrations that became the great revolutions of the late eighteenth and nineteenth centuries, working-class women of the cities joined and sometimes initiated protests. Especially active in the first stages of revolutions, they seized the opportunity to create new organizations and institutions for their own interests. While they ultimately were defeated, often by the same male revolutionaries they had aided, women had marched and built barricades, joined unions and formed cooperatives, initiated strikes, and participated in violence. In 1789, 1848, 1871, and 1917 women, often in opposition to like-minded men, seized the initiative. Their actions and the institutions they created in their own interests testify to the ingenuity and will of the common women of the cities.

Urban Women in 1789, 1848, and 1871

Women took action and participated in the opening events of the French Revolution of 1789–1795, the Revolutions of 1848, and the uprising of the Paris Commune in 1871. In 1789, Frenchwomen signed petitions stating their grievances: the erosion of guilds, lack of police protection, male competition in their trades.[5] One revolutionary pamphlet called women "the Third Estate of the Third Estate."[6] Parisian market women sent a song to the Third Estate in the summer of 1789:

> If the clergy, if the nobility
> My good friends
> Treat us with such rudeness
> And disdain
> Let them all fancy themselves capable of running the State
> While waiting, we will drink
> to the Third Estate.[7]

Women joined in the storming of the Bastille on July 14, 1789; afterward Marguerite Pinaigre applied for a pension as a "conqueror of the Bastille," saying that she had run "to several wineshops to fill her apron with bottles, both broken and unbroken, which she gave to the authorities to be used as shot in the cannon to break the chain on the drawbridge of the Bastille."[8] In October, thousands of Parisian women marched first to the city hall to demand bread, and then to the Champs Élysées where one witness saw "detachments of women coming up from every direction, armed with broomsticks, lances, pitchforks, swords, pistols and muskets."[9] Following eight or ten women with drums, they—and a large group of men—marched twenty-odd miles to the royal palace at Versailles, to "bring back the baker, the baker's wife, and the baker's boy"—the king, Louis XVI, the queen, Marie Antoinette, and the dauphin. At Versailles, a delegation of women met with the king and eventually escorted the royal family back to Paris. The "October Days" confirmed the power of the revolutionary National Assembly over the French monarchy.

In the urban Revolutions of 1848, which erupted in Paris, Vienna, Milan, Rome, Berlin, and other European cities and towns, women continued to initiate and join demonstrations and fighting. They defended these short-lived republics and were drawn wearing improvised uniforms and carrying weapons.[10] In cities, the first defense was the barricade, built by massing stones, wood, and furniture across a street to provide fortification. The following description from Dresden in 1849 is typical of women's participation in such actions:

> Many women, who came from all ranks of society, took part in the struggle of the Saxon people in Dresden from May 3 to 9. Many helped build the barricades, dragging stones and furniture, others supplied the fighters with meals on the streets which they had cooked. Still others took care of the wounded, bandaging their wounds in the rain of bullets on the open street or dragging them into their houses. A maiden whose fiancé, a gymnast, had fallen on the first day, defended a barricade for three days with the courage of a lion, shooting down many soldiers before she herself was felled by an enemy bullet.[11]

In Paris, a group of women "performers, workers, writers, and teachers" petitioned for the vote in March of 1848; others formed political clubs and working associations: the United Midwives, the Fraternal Club of Laundresses, the Association of Female Servants.[12]

Women took an equally active role in the uprising of the Paris Commune in 1871. In the opening days, mobs of women and children surrounded the government's troops, eventually convincing them to arrest their own general. "We were greatly mistaken in permitting these people to approach our soldiers," another general stated after the event, "for they mingled among them and the women and children told them: 'You will not fire upon the people.' "[13] The market women of Les Halles built a 65-foot-long barricade in half a day.[14] Other women joined revolutionary clubs and formed their own. A laundress, Mme. André, was secretary of the Club des Prolétaires, and in May *citoyenne* Valentin addressed the club, urging that women guard the gates of Paris as well as distributing clothes to poor children.[15] In April, women formed the Women's Union for the Defense of Paris and the Care of the Wounded; among its 128 members were seamstresses and dressmakers, sewing machine operators and laundresses.[16] As women participated in the early stages of revolution, so their participation came to signal the start of revolution: large numbers of women taking to the streets in protest embodied a government's loss of control.[17]

Once the first battles had been won, women organized groups and institutions to represent their interests. In France in 1790 and 1791, women petitioned the government: over three hundred signed one petition requesting that they be allowed to bear arms, declaring, "We are *citoyennes,* and we cannot be indifferent to the fate of the fatherland."[18] They rioted for bread, for the overthrow of the government, for price controls. They initiated episodes of *taxation populaire,* seizing and looting food and supplies. As "passive citizens" in the new French republic, women became part of the body politic, and as members of crowds, in the public revolutionary tribunals, and from the spectator galleries of the revolutionary clubs, they made their presence felt. On Sunday, February 25, 1793, for instance, women petitioned the National Convention for lower bread and soap prices. The convention adjourned until Tuesday. "Far from calming and satisfying [the women]," reported an eyewitness,

> this resolution embittered them still more, and upon leaving . . . the women in the corridors said aloud, to whoever was willing to listen: "We are adjourned until Tuesday; but as for us, we adjourn until Monday. When our children ask us for milk, we don't adjourn them until the day after tomorrow."[19]

In February 1793, several hundred radical Parisian women formed the Society for Revolutionary Republican Women. They stated that their purpose was "to be armed to rush to the defense of the Fatherland" and declared that "all the members of the society make up a family of sisters."[20] Founded by Claire Lacombe (b. 1765), an actress, and Pauline Léon (b. 1758), a chocolate manufacturer, the society included poor working-class women and spoke for their interests. Associated with members of the radical *Enragés* group, who wanted the revolution to continue further, the society aided in the overthrow of the moderate Girondins, petitioned for price controls and other revolutionary legislation, and demanded that all women wear the revolutionary red, white, and blue cockade. The Society for Revolutionary Republican Women was one of the first organized interest groups of working-class women.

In 1848, such women formed equivalent political clubs in France: the Society of the Voice of Women, the Women's Union, the Committee for the Rights of Women.[21] Others joined men's clubs where they were allowed to speak. The records of the Club Lyonnais mention "a simple proletarian woman, born from a poor family, wife to an honest republican," who asked that women be given rights and enough wages so they could live without selling their bodies. She also asked that "the seduced and abandoned young girl be able to keep her child without dishonor and that the shame should fall on her seducer."[22] A group of working-class women called the political club they formed the *Vésuviennes* after the volcano, explaining that their "lava [although] contained for so long" was "not at all incendiary, but completely regenerative." Like their predecessors of 1789, they too claimed the right to fight for the republic, and a number of women enlisted for a year.[23]

Women also organized working associations, ranging from benefit societies to workers' cooperatives. Some formed because equivalent male associations excluded women; other groups organized with male encouragement. In May 1849, for instance, "the associated male workers of Berlin" announced that they had formed a "Stocking-Cooperative" to "help the women work-ers." They asked for donations of wool which they proposed to distribute to women of their committee, who would then parcel it out to the female stocking-knitters. The knitters would complete the stockings in eight days and receive payments higher than usual in hopes "of raising the retail price a little."[24] By the summer of 1849, 104 of these women's work associations in France had joined together in a federation. French laundresses had success-fully petitioned the revolutionary government to shorten the workday, and the law now limited working hours to ten a day for both sexes. Workshops for extremely poor women to earn payment for piecework sewing opened, as they had in 1790 in Paris, where women received payments for piecework spinning.[25]

In the few months that the Paris Commune existed in 1871, women demanded even more. In a variety of ways, they insisted that they were and should be men's equals. Claiming an equal role in the defense of the city, some asserted that women were stauncher Communards than men. "The men are cowardly; they say they are the masters of creation, but they're only imbeciles," stated the call to arms issued by the Clubs of Female Patriots.[26] A meeting of the Commune Commission for Education decided to raise the salaries of women teachers to equal those of men, "seeing that the necessities of life are as imperative for women as for men, and, as far as education is concerned, women's work is equal to that of men."[27] The day after this resolution passed, triumphant government troops entered Paris, and in the last bitter days of fighting, women played a paramount role. "Our male friends are more susceptible to faintheartedness than we women are," wrote Louise Michel (1830–1905), the radical teacher who fought for the Com-mune. "During Bloody Week, women erected and defended the barricade at the Place Blanche—and held it till they died."[28] Thousands of women died during Bloody Week, 1,051 other women lived to be arraigned by the govern-ment after the Commune's defeat. A study of 115 of these women revealed that most were between thirty and fifty years of age, 75 percent were married women or widows, and 83 percent held working-class occupations, half in the clothing trades.[29] Testimony at their trials revealed some of their actions. Numerous witnesses, for instance, heard Florence Wandeval, a twenty-three-year-old day laborer, scream, "I've just set the f—— Tuileries on fire. Now a king can come; he find his palace in ashes [sic.] . . . and from tonight on, many others will burn!"[30] Wandeval was a cantinière to a revolutionary batallion; as cantinières, women cooked for, fed, nursed, and sometimes fought beside male troops. Requesting recognition of the heroism of their cantinière for bandaging their wounds under fire, the soldiers of the 66th Batallion wrote to the Commune stating that "she displayed conduct beyond all praise, conduct of the greatest virility."[31] Virtues like courage continued to be associated with men, and this *cantinière* was praised for acting like a man even while performing traditionally female tasks. Even revolutionary men tended to limit women to traditional roles; counterrevolutionaries saw women's participation in revolution as proof that revolutions were evil and wrongheaded.

The abrupt defeat of the Paris Commune of 1871 gave women a rough equality in fighting and death, and it is impossible to know if that equality would have continued if the Commune had survived. Earlier male revolutionaries had moved to exclude women from participating in politics and revolutionary activities even before the victory of conservative forces. In October of 1793, the Jacobins ruled that *all* women's clubs and associations were henceforth illegal. Seizing on a quarrel between the Society for Republican Revolutionary Women and some Parisian market women over the wearing of the revolutionary cockade, a representative of the Committee of General Security declared that, "in general, women are ill-suited for elevated thoughts and serious meditations. . . . We believe, therefore, that a woman should not leave her family to meddle in affairs of government."[32] A little over two weeks later, all women's deputations were barred from attending sessions of the Paris Commune. "It is horrible, it is contrary to all laws of nature for a woman to want to make herself a man," declared a revolutionary orator in the speech which convinced the Commune to vote unanimously to exclude women.

> The Council must recall that some time ago these denatured *viragos* wandered through the markets with the red cap to sully that badge of liberty and wanted to force all women to take off the modest headdress that is appropriate for them [the bonnet]. . . . Is it the place of women to propose motions? Is it the place of women to place themselves at the head of our armies? If there was a Joan of Arc, that is because there was a Charles VII; if the fate of France was once in the hands of a woman, that is because there was a king who did not have the head of a man.[33]

Women who acted as men rejected the traditional roles considered appropriate for them and thus became quasi-men, "denatured *viragos*" unworthy to be called women. We were "surrounded by 20,000 armed men and 40,000 furies—for they cannot be referred to as women," declared General Kilmaine in justifying the suppression of a revolutionary uprising in 1795. The conviction that all women, even revolutionary women, should remain at home out of public life united men who agreed on no other issue. The most radical joined with the most conservative: the French revolutionaries Babeuf, Marat, Hébert, and Robespierre all condemned women's public participation, as did Edmund Burke, the English critic of the revolution, who referred to "the revolutionary harpies of France, sprung from night and hell." In the Code Napoléon of 1804, which consolidated many revolutionary gains for men, women lost ground and were classified with children, criminals, and the insane as legal incompetents.

This unfortunate pattern repeated itself in the Revolutions of 1848. The French revolutionary provisional government delayed giving women the right to vote; when Pauline Roland (1805–1852), a radical teacher, tried to vote in the elections for the National Assembly, she was prevented from doing so.[34] Revolutionary men never let women vote even in their political clubs, and the revolutionary press made great sport of the women's clubs in general and the *Vésuviennes* in particular. Early in June, before the overthrow of the revolutionary government, police closed the Club des Femmes. In July, the Second Republic ruled that women could neither belong to clubs nor aid them. In 1850, police raided a meeting of the Federation of Working Women's Associations and arrested thirty-eight women and nine men. The leaders were tried for "plotting against society" and received jail sentences, the men's much longer than the women's because they were judged more "responsible."[35]

The defeat of republican governments only intensified the exclusion of women from politics. After 1851, in France and the German states, women were forbidden by law to participate in political activities or to attend meetings where politics was discussed. Only men could now be editors of newspapers, and Louise Otto-Peters, among others, was forced to cease publishing her *Women's Newspaper*, which had appeared weekly for two years. In the 1860s, male socialists advocated that women remain in the home, and even the men of the Paris Commune did not consider granting women political rights, like their predecessors in 1789–1795 and 1848.[36] Working-class women supported and died for the revolutions of the nineteenth century. They created political and economic groups to represent their own interests and needs. While they were ultimately defeated, they left a legacy of women's action which has never been completely forgotten.

The Impact of a Rising Standard of Living

From the 1870s to the 1930s, the lives of urban European women slowly improved. A recognition that infectious diseases, including cholera, did not respect neighborhood boundaries and social divisions in the crowded cities led to the provision of uncontaminated supplies of water to all a city's districts. Sewage, including human waste, no longer infected drinking water and homes. Cities now arranged to have streets cleaned, roads and sidewalks paved, gas and electricity provided. Housing and food codes slowly began to improve public hygiene, and it was hygiene, rather than medicine, which led to the decline of the urban death rate—which had always surpassed the urban birth rate—from the 1870s on.[37] By the turn of the twentieth century, increasing numbers of poor urban women began to live better than their predecessors. Running cold water, indoor flush toilets, sinks, and bathtubs appeared in new working-class dwellings. New building codes mandated new standards of ventilation, heat, and room size, and the two-room or larger apartment and row house began to replace the traditional single room. Pasteurized milk, now bottled or canned, gave the mother who needed to earn income outside the home a safe way of feeding her infant. Child labor began to end, and most nations mandated that children now spend eight to ten years at school. Illiteracy virtually disappeared in Europe, except for nations on the periphery: Russia, the Balkans, Spain, Portugal, Italy, Greece, and Ireland. While poor women usually were the last to benefit from an improved standard of living because they gave their families the best food and the most leisure, real progress had been made in raising the standard of women's lives as well. In 1890, the average life expectancy of women in Great Britain was 44 years; by 1910, it reached 52.4 years; by 1920, almost 60 years.[38]

Infant mortality, child mortality, and maternal mortality all decreased dramatically. By 1910, women who had faced high rates of death in childbirth a few decades earlier now risked only a 1.3 in 1,000 death rate in both home and hospital births.[39] In 1900, the death rate of infants in Europe ranged from a high of 252 per 1,000 live births in czarist Russia to a low of 91 in Norway. By 1925, only 50 Norwegian infants died for each 1,000 born alive; in the Soviet Union, the rate was down to 198 per 1,000. Many European nations almost halved the numbers of infants who died in their first year.[40]

Increased rates of survival meant that women could bear fewer children to achieve traditional family size, but in addition, between 1870 and 1930 women bore fewer children. Generally, couples reduced the size of their families when it became economically advantageous for them to do so, as they had always done. As industrialization spread, European family size began to decrease, starting from the top of the socioeconomic pyramid down. While much of this transformation remains mysterious, the contraction of the birth rate from the upper classes down is clear.[41] In Great Britain, for instance, an average working-class family had 5.11 children in the 1890s; an average professional family, 2.80. By the 1920s, the working-class family was down to 3.05 children; the professional family, to 1.69. These differences continued to lessen as the twentieth century progressed.[42] With the exception of a brief postwar "baby boom" from 1919 to 1921, family size in Europe continued to decline in the 1920s.[43] By 1914, all European nations except Ireland (whose demography was exceptional) showed substantial declines in the birth rate: down from 51.5 per 1,000 to 43.1 per 1,000 between 1875 and 1913 in largely peasant Russia; down from 35.4 per 1,000 to 24.1 per 1,000 in largely industrialized England and Wales during the same period. France again "led" this transition: in 1875, the French birth rate was 25.9 per 1,000; in 1913, 18.8 per 1,000. This decline caused grave national concern, especially when contrasted with the growing might of the German Empire, but even in Imperial Germany, the birth rate dropped from 40.6 per 1,000 in 1875 to 27.5 per 1,000 in 1913.[44]

This European-wide decline in the birth rate resulted from the increased use of contraceptive techniques, but these remained traditional. The restriction of family size preceded any widespread distribution of new contraceptive devices, like the condom or the diaphragm. Couples used abstinence or withdrawal, and if necessary, abortion.[45] Abortions almost certainly increased in the last half of the nineteenth century, as techniques became safer and more families sought to limit their size. By the last quarter of the century, hospital statistics registered the increasing numbers of women hemorrhaging or infected from abortions.[46]

European nation-states sought to curtail abortions in order to raise the birth rate. The claims of nationalism seemed to many to imply that women should bear as many children as possible.[47] Abortions remained illegal in Europe well into the twentieth century, and France even strengthened penalties in the 1920s in an attempt to raise the birth rate.

When more certain methods of birth control began to be available, poor women tried to obtain them. In 1921, Marie Stopes (1880–1958) opened the first birth-control clinic in London and was deluged with appeals for information. The following letter, received from a working-class woman in 1926, was typical:

> I have had 7 children. . . . my husband is only a labur and as been out of work 4 years. I dont want any more children and it seems how careful I try I seem to fall wrong I am just three days past my time and I feel worried I want to be on the safe side so I thought I would write and ask you think it is a shame that pore people should be draged down with families fed up with life keep having children hoping you will oblige me by writing return of post Yours sinculy[48]

Strategies tried by one woman might be impossible for another. "My mother had fourteen children and I didn't want that," explained Aida Hayhoe, born in the 1890s in an English village. "So if I stayed up mending, my husband would be asleep when I come to bed. That were simple, weren't it?" Her contemporary and fellow villager Maggy Fryett found such an evasion impossible: "I had nine. No way of stopping them. If there had been, I would have done. We never took nothing, nothing at all," she recalled bitterly.[49] A woman who tried to have fewer children in the face of her husband's opposition had a difficult time. "What is a woman to do when a man's got a drop of drink in him, and she's all alone?" a London working-class woman complained around 1911.[50] Given the steep decline in birth rates, however, working-class wives and husbands most likely collaborated in reducing their family size: many of the letters received by Marie Stopes in England in the 1920s are from working-class husbands concerned about their wives' health and the family's prosperity.

As family size and the birth rate declined, European nations responded by instituting legislation and benefits designed to make motherhood easier for poor women. Laws were passed forbidding women to work for a few weeks after childbirth, but women evaded the laws because they could not afford to stop earning income. Between the 1880s and the First World War, most European nations instituted maternity benefits—payments or aid to enable the mother to stop working for a while. These benefits usually began as voluntary insurance systems, but rapidly became attached to national health insurance programs and were made compulsory for both workers and employers. The most limited maternity benefits were those in the least industrialized nations: czarist Russia's program was only sporadically enforced, and Italy's applied only to the relatively small number of women who worked in industry or rice production.[51] The most generous programs, like that of the German Empire, gave pregnant workers up to eight weeks off with pay, two weeks before childbirth, six weeks after. Austria paid women for breast-feeding their own children and gave them free medical care before, during, and after the delivery. The motive for these programs was pro-natalist in all nations. "The money is meant for the mother," Lloyd George, the English chancellor of the exchequer, explained in 1911, "to help her in discharging the sacred function of motherhood, by proper treatment, fair play, so as to put an end to the disgraceful infantile mortality of this country."[52] Maternity benefits, although often small and rarely available to even a majority, were an important gain for working-class women in this era.

In addition to maternity legislation, nations instituted "protective" legislation for women who worked outside the home. The laws forbidding mothers to work immediately after childbirth constituted part of this movement, which sought both to make working conditions easier for women and to preserve women's role as mother and homemaker in an increasingly industrial world. Protective legislation for women first appeared in Victorian England.

Modeled on earlier laws limiting the hours and types of children's work, the first act in 1842 forbade women, girls, and boys under ten from all work underground. Public opinion had been shocked by parliamentary investigations into mining in the early 1840s, which revealed that women routinely worked half-naked with men in brutal conditions. Women usually worked as drawers or hurriers, loading carts with coal and hauling them through the mine tunnels. "I am a drawer, and work from six o'clock in the morning to six at night," testified Betty Harris, a thirty-seven-year-old mother of two.

> I have a belt round my waist, and chain passing between my legs, and I go on my hands and feet. The road is very steep. . . . it is very hard work for a woman. The pit is very wet where I work, and the water comes over our clog-tops always. . . . I have drawn till I have had the skin off me; the belt and chain is worse when we are in the family way. My feller has beaten me many a time for not being ready. I were not used to it at first, and he had little patience; I have known many a man beat his drawer. I have known men take liberties with the drawers, and some of the women have bastards.[53]

In mining, as in other heavy labor, women were prized for their "docility." "Females submit to work in places where no man, or even lad could be got to labour in," testified one mining foreman. "They work in bad roads, up to their knees in water, in a posture nearly double."[54]

The solution arrived at in England—and in all European nations by 1914—was to prohibit women from working underground. Further protective legislation followed. In 1844, an English Factory Act limited women's and children's hours in industrial employment: children under thirteen could work only six and a half hours a day; women, twelve. This measure was followed in 1847 by the Ten Hours Act, which limited the factory day for all workers. In 1867, an act passed limiting *all* women from working more than twelve hours a day, but this proved completely unenforceable.

Early English protective legislation for women combined the same mixture of motives as later European laws: the desire to make hard labor easier for women and the wish to keep women at home in their approved roles of wife, mother, and homemaker. Other nations legislated that women could not work outside their homes at night, although pay was higher then.[55] (Piecework in the home was not regulated.) If women stayed at home, it was hoped, they would be with their families where they "belonged," rather than "out" where they might have sex with other men. "Night work" became almost a code for prostitution, and forbidding it to women made the woman out at night more anomalous and more marginalized.

In addition to forbidding women from working outside the home at night, some nations gave female industrial workers time off so they could perform their duties at home. The German Industrial Code of 1891 mandated that women not work after 5:30 P.M. on Saturdays and before holidays, so they could do their housework.[56] In 1892, the French Union of Tobacco Workers demanded the same for its female workers:

> After having worked incessantly for 10 hours, our women, after having quickly made dinner, have to think about the kid's breakfast for the next day, have to mend their clothes, etc. Going to bed at 10 or 11 at night, getting up at 5, on Sundays they have to wash the family's laundry. They rarely have the time to change their own clothes, after having cared for their near ones.[57]

Unions and government arrived at the same solution: give women a few hours off from factory work so they could do their housework. Housework was a woman's responsibility, whether she earned income outside the home or not.

Protective legislation for women united men who agreed on few other subjects. Conservatives believed protective legislation would prevent employers from hiring women and thus keep them out of the labor market; radicals hoped protective legislation would enable women to become more independent by working outside the home. Protectionism appealed to those who thought women weak and in need of male protection and to those who thought the exploitation of women demonstrated the horrors of capitalism. "Woman," declared the French Radical party in 1907, "who must be protected by law in every circumstance of her life."[58]

Women divided sharply on the issue of protection. Some socialist women thought it patronizing, like Nathalie Lemel (1827–1921), the French bookbinder and Communard. "Women do not wish to be protected," she asserted at the Second International in 1889, adding that men had no right "to make protective laws for women assimilating them to children."[59] Others saw protection as the crucial issue for working-class women, and one which separated them from the largely middle-class voting rights movement. "Has not the star of the [English] women's rights movement, Mrs. Fawcett, declared herself expressly in opposition to any legal reduction of working hours for female workers?" wrote Eleanor Marx (1856–1908), Karl Marx's daughter and an active participant in English labor circles, in 1892.

> We see no more in common between Mrs. Fawcett and a laundress than we see between Rothschild and one of his employees. In short, for us there is only the working-class movement.[60]

From 1869 on, August Bebel and the German Socialist party began to introduce protective legislation for women in the Reichstag; in 1889, the Second (Socialist) International declared that women should be forbidden from night work and work "particularly damaging to the female organism."[61]

The image of the exploited female worker, whether a toiling child, pregnant laborer, or nursing mother, became a powerful symbol in socialism's struggle against capitalism. "Truly, it is no inspiring sight to see women, and even pregnant ones, at the construction of railroads," wrote Bebel in his influential book *Women and Socialism* of 1879,

> pushing heavily laden wheelbarrows in competition with men; or to watch them as helpers, mixing mortar or cement or carrying heavy loads of stone at the construction of houses; or in the coal pits and iron works.[62]

By 1913, most European nations had passed protective legislation for women. Generally, women were forbidden to work underground, at night, or for a few weeks after childbirth, and the hours they worked each day were limited. Shorter hours was one of the triumphs of nineteenth-century labor movements, and was the only area of protective legislation where gains for women and children led directly to gains for men. In 1893, the Second International demanded the eight-hour day for women; by 1900, it demanded it for all workers. The shortening of the working day in industry and some branches of manufacturing improved the lives of those women who continued to earn income outside the home.

Both shorter hours and protective legislation for women succeeded in many cases because of trade union support. By the end of the nineteenth century, women entered unions in large numbers, and male trade unionists, especially socialists, were making major efforts to unionize at least female factory workers. This was a reversal of earlier policy. The first workers' associations including women were all-female and resembled the older guilds. The Women's Benefit Society formed in Birmingham, England, in 1795 was typical of these groups, which originated as benefit and friendly societies. The union's charter called for a "mother and treasurer" of the society and stated that she should

> keep a good fire during the winter season on the nights the society meet, and shall take care that all provisions, ale, etc. that may be wanted for the entertainment of the members at the feasts, etc. shall be of the best, or for every neglect shall forfeit One shilling to the fund.

The society wanted "respectable" members under age forty; new members had to be voted in and contribute 2 shillings to a fund used to pay members if they were "sick, lame, or blind."[63] As a consequence of the French Revolution, "combinations" of workers became illegal in both England and France. Union activity revived in England in the early 1830s, with the Owenite attempt to enlist all workers into "one big union," the Grand National

Consolidated Trades Union. Women's unions were formed, and groups of stocking knitters, lace-makers, laundresses, bonnet-makers, shoemakers, and milliners sent delegates and held meetings.[64] The collapse of the GNCTU under government pressure retarded women's unionization. In 1840, a group of Nottingham "lace-runners"—women who embroidered lace on a piece-work basis in their homes—met to try and organize in opposition to a payment cut. A committee, composed of Mary Smith, Hannah Weatherbed, Mary Chapman, and Ann Davis called on their "sister" lace-runners to "turn out on Monday next" to protest; they also hoped "that the male portion of the society will assist us, as it is the cause of the poor working man as much as us females."[65]

But the "turn out" failed, and its failure embodied the reasons for women's relatively low rate of unionization in the middle decades of the nineteenth century. The women workers were afraid to demonstrate because they needed the income to survive; male workers did not join in and support their cause. From the mid-nineteenth century on, when trade unions became legal in most European nations, skilled male workers began to unionize in large numbers. At pains to distinguish themselves from unskilled male workers, they also sought to separate themselves from female workers, whether skilled or unskilled. "The working man will not advocate the admission of women into the representation," declared an English working-man's newspaper in 1842, "lest it should delay their own."[66] A few groups of skilled women workers formed—the milliners, the bookbinders, the dressmakers—but the large majority of women who earned income did so as domestic servants, pieceworkers in the home, or "unskilled" factory workers.

Servants and pieceworkers were and remained exceptionally difficult to unionize: working in individual homes, they had little contact with other women workers, and pressures from both the economy and employers prevented them from organizing. The chief gains for working women from unions came in factory work, and they occurred in the last decades of the nineteenth century. By then, it was apparent that if women in industry did not unionize, they would hold back men by acting as a surplus labor force and taking men's jobs during strikes and protests. As Marie Grubinger, a clothing worker and co-founder in 1890 of the Viennese Working Women's Association, put it, "As long as women are not in the movement, the men will gain little or nothing at all."[67] Although many male trade unionists blamed women workers for failing to unionize, female unionists emphasized the abysmal conditions of female labor. "Women are badly paid and badly treated because they are not organized," declared Mary MacArthur (1880–1921), "and they are not organized because they are badly paid and badly treated."[68] Undaunted by this vicious circle, MacArthur and others began working successfully to recruit women to unions. By the 1880s, men in occupations classified as unskilled began to strike and unionize, and women followed their example. The successful strike of the London dockers in 1889, for instance, led seven hundred East End women to strike for higher wages and better working conditions in the factory where they made matches, an unhealthy and poorly paid job. By the late 1880s, almost thirty-seven thousand Englishwomen belonged to unions, thirty thousand in textile unions of the cotton industry alone.[69] Instead of attempting to exclude women from unions and trying to get them to leave the job market, existing male unions began to admit women or aid women in forming their own unions. The effort of recruiting women was seen to be worthwhile, both for women and for men. "Of course, in common with everyone who has had much experience in the struggle of organizing women on trade union lines I consider the struggle a most disheartening and painful one," wrote Isabella Ford, secretary of the Leeds Society of Workwomen in 1900.

> But I hold very strongly that the fault does not lie with the women themselves. Those women who really grasp the aim of trade unionism, grasp it, I think, more firmly than men, because more religiously. Trade unionism means rebellion, and the orthodox teaching for women is submission.[70]

From the late 1880s on, both English and Continental unions began to recruit women in massive numbers, although women continued to be underrepresented in unions.

Male workers in "skilled" positions unionized first; women workers were most likely to unionize where they predominated: in the "unskilled" sectors of textile and food industries. The growth of socialist political parties aided the growth of women's unions. By 1912 in Germany, for instance, 222,809 women belonged to Social Democratic party trade unions, 28,008 to Catholic Center party unions, and 4,950 to Liberal party unions.[71] Roughly 233,000 German women were unionized, as were 90,000 Frenchwomen and 433,000 Englishwomen. In each of these nations, women comprised about 9 percent of the unionized labor force although they usually were more than 30 percent of the labor force as a whole.[72]

One issue which remained troublesome and often divided female and male workers was that of equal pay. Traditionally paid less than men, women feared asking for equal pay. Some male unionists advocated equal pay for women in hopes that employers would no longer hire women if they could employ a man for the same wages.[73] But if women accepted lower wages, they undercut men who worked at the same jobs. A few unions solved the problem by demanding equal pay for all: the Danish Tobacco Workers Union, which had originally expelled any member who taught the trade to a woman, admitted women in the 1880s. By the turn of the century, the Danish tobacco workers were demanding that "the female worker doing the same work as a man must consequently be paid the same wage," making it one of the first European unions to advocate equal pay for equal work.[74] Despite the justice of equal pay (as a Russian woman stated in 1905, "When I get hungry and go buy myself a pickle, they don't charge me half a kopek for it, do they? No, I pay the same kopek a man does")[75] unionized women continued to earn one half to two thirds the income of men.

Unions did provide some working-class women with a means of social mobility, and by 1914, socialist parties had become a working-class woman's chief avenue upward. A woman of energy and ability could organize other women, become a union or party official, and eventually achieve power through the party. Miina Sillanpää (1866–1952), for instance, started factory work at twelve and continued until she was nineteen. She then became a domestic servant and in the 1890s joined the Finnish Social Democratic party and helped to found its women's branch. She managed a home for domestic servants in Helsinki from 1900 on. She edited the *Servants' Journal* from 1905 to 1906 and *The Working Woman*, the party's women's publication, from 1907 to 1916. In 1906, Finland became the first European nation to allow women to vote, and by that year there were close to nineteen thousand women in the Social Democratic party, about 22 percent of the total membership. In 1907, Sillanpää became one of the nineteen women elected to the new, two-hundred-person Finnish parliament. In 1926, when she was sixty, Miina Sillanpää served as assistant minister for social affairs in the government; the next year, she was appointed minister in her own right.[76] Socialist women in other nations matched Sillanpää's achievements. In England, Margaret Bondfield (1873–1953) rose from shop assistant to become the first female English cabinet minister, in the second Labour government of 1929. In Austria, Adelheid Popp-Dworak, who had crocheted in bed at night as a child to earn income, headed the Austrian Socialist Women's Movement and became a parliamentary representative in the 1920s. Luise Zietz, the German woman who recounted how she had spun endlessly as a child at home, became a member of the Reichstag in the early 1920s. Klavdiya Nikolaeva and Aleksandra Vasilevna Artyukhina, one a typesetter and the other a textile worker, became heads of the Soviet Women's Organization in the 1920s.[77] While such women remained exceptional, unions and political parties— especially on the left—enabled some working-class women to attain influence and political power.

Most working-class women, however, remained un-unionized and began to leave the labor force if they could afford to do so. The rising standard of living of the period 1870 to 1930 enabled increasing numbers of women to stop earning income for at least a portion of their lives, usually when they married and when their children were young.[78] During this era, however, more single women earned income outside the home. By the twentieth century, they began to leave agricultural work and domestic service in favor of the new jobs which opened in offices and stores. A survey of Englishwomen looking for work in 1919 revealed that 65 percent stated they would not work as servants no matter what the pay; the rest would work as servants only if the pay was about double the going rate.[79] The numbers of female domestic servants rose only once in the twentieth century: during the Great Depression, when unemployment forced women to reenter the field.[80] Instead, women replaced men as clerks and typists (and were preferred by employers because they accepted lower wages) and increasingly staffed the new shops and department stores, the banks and telegraph offices, the phone switchboards and government offices of twentieth-century cities. Generally young and preferably good-looking, these women saw their work as temporary, as did their employers; many insisted that women leave if they married.[81] Half the female clerks in Germany in the 1920s were under twenty-five, compared to one quarter of the men.[82] As in other fields, men predominated in managerial and supervisory positions: although German women constituted one fourth of the white-collar workers, they held only one twentieth of the top positions.[83] Women worked as sales clerks, not managers; secretaries, not bosses; tellers, not bankers; nurses, not doctors.

Young women flocked to these new positions. In white-collar work, young working-class women seeking to escape the factory or domestic service in favor of "clean" work found themselves laboring beside young middle-class women who needed to earn income. To help support her widowed mother, Françoise Giroud (1916–   ) sought to become a clerk-typist when she was sixteen. "I put on lipstick and high heels to try and look older, and apply in person," she remembered. "I say I'm eighteen. It works." Giroud handled the correspondence and waited on customers. "I was pleased and proud to earn 500 francs a month. My sister, who was a salesgirl in a department store, earned only 300 francs and didn't have the right to sit down."[84] White-collar work, although poorly paid and often temporary, seemed prestigious and modern, and magazines and films cultivated the fantasy that a beautiful salesclerk or typist could marry her boss. By the mid-1930s almost 30 percent of all European women who earned income outside the home worked in this white-collar service sector of the economy.[85]

Women in World War I and the Russian Revolution

The improved standard of living, smaller family size, maternity benefits, protective legislation, unions, and new jobs comprised the most important changes in the lives of urban working-class women between the 1870s and the 1920s. Compared to these changes, the impact of World War I (1914–1918) on these women's lives was relatively minor. While middle- and upper-class women often reported that the war freed them from nineteenth-century attitudes limiting both work and personal life, working-class women's lives changed relatively little.[86] Unlike more privileged women, working-class women were used to earning income outside the home, and their entry into war work was more likely to be exploitative than liberating. Unlike more privileged women, working-class women and girls had rarely been shielded by a "double standard" of sexual behavior for women and men; rather, working-class women made the maintenance of the double standard possible for men of property. For working-class women in the cities, the growth of the new white-collar jobs was the one new trend fostered by the war which was not reversed afterward. Otherwise, World War I brought only a temporary suspension of the normal conditions of work outside the home, and traditional patterns returned in the postwar era.

As soon as the war broke out, European governments moved to suspend protective legislation for women for the duration.[87] Just as nations expected working-class men to serve in the military, so they exhorted working-class women to serve in the factories, taking the places of the men who had joined the armed forces. Drawn by high wages as well as patriotism, women thronged into these new, previously male jobs. As many as 684,000 French-women worked in munitions factories during the war, as did 920,000 English-women.[88] Entry was often difficult: male workers opposed the women even when women joined a union, worked for equal pay, and promised to give their jobs back to men after the war. "I could quite see it was hard on the men to have women coming into all their pet jobs and in some cases doing them a good deal better," remembered Joan Williams, an Englishwoman who worked in a munitions factory during the war. "I sympathized with the way they were torn between not wanting the women to undercut them, and yet hating them to earn as much."[89]

Governments initially insisted that women receive equal pay for doing a job formerly done by a man, but this policy was largely ineffective: factories tended to divide up jobs into smaller operations and pay women at a lesser rate.[90] Women's industrial wages rose during the war, both relative to men's and absolutely, but they still remained measurable as a percentage of male earnings. In Paris, women in metallurgy earned only 45 percent of what men earned before the war; by 1918, the women earned 84 percent of what men earned.[91] In Germany, women's industrial earnings relative to men's rose by about 5 percent.[92] Both women and men seemed to view the changes brought by the war as temporary. After the war, the men would return to their jobs, the women would leave men's work, and all would return to normal. Jessie Pope, an English poet, articulated this attitude in a 1916 poem entitled "War Girls":

> There's the girl who clips your ticket for the train,
>   And the girl who speeds the lift from floor to floor,
> There's the girl who does a milk-round in the rain,
>   And the girl who calls for orders at your door.
>   Strong, sensible, and fit,
>   They're out to show their grit,
>   And tackle jobs with energy and knack.
>   No longer caged and penned up,
>   They're going to keep their end up
>   Till the khaki soldier boys come marching back.[93]

As soon as the war was over, all belligerent governments acted quickly to remove women from "men's" jobs. In England, these women were made "redundant" and let go; in France, they were offered a bonus payment if they left factory work; and in Germany, the government issued regulations calling for women to be dismissed before men if necessary.[94] These policies were effective: by 1921, fewer French and English women worked in industry than had before the war.[95] Women's earnings decreased to return to lower percentages of men's, and the promise of "equal remuneration for work of equal value" made in the Versailles Treaty of 1919 remained a dead letter. Mass media concentrated on the relatively superficial changes in women's clothing, hair styles, and use of cosmetics and ignored the deeper continuities which structured most women's lives.

One important continuity was the urban working-class women's tradition of protests. Before and during the First World War, city women demonstrated and marched to protest intolerable conditions. They continued the tradition of the bread riot and added onto it the tactics of the industrial labor movements: the strike, the picket line, the invocation of revolution. In 1911, for instance, working-class housewives in the industrial Nord of France protested high food prices by seizing food supplies and marching carrying red banners, singing the "International," the hymn of the socialist movement.[96] The red flag became a symbol of working-class revolt during the French Revolution of 1789; by the twentieth century it was the emblem of socialist revolutionaries and was often carried by women demonstrating against high prices, bad working conditions, or unjust laws. French prostitutes rioted from 1906–1909 to protest the arrest of prostitutes under eighteen: they demonstrated and, when arrested, insulted the police and sang the "International."[97] Parisian police confiscated red banners carried by women clothing workers during a strike in Paris in 1917; about fifty thousand French women factory workers struck in 1917 and 1918.[98] In addition to riots and strikes, urban women invented new methods to protest new situations. In Barcelona in 1908, women of the flower market held a communal funeral for women killed in a bomb explosion.[99] Women in Glasgow of 1915 organized tenants to withhold their rent and held large, and ultimately successful, demonstrations to prevent the eviction of rent strikers.[100] Working-class Barcelona housewives convinced women textile workers in the factories to come out on strike and protest high prices with them in 1918; they carried signs reading:

> Down with the high cost of living. Throw out the speculators. Women into the streets to defend ourselves against hunger! Right the wrongs! In the name of humanity, all women take to the streets.[101]

Women in Turin and Paris in 1917 called for an end to the war as well as for lower food prices.[102]

While working-class women made their presence felt in many nations during the war, in Russia they helped to change history. For it was in Imperial Russia, strained by the efforts of the war, that a women's demonstration became a revolution.

In February of 1917, the Russian government was forced to institute bread rationing in the capital, Petrograd.[103] Bread had sold for 3 kopeks for a 1 pound loaf of rye bread in 1913; by the end of 1916, the price reached 18 kopeks. It was women who stood in the bread lines, many of them after putting in a twelve-hour day in the factories. By 1917, twice as many women worked in Petrograd factories as had before the war, and protective laws limiting their hours had been suspended. Women waited in bread lines because it was the only way to feed their families. "Men consider it better to die of hunger than to stand in a line," stated a Petrograd woman. "A woman can't act that way because she is the mother of her children and she needs to feed them."[104]

The Russian government realized it had a difficult situation on its hands. A police report of January 1917 warned of trouble:

> Mothers of families, exhausted by endless standing in line at stores, distraught over their half-starving and sick children, are today perhaps closer to revolution than [the liberal opposition leaders] and of course they are a great deal more dangerous because they are the combustible material for which only a single spark is needed to burst into flame.[105]

Bread rationing galvanized women. Disobeying the policy and advice of male political leaders of all parties, Petrograd women decided to demonstrate on International Women's Day, 1917.[106] Celebrated as March 8 in the West, the date fell on February 23 in the Russian Orthodox calendar, still in use in 1917. About ten thousand women, many of them workers in the textile factories, marched in Petrograd that day, carrying large red banners proclaiming, "Down with Autocracy," "Down with War," "Our Husbands Must Return from the Front," and "Peace and Bread."[107] Chanting for bread, the women were joined by other workers, and their call for a general strike succeeded. Alexandra Rodionova was then a twenty-two-year-old streetcar conductor in Petrograd. A wage earner since eleven, she had become a conductor—a job normally held by a man—during the war. "I remember how we marched around the city," she wrote later.

> The streets were full of people. The trams weren't running, overturned cars lay across the tracks. I did not know then, I did not understand what was happening. I yelled along with everyone, "Down with the tsar!", but when I thought, "But how will it be without the tsar?" it was as if a bottomless pit opened before me and my heart sank. Nevertheless I yelled again and again, "Down with the tsar!" I felt that all of my familiar life was falling apart, and I rejoiced in its destruction.[108]

Rodionova was elected to a citywide committee of streetcar workers on February 26, three days after the strike began. By then, some 200,000 people were out on strike, about 10 percent of the city's population. "This is a hooligan movement," the Czarina Alexandra wrote Nicholas II, who was at the battle front;

> young people run and shout that there is no bread, simply to create excitement, along with workers who prevent others from working. If the weather were very cold they would all probably stay at home.[109]

The commander of the Petrograd military garrison took the situation more seriously: by then, his soldiers were refusing to fire on the crowds, which the czar had ordered. "When they [the women] said, 'Give us bread!' we gave them bread and that was the end of it," he wrote. "But when the flags said, 'Down with the autocracy!'—how could you appease them then?"[110]

A few days later, the revolution had spread to Moscow and other Russian cities and the czar abdicated. Women continued to take an active role in the new Russian republic, just as they had in the early stages of previous revolutions. In late March, forty thousand Petrograd women marched to demand the vote, carrying banners reading, "The Woman's Place—Is in the Constituent Assembly" and "War to Victory."[111] Other women protested the war: in April 100,000 soldiers' wives marched asking for higher support payments, bread, and peace. Thirty-five of them formed a committee to press for their interests.[112] All political parties competed for women's support and appointed prominent women to their central committees. The provisional government included two women—Ariadna Tyrkova and Countess Sofya Panina—and a few women ran successfully for seats in the Moscow and Petrograd assemblies. In July, the government gave all citizens over twenty the right to vote, becoming the first European nation outside Scandinavia to enfranchise women.

During the Bolshevik Revolution of October 1917, women fought on both sides. Some organized into fighting batallions, others acted as nurses and messengers, guards and sentries. More armed women are estimated to have fought for the Bolsheviks, both in October and the ensuing civil war.[113] In October, Alexandra Rodionova was assigned to ensure that the streetcars kept running on the crucial nights of the Bolshevik takeover. She carried a pistol and dispatched two flatcars armed with machine guns to aid in the attack.[114]

During the civil war, which lasted from 1917 to 1921, women fought for and supported every faction. Lenin appointed Alexandra Kollontai (1872–1952), a longtime Bolshevik and feminist, as commissar for public welfare. She had helped to organize the committee of soldiers' wives, and in November she arranged for the First Congress of Women Workers and Peasants, attended by 1,147 women. Lenin addressed the congress and emphasized the need for women's participation. "Root out old habits, every cook must learn to rule the state," he declared that year, and the congress created Zhenodtel, the women's organization of the new government.[115] Enlisting women as delegates to Zhenodtel brought many peasants and factory workers into new positions of responsibility. "It seemed to me that I had lifted off from the ground and was flying into a dizzying unknown," Rodionova, the former streetcar conductor, recalled. "The former illiterate working girl had been transformed into a person, powerful with the knowledge of her own rights, a consciousness of responsibility for everything happening in the country."[116]

In contrast to earlier revolutions, the new Soviet Bolshevik government actively legislated to transform women's lives. In 1918, a new marriage law made marriage a civil ceremony and divorce easy to obtain. Kollontai issued a decree guaranteeing state protection for mothers and children and making maternity hospital care free. In 1920, abortion became legal for the first time in modern Europe. Lenin repeatedly raised the issue of housework, declaring in 1919 that notwithstanding all the laws emancipating woman [in the U.S.S.R.] she continues to be a *domestic slave*, because *petty housework* crushes, strangles, stultifies and degrades her, chains her to the kitchen and the nursery, and she wastes her labor on barbarously unproductive, petty, nerve-racking and crushing drudgery.

The answer, he argued, lay in the transformation of housework into a "large-scale socialist economy" through the creation of "public catering establishments, nurseries, kindergartens."[117] He and Kollantai envisioned a communist network of institutions like crèches and day-care centers, public restaurants and laundries, to free women from the labor of motherhood and housework.[118]

Attempts to implement these plans on a large scale failed in the face of the chaos caused by the four years of civil war which followed the Russian Revolution. Most women, faced with the problems of surviving in a society in which millions had died, opted for the protection of the traditional institutions of the family and village community.[119] Attempts to revolutionize the family and housework ended after 1922: Kollantai fell from favor, Lenin died, and by the late 1920s, the government's priority had become increased productivity and industrialization. Independent women's organizations were abolished: the International Women's Secretariat in 1926, Zhenodtel itself in 1930. Women were removed from their positions in the Red Army and replaced by men; the upper echelons of the Communist party remained overwhelmingly male.[120] The major gains for women were educational, as maternity centers opened, reading became widely taught, and girls trained for new types of work. "Young women need to study not the weaving of lace, not the embroidery of handkerchiefs," wrote Nadya Krupskaya, a leading Bolshevik and Lenin's widow, in 1926,

> but agronomy, animal husbandry, sanitation, technology, and so on. It is necessary for them to study those fields of production where a shortage of skilled workers threatens to have serious repercussions for the republic of workers and peasants.[121]

Aided—or hampered—by protection laws, maternity benefits, and an increase in the standard of living, Russian women began to leave the labor force: by 1928, only 24 percent worked outside the home, a figure lower than that in other nations.[122] The birth rate fell continuously in these years, and literacy rose, especially among women. By the end of the 1920s, Soviet working-class women had achieved a mixture of progress and tradition similar to that of women in capitalist societies.

Women in the 1930s

In the 1930s, European women's lives differed greatly depending on which nation-state they lived in. With the accession of Stalin to power in 1927, the lives of Soviet women began to diverge sharply from those of other Europeans. First, events like the forced collectivization of agriculture in the late 1920s and the resulting famine, the insulation of the Soviet economy from the Great Depression, and the purges of 1936–1938 produced a uniquely Soviet experience in the 1930s. Second, Stalin's decision to industrialize rapidly structured the lives of both Soviet women and men during these years. Industry took precedence over consumer goods; productivity over women's rights. Protective labor laws and maternity legislation were cut back to enable women to be hired by factories, and by 1940, Soviet women constituted 45 percent of the labor force.[123] Communist doctrine now stated that Soviet women had achieved equality with men and needed no special legislation or institutions to safeguard their interests. In 1936, Stalin decided that the family as a social unit should be strengthened and outlawed abortion in the Soviet Union; in 1943 and 1944, for similar motives, coeducation and divorce were also abolished.

Outside the Soviet Union, European nations responded to adverse economic conditions by attempting to force women out of jobs believed more appropriate to men. Governments of very different political parties—socialists, conservatives, liberals, and fascists—united on this policy. In England, where conditions remained depressed throughout much of the 1920s, women who married routinely lost their jobs, whether in teaching or in factory work. "I worked until I got married, in 1927," remembered Minnie Ferris, then employed as a steam presser in an East London laundry.

> Women lost their jobs when they got married and couldn't get another one. I had to stay home as a housewife. A friend of mine who worked at the biscuit factory, got the sack because she got married, too. . . . Today you can take a job anywhere, if you want to work after you're married, but then you couldn't. You'd have to do cleaning for someone or take in washing.[124]

Governments excluded women from unemployment benefits in a variety of ways. In 1920, England extended unemployment payments to almost two thirds of British workers, but domestic servants, cleaners, and those who worked in the home were not included. Anyone who entered domestic service lost all insurance eligibility, regardless of her previous employment. Unemployed women might be offered jobs in domestic service; if they refused them, they were classed as "not genuinely seeking work" and so lost all claim to benefits. In 1922, the government ruled that all married women workers could be automatically excluded from unemployment benefits unless their income was 10 shillings a week or less "when hardship was deemed to have resulted."[125]

In the grimmer conditions of 1931, Margaret Bondfield, then minister of labor, implemented an act which effectively excluded most working married women from unemployment benefits. Unless a woman had made a certain minimum number of contributions to the fund after her marriage (the ones she made before were not counted), unless she could prove that she was normally employed, was seeking work, and "could reasonably hope to obtain work in the district" (an impossibility in much of Great Britain in 1931), she was not eligible for the dole.[126] "The Labour Party itself was only lukewarm on such matters as 'equal pay,' " remembered Hannah Mitchell, a working-class activist, of these years,

> while on the employment of married women most of them were definitely reactionary. But they dressed up their objections, either in admirable sentiments about the "domestic hearth" with "mother's influence" thrown in as a tear-raiser, or else they went all Marxian and stressed the bad economics of two incomes going into one house, while men with a capital M were unemployed.[127]

Even for women who managed to hold onto a job, life became increasingly difficult as the Great Depression intensified. In 1936, when she was twenty, Nellie Priest worked in a London shoe factory. "Life was hard," she remembered.

> The wages were very low and we were on short time from Easter to October. . . . I couldn't even afford a ride to work. . . . If it rained, you would keep on pulling your skirt down to try and stop it shrinking—it was terrible. Nothing was ever pre-shrunk. Your coat would still be wet the next day, all round the shoulders and the bottom, but you still had to wear it, it was the only one you had. You would get colds upon colds upon colds. I would never like that over again. It was awful.[128]

The relative effect of the Great Depression on women and men cannot yet be gauged. Unemployment statistics show many more men put out of work, but women's work was often not counted in the first place. Doing a bit of cleaning, taking in laundry, performing piecework in the home were all traditional methods by which women earned income, and during the 1930s, when governments usually required proof of need for benefits, such work became all the more necessary and was often hidden. A study of Marienthal, an Austrian community where the closing of a local factory put virtually the entire population out of work by 1931, revealed that women worked even harder when unemployment grew. "The term 'unemployed' applies in the strict sense only to the men, for the women are merely unpaid, not really unemployed," the report concluded. "They cook and scrub, stitch, take care of the children, fret over the accounts, and are allowed little leisure by the housework that becomes, if anything, more difficult at a time when resources shrink."[129] An English study of 1,250 married working-class women done in 1939 found that the husbands' unemployment was a major factor in determining the women's health. Over and over, the women complained of being "run down through worry," and a statistical comparison of their household budgets revealed that ill-health and low income were inextricably connected.[130]

In Germany, the Nazi takeover of 1933 transformed women's and men's lives.[131] The Nazi's early pronouncements on women stressed the need for women to stay within the "woman's world" of "her husband, her family, her children, and her home," as Adolf Hitler declared to a meeting of the Nazi Women's Association in 1934.[132] Actively opposing the recent changes in women's rights as a "product of the Jewish intellect," the Nazis promised German women "emancipation from emancipation."[133] "Woman in the workplace is an oppressed and tormented being," stated a propaganda pamphlet from the 1920s.

> Day in and day out, she sits for hours at the typewriter or holding her shorthand pad . . . day after day the same misery. . . . National Socialism will restore her to her true profession—motherhood.[134]

In these attitudes, the Nazis differed little from other right-wing parties, and in the last elections of the late 1920s and early 1930s, women tended to vote for the Nazis (and communists) in lower percentages than men, and for the Catholic Center party and the conservative German National People's party in greater percentages than men. In the presidential election of 1932 between Hitler and Hindenburg, women voted for Hindenburg in far greater numbers than men, and in lesser numbers for Hitler.[135]

Often blamed at the time and later by political commentators and historians for Hitler's triumph, German women were the subjects of Nazi policy, not its agents. Even those women who had supported the Nazis before their rise to power were quickly ousted from positions of authority and often replaced by men in a regime which insisted that women should have no role in governing.

The accession of the Nazis to power meant that traditional attitudes about women's role and functions were reasserted and became laws. In the early years of the Third Reich, the Nazi government sought to structure women's lives in the crucial areas of marriage, motherhood, and paid labor.

"You have to start influencing women in their daily lives," Gertrud Scholtz-Klink (1902–   ), the former head of the Nazi Women's Association, declared in an interview in the early 1980s.

> Take it from me, you have to reach [housewives] where their lives are—endorse their decisions, praise their accomplishments. Start with the cradle and the ladle. That's what we did.[136]

Some of the earliest Nazi legislation began with the "cradle" and attempted to have women bear more children by a mixture of incentives and coercion. A government-subsidized Marriage Loan became available to couples in which the wife promised to leave the job market upon marriage. The birth of a legitimate child canceled one quarter of the loan; four children erased it. Mother's Day, established in Germany in the 1920s by a coalition of florists and conservative political groups, became an important Nazi holiday celebrated on Hitler's mother's birthday.[137] In 1939, the regime followed the examples of France and Italy and instituted the "Honor Cross of the German Mother," which bore the slogan "The child enobles the mother" and was given to mothers of four or more children.[138] Mothers of nine children or seven sons, whichever occurred first, received a special award, and members of the Hitler Youth and the armed forces were instructed to salute a woman decorated with the motherhood medal.[139]

To coerce a rise in the birth rate, the Nazis closed the sexual counseling centers, which had been the main source of contraceptive information, and prosecutions for abortion increased until they reached the rate of seven thousand a year.[140] Groups considered "racially undesireable" by the Nazis— like the Jews and Gypsies—were allowed to abort legally in the hopes they would commit "racial suicide." From 1934 on, Nazi legislation called for the sterilization of "lives unworthy of life," as they called them: those who belonged to undesired groups, ranging from the mentally ill to the criminal.[141]

The effect of these Nazi policies on women's and men's decisions to marry and have children is difficult to assess. Marriage and birth rates rose from 1934 to 1940, but not dramatically. It was not until 1939–1940 that the birth rate returned to the levels of the mid-1920s, under "decadent" Weimar.[142] Women's organizations, like other German institutions, "nazified" in the early 1930s, declaring allegiance to the Nazi state and expelling their Jewish members.[143] In 1934, Gertrud Scholtz-Klink became head of the new Nazi Women's Association and began a campaign to give women and girls a "general education," which included Nazi views of population and race, household management, mothering, and health care.[144] The Nazi Women's Association urged women to use their traditional roles to foster the goals of the Third Reich: as mothers shaping their children, as wives influencing their husbands, as consumers deciding where and what to buy. "Though our weapon in this area [the battle for German economic self-sufficiency] is only the ladle," declared Scholtz-Klink in 1937, "its impact will be no less than that of other weapons."[145] By 1941, one out of every five German women belonged to the Nazi Women's Association.[146]

Men joined Nazi organizations in far greater numbers than women. In part, Nazis put less effort into organizing women and girls than boys and men, but women also seem to have resisted Nazi attempts to enlist their labor. Despite their own early propaganda about women's "world" being the home, the Nazis found that they needed a reliable source of "unskilled" labor for their factories and farms just as other modern economies did. In 1937, the terms of the Marriage Loan were reversed: now a couple received the loan only if the wife promised to *continue* earning income outside the home.[147] As Germany rearmed, the need for labor grew, and by 1939 the number of women counted in the labor force grew to 14.6 million, up from 11.6 million in 1933. But that was the limit of growth. Throughout World War II (1939–1945), just under 15 million German women participated in the labor force. Unlike Great Britain and the Soviet Union, Germany never mobilized its women for war.

The reasons were both practical and ideological. German women ignored the regime's appeals for their voluntary labor far more than men: between 1935 and 1938, ten times as many men as women had enlisted in the Reich's Labor Service.[148] In 1938, the government instituted a "service year" for young women entering the job market; widely evaded, the "service year" did not attract any woman who could manage to avoid it.[149] In addition, a regime which had preached the sanctity of women's life in the home found it ideologically difficult now to insist on their forced labor outside it. Despite tremendous labor shortages, demands for women's mobilization were never enforced, and German women who refused to register for labor were not prosecuted. Instead, the regime turned to slave and contract labor by foreigners and prisoners of war—some of them women—to fill the gap: just under 8 million slave laborers worked in Germany in 1944.[150] By resisting Nazi demands for their labor, some German women were able to resist the regime passively.

Active resistance to the Third Reich, however, was as difficult for German women as for men. Hiltgunt Zassenhaus, a seventeen-year-old student in 1933, recalled how she initially refused to give the "Heil Hitler" salute then required at the beginning of each class. Her sympathetic teacher, Fraulein Brockdorf, urged her to give in. "Please don't make it so difficult for yourself—and for me." The next day, Zassenhaus was brought before the principal by her teacher and told she would be expelled if she did not salute. Her parents warned her of the consequences of her opposition, and the third day, she found the principal in her class waiting to watch her salute. She did so: "all eyes were on me; and while I again heard 'Heil Hitler' loudly and distinctly from the other girls, I made a desperate movement with my left arm into the air."[151]

Women active in socialist, communist, or labor circles were at risk. Although the initial arrests were mostly of men, enough women were included to necessitate the founding of the first women's concentration camp in Germany, at Moringen, in 1933. An estimated fifteen hundred women prisoners passed through Moringen between 1933 and 1938, about three quarters of them political prisoners.[152] Some women were imprisoned as hostages to coerce male relatives; others were jailed in their own right. Käthe Pick Leichter (1895–1942), for instance, an Austrian socialist and a Jew, had long championed the cause of working-class women and worked in the government on their behalf. "The issue was not opportunity for a few privileged women," she wrote, "but the raising of the miserable conditions of working women."[153] In 1934, when the Nazis came to power in Austria, she joined the underground resistance; her husband and children escaped to the United States. Arrested by the Gestapo in 1938, she was sentenced to the German women's concentration camp at Ravensbrück in 1939. While there, she wrote her memoirs and some poems, which survived by being memorized by fellow prisoners. "Brother, do you also stand all day with a shovel in your hands," she wrote in a poem addressed to her brother, also imprisoned in a concentration camp;

> Isn't it noon yet? Is there no end to sand?
> Or do you, like I, drag great heavy stones?
> Does your back hurt too, do your legs burn like mine?
> See, you are already a man, used to hard blows,
> I am weaker and my body has already borne children.
> What do you think about our children's lives?
> Will blows, punishment cells always hover as a threat?
> And then we go on—in the heart, hope and firmness;
> I in Ravensbrück, you in Sachsenhausen, in Dachau, or in Buchenwald.[154]

Leichter survived three years in Ravensbrück, where she performed heavy labor. "Käthe had to execute the heaviest street work and was forced to load ships in the harbor with bricks [used for ballast]," a cell-mate recalled. "Her hands were bloody, infected, and torn." In 1942, Leichter was gassed with fifteen hundred other Jewish women prisoners after having been shipped from Ravensbrück to Magdeburg.[155]

The success of German foreign policy and German armies between 1936 and 1942 ensured that millions of Europeans would have to live under Nazi rule and occupation. Women and men from France to the Soviet Union, from the Netherlands to Greece, were forced to deal with the war and occupation, with mass murder and genocide, with resistance and survival. In battles and bombardments, in factories and homes, in besieged cities and concentration camps, European women faced a new kind of warfare: the mass mobilization of industrial nation-states. While many continuities remained from earlier wars, World War II initiated armed conflict on a new scale in the lives of most European women.

4

CONTINUITY AND CHANGE:

WOMEN IN WORLD WAR II AND AFTER

❧

Women in World War II

From the eighteenth century through the First World War, wars in Europe were often "limited" with regard to women and children, who were supposed to be spared if possible. "If war is properly conducted," declared Frederick the Great of Prussia in the mid-eighteenth century, "the civilian population should not know it has taken place." While such rules broke down occasionally, by and large European women remained protected during World War I: casualties were overwhelmingly male. When people spoke of a "lost generation" after 1918, they meant a lost generation of young men, whose absence increased the proportion of women in the population.[1]

All this changed in the Second World War. Women and children were no longer protected, but as members of the civilian population were subject to bombardments and sieges. In concentration camps, Nazis often killed women and children first. World War II was fought in and over cities and towns. Starvation, rape, pillage, and mass murder reappeared as armies fought in all-out warfare. In 1941, for instance, a war correspondent described a scene in the Russian town of Dorogobuzh which could have occurred in any era of European history when war was not limited. The town, which had housed ten thousand people before the German invasion that summer, stood empty, except for

> an old woman; a blind old woman who had gone insane. She was there when the village was shelled and had gone mad. I saw her wandering barefooted around the village, carrying a few dirty rags, a rusty pail and a tattered sheepskin.

She survived on handouts from soldiers passing through, never speaking; she "just stared with her blind white eyes and never uttered any articulate words, except the word 'Cherti'—the devils."[2]

As in earlier warfare, women speedily improvised and organized the support services essential to keep society functioning. They provided child care, food, nursing aid, and refuges, both for their own families and for others. A postwar tribute gave credit to "Mrs. B.," a London beetroot-seller, who organized her neighborhood's air-raid shelter:

> When the [air] raids started [in 1940], she left the first aid post where she was a part-time volunteer, walked into the Ritchie Street [air raid] rest centre and took charge. She found a supply of milk for the babies, bedded them down early with their mothers, and administered [sleeping] powders. . . . Then she put the oldest and feeblest on the remaining beds and benches and had the whole household, 100 to 300 in all, asleep or quiet as the bombs came whining down. In the morning, she organized the washing, bathed the babies, swept the floors, supervised breakfast and went home about 11 o'clock. . . . In the evening she was back again. She made one rest centre a place of security, order and decency for hundreds of homeless people.[3]

In addition to resuming their traditional wartime roles, European women assumed new ones under the conditions of modern industrial warfare and the Nazi occupation and mass murders. Far more than during World War I, women enabled industries to increase production and keep functioning. In Great Britain, for instance, about twice as many women worked in war industries or the armed forces in World War II as in World War I.[4] In Germany, where the number of women in industry did not rise appreciably after 1939, half a million women worked as support staff for the military. Called "assistants," they functioned primarily as clerical staff and telephone operators in the early years of the war; by the end, they aimed searchlights and fired antiaircraft guns, drove trucks and ambulances, nursed the wounded and became fire fighters.

Germany and Italy never drafted their women; Great Britain and the Soviet Union did, and their ultimate victory may be due in part to their use of womanpower. In the Soviet Union, women replaced virtually all men in agriculture and, by 1945, comprised a majority of all industrial workers.[5] While there was never a formal call-up of Russian women, they participated of necessity, swept up in the German invasion of 1941. "Those were dreadful days," remembered Olga Sapozhnikova, a Moscow factory weaver.

> I was ordered, like most of the girls at the factory, to join the Labor Front. We were taken some kilometers out of Moscow. There was a large crowd of us and we were told to dig trenches. We were all very calm, dazed, and couldn't take it in. On the very first day we were machine-gunned by a Fritz who swooped right down. Eleven of the girls were killed and four wounded. . . . We went on working all day and the next day; fortunately no more Fritzes came. But I was very worried

about father and mother [both retired textile workers], with nobody to look after them.[6]

Allowed by her commander to return to Moscow, Sapozhnikova found food for her parents and went back to work at her factory when it reopened a few days later.

The women themselves had to figure out how to juggle war work and family responsibilities, as governments moved to draft women without providing support services for child care or housework. In Great Britain, the government registered all women between eighteen and fifty, and from 1941 drafted single women between twenty and thirty, giving them the choice of war work or military service. A large-scale 1942 poll found 97 percent of Englishwomen in favor of this policy, and by 1943, 90 percent of single women between eighteen and forty worked in industry or the armed forces, as did 80 percent of the married women of the same age.[7] "She's the girl that makes the thing that drills the hole that holds the spring," went a popular song of 1942,

> That drives the rod that turns the knob that works the thingumebob. . . .
> And it's the girl that makes the thing that holds the oil that oils the ring
> That works the thingumebob THAT'S GOING TO WIN THE WAR![8]

Women in the factories worked long hours under dangerous conditions, compensated less than men doing the same jobs. Under a law of 1939, women who were wounded or lost limbs in a war-related accident were compensated at rates 25 to 50 percent lower than those for men; public outcry forced a change in 1943.[9] Protests did not win women equal pay, however, even when they did the identical job as men. In government training centers, women were paid at a rate slightly less than two thirds of the men's. The railroad industry, resisting a measure to pay women workers equally, argued that "since the managers had been unable to find any industry where the principle of equal pay for equal work was applied, they did not see why they should apply it on the railways."[10] While women in some factories won equal pay—as the women at Rolls-Royce did in 1943 when the men went out on strike with them—the wartime coalition government strenuously resisted any large-scale attempt at equalization. In 1944, women teachers campaigned for equal pay and won a majority in Parliament on the issue. Calling the move "impertinent," Churchill made the issue a vote of confidence in his government and so defeated it.

Women also demonstrated for government nurseries, but were largely unsuccessful in gaining them. Wheeling children in carriages and carrying them in their arms, they argued that they could not contribute much to the war effort without nurseries.[11] The government urged private arrangements, as in this Ministry of Health poster:

> If *you* can't go to the factory
> help the neighbour who can
>
> How you can help
>
> Arrange *now* with a neighbour
> to look after her children when
> she goes to her war work—
> Or give your children to her when you go
>
> Caring for War Workers' Children Is A National Service[12]

During World War II, as in the decades following it, women's demand for public child care far outstripped what was actually provided.[13] Women factory workers remained responsible for food provision as well as child care, and rationing made marketing a lengthy and laborious task. Queuing in lines for scarce items, improvising meals from powdered eggs and dried milk, cooking with more vegetables and grains, Englishwomen found their ingenuity taxed by government food policies. "Whether you are shopping, cooking, or eating," declared the Ministry of Food in 1941, "remember, FOOD IS A MUNITION OF WAR, DON'T WASTE IT."[14] Women were exhorted to salvage materials in short supply: by 1944, the government reported that women had collected "1,117,788 tons of waste paper, 1,334,171 tons of metal, 82,889 tons of rags, 25,298 tons of scrap rubber, 43,948 tons of bones and thousands more tons of kitchen waste."[15] Household goods and civilian clothing were in short supply, and clothes and furniture were rationed. Women established clothing and furniture exchanges to assist those who lost all their possessions in the bombings.

Women in the military, who almost without exception were unmarried and without children, were spared the labor of queuing and finding ration points, making clothes, and looking after children. Instead, they either worked on the land or provided support services for the male armed forces. Rural women were almost always assigned to the "Land Army," where they had only eight days off a year and labored to produce larger harvests without male farm workers. City women worked in factories or entered the women's army, navy, and air force units. There, they learned to operate radios and transmit in code, fire antiaircraft guns and aim searchlights, ferry male soldiers in planes or trucks. "All these very young products of the dole, then the war, of white bread, 'marge' and strong tea, of a hard, city life already had the shrunken upper body, the heavy-set thighs, white and doughy, of mature women," recalled Mary Lee Settle, who enlisted as an airwoman in 1942.[16]

After a few weeks of basic training, "the girls around me had begun to fill out and glow," remembered Settle. "Air, exercise, regular meals, and the very act for some of them of sleeping above ground for the first time in years were making the blood run better in their veins."[17] Such women served as support staff, away from actual combat.

In the Soviet Union, the other belligerent nation which relied on women's direct contribution to winning the war, women in the armed forces saw active combat. Thousands of Russian women served in artillery and tank units and staffed regiments in the air force. "Anti-aircraft battalions, air-force regiments and signalling units were made up entirely of women," remembered Galina Utkina of the Battle of Stalingrad. "When the battle reached the city itself, women fought alongside the men outside the tractor and metallurgical plants, at the Mamai Barrow and on the streets."[18] Utkina asked to be sent to the front lines after her husband's death. Women at the front were usually widowed or single. Most Soviet women who fought moved from factory work to combat and sometimes back again. Over 100,000 Soviet women won military honors; eighty-six were awarded the coveted rank of "Hero of the Soviet Union." Most current knowledge of Soviet fighting women is based on the publicity given outstanding women: the case of Maria Oktyabr'skaya (1902–1944) is typical of this group. A telephone operator when the Germans invaded, she lost her husband in 1941. She then went to work in a Siberian armaments factory (much of Russian industry relocated east of the Ural Mountains). Like many workers, she contributed part of her wages to pay for a tank, and in 1943 she volunteered for training as a tank mechanic. In October, she was assigned to drive a tank under a male commander and given the rank of sergeant. She saw action in November at Novoye Selo and destroyed a German gun embankment by running it over with her tank, nicknamed Fighting Woman Friend. In January, her tank blew up crossing a mine field, mortally injuring Oktyabr'skaya—she died of her wounds in March. Posthumously awarded the Order of Lenin, her story was publicized, and a tank named Fighting Woman Friend in her honor took part in the invasion of Berlin in 1945.[19]

Women who knew aircraft staffed the all-female 122nd Air Group, formed in 1941. Comprised of three regiments—the fighters, the bombers, and the night bombers, the group flew 4,419 operations, participated in 125 air combats, and scored 38 aerial victories. Soviet airwomen won numerous decorations for their heroic exploits. One of the most famous was Lily Litvak (d. 1943), a teenage flying instructor who became the chief Soviet female air ace. Blond and barely over 5 feet tall, she enlisted in 1941. Just before her unit was sent to Stalingrad, she wrote her mother, who kept a shop in Moscow, to send her warm clothing, toothpaste, notebooks, and handkerchiefs. "Please, Mamenka, don't address your letter to pilot Litvak. Just make it L. Litvak. The names of the pilots are supposed to be a secret."[20] Famous for her success in aerial combat, she shot down twelve German planes. Known as "the White Rose of Stalingrad," she returned to the front after recovering from a leg wound, only to be shot down herself in 1943. Her last letter to her mother spoke of how the war had consumed her:

> Battle has swallowed me completely. I can't seem to think of anything but the fighting. . . . I love my country and you, my dearest mother, more than anything. I'm burning to chase the Germans from our country so that we can live a happy normal life together again.[21]

Women in regions occupied by the German armies faced even harsher conditions, especially in Eastern Europe. If they were Jewish, they faced extermination, as the Nazis sought to eliminate the European Jews. In 1940, as the German armies conquered France, Belgium, the Netherlands, Norway, and Denmark, brutalities calculated to intimidate the local population were standard. In the French village of Luray, for instance, Mme. Bourgeoise, an elderly woman,

> was so infuriated when the Germans came to requisition and occupy her house that she shouted and shook her fist at them. Two soldiers grabbed her and tied her to a tree in the garden and assassinated her before the eyes of her horrified daughter. They told her daughter to leave the body tied to the tree for twenty-four hours as a warning to all.[22]

Generally, women who were not Jewish in the occupied nations of Western Europe escaped the harshest consequences of Nazi rule. Unlike men, they were not drafted for forced labor in Germany or shot as hostages. But by the end of 1944, as the Nazis began to lose, they struggled to maintain their hold on Western Europe at all costs. The village of Oradour in southeast France, for instance, was destroyed in 1944 to discourage others from supporting the Allied armies. The Germans shot the men in the town square, herded the women and children into the church, and set the church on fire. Those escaping were shot. Marguerite Rouffranche got out through a window; her two daughters and grandson died. "A woman who had been near me sought to escape as I did," she remembered, "and she had a 7-month-old baby with her. She threw the baby out the window . . . the Germans opened fire and killed her." The baby also died, and Rouffranche was the only survivor.[23] Women in occupied Western Europe also suffered from the hunger and chaos caused by armies on the move. Catherina Barnes, then eighteen, remembered the winter of 1944–1945 in Rotterdam:

> For a week you could not come out and all that we had was one big pack of dried peas. We had three people that died and we could not even bury them, you know, they were in sheets, wrapped up in sheets somewhere because of the fighting.[24]

The challenge of living under conditions of severe social breakdown affected women in Western Europe at the end of the war; in Eastern Europe, unlimited warfare came earlier. The battle over Leningrad, which lasted from 1941 to 1943, subjected a city of 3 million people to conditions of siege warfare. The Germans attempted to bomb and starve the city into submission. Bombs fell almost every day during the autumn of 1941, and no food came in. By the end of the first year of the siege, almost a million people had died of starvation.[25] Women rallied to the defense of the city, digging trenches, stringing barbed wire, building bunkers. "I love you with a new love," wrote the poet Olga Berggoltz from Leningrad in 1941,

> Bitter, all-forgiving, bright—
> My Motherland with the wreath of thorns
> And the dark rainbow over your head.[26]

Another poet, Vera Inber, remembered that hunger came to erode the normal human decencies. One evening during the first bad winter, she and her husband were walking through the snow during a period of shelling. An old woman called to them to help her: "Darlings, I lost my bread card. Do help me. I can't find it without you." Inber was horrified at her own reply: "Find it yourself. We can't help." Her husband retrieved the woman's card; to lose one was a sentence of death by starvation.[27]

As conditions worsened, people stayed indoors and died of hunger. Tanya Savicheva (1931–1943) chronicled the deaths of her family by pasting sheets of paper on her bedroom wall:

> Z—Zhenya died 8 December, 12:30 in the morning, 1941
> B—Babushka died 25 January, 3 o'clock, 1942
> L—Leka died 17 March, 5 o'clock in the evening, 1942
> D—Dedya Vasya died 13 April, 2 o'clock at night, 1942
> D—Dedya Lesha died 10 May, 4 o'clock in the afternoon, 1942
> M—Mama 13 May, 7:30 in the morning, 1942
> S—Savichevs died. All died. Only Tanya remains.[28]

Tanya Savicheva was evacuated from the city in the spring of 1942, suffering from chronic dysentery. She died in the summer of 1943. March 8, 1942, International Women's Day, was celebrated that year by women beginning to clean up the city, burying the bodies still covered by the winter snow.

Under extreme conditions, survival often became an act of resistance—especially for the European Jews. But in addition to attempting to survive, Jewish and other European women also actively resisted Nazi rule. Women organized the resistance networks which smuggled information and people out of occupied Europe.[29] In France, Marie-Madeleine Fourcade directed a three-thousand-person network.[30] In Yugoslavia, women partisans fought the Germans, and at least one village was "commanded" by a Jewish woman. A childhood friend from Belgrade describing her in 1944 said that "the years of terror had left their mark on her face, and despite her youth her hair was streaked with grey."[31] Women fought in the Warsaw ghetto uprising of 1943 and provided a majority of the couriers between ghettos and the "Aryan" world outside: if searched by police, they could pass as Christians more easily than Jewish men, whose circumcision made them identifiably Jewish.[32]

Women's activities in the resistance ranged from providing "safe houses" for escaping refugees to carrying messages, from publishing underground newsletters to taking part in assassination attempts against the Nazis.[33] Some gave momentary assistance; others dedicated their lives to undermining fascism. The history of women in the resistance is just beginning to be written. Most testimony describes one discrete action: a French servant hiding a radio under her skirts from the Germans; a Danish fishmonger helping to smuggle Jews to neutral Sweden; Polish partisans living in a forest outside Warsaw, "sunburned and disheveled, clad in odds and ends of old clothing. . . . some were armed with revolvers, others with carbines."[34] One life about which more is known is that of Hannah Senesh (1921–1944), a Hungarian Jew whose diary and poems record her conversion first to Zionism and then to active resistance.

Senesh grew up in a comfortable Jewish home in Budapest; her father, a well-known playwright, died when she was six. Her mother sent Senesh to a fashionable Protestant girls' school where she first experienced overt anti-Semitism. Elected head of the school's literary society when she was fifteen, Senesh was forced to yield to another girl because the student government ruled that the head had to be Protestant. "Only now am I beginning to see what it really means to be a Jew in a Christian society," she wrote in her 1937 diary.[35] Like many European Jews in the twentieth century, the Senesh family "assimilated": they "did not feel it important to observe the outer formalities of religion" and gave their children Christmas presents and a Hungarian education.[36]

Senesh's diary, which she began at thirteen, records the rise of violence and anti-Semitism in Europe. By the time of the Munich crisis of 1938, Senesh stated, "I've become a Zionist. . . . To me it means, in short, that I now consciously and strongly feel I am a Jew and am proud of it."[37] She left Hungary for Palestine in September 1939, a few days after the outbreak of World War II. In January 1940 she wrote, "1939—brought so many changes in my way of life and within me. It was a year filled with constant tension, excitement, fear, and last autumn World War II began. . . . Due to these outer pressures, plus my personal leanings and abilities, I became a Zionist, and a real Jew."[38] From 1939 to 1943, when she was twenty-one, Senesh worked to become a worthy Palestine immigrant: learning Hebrew, working in agricultural communes (kibbutzim), reclaiming her Jewish identity. Word reached Palestine of the deportation and mass murder of Europe's Jews. In 1943, Senesh conceived her plan to help rescue Hungarian Jews. "I was suddenly struck by the idea of going to Hungary," she wrote in her diary, "to help with the youth emigration, and also to get my mother out. Although I'm quite aware how absurd the idea is, it still seems both feasible and necessary to me, so I'll get to work on it and carry it through."[39] Already recruited into the Palmach, the attack force of the Jewish underground army in Palestine, she contacted the British, who made her an officer in their air force and agreed to parachute a small group of resistance fighters into Yugoslavia in March of 1944. A companion remembered her "smart in her army uniform, pistol strapped to her waist."[40] There, she wrote her most famous poem, in Hebrew:

> Blessed is the match consumed
> in kindling flame.
> Blessed is the flame that burns
> in the secret fastness of the heart.
> Blessed is the heart with strength to stop
> its beating for honor's sake.
> Blessed is the match consumed
> in kindling flame.[41]

She would be severely tested: in June, the group crossed the border into Hungary and were captured by the Germans.

Tortured and beaten, Senesh refused to reveal the radio code which linked the group to the British. The Germans brought her to a Budapest jail, where they confronted her with her mother, threatening both women with torture if they did not yield the code. "Her once soft, wavy hair hung in a filthy tangle, her ravaged face reflected untold suffering, her large, expressive dark eyes were blackened, and there were ugly welts on her cheeks and neck," her mother recalled of this meeting. Senesh did not reveal the code, and mother and daughter were imprisoned in the same jail, separated from each other. Although in solitary confinement, Senesh communicated with other prisoners by cutting out paper letters and then piling her table and chair on her bed so she could climb up and flash messages out her window. She wrote her last poem in this cell, returning to her native Hungarian language:

> I could have been
>> twenty-three next July;
> I gambled on what mattered most,
> The dice were cast. I lost.[42]

In November 1944, Hannah Senesh was executed in prison by a German firing squad.

Senesh chose her fate. Millions of other Jewish women had little choice in shaping their futures, especially after 1941, when the Nazis began to systematically exterminate Europe's Jewish population. Forced to wear the yellow six-pointed Star of David prominently on their clothing, Jews were quickly differentiated from the rest of the population.[43] Women sewed the stars on clothing, including their children's. Some remember the adult-sized stars glaring from small chests; others sewed on baby-sized markings. "Dvortche, Hayim-Idl's wife, embroidered a tiny yellow hexagram, like a star of David, and sewed it to Bella's dress over the heart," recalled a woman who identified herself only as Ka-Tzetnik (concentration camp inmate) 135633, of life in the last days of the Warsaw ghetto. "And the tiny patch really had poignant charm, like the first wee white shoes of an infant."[44] In larger cities, Jews were first moved into a ghetto area where they could be easily controlled. From 1942 on, these ghetto populations began to be shipped to the death camps in the East which were kept secret. Women and children formed a disproportionate amount of the ghetto population, since men had often been rounded up or captured earlier.[45] As the pace of the shipments intensified, normal kindness toward children and women disappeared. "Running behind that last wagon, a lone woman, arms outstretched, cried, 'My child! give me back my child!' " remembered Vladka Meed of an early deportation from the Warsaw ghetto.

> In reply, a small voice called from the wagon, "Mama! Mama!" The people in the street watched as though hypnotized. Panting now with exhaustion, the mother kept running after the wagon. One of the guards whispered something to the driver, who urged his horses into a gallop. The cries of the pursuing mother became more desperate as the horses pulled away. The procession turned into Karmelicka Street. The cries of the deportees faded away; only the cry of the agonized mother still pierced the air. "My child! Give me back my child!"[46]

Crammed into railway cattle cars, some deportees still believed they were headed to labor camps, although the conditions of the journey often ended that illusion. "No air can enter or escape . . . 75 to a car . . . no toilets . . . no doctors . . . no medication," remembered Isabella Katz of her passage from Hungary to Auschwitz in 1944. "I am menstruating. There is no way for me to change my napkin . . . no room to sit . . . no air to breathe."[47] Katz's mother died on the journey; she and her sisters survived the following year at Auschwitz in Poland, the largest of the death camps, where 3 to 4 million died.

In the death camps, women and children often perished on arrival: Nazi policy sent women with small children to the gas chambers and ovens. Charlotte Delbo, a French survivor of Auschwitz, described this selection process:

> MARIE
> Her father, her mother, her brothers and her sisters were gassed on arrival.
> Her parents were too old, the children too young.
> She says: "My little sister was beautiful.
> You cannot imagine how beautiful she was.
> They must not have looked at her.
> If they had looked at her, they would not have killed her.
> They would not have been able to."[48]

Those picked to survive, and aid in the labor of the camps, tended to be young, healthy, and without children. Survivors describe the world of Auschwitz and the other death camps as "hell"—a place where inhumanity became normal. "You were speaking of being dead. I had that feeling when I first came from the ghetto, and after the cattle train came to the camp," remembered a woman in 1983.

> They told us to undress. I was young. There were men standing . . . I just couldn't help myself. I wanted to hide myself. What do I hide first? My whole body? You can't hide your body. And then they shaved us. . . . And we were naked—no hair. In the meantime, they were burning our families in the crematorium. By the time we'd gotten into camp, it was at night and all the lights were on and I thought I was dead and in hell. That was the only explanation for the experience and for the sight. Can you imagine men watching and telling you to get undressed and trying to push? Everyone was trying the same thing, to get back, but you can't go further back than the barrack walls, right? And the lights on and the electrified barbed wire. I thought I was dead and I'm in hell, only in hell they burn. This was how I associated the situation.[49]

Guards shaved women's and men's head, armpit, and pubic hair and then tattooed their camp number on their right forearm. "The branding demoralized Clara," remembered Fania Fénelon, a French Jew. "Dazed, still incredulous, she gazed at her round white arm: 'Why are they treating us like this?' . . . I see us now . . . demoralized, shivering, tattooed and hairless. It was odd, but that was the real humiliation, having no hair."[50] Other women recalled the feeling of being desexed which a shaven head produced. "Our heads are shaved. We look like neither boys nor girls," remembered Isabella Katz. Soon, the starvation diet caused menstruation to stop: a blessing in the camps where there were no extra rags to use as napkins. Soon, women and men were reduced to the half-life of the starved slaves of the camps:

> We haven't menstruated for a long time. We have diarrhea. No, not diarrhea—typhus. Summer and winter we have but one type of clothing. Its name is "rag." Not an inch of it without a hole. Our shoulders are exposed. The rain is pouring off our skeletal bodies. The lice are having an orgy in our armpits, their favorite spots. . . . We're hot at least under our armpits, while our bodies are shivering.[51]

Women and men survived by laboring: either useless work calculated to wear them down, or the horrible jobs of maintaining the camps: collecting the clothes, false teeth, hair, and belongings of the dead; hauling bodies out of the gas chambers and ashes out of the ovens; working in camp hospitals and brothels. "I am terribly ashamed to have experienced all this," wrote Hannah Lévy-Hass in a diary she managed to keep in Bergen-Belsen in the spring of 1945.

> It is abominable what they had made of human beings. The darkest scenes of the Middle Ages and of the inquisition were repeated here, multiplied to the utmost. Their monstrous repetition will mark "civilized" and "cultured" Germany of the twentieth century forever with the sign of shame.[52]

Women's camp experiences differed from men's, but not as outsiders expected. "The question I was asked was always the same," remembered Micheline Nauriel about her release, " 'Tell me, were you raped?' "[53] Rape was infrequent, and some camps had brothels to service German soldiers and SS guards. At Auschwitz, the notorious "House of Dolls" brothel used female guards to control the women, who were inspected for venereal disease and had "Field Whore" and their identification number tattooed on their chests.[54] Female guards could be just as sadistic as men, as were other women in authority: Ilsa Koch, wife of the commandant at Buchenwald, was notorious for her brutal and heartless treatment of prisoners.

While women suffered more sexual abuse in the camps than men, both inmates and survivors agreed that women coped with the camp ordeal better than men.[55] Unlike men, who tended to be "lone wolves," women formed surrogate families, becoming each other's "sisters," "daughters," or "mothers" as their real families perished.[56] Women shared food, cleaned their quarters together, and celebrated holidays as a group. Almost all female survivors testify that they would not have lived without the help and support of other women. "Women by the thousands were walled-in," wrote the Greek poet Victoria Theodorou of the concentration camp she survived,

in this empty, nameless island
—officially declared unfit for human habitation
infested with yellow fever and typhus—
We were the first to camp here
on this meager soil we worked, we gave birth
we buried, we sang
we abolished emptiness
we built kilns and workshops
wells and windmills
here we lived out the clay age
we dug for roots
we coaxed the music from the reed
we made a lyre from the turtle shell[57]

Millions of Europeans, women and men, perished in the camps and on the battlefields, in the ghettos and the bombed cities—perhaps as many as 50 million dead. Six months before the war's end in May 1945, Fania Fénelon was shipped from Auschwitz to Bergen-Belsen, part of the German attempt to conceal the extermination camps. Tens of thousands of women slept in the mud, under a canvas which collapsed in the rain. "I was later to learn that, a few yards from me, Anne Frank was lying under this same canvas," Fénelon remembered.[58] Fénelon survived; Frank did not. Her death epitomizes the losses from World War II.

Anne Frank (1929–1945), born in Frankfurt, moved with her family to Amsterdam in 1933. In 1940, when she was ten, the Germans occupied the Netherlands; the next year saw the first roundups of Amsterdam's Jews and many regulations aimed at separating Jews from the rest of the Dutch. Anne Frank and her older sister Margot transferred to a Jewish high school; the next year, the family went into hiding in an apartment behind a warehouse maintained by sympathetic Christians. Anne Frank began a diary the month before she went into hiding and kept it throughout her stay. She recorded the passionate feelings of adolescence pent up in a life of secrecy, where movement and noise had to be curtailed all day. She "hated" her mother, "loved" her father, chafed at the lack of privacy. In these pages, her unique voice emerged:

I want to get on; I can't imagine that I would have to lead the same life as Mummy and Mrs. Van Daan and all the women who do their work and then are forgotten. I must have something besides a husband and children, something that I can devote myself to!

She longed to become "a journalist or a writer" and to write something "great."[59] In addition to recording her feelings, Anne Frank recounted what she called "the march of death": the roundups of Jews and their shipment to the East. The family listened avidly to the radio, following the course of the war whose outcome would determine their survival. In July 1944, Frank wrote,

> in spite of everything I still believe that people are really good at heart. . . . I see the world gradually being turned into a wilderness, I hear the ever-approaching thunder, which will destroy us too, I can feel the sufferings of millions, and yet, if I look up into the heavens, I think that it will come out all right, that this cruelty too will end, and that peace and tranquility will return again. In the meantime, I must uphold my ideals, for perhaps the time will come when I shall be able to carry them out.[60]

One month later, the Frank family was betrayed, arrested, and sent to Auschwitz. Anne Frank died in Bergen-Belsen in March 1945—two months before the end of the war in Europe, three months before her sixteenth birthday. The loss of lives like hers, whether young or old, promising or fulfilled, can never be fully measured.

Working-Class Women Since 1945

For European women, the postwar era divides at the end of the 1950s. Until then, women and men alike had to cope with the ravages of war. In much of Eastern Europe, there are no population statistics for the years 1944 and 1945: the disruption was too great. Bombed-out, displaced, homeless, women and men struggled to restore life to normal. "As stake / for a new start only the fact / that scorched earth can still be used," as the Dutch poet Judith Herzberg put it.[61] For many, the hardest times came immediately after the war: even in victorious England, the most stringent rationing was from 1946 to 1950, not during the war. In 1947, the popular English magazine *Picture Post* devoted an issue to "Mrs. Average," who still spent at least an hour a day queuing for scarce goods. "Mrs. Average" wants "more leisure, more colour, more food and clothing, less weary work," concluded the magazine. "But most of all she wants hope. That is why she is sadder now than during the war."[62]

Disparities were immense. In the West, aid came from the United States; in the East, material resources often went to Soviet Russia. The gap between those who had survived violence and danger and those who had been relatively protected loomed large. In 1946, for instance, Hildegard Knef worked as a film actress in bombed-out Berlin. Suffering from lice, dysentery, and malnutrition, she and her colleagues produced films in unheated warehouses. That year, Knef traveled to Zurich, in neutral Switzerland, where she was shocked to hear a Swiss woman complain, "The war was dreadful, we could not get rice or cocoa." Knef remembered that the remark made her "want to go back to a town without houses, without widowpanes, without roofs; holes in the asphalt, rubble, rubbish, rats. I wanted to be poor among the poor."[63]

By 1960, the worst was over, and the standard of living rose again. From the mid-1960s on, Europeans lived better than they ever had before. Not only running water, but hot running water became standard in newly built working-class dwellings, which were no longer only one or two rooms. Flush toilets, baths, central heating, and electricity could often be taken for granted, especially in new buildings in the cities. Appliances previously restricted to the wealthy—refrigerators, ovens, televisions, and even automobiles—came increasingly within the means of the average person. In every category— longevity and health, *per capita* and disposable income, level of education— Europeans were better off than they ever had been before.

Women shared in these gains, but for European women the period between 1960 and 1985 brought continuities as well as changes. For women, the greatest changes have been in life expectancy and control of reproduction. By 1980, life expectancy at birth for a European woman was close to eighty, about five years longer than for men. Maternal mortality rates have declined so drastically that they are now measured not per every thousand women, but per every ten thousand or one hundred thousand, and rates of less than 2 per 10,000 are common. Death in childbirth has become rare rather than a normal part of life.

Childbearing and child rearing take up a far smaller portion of women's lives. In 1900, the average European woman lived fifty years; over half her life was spent raising children. By 1970, a European woman could expect to live to seventy-five, and only eighteen of those years—usually from twenty-two to forty—were devoted to children.[64] Prosperity enabled women and men to marry young, in their early twenties.[65]

European women can control their fertility far better than ever before. Contraception and abortion have made it possible for most women to bear only the children they want and to have sexual intercourse without much risk of pregnancy. Birth rates confirm this new control over reproduction. In the years immediately following the war, birth rates rose, but since 1950, they have consistently declined throughout Europe. By 1970, the average European birth rate was 16 per 1,000, lower than possible before access to cheap, reliable contraception. Even in Catholic societies, where contraception and abortion were illegal, couples restricted their fertility: birth rates declined to equal those in societies where such remedies were legal.[66] In 1969, 53 percent of Italian women polled approved of the "artificial" contraceptive techniques opposed by the Roman Catholic church.[67] Only extremely severe state policy could thwart this decline: in Rumania, the birth rate was 14.3 per 1,000 in 1966; the following year, it rose to 27.4 because of a stringent anti-abortion policy.[68]

Birth-control techniques vary according to custom and available options. Generally, women in Western Europe have relied more on contraception; women in Eastern Europe, more on abortion.[69] Western European women use the birth-control pill; Eastern Europeans often distrust it. "We have the I.U.D., although it isn't used very much," a Moscow woman told a Swedish reporter in 1980.

> Women who want one have to stand in line a long time. We try not to use the pill because of the possible danger to a future fetus. Some also use aspirin, but that's also considered to be dangerous. But above all we have many abortions.[70]

In most European nations, east and west, maternity benefits are generous by United States' standards. Most nation-states, concerned about low birth rates, seek to counter them by trying to make motherhood more attractive to women. Prenatal care is usually free, comprehensive, and universal; most nations pay the mother a maternity allowance; and all have maternity benefit programs. The most comprehensive as of 1985 was that of the German Democratic Republic (East Germany): a woman is guaranteed a six-week pregnancy leave at full pay before giving birth, and she can take up to a year off at full pay after the birth (the "baby year"). If the mother cannot find a nursery for her child, the leave can be extended for two more years.[71]

With regard to the absolute standard of living, life expectancy, control of reproduction, and childbirth, European women have never been better off. "If I compare my own situation when I was a young stenographer [in the 1930s] with a girl of today earning the same salary, there is simply no comparison," wrote Françoise Giroud in 1974.

> A month of paid vacation; I never knew what it was to have a vacation. Charter trips and organized travel; in my day that didn't even enter into your wildest dreams. Paperback books, the lithograph available at the local ten-cent store, blue jeans and the T-shirts, instant mashed potatoes, the transistor radio you can buy for next to nothing, the boyfriend who has a Citroen 2CV he bought secondhand, and off you go to the country. And the pill! It's not a better world; it's another world altogether.[72]

Despite such general improvements caused by the rise in the standard of living, there was little change for European women in the area of paid labor outside the home. There, traditional attitudes and patterns continued to shape women's lives. In the types of paid labor available to women, in the ways in which women's labor is counted and rewarded, in women's obligation to care for their families in addition to earning income, continuities outweigh changes. More European women earned income outside the home—and so are considered to be part of the labor force—in 1980 than in 1960. The patterns of difference in women's participation in the labor force between Eastern and Western Europe have changed. In 1960, far many more Eastern European women were counted as part of the labor force; by 1980 national patterns had become more complex.

PERCENTAGE OF WOMEN IN THE LABOR FORCE 1960–1980[73]

Nation	1960		1980	Change
Netherlands	22 %		29 %	+7 %
Ireland	26 %	(1961)	28 %	+2 %
Sweden	30 %		71 %	+41 %
Great Britain	32 %		41 %	+9 %
France	33 %	(1962)	39.4 %	+6.4 %
Denmark	37 %		60.3 %	+23.3 %
F.R.G. (West Germany)	37.2 %		38 %	+.8 %
Austria	40 %		36.6 %	−3.4 %
Poland	44 %		43 %	−1 %
G.D.R. (East Germany)	45 %		50.2 %	+5.2 %
U.S.S.R.	53 %		51 %	−2 %

However, more women earning income outside the home has changed neither women's occupational segregation nor their low pay relative to men. In no European nation do women earn equally to men. The traditional disparity, where women earn between one half to two thirds what men earn, has shifted slightly: now figures range from nine tenths to three fifths. But even in the socialist nations, which have enforced equal pay for equal work, women do not earn equally to men because women still cluster in poorly paid "women's jobs," and because women still carry the "double burden" of performing paid labor in addition to housework and child care, which cuts into their earning capacity.

In socialist and capitalist societies, in Eastern and Western European nations, girls and women still enter traditionally female jobs. "*Every* girl now thinks in terms of a job. This *is* progress," Eva Johansson, a Swedish school vocational guidance counselor, answered when asked if she thought there had been change in 1975. "They want children, but they don't pin their hopes on marriage. They don't intend to be housewives for some future husband. But there has been no change in their vocational choices."[74] In 1980, a United Nations seminar reported that an analysis of twenty-five European nations showed that women still comprise more than 80 percent of their nations' "stenographers and typists, nurses and midwives, tailors and dressmakers."[75] In Eastern Europe, more women continue to work in agriculture, but there, as in previous centuries, their labor is largely unmechanized, classified as "unskilled," and paid less.[76] Even in the Soviet Union, which prides itself on having broken down stereotypes in labor, women still are over 95 percent of the nurses, typists, day-care personnel, pediatricians, secretaries, and librarians.

Women's traditional jobs remain in modern Europe. Most women enter the service sector of the economy, working in offices or restaurants, in shops or as cleaning women. In service positions, they remain the subordinates: the secretaries, not the managers, the "unskilled" workers, the least well paid. Most women still have to find ways of earning income which are compatible with child care. Despite education, many women find, as Linda Peffer did in England in the 1970s when she entered the job market after having gone to school through ninth grade, that "there was nothing that I could do, because I had no qualifications."[77] Over half the women in England who earn income outside the home do so in the "service industries": office work, retailing, cleaning, child care, teaching, hairdressing.[78] For women with small children, cleaning remains a reasonable choice, despite its low pay and hard labor. In the second half of the twentieth century, however, the employer is more often a corporation than an individual family.[79] May Hobbs, who eventually organized her fellow office cleaners, began when her children were little and she needed to find night work. "I could never go out during the day when the kids were growing up, they was all different ages," stated Jean Mormont in the late 1970s. She worked in a textile factory until she had children in the 1950s. "The only jobs that were going were cleaning jobs in them hours that I wanted to suit me like. It's all you could do."[80] Domestic service, however, has declined since the 1930s. Urban European women now tend to find jobs in offices or in factories.

About one quarter of employed Englishwomen work in factories. As in service jobs, women cluster in a few areas: food and drink manufacture, clothing and shoe manufacture, textiles, and electrical engineering.[81] In a study of the differences between women's and men's factory work in the French metallurgical industry in the 1960s, Madeleine Guilbert found eight chief differences: women's work was physically easier, less complex, more repetitive, more manual, and involved a longer series of operations done at a faster tempo. Jobs that carried responsibilities were generally restricted to men, as were chances for promotion, which were "practically zero for the women."[82] Women's factory work classified as "skilled" even declined in the twentieth century: in 1911, 24 percent of female English factory workers were in this category; in 1977, only 14 percent.[83] In the Federal Republic of Germany (West Germany) in the mid-1970s, 94 percent of female factory workers were classified as "unskilled" or "semiskilled." Seventy percent worked at a piecework rate.[84] "Skilled" work paid higher rates, and classifying women's factory work as "unskilled" lowered women's rates of pay. In the F.R.G. until the 1950s, for instance, special "women's wage categories" set women's wages as a percentage of men's. When courts ruled this illegal, the women's wage categories were instead called "light wage categories," but the reality remained the same: women earned less.[85]

Work classified as unskilled or semiskilled can be exacting and laborious. As in past eras, however, if women do such work it pays less simply because it is done by women.[86] In the 1970s, a Frenchwoman named Alice described her "semiskilled" factory job:

> I made tiny parts for the batteries of electric clocks. I know that because I made a point of finding out. I'd never have known otherwise, and I don't want to die a complete idiot. I cut platinum strips into small 1-millimeter pieces—7 to 8000 an hour. . . . Then I would place each individual part on the plate, rub it smooth, pick it up with a kind of needle-point and put it into a shaping machine. If they were not clean enough I'd have to take them out again, rub them a little more and repeat the process. Then I'd be finished. On the average, I had to complete 800 an hour. It was awful how quickly you had to move your fingers.[87]

Alice called her work "uninteresting and exhausting." Women factory workers have consistently used such words to describe their jobs; the compensations are the rates of pay, which are high for women, and the companionship of other workers. A 1972 survey of women in the English clothing industry found that 43 percent of the women leaving "agreed that only a cabbage could get satisfaction from the job. Well over half find their job boring and repetitive, though 60 percent found the atmosphere friendly."[88]

As in earlier periods, many women who earn income, in their homes or out, try to conceal this income from governments. Women mind children; they do piecework; they work part-time for a few hours a week. They do telephone sales, home sewing, and run secondhand stores. And, as in earlier periods, some work as prostitutes. In all European cities, prostitution still exists, although the number of male clients is declining.[89] Women work on their own, in brothels, or in the new "Eros Centers," which originated in Hamburg in 1967 but rapidly spread to other European cities. The first Eros Center was a drive-in motel with 130 prostitutes, each in an identical room.[90] "Eros Centers? Worse than science fiction," declared F——, a Lyons prostitute who had spent a number of years in one. "Everyone, including the State, scientifically dividing up the cake [the government taxes both the prostitutes

and the man who runs the center], and all on our backs. A nightmare." F——
remembered she and the other women clustered on the stairs soliciting
clients. They wore only bras and panties and had to pay for their rooms on
a daily basis.[91]

F—— was one of a group of Lyons prostitutes who stopped work and
occupied a church in 1975 to protest the dangers and exploitation of their
job. In round-table discussions, they testified that most had started because
they needed money. "It's always need, money that's behind prostitution,"
stated A——. "There comes a time when you're in a fix. And for a girl, a
woman, when she's in a fix, she's always got one thing left that's worth
anything—her body."[92] None saved money: they found their expenses—for
rooms, for clothes, for police protection—outran their earnings.[93] Many had
children whom they saw as the center of their lives and the chief reason they
kept working.[94]

Along with the relative lack of change in jobs available to women, another
constant in the lives of European women who need to earn income is the
double burden of earning income on the one hand and raising a family and
maintaining the household on the other. Whether they earn income outside
the home or not, women remain primarily responsible for child care and
housework. "Our parachutes are the same as theirs, we jump from the same
planes, we've got guts, and our performance isn't much different from the
men's," stated a member of the Czech women's national parachute-jumping
team, "but that's where emancipation ends. I'm married, I'm employed, and
I have a daughter. And a granny. If I didn't say, 'Granny, keep an eye on her,'
my sporting career would be at an end."[95] Although many European nations
provide public day-care, demand tends to outstrip supply.[96]

Recent time-budget studies of working-class families in which both adults
earn income outside the home show women consistently doing more
housework and child care than men, usually six times as much.[97] A 1973
survey of seven thousand women in the European Economic Community in
such families showed almost half the Italian men, one third of the West
German and French men, and one quarter of the men in the Low Countries
doing no housework at all.[98] When women who earn outside the home
complain, they speak not so much about work itself as about the lack of time
the double burden produces in their lives. "It's not work that alienates you,
it's time," an Italian woman in the metallurgy industry told an interviewer
around 1980.[99] In 1969, Natalia Baranovskaya published a short story called
"A Week Like Any Other" in *Novy Mir*, a Soviet literary magazine. Also
known as "An Alarm Clock in the Cupboard," the piece portrayed the hectic
life of Olga Voronkova, an engineer, wife, and mother. On the surface,
Voronkova has achieved the good life denied her foremothers. She is "a real

Soviet woman, proud to be a good mother and a good worker," as a co-worker states.[100] Voronkova is happily married, the mother of two healthy children, and an engineer at a scientific institute. But the opening words of the story are, "I'm running, running . . ." and the piece forcefully conveys the hectic and fatiguing life motherhood, household responsibilities, and earning income can produce. "My favorite sport is racing," Voronkova muses ironically, "from home to bus, from bus to metro, bus again. Metro again. To work and back. Shopping. I'm a qualified long distance runner."[101] Despite her sympathetic husband, who helps her a bit with the children and housework, Voronkova lives in a state of chronic exhaustion, her mind always calculating how she can do all she has to do, her body so tired she cries on Saturday from sheer fatigue. The story ends with no solution, with Voronkova fretting about time in the middle of Sunday night:

> I wake up and I don't know why. I am disturbed. . . . I lie in bed with open eyes, listening to the night's silence. The radiators bubble, there are footsteps in the apartment upstairs. And the alarm clock is ticking away—unceasingly, mercilessly.[102]

The double burden makes it difficult for European women to earn equally to men; it also makes it harder for women to unionize to change their situation. In 1970, a Leeds textile worker explained why she found it difficult to join a union or a strike:

> A woman *has* to go out to work, and she works. And a man can stick up for his rights and all stick together, but a woman has a lot more other things on her mind. Feeding her family, you know . . . by the second hour [of a strike], you're thinking—"Oh, I could be doing my washing. . . . I could be at home, doing my shopping." And that's the whole difference.[103]

The processes by which women choose to work in largely female jobs and accept lower pay and a subordinate status are beginning to be studied. A continuing survey of 450 kindergartners in the German Democratic Republic (East Germany) in the early 1970s found that while children tended to champion their own sex at age four and proclaim that it was the "best" and could do anything, by five and a half they had learned and internalized traditional stereotypes about female and male roles. When asked who performed certain activities, the five-year-olds answered that it was Mother who cooked the food, went marketing, dusted the rooms, did the laundry, and sewed buttons on. Father read a book, sat in front of television, read the newspaper, drank beer, and smoked cigarettes.[104] By grade school, most girls opt for traditional women's roles, like these English girls from the village of Gislea in the mid-1970s:

Debbie, aged ten: "Both my sisters are older than me and one works in a factory. She likes it, so I might work in a factory when I grow up. But best I want to be a nurse, and if I couldn't be a nurse, then I want to be a schoolteacher."

Catrina, aged six: "When I grow up, I'm going to be a nurse, and if not, a hairdresser."

Beverley, aged ten: "I'm going to be a hairdresser when I grow up, and if not, a nurse."

Lynne, aged ten: "I don't think much about what I want to be when I grow up. Until I get a job, I'll be a normal housekeeper or something. I don't mind housework. I think I want to be a housewife, until I think of another job."[105]

Throughout Europe, most girls continue to aspire to traditional lives and envisage a far narrower range of future occupations for themselves than boys.[106]

But change in traditional patterns has taken place and is becoming increasingly possible. May Hobbs was able to organize the office cleaners, and they won important benefits. More women work in nontraditional jobs and are beginning to pressure and hope for change. Governments are increasingly aware of the costs of women's double burden to them in terms of productivity and are making some efforts to make it easier for men to share housework and child care: in Sweden, France, and Norway, for instance, fathers as well as mothers have the right to stay home from work if their child is sick. Such change comes primarily from women saying "no"—from women rejecting the traditions which have oppressed them throughout the centuries. "Do please teach [my daughter] the right things," a Yugoslav peasant woman begged the village's female teacher in the 1930s. "She's a girl, you know yourself the position of women today. Martyrs. I want to make sure she doesn't have that fate."[107] Ultimately, the efforts of European women to ensure a better life for themselves and their daughters necessitated that many traditions governing women's lives be rejected. Rejecting these traditions was immensely difficult: European women who did so had to rebel against and say "no" to all the age-old strictures and limits on women's behavior. When they did so rebel, they became feminists, whether working-class, peasant, bourgeois, or aristocratic. The working-class women who struck for better pay for women, the middle-class women who mobilized to win the vote for women, and the thinkers and writers who demanded equal rights for women united in their rejection of traditions subordinating women and in their creation of a new tradition empowering women: feminism.

IX

TRADITIONS REJECTED

•

A HISTORY OF FEMINISM IN EUROPE

1

FEMINISM IN EUROPE

✿

FEMINISM ASSERTS that women are first of all human beings and, as such, deserve justice. What justice for women involves has changed over the centuries, but feminists are united in their conviction that women are oppressed and that that oppression can and should be ended. These seemingly simple demands have necessitated opposing, rethinking, and revaluing much of Western culture. To assert these claims for women, feminists have had to reject many of Europe's most ancient and sacred traditions in the face of much opposition and hostility. They have had to envision and believe in the benefits of wider roles and options for women in the hope of creating a better world for all. While feminists have differed on many issues—on whether they want women treated equally or differently from men, on how much they have been able to demand for women—they all begin by claiming full humanity for women. "There is not the slightest doubt that women belong to the people of God and the human race as much as men and are not another species or dissimilar race, for which they should be excluded from moral teachings," declared the French courtier and writer Christine de Pizan in 1405.[1] Almost five hundred years later, and in very different social circumstances, Hannah Mitchell described how she became aware of injustice as a child in a poor English farming family in the 1870s:

> At eight years old my weekly task was to darn all the stockings for the household, and I think my first reactions to feminism began at this time when I was forced to darn my brothers' stockings while they read or played cards or dominoes. Sometimes the boys helped at rugmaking, or in cutting up wool or picking feathers for beds and pillows, but for them this was voluntary work; for the girls, it was compulsory, and the fact that the boys could read if they wished filled my cup of bitterness to the brim.[2]

Claiming justice and humanity for women has meant rejecting much, from basic cultural tenets to casual everyday behavior. Given ancient European traditions which insist that women defer and subordinate themselves to men, given traditions which define women only by their relationships with men, given traditions which undervalue women and take men as the standard, the only valid way for women to claim full humanity has been to reject those traditions. It is ultimately impossible to see oneself *both* as dependent upon, subordinate, and inferior to men *and* as a full human being. From the fifteenth century on, some European women have written asserting their claim to full humanity and the subsequent right to work out their own lives. Feminists have consistently rejected the assumption of women's inferiority. They have asserted that what may seem to be inferiority is really inequality, created by centuries of male dominance. "The history of women is solely a history of their persecution and lack of rights, and this history says: men have oppressed women all along," declared Hedwig Dohm, the German writer, in 1876.

In addition to rejecting traditions which subordinate and denigrate women, feminists have also created new visions of society not only for women but for all humanity. For if the traditions which subordinate women are cultural and social, rather than natural and innate, they can be changed and the lives of women transformed. "Sooner or later you are going to see land," Dohm promised other women, "the country that, for centuries, indeed for millennia, you have searched for in your heart—the country where women belong not to men but to themselves."[3] Over a century later, an English women's group echoed this hope:

> A world freed from the economic, social and psychological bonds of patriarchy would be a world turned upside down, creating a possibility for the development of human potential we can hardly dream of now.[4]

Central to feminists' vision of the future is identification with other women. "I felt myself linked with my sex by too close a solidarity ever to be content to see myself abstracted from it by an illogical process," declared the French feminist Jenny d'Héricourt (1830–1890) in 1857.

> I am a woman—I glory in it; I rejoice if any value is set upon me, not for myself, indeed, but because this contributes to modify the opinion of men with respect to my sex. A woman who is happy upon hearing it said: *"You are a man"* is, in my eyes, a simpleton, an unworthy creature, avowing the superiority of the masculine sex; and the men who think that they compliment her in this manner are vainglorious and impertinent boasters. If I acquire any honor, I thus pay honor to women. I reveal their aptitudes.[5]

It is views like these that distinguish feminists from women of achievement who overcame limitations for themselves but saw no link between the obstacles and their gender and had no interest in winning such rights for other women. "I considered myself to be unique—the One and Only," recalled Simone de Beauvoir of her college days at the Sorbonne in the 1920s, before she became a feminist. She took pride in an evening of intellectual discussion in which "all the girls retired . . . but I stayed with the young men."[6] Many years later, in 1972, de Beauvoir stated that she became a feminist when she acknowledged her solidarity with other women rather than her separation from them and realized that she had to "fight for an improvement in woman's actual situation."[7] Feminists differ on whether common cause can be made with all women or only with some of them, but all feminists identify as women and with women, working to extend benefits to others. "It is time that women should arise and demand their most sacred rights in regard to their sisters," as the English reformer Josephine Butler declared in 1870.[8]

While feminists identify with other women, they have been a small minority. The ideology of female inferiority and subordination has been so deeply integrated into the fabric of both women's and men's lives that most have accepted it without noticing that it exists. Within the framework of male dominance, most European women mastered the strategies of those in subordinate positions. They valued what they had and worked for the survival of their families. Operating within a male-dominated family and a male-dominated world, they learned to manipulate and please, to adjust and endure. Many succeeded in living full and productive lives within traditions that subordinated women: Glückel of Hameln, a seventeenth-century German-Jewish wife who ran a trading business; Mary Collier, an eighteenth-century English laundress who wrote poetry about her life; Marie Charpentier, a Parisian woman who took part in the storming of the Bastille in the French Revolution of 1789.[9] Records of women's lives and women's rebellion are still being discovered, but most women lived within the framework of traditional female roles and behavior. By rejecting these traditions, feminists separated themselves from the majority of women. They have most in common with each other, rather than with nonfeminists of the same social group, nation, or era.

United across the centuries by their rejection of traditions, their identification with other women, and their hopes for a transformed future, European feminists have consistently claimed their right to determine their own destiny. Rejecting the dictum widely repeated in the nineteenth century that a woman had to choose between being a housewife or a prostitute, the French teacher Hélène Brion declared in 1916 that woman "wants the right to make a third way for herself, a free life in which she herself will choose the elements."[10] "The Freewoman wants no ready sphere . . . the Freewoman wants the whole round earth to choose from," as the Englishwoman Caroline Boord wrote in 1911.[11] This rejection of man-made and male-centered rules for their lives links feminists who differ in almost every other regard. It is the premise underlying both the feminist writings of the fifteenth-century *querelles des femmes* (debate over women) and the movement for women's liberation of the 1970s. While the specific traditions rejected and the methods for rejecting them have varied over time, the starting point for all European feminists has been their refusal to accept the prevailing views of women's nature, function, roles, and relative worth.

Much had to be rejected. Most of the European traditions about women advocated their subordination to men on the grounds of women's inferiority. The major thinkers and writers, the holy books and sermons, the popular tales and sayings stressed the necessity of female subordination. In a section of *Politics* arguing that Sparta had declined because its women had too much influence, Aristotle declared that "the courage of the man is shown in commanding, of a woman in obeying."[12] In the Old Testament, women were ranked as less valuable than men, were excluded from genealogies, and were told that "your desire shall be for your husband, and he shall rule over you."[13] In the New Testament, women were instructed to be silent in church and learn from their husbands, since "the man is not of the woman, but the woman is of the man. Neither was the man created for the woman, but the woman for the man."[14]

While these pronouncements were not all that was said of women, they did become central tenets of European culture. Women's subordination and inferiority became traditions, woven into the fabric of everyday life as well as the writings of scholars. The male ideal of a woman continued to be a loving, faithful, supportive, and subservient wife; the male nightmare continued to be the "wicked women, wilful, and variable, / Right false, fickle, fell [fierce], and frivolous," as a fourteenth-century English poem put it.[15] Peasant proverbs warned men to keep control: variants of "A woman, a horse, a walnut tree / the more they are beaten, the better they be" appeared in many languages. Churchmen and scholars stressed the role of Eve in causing the "original sin"; fables and poems often presented deceitful and unfaithful wives. In every level of European society, women's subordination was preached. In practice, a few women had always been able to achieve relative power and freedom within the framework of female subordination. But over time, their situation worsened. Noblewomen, nuns, and craftswomen had more options and opportunities in 1200 than in 1500.

With the Renaissance, many European traditions came into question.

The growth of commercial capitalism, the transition from aristocratic feudalism to dynastic monarchy, the spread of education and printing, the Protestant Reformation, and the Scientific Revolution changed a great deal in the lives of men. And yet for women, little improved, and in many ways they lost ground in this era.[16] The old pronouncements about women's inferiority and the desirability of their subordination continued and were added to, even by those men who advocated that women be educated and treated more courteously. "Since nature always intends and plans to make things most perfect, she would consistently bring forth men if she could," wrote Baldassare Castiglione in his important treatise on manners, *The Book of the Courtier* (1518). "When a woman is born, it is a defect or mistake of nature and contrary to what she would wish to do."[17] Discussing the education of girls in an influential seventeenth-century treatise, the French archbishop François de Fénelon remarked that

> a woman's intellect is normally more feeble and her curiosity greater than those of a man; also it is undesirable to set her to studies which may turn her head. Women should not govern the state or make war or enter the sacred ministry. . . . Their bodies as well as their minds are less strong and robust than those of men.[18]

The persistence of traditional attitudes about women in a period in which so much changed for men widened the disparity between women's and men's lives. The new values of humanism—education, individuality, civic virtue—seemed to include all and yet remained restricted to men. The Renaissance ideal of "man" excluded women and perpetuated traditional views of male dominance.[19]

It was under these circumstances that women first wrote opposing and challenging female subordination. The increasing disparity between women's and men's lives, the new visions of human potentiality of the Renaissance, and the growing commercial economy which enabled more women to learn how to read and write led some to feminism. Women almost certainly had felt the injustice of their situation before then—as Hannah Mitchell did when she had to knit as her brothers read—but no written record of this remains. The only evidence of women's rejection of subordinating attitudes is negative: all the strictures about the necessity of female obedience, all the complaints about female "impertinence," "lightness of mind," and lack of "proper modesty" signal the existence of women who did not conform to the traditional womanly ideals of deference, compliance, and silence. But from the fifteenth century on, some women found the confidence in their own judgment to assert their opposition to traditions advocating female subordination and inferiority.

Feminism was born in opposition. The first feminists wrote to overturn male arguments for female inferiority and subordination.[20] Educated themselves, they claimed the right to education for other women and argued that if women and men were educated equally, "the girl would be more perfectly instructed than the boy and would soon surpass him," as the Venetian poet Lucrezia Marinella wrote in 1600.[21] They criticized men's treatment of women in courtship and marriage, delineating the types of men who most oppressed women: the seducer, the bully, the wife-beater, the miser, the fop.[22] From the fifteenth to the eighteenth centuries, these constituted the major assertions of early European feminists.

The revolutionary political and economic developments of the seventeenth and eighteenth centuries—the English Civil War, the French Revolution of 1789, the Industrial Revolution—enabled feminists to claim more for women. New political rights for men led some to demand equal rights for women. Looking to the nation-state to redress women's grievances, eighteenth- and nineteenth-century European feminists worked for women's rights to equal citizenship, to equal political participation with men. The growing industrial capitalist economy enabled more women to cease earning income or laboring hard in the home, but increasingly, these women claimed more for themselves than the life of a traditional wife. In the nineteenth century, feminists organized other women and men to demand political and legal rights for women ranging from child custody to control of property, from equal public education to the vote. Between 1875 and 1925, these women's rights movements achieved many of their goals.

As the changing economic and political conditions of the nineteenth century led some women to work for new political rights, so they led others to work for improved economic circumstances. Feminist socialism appeared within the new socialist movements. From the 1830s, the feminist socialists—most of them women—questioned the conditions under which women worked. From then on, feminist socialists challenged the traditional patterns that had governed and delimited women's work over the centuries. They demanded equality for women in the workplace, access to better jobs, better pay, better working conditions, and better education. By the twentieth century, some insisted that women should have equal opportunities with men in their choice of employment, their access to training, and their chances of advancement. Through unions, socialist political parties, and their own women's organizations, these women extended the spectrum of feminist claims to the world of women's work, both paid and unpaid.

The years between the two world wars witnessed a temporary lull in European feminism. Told they had achieved all that was necessary for equality, feminists in capitalist and socialist societies ceased to make women's

1. Christine de Pizan (1365 – c. 1430) and Reason begin building the foundation of the City of Ladies.

2. Suzanne Voilquin (1801–c. 1876), the French Saint-Simonian feminist, in 1839.

3. Harriet Mill (1807–1858), the English feminist.

4. Four Russian "Amazons"—(clock
wise from top) Sophia Leshern vo
Gershfeld,   Alexandra   Kornilov,
Sophia Perovskaya, Anna Vilberg—i
the 1870s. Perovskaya was execute
for assassinating Czar Alexander II i
1881.

5. Danish suffragettes marching, c. 1900.

6. Clara Zetkin (1857–1933) the German feminist socialist, with (*right*) Nadezhda Krupskaya (1869–1939), the Russian revolutionary, c. 1929.

7. The French feminist Simone de Beauvoir (1908–1986).

8. Italian feminists demonstrate for the right to legal abortions, mid-1970s.

9. Englishwomen forming a human chain to blockade the missile base at Greenham Common, March 22, 1981.

movements their chief political activity. On the defensive for much of the interwar period, feminists worked for issues only peripherally connected to feminism: women's welfare, anti-fascism, pacifism. The upheavals of the Great Depression, the rise of fascism, the Second World War, and the postwar recoveries necessitated that Europeans grapple with issues of basic survival. In these circumstances, most women did not identify as feminists, but as Russians, Englishwomen, and Germans; as communists, fascists, and democrats; as Catholics, Protestants, and Jews.

Feminist concerns reappeared in the late 1960s as the movement for women's liberation. The women's liberation movement revived earlier dreams of political and economic equality, but it also went further, demanding a radical transformation of the basic institutions of society. These new feminists placed women at the center and rejected the ancient tradition that man was the measure of all things. They redefined women's lives in individualist terms, not in relation to men and the family. They insisted that previously taboo subjects be discussed and reformed. In earlier eras, women who raised issues concerning sexuality and fertility had been ostracized; their behavior and demands challenged the basic premises of male domination. But by the last quarter of the twentieth century, European feminists had overridden such prejudices. Tens of thousands of women organized to win divorce reform, birth control, abortion, freedom from sexual harassment. The women's liberation movement demanded new rights for previously outcaste women: victims of rape, prostitutes, single mothers, lesbians.

In the 1980s, European feminism both engaged in feminist scholarship and created a feminist perspective on contemporary political issues. In intellectual discourse, women's studies reshaped many traditional disciplines. Women's history, women's literature, women's art and music have become legitimate areas of study, and their findings reject former assumptions about women. Women's studies has raised basic questions about the limits and prejudices of male-focused language, science, and philosophy. In politics, feminists are currently active in the European ecological and anti-nuclear movements. Arguing that "the future will be female or it will not be," they link male dominance, ecological destruction, and nuclear war.[23]

While different historical circumstances enabled feminists to assert different claims for women, all feminists are united in their common rejection of European traditions subordinating women and in their common goal of a better life for women. Conscious of their identity as women, they have sought not to deny their womanhood, but to expand its claims for other women as well as themselves. Feminists have already transformed seemingly eternal traditions in Europe, winning women education, political and economic rights, and new powers and opportunities impossible in earlier ages.

They have always coupled their discontent with the present to their hopes for the future, asserting as the French socialist Louise Michel did in 1886 that "beyond our tormented epoch will come the time when men and women will move through life together as good companions. . . . It is good to look to the future."[24] Providing wide-ranging criticism of ancient traditions of European culture, feminists seek eventually to change the entire world, not only for women, but for everybody.

2

ASSERTING WOMEN'S HUMANITY:

EARLY EUROPEAN FEMINISTS

✿

Christine de Pizan

The feminist movement which transformed the lives of European women began in the early fifteenth century with the writings of the French courtier Christine de Pizan (1365–c. 1430). Pizan is the first woman known to have participated in the *querelles des femmes*, the literary and philosophic debate about the value of women. The *querelles des femmes*, which continued sporadically from the early fifteenth century until the eighteenth century, opposed women and men, just as another standard topic of argument opposed the ancients and the moderns. From the fourteenth century, learned men debated whether women were human, what their nature was like, whether they could be educated, if they were good for men. While men took all positions in this debate—some arguing women's inferiority to men, others asserting women's equality or superiority—women uniformly championed their own sex, following Christine de Pizan.

In her *Book of the City of Ladies*, completed in 1405, Pizan described how she first became aware of the injustice of men's views of women and decided to oppose them.[1] By 1405, Pizan was over forty, a writer at the French court whose works circulated widely in manuscripts. Widowed at twenty-three, she had successfully supported herself, her two young children, her mother, and other relatives by writing. *The Book of the City of Ladies* begins with the author alone in her study, "surrounded by books on all kinds of subjects, devoting myself to literary studies, my usual habit."[2] She picks up one and begins to read, realizing that this work, like so many others, denigrates women. "Just the sight of this book, even though it was of no authority, made me wonder," she wrote,

how it happened that so many different men—and learned men among them—
have been and are so inclined to express both in speaking and in their treatises
and writings so many wicked insults about women and their behavior. . . . it seems
that they all speak from one and the same mouth. They all concur in one
conclusion: that the behavior of women is inclined to and full of every vice.[3]

Pizan describes how she tried to counter these assertions at first, but found
herself agreeing. She began to argue "vehemently against women, saying it
would be impossible that so many famous men—such solemn scholars, possessed of such deep and great understanding, so clear-sighted in all things,
as it seemed—could have spoken falsely."[4] Unable to oppose her culture's
traditions denigrating and subordinating women, Pizan feels defeated: "As I
was thinking this, a great unhappiness and sadness welled up in my heart, for
I detested myself and the entire female sex, as though we were monstrosities
in nature."[5]

A woman's acceptance of traditions subordinating women leads ultimately to this sense of defeat. Either she must accept her culture's assessment
of women and join an inferior caste, or she must try to identify herself as other
than a woman and so become false and inauthentic. What makes Christine
de Pizan a feminist is her refusal of these alternatives. Instead, she decides
to trust herself over male authorities. "Occupied with these painful thoughts,
my head bowed in shame, my eyes filled with tears, leaning into the pommel
of my chair's armrest, I suddenly saw a ray of light," she wrote.[6] Shifting to
the allegorical mode, Pizan describes how three crowned ladies—"Reason,
Rectitude, Justice"—appear to "Christine," as she calls herself in this work,
and urge her to trust her own judgment. "We have come to vanquish from
the world the same error into which you had fallen," they declare, "so that
from now on, ladies and all valiant women may have a refuge and defense
against their various assailants."[7] The ladies chide Christine for having
doubts about women's worth and demolish men's criticisms of women, calling them contradictory, untrue, and even blasphemous. Christine revives
from her despair and asks what she can do; Reason replies, "Get up, daughter!
Without waiting any longer, let us go to the Field of Letters. There the City
of Ladies will be founded."[8]

The walled city forms the prevailing metaphor of this work. Christine
must shovel away the "dirt" of men's false views of women before she can
build the city on a firm foundation. The "stones" forming the city's walls are
the lives of exemplary women, which provide a refuge to protect other women
from male slander. Christine reviews various male disparagements of women;
Reason, Rectitude, and Justice rebut each one by telling of noble and virtuous
women who are equal if not superior to any man. Beginning with the Amazons—"ladies of great courage who despised servitude"—Lady Justice and

her sisters use women's lives to demonstrate women's virtues and strengths, showing that "all things which are feasible and knowable, whether in the area of physical strength or in the wisdom of the mind and every virtue, are possible and easy for women to accomplish."[9]

The Book of the City of Ladies insists that women are not innately inferior to men; their inferior education and training have created the illusion of inequality. Pizan argued that "if it were customary to send daughters to school like sons, and if they were then taught the natural sciences, they would learn as thoroughly and understand the subtleties of all the arts and sciences as well as sons."[10] She asserted that marriage was harder on women than men, "for men are masters over their wives, and not the wives mistresses over their husbands." She accused men of cruelty to their wives:

> How many harsh beatings—without cause and without reason—how many injuries, how many cruelties, insults, humiliations, and outrages have so many upright women suffered, none of whom cried out for help? And consider all the women who die of hunger and grief with a home full of children, while their husbands carouse dissolutely or go on binges in every tavern all over town, and still the poor women are beaten by their husbands when they return, and *that* is their supper.[11]

Some of her arguments sound unusual for her time. For example, she asserted that women "take absolutely no pleasure in being raped" and argued that it is in men's interest to believe that they do.[12] This different perspective comes from Pizan's feminism. By rejecting her culture's traditions subordinating women and "building" women a refuge where they would be safe from their "enemies and assailants," Pizan created an ideology which united the women who embraced it in subsequent centuries. Isolated from much of their own culture, early feminists, regardless of their specific nation or era, most resembled each other.

Early European Feminists

From the early fifteenth century through the eighteenth century, some European women wrote as feminists, assuming the same critical stance as Pizan had. Some of these women wrote alone, in isolation; others, like Pizan, participated in the *querelles des femmes,* the debate over women's worth relative to men's which ranged from fifteenth-century Paris, to late-sixteenth-century Venice, to seventeenth-century London and Paris. All championed the female sex, arguing that the traditional views of women as inferior and evil must be rejected. Like Pizan, they asserted women's basic humanity and called for better education and kinder treatment in marriage for the female sex. Arguing, like Pizan, that their discontent was valid, they insisted that women's subordination was neither natural nor innate, but man-made, and thus capable of transformation.

"Happy are you, reader, if you are not of that sex to whom they forbid all good things," wrote the Frenchwoman Marie de Gournay (1566–1645) in her pamphlet *Le grief des dames (The Ladies' Grievance)* of 1626,

> forbidding it liberty, yes, gradually even forbidding it all virtues, taking power from it . . . [leaving women] to be ignorant, to play the fool, to be a servant.[13]

Four years earlier, Gournay had published *The Equality of Men and Women*, in which she asserted like Pizan that women's basic humanity made them innately equal to men. These early feminists based their arguments about equality on virtue: woman's virtuous nature and moral capacity proved her human equality, in the sight of both God and man. Pizan's City of Ladies is built "entirely of virtue, so resplendent that you may see yourselves mirrored in it."[14] Pizan's crowning proof of women's virtue was the Virgin Mary, whom she portrayed leading all the other exemplary ladies into her city.

Feminists after Pizan continued to claim equality based on virtue. "GOD has given Women as well as Men intelligent Souls," argued the English feminist Mary Astell (1666–1731) in 1694.[15] Almost a hundred years later, in 1792, her countrywoman Mary Wollstonecraft reasoned the same way. Since virtue was characteristic of humans and women were created human by God, they must "be considered either as moral beings, or so weak that they must be entirely subjected to the superior faculties of men."[16] Calling that latter position unchristian, since it implied that women did not have souls, Wollstonecraft like her predecessors based female equality on Christian doctrine.

Like Pizan, these early feminists rejected female inferiority by citing examples of superior women. Women from the Bible, women from fables and legends, women from history and the present proved that women could function as well as or even better than men. Pizan concluded her second section of *The City of Ladies* by having Lady Justice extol "the princesses and ladies of France" of Pizan's own day as proof of women's equality. By the early seventeenth century, when Marie de Gournay wrote her pamphlets, a list of female exemplars had become standard in works defending women.[17] The author's case was strengthened if she could point to a female ruler, preferably from her own nation. Gournay cited female regents in France as proof of women's capacity to govern and dedicated her pamphlet on equality to Anne of Austria, Queen of France.[18] Mary Astell dedicated her proposal for a women's college to the future Queen Anne of England and later argued that the presence of Anne on the English throne not only set an example for others but also made a belief in "the Natural Inferiority of our Sex . . . Sedition, if not Treason . . . in this Reign."[19]

If women's virtue and capacity for rule demonstrated their equality, where had their subordination come from? Early feminists agreed that it was the fault of male jealousy and custom. Lady Reason in Pizan's *Book of the City of Ladies* asserts that men have attacked women because "of their own vices," "out of jealousy," and because they "are naturally given to slander."[20] Once women had been subordinated, went this argument, the tyranny of custom ensured that subordination would continue. "Custom, when it is inveterate, hath a mighty influence: it hath the force of Nature it self," wrote the English educator Bathsua Pell Makin (c. 1608–c. 1675) in a pamphlet urging women's education.

> The Barbarous custom to breed Women low, is grown general amongst us, and hath prevailed so far that it is verily believed . . . that Women are not endued with such Reason as Men; nor capable of improvement by Education, as they are. It is lookt upon as a monstrous thing, to pretend the contrary.[21]

Catharine Macaulay (1731–1791), the English historian, repeated this argument at the end of the eighteenth century. People like to believe their "fond prejudices," and "it is from such causes that the notion of a sexual difference in the human character has, with very few exceptions, universally prevailed from the earliest times," she concluded. It is "the pride of one sex, and the ignorance and vanity of the other" which "have helped to support an opinion which a close observation of Nature, and a more accurate way of reasoning, would disprove."[22]

Challenging and rejecting tradition, feminists invoked reason, literacy, and education to prove women's capacities. Pizan built the City of Ladies on the "Field of Letters," and both she and later writers argued that women's supposed inferiority was caused by their lack of a decent education. From the beginnings of European culture, some had argued that girls be educated to make them more pious, either within the traditional confines of the family or the protected cloisters of religious orders of women. The seventeenth-century Dutch prodigy Anna Maria van Schurman (1607–1678) used this argument in her famous treatise of 1637, *Whether the Study of Letters is Fitting to a Christian Woman*. Van Schurman's thesis was that education would make women better and more obedient wives and Christians.[23] Feminists claimed more. They insisted that women should be educated not to enable them to function better in women's traditional roles, but to make them better human beings.

Writing in the humanist tradition of the Renaissance, these early feminists asserted that learning was intrinsically good and so should not be denied women. Knowledge "is the first Fruits of Heaven, and a glymps [sic] of that Glory we afterwards expect," declared Bathsua Makin in 1673.[24] Makin argued that education would benefit women themselves, their families, and their nations and founded her case on women's innate intelligence: "Had God intended Women only as a finer sort of Cattle, he would not have made them reasonable."[25] A generation later, Mary Astell proposed the creation of an "institution" for women's higher education in England on the same grounds.[26] Both Makin and Astell emphasized unmarried and widowed women's particular need for education, but both, like other early feminists, thought all women would benefit from a more equal education.[27] Given "books and teachers instead of . . . linen, embroidery hoops, and pillows," argued the seventeenth-century Spanish writer María de Zayas, women "would be just as apt as the men for government positions and university chairs, and perhaps even more so."[28] With better education, wrote Mary Lee, Lady Chudleigh, in her *Ladies Defence* of 1701, women could even gain independence from men. Educated women

> will respect procure,
> Silence the Men, and lasting Fame secure;
> And to themselves the best Companions prove,
> And neither fear [men's] malice, nor desire their love.[29]

Early feminists believed that the situation of women could best be improved by the private remedies of better education and kinder behavior from lovers and husbands.

Mary Wollstonecraft

Mary Wollstonecraft (1759–1797), the English radical, repeated these early feminist arguments in her *Vindication of the Rights of Woman,* published in 1792.[30] But Wollstonecraft also went further than earlier feminists, and she did so most clearly in her relatively democratic view of the nation-state and its role as an agent for improving the lives of women. Pizan and others wrote to advocate better private education for princesses and queens. Makin wanted to found a school for "gentlewomen"; Astell proposed that "ladies" create a college of their own. But Wollstonecraft dedicated her book to Talleyrand, who had just written the report on education for the new revolutionary French government, and she proposed that France institute a national system of free and universal primary education for both sexes. Thereafter, she planned that most working-class young people would receive traditional vocational training, while "the young people of superior abilities, or fortune," would pursue higher education together.[31] Daring France to "effect a revolution in female manners," she argued that this type of education would enable women to lead more useful and fulfilling lives. At the present, many fell "prey to discontent" who with education could have "practised as physicians, regulated a farm, managed a shop, and stood erect, supported by their own industry."[32]

Earlier feminists had envisioned more equal education leading to women's equality with men. What distinguishes Wollstonecraft was her demand that political legislation be used to reverse traditions subordinating women. By the late eighteenth century, when she wrote, the modern nation-state had become a powerful institution, reflecting the decline of dynastic monarchy and the rise of the middle class. Wollstonecraft was one of the first to call on a national government to create a new system of education for women. Like her predecessors, she insisted that what appeared to be women's inferiority was only their inferior training. "Men complain, and with reason, of the follies and caprices of our sex, when they do not keenly satirize our headstrong passions and grovelling vices," she declared. "Behold, I should answer, the natural effect of ignorance!"[33]

Despite its title and reputation as a feminist classic, the *Vindication of the Rights of Woman* concentrates on education. The thesis of Wollstonecraft's *Vindication* is that women are born human, but made "feminine," and thus inferior to men, through poor education. The organization of the *Vindication* focused on Wollstonecraft's criticism of "some of the writers who have rendered women objects of pity, bordering on contempt," because of their faulty theories of education, and on her own proposal for a system of equal national education.

By 1792, when she wrote the *Vindication,* Wollstonecraft had experienced many of the liabilities facing women. The daughter of a physically abusive, spendthrift father, she earned her living from age seventeen, working as a paid companion, a governess, a seamstress, and a schoolteacher. When she came to write the *Vindication of the Rights of Woman* (she had written a *Vindication of the Rights of Man* the year before, in 1791), she condemned the education currently offered because it made "women more artificial and weak in character than they would otherwise have been." It deformed women's innermost values with "mistaken notions of female excellence."[34]

Her chief enemy was Jean Jacques Rousseau, whose writings on education were especially popular with progressive reformers in the last quarter of the eighteenth century. While Wollstonecraft admired his improvements in boys' schooling, she deplored his view of girls. Rousseau had argued that since little girls were born to please men (demonstrated by their innate flirtatious-ness and love of dressing up), their education should suit them to their future role as wife. In rejecting Rousseau's views, Wollstonecraft, like other feminists, trusted her own judgment and experience over received male authority. "I have, probably, had an opportunity of observing more girls in their infancy than J.J. Rousseau," she wrote.

> I can recollect my own feelings, and I have looked steadily around me; yet, so far from coinciding with him in opinion respecting the first dawn of female character, I will venture to affirm that a girl, whose spirits have not been dampened by inactivity, or innocence tainted with false shame, will always be a romp, and the doll will never excite attention unless confinement allows her no alternative.[35]

She asserted that Rousseau's views weakened women and taught them only how to flirt and marry well:

> Women are told from their infancy, and taught by the example of their mothers, that a little knowledge of human weakness, justly termed cunning, softness of temper, *outward* obedience, and a scrupulous attention to a puerile kind of propriety, will obtain for them the protection of man; and should they be beautiful, everything else is needless, for, at least, twenty years of their lives.[36]

Rejecting this strategy as both demeaning and unsuccessful, Wollstonecraft wrote "to shew [sic] that elegance is inferior to virtue, that the first object of laudable ambition is to obtain a character as a human being, regardless of the distinction of sex."[37] More forceful in her arguments than many of her predecessors, Wollstonecraft also went further in her remedy, calling on the nation-state to reform girls' education.

Wollstonecraft's reliance on the state as an agent of reform is equally apparent in her writings on marriage. From Pizan on, early feminists had criticized husbands' treatment of wives. By the turn of the eighteenth century, such criticism was often extremely harsh. "Wife and servant are the same, / But only differ in the name," wrote Mary Lee, Lady Chudleigh, in her "To The Ladies" of 1703, reprinted many times.

> Like mutes, she signs alone must make,
> And never any freedom take:
> But still be govern'd by a nod,
> And fear her husband as a God.[38]

A few years earlier, in *Some Reflections upon Marriage*, Mary Astell raised the question of the connection between government and male dominance in the family. "If absolute Sovereignty be not necessary in a State, how comes it to be so in a Family?" she wrote, reminding her readers of the recent Revolution of 1689 which limited royal power in England. "If all Men are born Free, how is it that all Women are born Slaves?"[39] Astell raised these questions, but did not answer them; Wollstonecraft moved to the logical conclusion of state intervention to rescue women from cruel and abusive husbands.

In her last piece of writing, the novel *Maria, or The Wrongs of Woman* (1797), Wollstonecraft portrayed a nightmare marriage. Maria's husband, after trying to prostitute her, confines her in an insane asylum and takes away her infant—all legal in England at the time. Wollstonecraft died from the aftereffects of the birth of her second daughter before she could finish the novel, but among the last pages she wrote is Maria's plea to a judge for divorce. Revolutionary France instituted divorce in 1792; Wollstonecraft hoped for it in England as a way for women to escape their husbands' brutality. By looking to the state to reform marriage and education, by believing that legislation would end women's subordination, Wollstonecraft initiated a new era in feminist discourse. If women were not innately inferior, and if they could be educated to be the equals of men, then why should they be excluded by law from political life? "When, therefore, I call women slaves," she wrote in the *Vindication*, "I mean in a political and civil sense."[40] The very title of Wollstonecraft's chief work asserted that women had rights, and in a revolutionary challenge to the revolutionary government of France, she urged that women govern equally with men:

> Let an enlightened nation then try . . . allowing them [women] to share the advantages of education and government with men, see whether they will become better, as they grow wise and become free. They cannot be injured by the attempt.[41]

By demanding women's political and legal equality, Wollstonecraft helped to inaugurate a new era in European feminism. From the late eighteenth century through the first quarter of the twentieth, European feminists concentrated on winning these new demands.

3

ASSERTING WOMEN'S LEGAL AND POLITICAL EQUALITY: EQUAL RIGHTS MOVEMENTS IN EUROPE

❦

IN EUROPE, political subordination was countered by the doctrine of rights: inalienable civil liberties claimed first by small groups of men for themselves and, later, by feminists for women. These feminist groups were known in their own day as "equal rights" movements, because they claimed women should have legal and political rights equal to those of men. They rejected the ancient traditions which placed women in a special legal status, under male control and lacking rights given all adult males. In the seventeenth century, only the most radical claimed equal rights for everybody. In 1646, for instance, John Lilburne, a member of the Leveller party in the English revolution, asserted that

> all and every particular and individual men and women that ever breathed in the world, are by nature all equal and alike in their power, dignity, authority and majesty, none of them having (by nature) any authority, dominion or magisterial power one over or above another.[1]

Some Englishwomen asserted their right to petition the government for economic relief during the Civil War and "Glorious Revolution" of 1689, but they did not claim any universal theory of rights for women. One hundred years later, in France, women claimed rights for themselves.[2] From 1788 on, Frenchwomen petitioned the government, stormed the Bastille, and marched to Versailles to bring the royal family back to Paris. Active throughout the French Revolution, tens of thousands of women entered the political arena in these years.[3] Some of their writings took the new revolutionary formulations of rights for men and applied them to women and women's lives. "We are suffering more than the men," wrote the editors of the radical journal *The New National Ladies' Offering,* who with their declaration of rights leave us in a state of inferiority, or, to tell the truth, of the slavery in which they have kept us for so long. If there are husbands sufficiently *aristocratic* in their homes to oppose a division of patriotic duties and honors, we shall use against them the weapons which they have employed with so much success.[4]

Throughout the nineteenth century, the winning of political and legal rights by men who had not previously possessed them naturally raised the issue of such rights for women. The French Declaration of the Rights of Man and the Citizen, passed in the summer of 1789, inspired a number of women to claim similar rights for women. "You have restored to man the dignity of his being in recognizing his rights; you will no longer allow woman to groan beneath an arbitrary authority," declared Etta Palm d'Aelders, a Dutch-woman active in the revolution, in her address to the National Assembly in the summer of 1791 when she asked for equal education for girls as well as equal rights for women.[5] That same year Olympe de Gouges (1748–1793, born Marie Gouze), a playwright and revolutionary, composed *The Declaration of the Rights of Woman and the Female Citizen*, a revision of the Declaration of the Rights of Man written to include women. Demanding women's right to absolute political and legal equality, Gouges based her claim on the liberating powers of reason and revolution to overturn unjust traditions:

Woman, wake up; the tocsin of reason is being heard throughout the whole universe; discover your rights. The powerful empire of nature is no longer surrounded by prejudice, fanaticism, superstition, and lies. The flame of truth has dispersed all the clouds of folly and usurpation.[6]

Gouges called for better education and equal rights in marriage, including a sample "Social (Marriage) Contract between Man and Woman" in her *Declaration*. Individualistic herself (despite her working-class origins she remained a royalist and addressed her *Declaration* to Queen Marie Antoinette), she urged other women to individual action: "Regardless of what barriers confront you, it is in your power to free yourselves; you have only to want to."[7] This was the climate in which the English feminist Mary Wollstonecraft composed her *Vindication of the Rights of Woman*. The liberating possibilities of the French Revolution inspired European feminists to claim legal and political rights for women.

But the same revolution which raised those possibilities also defeated them. All women were barred from political activity in France in October 1793 on the ancient grounds that "a woman should not leave her family to meddle in the affairs of government."[8] That same year, Gouges was guillotined and d'Aelders forced to flee France. In the last years of her life, Wollstonecraft wrote of the dangers facing any woman who challenged accepted traditions of womanhood. Maria, the heroine of *The Wrongs of Woman*, concludes that "women [are] the *outlaws* of the world." Since they are excluded from the law's benefits, they really have no country.[9] Activist women of the French Revolution were disowned and forgotten, and for many years Wollstonecraft's reputation was besmirched by her liaisons with men and by having borne a child out of wedlock.

Feminists who first claimed political and legal rights for women lived isolated and estranged. First, feminism itself demanded a rejection of much of traditional European culture. Second, late-eighteenth- and early-nineteenth-century feminists demanded political rights in isolation, without political movements behind them. Third, simply looking at law and politics from the perspective of women's rights led to the conclusion that European women lived in servitude, subordinated to men. "Though the situation of women in modern Europe," wrote the English historian Catharine Macaulay in 1790,

> when compared with that condition of abject slavery in which they have always been held in the east, may be considered as brilliant, yet if we withhold comparison, and take the matter in a positive sense, we shall have no great reason to boast of our privileges. . . . For with a total and absolute exclusion of every political right to the sex in general, married women . . . have hardly a civil right to save them from the grossest injuries.[10]

Feminists who rejected political and legal traditions subordinating women used the words "slave," "outlaw," and "pariah" to describe both women's situation in general and their own sense of estrangement in particular. Flora Tristan (1803–1844), the French socialist, entitled her autobiography *Peregrinations of a Pariah;* her last written work was a pamphlet called *Women's Emancipation, or the Pariah's Testament.* Wollstonecraft consoled herself that she would find her audience in the future, and the Polish-born American feminist Emma Goldman wrote that Wollstonecraft and other "pioneers of human progress are like the seagulls: they behold new coasts, new spheres of daring thought, when their co-voyagers see only the endless stretch of water."[11]

The first feminists concerned themselves little with politics. Pizan, Zayas, Gournay, Astell, and others wrote as philosophers, hoping to encourage debate and the acceptance of new ideas, to challenge cultural traditions denigrating women. But in the late eighteenth and early nineteenth centuries, feminists wrote in a different world, in which democracy had become a political ideal and political rights a matter of debate. These feminists wrote as political activists, hoping to influence their nation-states to change laws in favor of women. Under these new conditions, national political and religious differences shaped the fortunes of both feminists and feminism in Europe.

Feminism and Liberalism

Women found it easiest to become feminists and organize women's rights movements where liberalism flourished. The two movements shared many positions and tenets. Like feminism, liberalism arose in opposition to traditional society and its beliefs. Like feminism, liberalism prized the individual and trusted individual judgment and reason over received authority. Like feminism, liberalism believed in the power of education and reform to eradicate age-old boundaries and hierarchies. To some liberals, feminism seemed the logical next step in the moral progress of society. "The social subordination of women thus stands out an isolated fact in modern social institutions," wrote the English philosopher John Stuart Mill in 1869,

> a single relic of an old world of thought and practice exploded in everything else. . . . This entire discrepancy between one social fact and all those which accompany it, and the radical opposition between its nature and the progressive movement which is the boast of the modern world . . . surely affords . . . serious matter for reflection.[12]

Women most often became feminists if they associated with liberal or radical groups, through familial or social connections. Involvement in causes ranging from the abolition of slavery to utopian socialism, from philanthropy to political revolution, led women to feminism. Feminism flourished during liberal revolutions and declined when conservative order was restored. In England, the success of liberal politics encouraged the early formation of a women's rights movement. In France, feminism grew during the revolutions of 1789, 1848, and 1871 and within revolutionary groups like the early socialists, but it declined during the periods of repression which followed the revolutions, especially the first and second Napoleonic empires. In Germany, feminism achieved its strongest early expression during the Revolutions of 1848; the period of conservatism that followed weakened feminism as well as liberalism in Germany. In Russia, the rise of a liberal movement in the 1850s led to the rise of feminism. The later repression of both moved many liberals and feminists into the revolutionary camp.

But while liberalism provided a more congenial climate for feminism, it also posed political problems for feminists. In nations where men first won the suffrage on a property franchise—by paying a certain amount in taxes or possessing a certain amount of land—feminists faced a dilemma. Should they work for the vote for all women, when all men could not vote, should they support a property franchise for some women, or should they work for men first? Feminist equal rights movements differed on their policies, just as liberals did in their views of how far the franchise should be extended.

In territories aspiring to nationhood in nineteenth-century Europe, women's activity in the liberal nationalist movements sometimes led to their winning political rights. "Women managed large sales of their work and gave the money to the national party," recalled the Finnish feminist Baroness Alexandra Van Grippenberg (1859–1913). "Thus it is natural that the nationalist work became an important means of development for the women of Finland."[13] When Finland gained independence from Russia, Finnish women became the first in Europe to win the vote on a national basis, in 1906.

Feminism and Christianity

Nineteenth-century liberals worked for nationalism, for the extension of political rights, for better education. They opposed absolute monarchy, government regulation of the economy, and, in many instances, the Roman Catholic church. Although both Protestantism and Catholicism opposed feminism and despised women activists, equal rights feminism arose more easily in Protestant nations than in Roman Catholic ones.

In an era of change, the Roman Catholic church embraced tradition and discouraged reform, particularly with regard to women. For women, the Catholic church offered its traditional solaces. The Virgin Mary and the female saints remained important figures. The age-old, approved female roles of wife and mother within a family or nun within a holy order remained intact. Active women who did not marry continued to find a place within the Catholic religious orders. Many new orders were founded in the course of the nineteenth century, and nuns led missions, ran hospitals, and directed schools.

In Catholic nations, nuns educated most girls well into the twentieth century—long after most boys were being educated by the state. Nuns inculcated girls with the traditional female virtues of obedience, deference, modesty, and self-sacrifice, and Catholic girls learned they would be respected and honored for these virtues. "I often imagined that I was Mary Magdalene, and that I was drying Christ's feet with my long hair," wrote Simone de Beauvoir of her girlhood during World War I.

> The majority of real or legendary heroines—Saint Blandine, Joan of Arc, Griselda, Geneviève de Brabant—only attained to bliss or glory in this world or in the next after enduring painful sufferings inflicted on them by males. I willingly cast myself in the role of victim.[14]

Successive nineteenth- and twentieth-century popes extolled women's traditional roles and praised women for their acceptance of subordination. Most Catholic women remained practicing Catholics at a time when men's participation in the Church declined drastically.

Since Catholicism enshrined ancient traditions concerning women, women who rejected those traditions to demand equal political and economic rights often had to break entirely with the Church. De Beauvoir called her loss of faith "my conversion to the real world," stating the opposition between the two realms expressed by many Catholic women who became feminists. "They wanted the three of us to sit in parlors, patiently embroidering our days with the many silences, the many soft words and gestures that custom dictates," wrote the Portuguese feminists known as "the Three Marias" about their Catholic girlhood. "Since we have broken out of the cloister, we have broken out once and for all."[15]

Until the mass women's movements of the 1970s, women who "broke out" of Catholicism found themselves relatively isolated in their own cultures, especially from other women. Women who opposed the Catholic church usually broke with most other women and joined with like-minded men in anticlerical groups like the Freemasons, republicans, or socialists. But in these circles, feminism was subordinated to other causes. Women's rights movements in Catholic nations were very small and often dominated by men. Women attracted to them were usually not Catholics, but Protestants, Jews, or atheists. By providing women with its traditional support and allowing little room for dissent, Catholicism offered many women stability and security. Few women became feminists or joined women's rights movements in Catholic European nations until very recently.

Protestantism, while no more favorable to women in its ideology or institutions, created an atmosphere more conducive to feminism. Protestantism narrowed women's roles while widening room for dissent and the rejection of tradition. The models of Mary and the saints disappeared, as did religious women's convents. While some Protestant sects allowed women a wide range of roles at the start of the Reformation, all subsequently imposed a new uniformity on women. Protestant women were supposed to be traditionally subordinate wives and mothers. Seventeenth-century feminists found no more support in Protestant England than in Catholic France.

But Protestantism, especially its more radical sects, provided feminists with a basis for rejecting traditions about women. The concept of spiritual equality and the conviction that all individuals had access to God encouraged independence of mind. Protestantism taught both girls and boys to read the Bible so that both women and men could work for their own salvation. At the same time, it preached the traditional wifely role for women. These contradictions between spiritual equality and female subordination led some women to question women's traditional roles. The Protestant emphasis on individual spiritual responsibility could provide a strong-minded woman, like Mary Wollstonecraft or Josephine Butler, with the courage of her own convictions, enabling her to reject traditions subordinating women and to act outside the prescribed circle of the family. In the nineteenth and early twentieth centuries, Protestantism provided the most hospitable climate for women's rights movements. It is why European women won the vote in Protestant nations a generation earlier than in Catholic nations.[16]

Even though Protestantism and liberalism were more conducive to feminism than Catholicism and conservatism, the difference was only one of degree rather than kind. No Protestant nation, however liberal, welcomed women's demands for equal higher education, much less legal and political equality. In every Protestant nation, from Great Britain to Germany, from Sweden to the United States, feminist claims met with massive opposition, and women had to organize themselves to fight for their rights. In every European nation, feminists were a small minority in opposition. This isolation makes their success in building movements which won women legal and political equality all the more impressive.

Women's Rights Movements

Early feminists of the fifteenth to eighteenth centuries came from a wide range of social backgrounds. Some were from the old, landed aristocracy, like Mary Lee, Lady Chudleigh, others from the newly ennobled court aristocracy, like Margaret Lucas, Duchess of Newcastle. Christine de Pizan's father was a court astrologer. Some feminists were middle class, but poor, like Marie de Gournay or Bathsua Makin; others were middle class but wealthy, like Mary Astell or Catharine Macaulay. Some came from the urban "demimonde" where authors, theatrical people, free-lance workers, prostitutes, and criminals mingled: Olympe de Gouges was a butcher's daughter who married a junior army officer. In Paris, she became known for her love affairs as well as her theatrical comedies before she wrote her feminist works.

In contrast, the women's rights movements of the nineteenth and early twentieth centuries were overwhelmingly composed of women from the middle classes—women whose families had moderate wealth from land, trade, the professions, or industry. Women of other classes faced different situations and problems. Aristocratic women—like aristocratic men—still retained many privileges of rank and wealth; working-class and peasant women—like working-class and peasant men—still spent much of their lives laboring for survival. But in the growing middle classes of the nineteenth century, a great discrepancy between women's and men's experiences developed. Middle-class women most strongly experienced the deprivation of rights their men had won or were winning.[17] Middle-class women, whose men had won political, educational, and economic rights, most often demanded such opportunities for themselves. Middle-class girls desperately needed better education in order to live at a level above that of a servant if they ever had to earn income. Middle-class women, usually neither earning income nor owning property themselves, depended totally on the charity of their male relatives for their financial existence. Middle-class women, many of whom supported philanthropic and social causes, were most conscious of the relative deprivation they suffered by having no independent legal existence nor any political power. In every nation where women's rights movements formed, women from the middle classes provided most of their early membership. Thus the largest and best-known women's rights movement of the era arose in England, where the middle classes first achieved numerical and political power.

The English Women's Rights Movement: 1832–1928

For decades the English women's rights movement provided a model to other European equal rights movements. "Truly these Englishwomen *are* fighters, real fighters, for they put their all into it," wrote Minna Cauer, a German feminist, in 1906. "We on the other hand are infinitely gentle, and unfortunately, so infinitely patient."[18] The rejection of legal and political traditions subordinating women can be seen most clearly in the development of the women's rights movement in England. There, tens of thousands of women struggled over a seventy-year period to win legal and political rights. They worked for more equitable child custody and divorce laws, for laws allowing married women control over their own wages and property, for higher education and the right to vote and participate in politics. From the 1850s to the 1920s increasing numbers of women organized to demand their rights, confident, as the Edinburgh branch of the National Society for Women's Suffrage put it in 1878, that "like stroke on stroke which eventually overthrows the largest forest tree, so these strokes, dealt week by week, year by year, will eventually overthrow all opposition to the accomplishment of the just object we have in view."[19]

Englishwomen first demanded the vote. In 1832, the "Great Reform Act" broadened suffrage qualifications, giving the vote to about one out of every five men on a property basis. The year before, articles had appeared in radical journals urging women's suffrage. In response, the House of Commons inserted the word *male* into voter qualifications for the first time in English history: previous legislation had been based solely on property requirements. Mary Smith, a single Yorkshire property owner, petitioned Parliament urging that "every unmarried female possessing the necessary pecuniary qualification should be entitled to vote for Members of Parliament."[20] The reformed House of Commons laughed in response to the reading of the petition.

In the 1830s and 1840s, Englishwomen active in politics joined other movements rather than organizing to win rights for themselves. They worked for the abolition of the slave trade, the abolition of laws keeping the price of grain high, the passage of the "Charter," which sought to broaden political rights to include working-class men. Feminists in the United States later said the exclusion of female delegates to the World Anti-Slavery Convention in London in 1840 fired them to organize the first Women's Rights Convention at Seneca Falls, New York, in 1848. Some Englishwomen were equally outraged. "Your noble spirits lighted a flame which has warmed . . . [English-women who had] thought *not* of our bondage," wrote Anne Knight (1786–1862) to the feminist Grimké sisters of South Carolina.[21] A Quaker long active in the Chartist and antislavery movements, Knight joined with seven Chartist women in Sheffield in 1847 to form the first English "Female Political Association" to work for women's suffrage. She wrote a pamphlet advocating women's suffrage and designed stickers bearing women's rights slogans to seal her letters.[22] In 1851, Knight and the Female Political Association organized a meeting which led to another parliamentary petition, this addressed to the House of Lords. Harriet Taylor Mill, the English feminist, concluded her 1851 essay on the "Enfranchisement of Women" by referring hopefully to this petition, which in the end received little support.

In her essays, Harriet Hardy Taylor Mill (1807–1858) rejected legal and political traditions subordinating women more firmly and confidently than her predecessors. Her second husband, the English philosopher John Stuart Mill, always credited her with the development of his ideas concerning women. (Mill and Taylor met in 1830, when she was married to John Taylor. They married in 1851, after Taylor's death.) Basing her views solidly on liberal principles and the recent winning of some rights for women in the United States, Harriet Mill published an essay advocating women's suffrage in a radical journal in 1851. Beginning by reporting on women's rights conventions in the United States, Mill claimed full legal and political citizenship for Englishwomen: "What is wanted for women is equal rights, equal admission to all social privileges; not a position apart, a sort of sentimental priesthood."[23]

Like contemporary feminists in the United States, Harriet Mill compared men to slave-owners and women to slaves. Just as "the prejudice of custom" denied rights to slaves for centuries, so it denied them to women. Just as slavery corrupted both master and slave, so women's oppression corrupted both sexes: "in the one it produces the vices of power, in the other, those of artifice."[24] And just as the abolition of slavery was both a moral and political issue, so was the abolition of women's oppression:

> The real question is, whether it is right and expedient that one-half of the human race should pass through life in a state of forced subordination to the other half.[25]

Mill's remedies were education, law, and politics. She advocated equal higher education, arguing that "high mental powers in women will be but an exceptional accident, until every career is open to them, and until they, as well as men, are educated for themselves and for the world—not one sex for the other."[26] She advocated new laws to allow women to keep their own wages and to protect them from male violence and brutality. And she urged women's rights as citizens as a matter of simple justice: "That women have as good a claim as men have, in point of personal right, to the suffrage, or to a place in the jury-box, it would be difficult for anyone to deny."[27] Like earlier feminists, she invoked illustrious female rulers to prove her case, arguing that "if there is any one function for which [women] have shown a decided vocation, it is that of reigning."[28] While Harriet Mill's poor health prevented her from political activity, her husband, John Stuart Mill, based his classic expression of feminist thought, *The Subjugation of Women* (published in 1869, eleven years after Harriet Mill's death), on her ideas. "In what was of my own composition," he wrote in the introduction, "all that is most striking and profound belongs to my wife . . . [from] our innumerable conversations on a topic which filled so large a place in our minds."[29]

While other European women's rights movements often called Mill's *Subjugation of Women* the spark which began their political activity, Englishwomen organized to claim the legal and political rights Harriet Mill championed earlier, in the 1850s. In 1848, Queen's College was established to provide women with college-level lectures; the following year Bedford College, administered in part by women, did the same. At these first colleges, both in London, middle-class women not only gained an education, they met others like themselves, forging the connections which made the creation of a women's rights movement possible.

The women who created this movement came from liberal middle-class, often clerical, families. Many had been active in earlier reform movements and most were born in the 1830s. The organizing spirit of the group was Barbara Leigh Smith, later Barbara Bodichon (1827–1891). Her father, a wealthy radical member of Parliament, raised her and her brothers in the same way, providing them with an identical education and giving each an independent income of £300 a year when they turned twenty-one. After attending Bedford College, Leigh Smith opened her own school and taught for several years. In 1854, she published a feminist pamphlet, *A Brief Summary in Plain Language of the Most Important Laws Concerning Women, Together with a Few Observations Thereon.* Collecting these laws convinced her of the need for action, especially for married women, who had virtually no legal rights. "A woman is courted and wedded as an angel," she wrote, "and yet denied the dignity of a rational and moral being ever after."[30] The next year, she organized a committee of like-minded women to gain married women control over their earnings and property.[31] Most were Unitarians and knew each other socially. Committee members included Bessie Raynor Parkes, Leigh Smith's childhood friend who published a book on girls' education in 1854, and Mary Howitt and Anna Jameson, both well-known writers. The committee held public meetings to collect signatures for a petition to Parliament, a novel and unusual activity for women then. By March 1856, when the petition was presented, twenty-six thousand names had been signed and the list was headed by a number of prominent women: Elizabeth Barrett Browning, the poet; Jane Carlyle, wife of the famous author; Charlotte Cushman, the actress; Elizabeth Gaskell, the novelist; Harriet Martineau, the political reformer.[32]

Despite this mobilization to gain married women control over their own property, the attempt failed. In 1857, Parliament passed a slightly liberalized Divorce Act. Previously, each divorce had required a separate act of Parliament; now, legal separation was provided for and couples could divorce if the wife committed adultery or if the husband committed adultery and another crime, like desertion or extreme cruelty. Since the most abused wives could now divorce or separate, both Parliament and public opinion generally reasoned that other wives had no need to control their own wages or property.[33]

This defeat transformed the women's committee into a women's movement. Instead of disbanding, they increased their efforts. By the early 1860s, the group had purchased a building on Langham Place in London, where they initiated a number of ventures designed to promote the winning of women's rights. They published the *Englishwomen's Journal,* a feminist magazine edited by Bessie Parkes. They established the Victoria Press to print the *Journal,* and both press and journal were directed and staffed entirely by women. They founded the Society for the Promotion of the Employment of Women to train women who needed to earn income to work as bookkeepers and clerks instead of governesses and companions. They opened the Ladies' Institute, which contained a club, a reading room, and classrooms, providing comfortable spaces for women outside their homes, not previously available in Victorian London. The original group—Bodichon, Jameson, Howitt, and Parkes—soon attracted other women interested in women's rights: Adelaide Anne Proctor, the poet; Emily Davies, the educator; Elizabeth Garrett, who became the first female English physician; Emily Faithfull, the Scottish printer who ran the Victoria Press. Similar groups coalesced in other English cities and a political network formed.

In the 1860s, the English women's rights movement launched four campaigns. They organized to lobby Parliament into passing legislation allowing married women control over their earnings, which they achieved in 1878, and over their property, which they won in 1882. They founded women's colleges at the major universities. They trained students to pass the Oxford and Cambridge examinations, raised money for buildings, and convinced college authorities to allow female students to attend some classes. Women began to enter Oxford and Cambridge in the 1870s; in 1878, the University of London granted them degrees.[34] They campaigned to abolish the Contagious Diseases Acts, which allowed police inspection for venereal disease of women suspected of being prostitutes, and they succeeded in 1884. All the while, they continued to work for the vote.

In 1865 Bodichon, Davies, and Parkes, among others, campaigned for John Stuart Mill in a carriage covered with election placards. Mill won his election to Parliament, where he raised the issue of women's suffrage. That same year, Elizabeth Garrett spoke about the suffrage to these women and about fifty others at the Ladies Discussion Society at Langham Place. Realizing they all supported women's right to vote, they organized as the Women's Suffrage Committee, later the London National Society for Women's Suffrage. These equal rights feminists often based women's claim to the vote *not* on women's similarity to men, but on their differences. "I advocate the extension of the franchise to women because I wish to strengthen true womanliness in women, and because I want to see the womanly and domestic side of things weigh more and count for more in all public concerns," declared Millicent Garrett Fawcett in 1878.[35] (Fawcett was Elizabeth Garrett Anderson's younger sister and the future leader of the English women's rights movement.) In 1866, 1,499 eminent women, including Florence Nightingale, Harriet Martineau, and Mary Somerville, the mathematician, signed a petition for the vote which was presented to the House of Commons, where eighty members voted in favor of women's suffrage.

By the early twentieth century, Englishwomen had achieved much. They won the right to sit on town councils and school boards. They could become poor-law officers and factory inspectors. They could vote in municipal and county elections if they had the requisite property. They could even be mayors, and ten, including Dr. Elizabeth Garrett Anderson (1836–1917), were elected in 1907. But they had still not won the right to vote in national elections.

This delay in winning the vote was caused by a political stalemate which stymied not only Englishwomen, but all women's rights movements of the period. Feminists' natural political allies were men open to change: liberals, radicals, socialists. But these men of the left generally feared the women's vote, which they assumed would go not to themselves, but to conservative, right-wing parties. In both Protestant and Catholic societies, women were judged to be more religious, and so, more reactionary, than men. Conservative parties had no interest in women's suffrage and upheld the tradition that women had no place in political life. Thus, feminists who wanted the vote for women could only ally with men and male parties which often resisted giving women the vote on the national level. "I believe that at the present time, it would be dangerous—in France—to give women the political ballot," argued Léon Richer, himself head of the French League for Women's Rights, in 1888. "They are, in great majority, reactionaries and clericals. If they voted today, the Republic would not last six months."[36] It would be "political folly" to work for votes for women before all men had votes, stated the Austrian socialist Viktor Adler in 1903.[37] In England, the women's rights movement postponed its own goals for forty years to remain allied with the Liberal party. The movement split from 1884 to 1897 over the question of whether to ally with the Liberals or remain an independent political entity. Liberal party leaders, from Gladstone in the 1860s to Asquith in the crucial period before the First World War, steadily resisted allowing women to vote. Votes for women, thundered Gladstone in 1884, when he was prime minister and a new Voting Act was under discussion, was

> one of the questions which it would be intolerable to mix up with political and Party debates. If there be a subject in the whole compass of human life and experience that is sacred, beyond all other subjects, it is the character and position of women.[38]

In 1897, the divided English women's rights movement reunited as the National Union of Women's Suffrage Societies. The new emphasis on the vote led to new, more militant tactics. Working-class women joined in the thousands, actively recruited by the previously middle-class suffrage societies in the large textile factories of the North. In 1900, over twenty-nine thousand female Lancashire factory workers signed a petition demanding the vote for themselves because in the homes, their position is lowered by such an exclusion from the responsibilities of national life. In the factory, their unrepresented condition places the regulation of their work in the hands of men who are often their rivals as well as their fellow workers.[39]

More women, both middle- and working-class, began allying with the newly formed Labour party and the smaller, more radical Independent Labour party. A deputation of 300 women, representing 50,000 textile workers, 22,000 women of the Women's Cooperative Societies, and 1,530 college graduates, as well as the suffrage and temperance organizations, met with the prime minister in 1906 to demand the vote. He counseled "the virtue of patience."[40]

From then on, feminists lost patience and turned to more aggressive tactics. The leader of the 1906 deputation was Emily Davies (1830–1921), who by then had been working for women's rights for fifty years. From 1906 to the outbreak of the First World War in 1914, the English women's rights movement concentrated on winning the vote, but split in these years into two rival factions: the National Union of Women's Suffrage Societies (NUWSS), led by Millicent Garrett Fawcett (1847–1929) and the Women's Social and Political Union (WSPU), led by Emmeline Goulden Pankhurst (1858–1928). The divergent tactics and personalities of these two feminists dominated this phase of the women's rights movement. Both came from politically active liberal families; both were introduced to the women's movement when they were young—Fawcett by her older sister, Pankhurst by her mother. Both were conventionally attractive by the standards of the day, and prided themselves on the taste and elegance of their dress. But they differed sharply on methods in the decade before World War I. Fawcett worked to expand NUWSS membership, to publicize the demand for the vote through speaking tours and distribution of their journal, *The Common Cause,* and to lobby Liberal politicians into voting in favor of women's suffrage. Fawcett continually urged liberal political tactics: democratically elected head of the NUWSS herself, she continued to believe that parliamentary democracy would eventually acknowledge the justice of women's claim to equal rights. "Let us prove ourselves worthy of citizenship, whether our claim is recognized or not," she advised the over fifty-three thousand members of her group in 1914.[41] The NUWSS took "Faith, Perseverance, Patience" as their motto and repudiated all use of violence but became increasingly impatient as the struggle to win the vote intensified. By 1912, the NUWSS had broken its alliance with the Liberals and was supporting Labour party candidates. NUWSS members braved hostile crowds to speak for the vote: in 1913, for instance, Selina

Cooper and Margaret Aldersley, two NUWSS members from industrial Lancaster, addressed a mob in the village of Haworth. "They threw rotten eggs and tomatoes and all sorts of things," recalled Cooper's daughter.

> My mother went out, and she stood on this cross, and she said, "I'm stopping here, whatever you throw, so go and fetch all the stuff you've got to throw, because," she says, "I'm going to speak to you, I've come here to speak. And" she says, "this blooming village would never have been known about but for three women—the Brontës."[42]

Large rallies and demonstrations, where Fawcett and others spoke to tens of thousands of women and men, occurred in 1910, 1912, 1913, and 1914.

By these years, Fawcett insisted that banners identify her as the president of the "law-abiding suffragists" and that the word *non-militant* appear on the cover of *The Common Cause*. "I can never feel that setting fire to houses and churches and letter-boxes and destroying valuable pictures really helps to convince people that women ought to be enfranchised," she declared in 1913, repudiating the violent tactics introduced by Pankhurst and the WSPU after 1906.[43] While her view was shared by most Englishwomen who worked for the suffrage in these years, there is no doubt that the movement as a whole was given publicity and new energy by the more "unladylike" tactics of Pankhurst and the WSPU.[44]

"The argument of the broken pane of glass is the most valuable argument in modern politics," declared Emmeline Pankhurst in 1912.[45] Years of increasing militancy lay behind that declaration. After the death of her husband, a feminist attorney, in 1898, Pankhurst earned income as a poor-law officer and volunteered her political energies to the Independent Labour party. In 1903, exasperated by party timidity on the women's vote, Pankhurst created the Women's Social and Political Union in Manchester, supported by her three daughters, Christabel (1880–1956), Sylvia (1882–1956), and Adela (1885–1961). From 1906 on, the WSPU raised the issue of votes for women whenever Liberal politicians spoke, and participated in local and national suffrage demonstrations, often at personal risk. The socialist men "tried to protect the women," remembered Hannah Mitchell of a suffrage rally in Manchester in 1906.

> The mob played a sort of Rugby football with us. Seizing a woman they pushed her into the arms of another group who in their turn passed her on. . . . two youths held on to my skirt so tightly that I feared it would either come off or I should be dragged to earth on my face. But my blood was getting up. . . . I turned suddenly, gave one a blow in the face which sent him reeling down the slope and pushed the other after him. . . . Realizing my umbrella was still in my hand, I ran after this man and hit him in the jaw with it.[46]

WSPU women like Mitchell continued to disrupt meetings and to organize their own demonstrations: an open-air rally in London in 1908 drew between 250,000 and 500,000, and 30,000 women joined the WSPU march which preceded it.[47] Continued government inaction led the WSPU to adopt the violent tactics of the Irish independence movement: breaking windows, pouring liquids down mailboxes, cutting telegraph wires, carving "Votes for Women" into golf courses. They courted arrest and began to go on hunger strikes when in prison. Faced with these tactics, the government responded with forced feeding. Methods were primitive and painful. "I have been fed through the nostril twice a day," wrote Mary Leigh in a statement to her attorney in 1909.

> The sensation is most painful—the drums of the ears seem to be bursting and there is a horrible pain in the throat and the breast. The tube is pushed down twenty inches. I have to lie on the bed, pinned down by the wardresses. . . . I resist and am overcome by weight of numbers.[48]

Suffragette militancy and government repression climaxed in 1913 and 1914. By 1913, over one thousand suffragettes had gone to prison for their beliefs, and suffrage marches were led by women "in white dresses, carrying long silver staves, tipped with the broad arrow, showing they had suffered imprisonment."[49] WSPU members now went on hunger-and-thirst strikes; the Liberal government retaliated by passing the "Cat and Mouse Act," which allowed prisoners to be released to recover their health and then rearrested so they would serve every day of their sentence. Justifying this act, and the government's decision to classify the militant suffragettes as criminals rather than political prisoners, the home secretary rejected all other responses. Letting the women die would only create political martyrs; incarcerating them in lunatic asylums was impossible because doctors would not declare them insane; exiling them might inspire their wealthy supporters to rescue them; granting the vote was out of the question in the "existing state of lawlessness."[50]

WSPU women responded with equal intransigence. They took "Deeds, Not Words" as their motto and increased attacks against property. Emily Wilding Davison (1872–1913), an Oxford graduate and WSPU militant who had been force-fed forty-nine times in prison, deliberately hurled herself in front of the king's horse in the popular Derby Day race. Sacrificing herself to aid the cause of votes for women, she carried a WSPU banner under her coat; her funeral occasioned another large suffrage demonstration. Early in 1914, another suffragette slashed Velasquez's painting of Venus. Her note explained that she had tried to destroy the picture of the most beautiful woman in mythological history as a protest against the Government for destroying Mrs. Pankhurst who is the most beautiful character in modern history.[51]

Emmeline Pankhurst was jailed and released twelve times in 1913; her daughter Christabel directed the WSPU from Paris. Only members totally loyal to these two remained in the WSPU: they had expelled most of their other supporters, including Sylvia and Adela Pankhurst, Emmeline's second and third daughters. All men now were seen as the enemy, in part because they opposed women's suffrage, in part because they caused the "great scourges" of venereal disease and prostitution. The WSPU chose "Votes for Women and Chastity for Men" as their new slogan.

Great Britain's entry into World War I in August 1914 abruptly ended the suffrage struggle. Emmeline and Christabel Pankhurst ceased their efforts for the vote and committed themselves to war work and fanatical patriotism: they changed the name of the WSPU journal from *The Suffragette* to *Britannia* and argued that all would be lost if the Germans won. The government granted amnesty to all suffragette prisoners and released them from jail. Millicent Fawcett dedicated the NUWSS to war work.

Before World War I, anti-suffragist groups had asserted that equal political rights for women was "against nature" and would weaken Britain's ability to defeat more "manly" nations.[52] However, women's conduct during the war swayed public opinion toward the vote.[53] Women's war work—toiling in the factories, or nursing soldiers and driving ambulances at the front—their staunch patriotism, and their ability to assume "men's jobs" all supported the feminists' case. Paying tribute to the heroism of Edith Cavell, an English nurse executed by the Germans for espionage in 1915, Prime Minister Asquith declared, "There were thousands of such women, but a year ago we did not know it."[54] The next year, his government moved to grant women's suffrage.

Englishwomen over thirty won the vote in 1918. (Women between twenty-one—the age at which men could vote—and thirty were considered too "flighty" and had to wait until 1928 for the suffrage.) By the end of World War I, the issue of women's suffrage had ceased to be radical in many nations. In a world which had witnessed a communist revolution in Russia, the collapse of the German and Austro-Hungarian empires, and the deaths of millions from war and influenza, equal political rights for women seemed an antiquated cause, of interest, as its detractors had often claimed, only to a few women. In 1919, even Pope Benedict XV came out for votes for women: it was widely thought that the women's vote would go to Catholic and conservative parties. In fact, women voted much the way men did, especially in

England.[55] The WSPU disbanded; the NUWSS became the National Union of Societies for Equal Citizenship (NUSEC) and worked to finish the granting of equal political and legal rights for women. In 1919, the Sexual Disqualifications (Removal) Act allowed women to be lawyers, jurors, judges, and members of Parliament. In 1923, grounds for divorce became equal; in 1925, mothers won equal rights to child custody.

To the feminists who had fought so hard for the vote, all else seemed anticlimactic. The winning of the suffrage in 1918 "was the greatest moment of my life," wrote Millicent Fawcett in her memoirs in 1924. "We had won fairly and squarely after a fight lasting fifty years. Henceforth, Women would be free citizens."[56] While rejecting the traditions of women's political and legal inequality was an important and necessary step in the development of feminism, by 1925 the limitations of focusing only on issues of citizenship and the need to move on to other demands for women was clear both in England and elsewhere in Europe.

Achievements and Limitations of Equal Rights Feminism

In the first quarter of the twentieth century, women won the vote in nations most similar to England. In Scandinavia, feminists allied with liberals, socialists, and nationalists, and when those groups attained power, women won the franchise.[57] But winning the vote showed the limits of equal rights feminism. Before the twentieth century, with its focus on winning the vote, equal rights feminists had tried to keep the issue in perspective. Women's suffrage, stated Millicent Fawcett in 1886, "will be a political change, not of a very great or extensive character in itself, based upon social, educational and economic changes which have already taken place."[58] Because of the bitterness of the struggle for the vote, this perspective was lost.[59] Liberalism argued that once people were enfranchised, they possessed the means to work out their own liberation. Equal rights feminism, strongly tied to and influenced by liberalism, tended to the same view. Both liberalism and equal rights feminism declined when the vote *alone* proved incapable of liberating groups which remained economically and culturally subordinated.

Rights of citizenship—the vote, the right to serve on juries, the right to hold political office—in fact meant relatively little to most women. In addition, these were the kinds of rights which, when won, were often taken for granted. Equal rights feminists, who had fought so hard for the vote, sometimes became disillusioned when elections failed to make a major difference in the lives of most women. "Nowadays, when it is often difficult to persuade women to come out and vote," wrote Hannah Mitchell, active in the WSPU before World War I,

I wonder whether these women, like all electors today, who have had the vote handed to them on a gold plate, so to speak, would not have been just as well left among the "infants, imbeciles, and criminals" [who had also been denied the vote under English law].[60]

Prior to women's suffrage, many equal rights feminists believed that women's vote would change the world. "Women's suffrage spreads culture!" asserted a 1908 manifesto of the German Suffrage Union, ten years before German women won the vote.

Women's suffrage encourages peace and harmony among different peoples.
Women's suffrage effectively promotes abstinence and thus prevents the ruin of a people through alcohol.
Women's suffrage opposes the exploitation of the economically and physically weak, it takes pity on children and tormented animals.[61]

These hopes did not materialize. Women did not vote in a separate bloc as women or feminists—they voted much as men of their classes did.[62] In Sweden, for instance, a "Woman's List" of candidates for the Stockholm City Council in 1927 received only 0.6 percent of the vote.[63]

Catholic nations resisted women's suffrage, and in Catholic nations, equal rights feminists remained disparaged and ignored. In 1934, the French feminist Louise Weiss complained that "peasant women remained open-mouthed when I spoke to them of the vote. Working-class women laughed, women clerical workers shrugged their shoulders, bourgeois women rejected me, horrified."[64] Fear among liberal and left-wing male politicians of hordes of Catholic women voting for Catholic and conservative parties delayed women's suffrage. In Catholic nations women did not vote until after the Second World War.[65]

In addition to having relatively little impact on the lives of most women, equal rights feminism succeeded only in those nations where class and political boundaries could be easily crossed. In both England and Scandinavia, feminists worked with liberal *and* socialist parties, and equal rights organizations united middle-class *and* working-class women. Elsewhere, political and economic disparities led to the development of two separate women's movements. One was an equal rights movement: middle-class, liberal, and focused on the vote, especially for women of property in societies where male suffrage was not yet universal. The other women's movement was socialist: composed of working-class women and those who identified with them, and focused primarily on issues of the economy and overturning capitalism. In France, Russia, Italy, Austria, and Germany, the two women's movements loathed each other, and the feminist unity achieved in England proved impossible in most other European nations.

Equal rights feminists and feminist socialists differed sharply on issues and tactics. First, they divided bitterly on the issue of protective legislation limiting women's working conditions and hours, even in England. "If every demand raised by these women [the equal rights feminists] were granted today," declared Eleanor Marx (1856–1908), Karl Marx's youngest daughter, in 1892,

> we working-women would still be just where we were before. Women-workers would still work infamously long hours, for infamously low wages, under infamously unhealthful conditions. . . . Has not the star of the women's rights movement, Mrs. Fawcett, declared herself expressly in opposition to any legal reduction of working hours for female workers?[66]

Millicent Fawcett's husband, Henry, a Liberal Member of Parliament, first opposed laws limiting working women's hours in 1873. Obsessed with literal equality, thinking in terms of the professions rather than factories, condescending toward the working class, equal rights feminists backed "equality" for women workers. This meant opposing all attempts made to restrict women's labor in industry, everything from shorter hours to prohibitions against night work and labor in mining. Equal rights feminists clashed with labor unions and socialist parties, and in most cases, this policy of opposing protective legislation for women led to or exacerbated the split between themselves and feminist socialists. In England, feminism had largely overridden such class divisions; elsewhere it could not. Feminist socialists repudiated a cross-class alliance; so did equal rights feminists.

Equal rights feminists and feminist socialists also differed on tactics. In Germany, for instance, the *Bund Deutscher Frauenverein* (*BDF*, the League of German Women's Associations) not only refused to link up with more activist or left-wing groups, they also repudiated the tactics which had given the English equal rights movement widespread appeal: rallies, marches, demonstrations, resistance both nonviolent and violent. "Through the imitation of men's revolutionary violence women themselves destroy the possibility of taking a place in public life," declared the *BDF* in 1913.[67] Socialists marched, therefore marches must be avoided. In 1912, propertied Munich suffragists organized a procession of eighteen private coaches, the very emblem of nineteenth-century wealth. Trimmed in the suffragist colors of green, purple, and white, these horse-drawn carriages in an automobile age embodied the limits which class could place on feminism.[68]

Indifferent to socialism and economic issues important to women, and satisfied with what seemed to be the full achievement of women's legal and political equality, some equal rights feminists embraced new causes once the vote had been won. Emmeline Pankhurst joined the Conservative party and worked for child welfare. Christabel Pankhurst became an evangelical Christian. More often, however, politically active women turned to socialism to solve women's problems. The novelist Virginia Woolf (1882–1941) was one: in her 1929 feminist essay *A Room of One's Own,* she declared that to create equally to men, a woman needed £500 a year and a room of her own. "The news of my legacy [an aunt left her an independent income] reached me one night about the same time that the act was passed that gave votes to women," she remembered. "Of the two—the vote and the money—the money, I own, seemed infinitely more important."[69] Woolf became active in Labour women's groups. Adela and Sylvia Pankhurst devoted themselves to socialism and attempted to transform the capitalist economy, an issue which they saw as far more important to most women than the vote. In this move, Virginia Woolf and Sylvia and Adela Pankhurst joined the many thousands of women, a few of them middle- and even upper-class in origin, who believed that socialism offered the best hope for feminism and its demands for women.

4

FEMINIST SOCIALISM IN EUROPE

❦

Feminism and Socialism

Feminists shared much with socialists. Feminists rejected European traditions subordinating women, argued that the system of institutionalized male dominance was the chief cause of women's oppression, and looked to build a new world where all would be free from the bonds of patriarchy. Socialists rejected European traditions subordinating workers, argued that the system of institutionalized capitalist ownership was the chief cause of human oppression, and looked to build a new world where all would be free from the bonds of private ownership and class struggle. Both philosophies involved a radical rejection of central traditions of European culture. Socialism rejected private ownership of the means of production, like land or factories, and advocated their social or public ownership. Feminism rejected male "ownership" of and dominance over women. In the nineteenth century, changed economic and political circumstances enabled both feminists and socialists to build large political movements, attracting thousands to their causes. Some linked feminism to socialism: the eccentric French philosopher Charles Fourier (1772–1837) insisted in 1808 that "as a general thesis: *social progress and historic changes occur by virtue of the progress of women toward liberty. . . . the extension of women's privileges is the general principle for all social progress.* "[1] By the 1830s, a number of women and men—the feminist socialists—concluded that the economic transformation of capitalism into socialism would inevitably liberate women. Committed to both socialism and feminism, they believed that the former would ensure the latter. "When the Revolution comes, you and I and all humanity will be transformed," declared the French socialist Louise Michel (1830–1905) in 1885.

Everything will be changed and better times will have joys that the people of today aren't able to understand. . . . Beyond our tormented epoch will come the time when men and women will move through life together as good companions, and they will no more argue about which sex is superior than races will argue about which race is foremost in the world.[2]

Michel and other feminist socialists insisted that only socialism, not the "bourgeois" equal rights movements, could improve the lives of the vast majority of women, to whom legal equality meant little. "The issue of political rights is dead," declared Michel. "Equal education, equal trades, so that prostitution would not be the only lucrative profession open to a woman—that is what is real in our program."[3] Michel and other feminist socialists focused on issues of women's labor, particularly in industry. They criticized equal rights movements for ignoring economic reality—the need for most women to earn income under exploitative conditions. They concentrated on improving women's working conditions, hours, and rates of pay. They recruited working-class women to socialist parties and trade unions.

While they worked to improve women's lives, however, they continued to believe that socialism, rather than feminism, would do more to transform the world. In any conflict between feminism and socialism, these women and men put socialism first, arguing, as the German socialist Clara Zetkin did in 1895, that

the proletarian woman cannot attain her highest ideals through a movement for the equality of the female sex, she attains salvation only through the fight for the emancipation of labor.[4]

Simultaneously, however, feminist socialists—most of them women—complained about the lack of feminism within socialist circles. "In theory comrades have equal rights," Zetkin wrote in a private letter, "but in practice the male comrades have the same philistine pigtail hanging down the back of their necks as do the best-wigged petty bourgeois."[5] Sensitive to issues of both male dominance *and* economic oppression, feminist socialists often performed a balancing act, proving their party loyalty by subordinating their feminism, but using party institutions to further feminist demands.

Feminist socialists often had to fight on two fronts: against governments which outlawed and attempted to crush socialism, and against socialist men who resisted both feminism and women's equal participation in the movement. These battles changed over time, as economic and political conditions as well as socialist theory evolved in the course of the nineteenth century. Initially, early male advocates of socialism excluded women. The most radical male leaders of the French Revolution of 1789 scorned women's political activity, even on behalf of their causes.[6] But by the 1830s, women participated in radical socialist groups, like the Saint-Simonians in France and the Owenites in England. In these circles, theory and practice were both expansive, and welcomed women and some feminist principles. The short lives of these movements limited women's gains in them.

By the 1850s and 1860s, Karl Marx and Friedrich Engels had reshaped socialism. Marx and Engels consistently argued that paid labor outside the home would ultimately liberate women by bringing about a new stage of social development. Under socialism, all would change, including relations between women and men, the form of the family, and women's economic oppression. "The first condition for the liberation of the wife is to bring the whole female sex back into public industry," Engels wrote in his *Origin of the Family, Private Property and the State,* in 1884.[7] In *Capital* (1867), Marx insisted that "modern industry . . . creates a new economic foundation for a higher form of the family and of relations between the sexes."[8] Although Marx and Engels never described exactly what shape the "higher form of the family" would take and rarely discussed "women's issues," they did appoint Harriet Law (1831–1897), an English schoolteacher, to sit on the General Council of the First International (originally the International Working Man's Association).[9] She was, however, the only woman present.

Women's absence from socialist circles in this era came from male prejudice and misinterpretation of Marx's and Engels's writings. Many socialist men hoped their wives would never have to earn income outside the home. In criticizing capitalism, Marx and Engels had deplored the disappearance of women's traditional roles: Marx called the dissolution of "the old family ties" "terrible and disgusting"; Engels early in his career declared that a wife earning wages outside the home "deprives the husband of his manhood and the wife of all womanly qualities."[10] Seizing on such statements, both French and German male socialists argued that women should remain in their traditional roles. "The rightful work of women and mothers is in the home and family," resolved the Lasallean General German Workers' Association in 1866.

> Alongside the solemn duties of the man and father in public life and the family, the woman and mother should stand for the cosiness and poetry of domestic life, bringing grace and beauty to social relations, and be an ennobling influence in the increase of humanity's enjoyment of social life.[11]

Socialists welcomed women's participation more in the last quarter of the nineteenth century, especially in Germany. The German Socialist party was virtually outlawed from 1878 to 1890, and in these circumstances, women's participation became crucial both in keeping the cause alive and in training the next generation of German socialists. When Clara Zetkin returned to Germany from exile in 1886, she noticed the difference.

> What gave me the most joy was that women are being drawn more and more into the movement. I met quite a large number of women . . . distributing brochures, flyers, and election appeals. Most comrades see participation and activity by women no longer as a nice convenience, but as a practical necessity. Quite a transformation in attitudes since I left Germany.[12]

In addition, Zetkin's countryman and fellow socialist August Bebel published *Women Under Socialism* in 1879. This immensely popular work went into hundreds of editions. Bebel argued that capitalism caused most of women's oppression, and he linked socialism firmly to women's liberation. "The woman of future [socialist] society is socially and economically independent," concluded Bebel. "She is no longer subject to even a vestige of dominion and exploitation; she is free, the peer of man, mistress of her lot."[13] While Bebel's own feminism was questionable (he preferred women to be "supporters" rather than equals), his book made it easier for women to participate in socialist movements from the 1880s on. Many testified they first encountered the oppression described by feminists in its pages. "Neither in the family nor in public life had I ever heard of all the pain the woman must endure. One ignored her life," remembered Ottilie Baader (1847–1925), then a seamstress. "Bebel's book courageously broke with the old secretiveness."[14] Baader credited Bebel's book with making her a socialist, and she later became one of the leaders of the German Socialist Women's Movement.

Socialist women also found readier acceptance in socialist circles in the last quarter of the nineteenth century, in part because of the heroic example of Russian women. In Russia, all political parties were outlawed, all liberal ideas suspect. In these circumstances, Russian reformers and intellectuals demanded rights for all, serfs as well as aristocrats, women as well as men.[15] Welcomed into revolutionary circles, Russian women participated in all kinds of revolutionary activities, from assassination to organizing, from writing political leaflets to serving lengthy sentences in Siberia. Women comprised over 10 percent of the Russian revolutionaries of the 1870s; this proportion increased to about 30 percent in the most extreme groups, which demanded total commitment from members.[16] The dedication and heroism of these women won them respect from other socialists.

By the twentieth century, women participated in and occasionally influenced socialist movements throughout Europe. The German Socialist party, for instance, which was the largest in Europe, included some 175,000 female members (16 percent of the total membership) and an additional 216,000 women in socialist trade unions by 1914. As many as 124,000 women subscribed to the party's women's paper, *Die Gleichheit* (Equality).[17] Two strong women, Clara Zetkin (who identified herself as a feminist) and Rosa Luxemburg (who did not), helped lead the party and determine its policies.

For some women, like Luxemburg, socialism was an end in itself—the only movement which could liberate women and men. Women's duty is to "adopt resolutely the masculine tactic," declared the French socialist Aline Valette (1850–1899) in 1893, and "work towards the *appropriation of the instruments of production by the collectivity.*"[18] Other female socialists, like Zetkin, became feminist socialists, determined to champion women's rights as well as socialism. They believed, as Louise Michel asserted in 1885, that

> a man, too, suffers in this society, but no sadness can compare to a woman's.
> . . . We know what our rights are and we demand them. Are we not standing next to you fighting the supreme fight? Are you not strong enough, men, to make part of that supreme fight a struggle for the rights of women? And then men and women together will gain the rights of all humanity.[19]

Feminist socialists subordinated their feminism to their socialism if there was a conflict between the two, but from the 1830s on they fought for women within socialist parties and trade unions. Despite male prejudice, socialism provided the most enduring, congenial, and supportive environment for feminists until the second half of the twentieth century. Feminist socialists added new demands about women's work to the feminist spectrum, while continuing to reject other legal and cultural traditions which still limited women's lives.

Early Feminist Socialists: Owenites and Saint-Simonians

Records of female feminist socialists date from the early 1830s in France and England. In the same years that middle-class Englishwomen first demanded the vote, working-class Englishwomen began to form unions and cooperative societies to further women's economic and political interests. The end of the repression which had outlawed trade unions and stifled radical politics until 1829 in England released a wave of female socialist activity, aided by the Owenite movement (named for Robert Owen, a radical manufacturer-philanthropist) which in these years sought to organize all workers into "one big union." Sixty London women formed the Society of Industrious Females in 1832 to create a cooperative in which they could ensure "a fair distribution of the products of our labour"; in 1833, another group of London women formed the Practical Moral Union of Women of Great Britain and Ireland to "combine all classes of women" and work for their rights.[20] These early socialist women, in England and in France, oscillated between including all women in their endeavors and including only women of the working class.

In 1834, the Owenist Grand National Consolidated Trades Union (GNCTU) formed, and thousands of women joined local branches. In the union and in the pages of the Owenite newspaper, *The Pioneer*, working-class women created a forum for themselves.[21] "It is time the working females of England began to demand their long-suppressed rights," stated an editorial written by Frances Morrison, the wife of the editor.

> Why should the time and the ingenuity of the [female] sex . . . be monopolized by cruel and greedy oppressors, being in the likeness of men, and calling themselves masters? Sisters, let us submit to it no longer. . . . unite and assert your just rights![22]

In these publications, Owenite women criticized both employers and husbands, linking the economy and marriage. "There is a jealousy in the men against female unions," wrote "A Woman" in 1834. "The men are as bad as the masters,"[23] went one typical complaint, while others focused on the need for women to unionize to protect their rights. "A great deal was said of the slavery of the working classes, and of the inadequate wages of the men," stated a woman "in a kind of whisper" from the back of the hall at a union meeting, "but never a word of the slavery of the poor women, who were obliged to toil from dawn to midnight for 7 or 8 shillings a week."[24]

After the outlawing of the GNCTU in 1834, some Owenites formed separate socialist communities, arguing as one woman did in 1840 that

> under the withering influence of competition for wealth, mammon worship and an aristocracy of birth . . . very little progress can be made in the attainment of true liberty for women. . . . [We should] organize small societies on a better system, as examples and patterns to the rest of the world—so that men and women may meet in equal communion, having equal rights and returns for industry.[25]

While these communities failed to survive for long, a few Owenite women continued to speak in favor of feminist socialism in England. "One great evil is, the depraved and ignorant condition of women," declared Emma Martin (1812–1851) in 1840;

> this evil can only be removed by Socialism. We love Socialism, because it is more moral—we love Socialism, because it is more benevolent—we love Socialism, because it is the only universal system of deliverance that man and woman can adopt.[26]

In early Victorian England, such militancy was doomed. Tainted by accusations of atheism and sexual license, weakened by the failure of the Revolutions of 1848, Owenism disappeared by the 1850s.

A similar pattern of the early appearance of feminist socialism and its subsequent demise evolved in France. There, working-class women entered the Saint-Simonian movement (founded by Henri Rouvroy, Count de Saint-Simon), which was dedicated to socialism, science, and the search for a "female messiah" who would "save the world from prostitution as Jesus saved it from slavery."[27] By 1830, about two hundred women regularly attended Saint-Simonian lectures; in 1831, 110 working-class women identified themselves as "faithful adherents" to the movement.[28] In 1832, a small group of these women published their own newspaper, *The Free Woman (La femme libre)*, declaring, "With the emancipation of woman / Will come the emancipation of the worker."[29] Identifying themselves as "proletarian women," the group only printed articles by women, and signed the articles using only their first names, arguing that women using men's last names signaled female slavery.[30] Like Owenite women, Saint-Simonian women oscillated between welcoming all women to their movement and focusing attention on working-class women in particular. "The woman question is fundamentally connected to that of women workers," wrote the editor, while the first editorial asked women generally for their support:

> We call on all women, whatever their rank, their religion, their opinion, provided that they feel the oppression of women and the people and that they wish to join with us, to associate themselves to our work and share our efforts.[31]

Like Owenite women, Saint-Simonian women complained about the men of their movement: "At bottom, the male Saint-Simonians are more male than they are Saint-Simonian," wrote Suzanne Voilquin (1801–1877?), one of the editors of the newspaper.[32] Voilquin worked as a laundress and an embroiderer prior to publishing *The Free Woman*, which ran for forty issues before funds gave out. "I now preach a new peaceful crusade against despotism, against the crushing yoke of prejudice," she wrote, "and above all, against this harmful belief which subordinates and declares our sex inferior to the other."[33]

Although the Saint-Simonian movement collapsed under government and social repression in the early 1830s, some Saint-Simonian women continued to advocate feminist socialism. Jeanne Deroin (1810?–1894), one of the editors of *The Free Woman*, continued to organize for a workers' federation and refused to take her husband's name when she married in 1832. Jailed as a radical in the 1840s, she reemerged during the Revolutions of 1848, when she and Voilquin joined other women to form a new women's club and publish a new journal, *Women's Voices (Voix des femmes)*. The first issue proclaimed, "It is a mistake to believe that by improving the lot of men, that by that fact alone, the lot of women is improved."[34] Although their club and newspaper were closed in 1848, and the revolutionary government forbade all women participation in political clubs, Deroin ran for the legislative assembly in 1849, declaring in her election manifesto that "a Legislative Assembly, made up strictly of men, is as incapable of making laws as would be an assembly that was entirely composed of privileged persons to discuss the interests of the workers, or an assembly of capitalists to uphold the honor of the country."[35] She and Pauline Roland, a fellow Saint-Simonian, concentrated on organizing a federation of workers' associations, until they were arrested by the republican government in 1850. Sentenced to lesser terms because they were women, Deroin and Roland sent greetings from prison in 1851 to feminists in the United States:

> Sisters of America! your socialist sisters of France are united with you in the vindication of the rights of woman to civil and political equality. We have, moreover, the profound conviction that only by the power of association based on solidarity—by the union of the working classes of both sexes to organize labor—can be acquired, completely and pacifically, the civil and political equality of women, and the social right for all.[36]

The theories and principles of these early feminist socialists outlived their authors, who often died young. One year later, Roland was dead and Deroin living in exile in England. In their letter, they had referred to "Woman, the pariah of humanity," echoing the phrase coined by another French socialist, Flora Tristan (1803–1844). Influenced by the English Owenites, Tristan worked in France in the late 1830s and early 1840s as an independent socialist, also trying to form workers' associations. It was Tristan who called women "the proletariat of the proletariat" in her last writing, *Women's Emancipation, or The Pariah's Testament.* She declared that "the most oppressed man can oppress one being, his wife."[37] While the work and writings of women like Tristan, Voilquin, Roland, and Deroin and their Owenite counterparts across the Channel in England remained neglected and ignored for many years, their influence persisted. Twentieth-century feminists rediscovered their writings and built on their demands. Their insistence on women's liberation and their expansive view of equality outlasted the opposition they encountered from their male comrades. As a girl, the eminent French painter Rosa Bonheur (1822–1899) attended a Saint-Simonian school. "The influence it had on my lifework cannot be exaggerated," she later declared. "It emancipated me before I knew what emancipation meant and left me free to develop naturally and untrammelled."[38] Early feminist socialists created a liberating alliance between socialism and feminism in Europe of the 1830s and 1840s.

Marriage, Sex, and Socialism

Early English and French socialists connected economic and political concerns to women's sexual oppression. These feminist socialists, like some equal rights feminists, insisted that male dominance in sexual relations and marriage oppressed women. In her *Pariah's Testament*, published in 1846, two years after her death, the French socialist Flora Tristan called for four measures which she thought would emancipate women. First came the familiar rights to equal education and professional training. The other three dealt with sexual relations. To be free, argued Tristan, women needed the "right to a free choice of a mate," the right to divorce and remarriage, and the "right of unwed mothers to respect and equality before the law," coupled with the right of illegitimate children to share in the father's estate.[39] Illegitimate herself, Tristan had married her employer at seventeen; when the marriage proved unhappy, she was unable to divorce. (Divorce had been legal in France from 1792 to 1816, but was completely outlawed from 1816 to 1884.) To Tristan and other women active in early socialist movements, property relations and sexual relations were connected subjects, and to question one was to question the other.

English and French feminist socialists often questioned traditional sexual behavior. Owenites denounced the sanctity of marriage, especially marriage based on property, and were early advocates of birth control: Owen's son, Robert Dale Owen, published *Moral Physiology*, an important early birth-control tract, in 1830. Five years later, Robert Owen published his *Lectures on the Marriages of the Priesthood in the Old Immoral World*, in which he extolled "marriages of Nature," untrammeled by law or religion. By the 1840s, the association of socialists with "free love"—sex without or outside of marriage—was fixed in the public mind, and a great deal of criticism of socialism was based on socialists' supposed sexual license.[40] An 1840 broadside widely distributed in Manchester equated a deserted single mother with "a SOCIALIST'S BRIDE," condemning socialists' criticisms of marriage. A Christian denunciation of the same year declared that under socialism, "No man is to be confined to one woman, nor is any woman to be confined to one man; but all are to yield themselves up to be governed by the unrestrained instincts of nature, in imitation of dogs and goats."[41]

While most Owenite women remained monogamously married, even in the Owenite communities, Saint-Simonian women often did not. In both England and France, both groups were denounced as sexually immoral. "The principles of both are the same," wrote a hostile observer in 1834,

> open profligacy and plunder, and they are . . . addressed, in the first instance, to the weaker sex, upon whom they hope to make a fatal impression, as the serpent succeeded with Eve.[42]

In France, the Saint-Simonians began preaching the doctrine of the "liberation of the flesh" in the late 1820s. Women involved in the movement questioned traditional sexual fidelity as well: in 1833, Suzanne Voilquin ended her childless marriage informally when she discovered her husband had fallen in love with another Saint-Simonian.[43]

Voilquin later declared that she advocated divorce only if the wife had no children, "for, even though in theory Saint-Simonianism was well ahead of the actual world, the family group that I was a part of was not powerful enough, nor was it organized strongly enough, to replace paternal protection."[44] Other Saint-Simonian women, who attempted to "liberate" sexual relations, discovered the heavy price a woman had to pay for violating traditional values of chastity and fidelity. The life of Pauline Roland, a Saint-Simonian, reads like a horrific cautionary tale about what happened to a woman who broke society's sexual rules.

Introduced to Saint-Simonian ideas by her tutor, Roland (1805–1852) went to Paris in 1832 to join the group. The following year, she wrote another Saint-Simonian woman that she wished to experience sex and motherhood without marriage:

> To abstain, always to abstain—is this living, is this doing well? So why should I continue to live celibate when the 12 most beautiful years of my life have already been consecrated to this crazy Moloch [a god who demanded child sacrifice] of the Christians? . . . I want to be a mother, but with a mysterious paternity. . . . I will be proud of my motherhood and my child will be proud of its birth.[45]

Roland bore four children out of wedlock; the fathers were two Saint-Simonians. She supported her children and herself by writing, but in the late 1840s found herself destitute. Appealing to the fathers for money, she received no replies and had to beg to survive.[46]

By the time of their trial in 1850, both Pauline Roland and Jeanne Deroin were notorious for their untraditional views of marriage and sex. Roland had recently proclaimed that "socialism, the new religion," would remake marriage so that it embodied "perfect equality."[47] The arrest order for her charged that "for many years she has been promulgating communist-socialist opinions. As an unmarried mother, she is the enemy of marriage, maintaining that subjecting the woman to the control of the husband sanctifies inequality."[48] At their trial, Deroin tried to explain their views as part of their defense:

> I want absolute equality of the two sexes. They pretend that I am dreaming of promiscuity. Heavens no! On the contrary, I dream of, I desire the realization of a state of society in which marriage will be purified, made moral . . .

The judge interrupted her: "It is impossible for me to let you continue. You are attacking one of the most respectable of institutions, you are attacking a section of the Civil Code."[49] Sentenced to six months in prison, Deroin left France on her release; Roland was rearrested the next year and sent to an Algerian prison. She died on the return voyage. From her last prison she reflected on her misgivings about her life of "free love": "Having recognized the error of this false theory that the mother alone is the family, I have been condemned to live this life. . . . I don't understand my complete abandonment by two good, noble, intelligent men, both so tenderly loved."[50] Roland found motherhood outside traditional marriage untenable: if men refused financial responsibility for their children, no social institutions existed to ease the mother's burden.

By 1848, when Marx and Engels wrote *The Communist Manifesto,* the association of socialism with sexual promiscuity had become so common that they addressed the issue directly:

> "But you communists would introduce community of women," screams the whole bourgeoisie in chorus. . . . Bourgeois marriage is in reality a system of wives in common and thus, at the most, what the communists might possibly be reproached with is that they desire to introduce, in substitution for a hypocritically concealed, an openly legalized community of women. For the rest, it is self-evident that the abolition of the present system of production must bring with it the abolition of the community of women springing from that system, i.e. of prostitution both public and private.[51]

Socialists continued to criticize prostitution, which they connected to capitalism. In their private lives, socialists remained relatively free in sexual matters. A number lived in *unions libres,* the working-class common-law marriage, where the couple remained faithful but did not go through a formal ceremony. Engels lived this way for most of his life. In the second half of the nineteenth century, as Eastern European governments forced socialists into exile, many couples based their decisions to marry, or not to marry, on citizenship requirements. Throughout Europe, a wife assumed her husband's citizenship upon marriage. Clara Eissner Zetkin never married the Russian Ossip Zetkin and was therefore able to retain her German citizenship; Rosa Luxemburg went through a fictitious marriage to obtain German citizenship.

By the 1870s, such fictitious marriages were commonplace in Russia: revolutionary women used them to escape parental control and to study in Russia or abroad. These fictitious marriages seemed to hostile observers to signal sexual promiscuity. In 1873, for instance, the czarist government ordered the return of Russian female students from Zurich where they were charged with practicing "communist theories of free love."[52]

For women, the association of socialism with sexual freedom cut two ways. Some women, like men, enjoyed freedom from formal marriage and had affairs, ending relationships when they wished. Others led extremely conventional lives, perhaps to overcome the stigma of being involved with such a supposedly licentious movement. For women, "free love" made membership in socialist movements more problematic than for men. Many pointed out that love could not be "free" for women as long as women bore unwanted children and had to raise them alone. Others argued whether topics like marriage, divorce, contraception, abortion, and child rearing were appropriate subjects for socialist discourse or mere "women's issues" which would cease to be problems when capitalism disappeared. Some feminist socialists responded by ignoring sexual issues and instead focused on labor legislation or the vote as concerns of primary importance to working-class women. Others insisted on raising such topics within their parties, braving the gibes and criticism which greeted their attempts. Until well into the twentieth century, topics concerning marriage, sexuality, and the family divided feminist socialists and continued to make the women who raised them in public notorious, derided rather than respected.

Feminist Socialism in Europe: 1875–1925

Between 1875 and 1925, women achieved much within the growing realm of European socialist politics and unionism, but success came gradually. The military defeat of the socialist Paris Commune in 1871 ended hopes of an imminent revolution. Women who fought for the Commune were either executed like their male comrades, or sentenced to lengthy prison terms: Louise Michel spent the 1870s in the French penal colony of New Caledonia in the South Pacific. Socialist parties and unions were small and overwhelmingly male, largely confined to "skilled" male workers and intellectuals. By 1925, all this had changed. Socialist women had held ministerial positions in Great Britain, Scandinavia, and the Soviet Union as socialist parties came to govern. Thirty-three women sat as members in the German parliament (Reichstag), nineteen of them representing socialist and communist parties.[53] Hundreds of thousands of working-class women belonged to socialist trade unions.

New, large trade unions actively recruited women from the 1880s on, to prevent lower-paid women from driving down male wages. Outside of Russia, where all unions were illegal, women found socialist union activity a new path to success. Adelheid Popp-Dworak (1869–1939), the leader of the Austrian socialist women's movement, rose in this way. The fifteenth child of a widowed factory worker, Popp labored at home as a child at piecework and herself entered the factory as a young teenager. In the early 1890s, she encountered socialism, taught her by a fellow worker who brought her to her first socialist meeting, called to organize women factory workers. Popp was one of nine women among three hundred men; when she spoke, the men "cried 'Bravo' before I opened my mouth; merely from the fact that a working woman wanted to speak."[54] Popp was quickly recruited:

> The applause in the meeting was boundless; they surrounded me and wanted to know who I was; they took me at first for a member of the Branch, and requested me to write an article addressed to working women for the Union paper on the line of my speech.[55]

A short time later, the party paid her "to devote all my time to the organization among working women, and to help to work at a newspaper for working women."[56] After World War I, Popp became a member of the Austrian parliament.

Women also acted as socialist revolutionaries as well as union organizers. In Eastern Europe, government repression and forced exile created an international socialist community. In St. Petersburg, in Warsaw, in Berlin, in Vienna, and in Zurich, women joined the ranks of the revolutionary exiles.[57] Proving their dedication and fortitude, these socialist women created a tradition of female participation within conspiratorial revolutionary groups. Some—like some female trade unionists—also developed into feminist socialists, realizing that women's issues must also be championed within socialist parties, where men rarely saw women as equals.[58] "I do not want anything exceptional or even special rights," declared the German socialist Luise Zietz (1865–1922) around 1900, "but—as we are already second-class citizens—we refuse to be degraded to second-class comrades."[59]

In these years, socialist women fought on two fronts: against governments who wanted to crush their movement, and against socialist men who did not want women in party positions of any authority. Between 1875 and 1925, socialist women largely won these battles, although some parties, like the French, remained resistant to their participation. But feminist socialists—socialist women who advocated women's rights—faced a third battle: to raise women's issues without being accused of having abandoned their socialism in favor of bourgeois women's rights movements. Their success was limited.

While they articulated and developed their position from 1875 to 1925, many of their demands for women remained unfulfilled, even where socialist revolutions succeeded.

THE RUSSIAN AMAZONS

In czarist Russia of the 1860s and 1870s, privileged young women by the hundreds left their families to study and pursue revolutionary political paths.[60] "The emancipation of the serfs in 1861 had given rise to the women's movement. Like a huge wave, the movement to liberate women swept over all the urban centers of Russia," remembered Elizaveta Kovalskaya (c. 1850–1933), who became a revolutionary populist. "I, too, was caught up in it."[61] Most Russian women who joined these groups came from propertied families and knew oppression only secondhand, by observing the treatment of serfs on their families' estates or of workers in factories. In the 1870s, these women joined women's study groups, which discussed the "woman's question": what women's proper role in society should be. Many of these women's groups then transformed themselves into conspiratorial revolutionary societies, convinced that only the total transformation of Russian society could liberate women.

Although atypical in her background, Elizaveta Kovalskaya's life followed this pattern. Kovalskaya's mother was her father's serf, and Kovalskaya's first fears were that she would be sold off the estate. She persuaded her father to educate her, and as his only child, she received an inheritance when he died in the 1860s. She used the money to organize "free courses for women seeking higher education," workers' study groups, and a group "exclusively for women who were interested in socialism."[62] When college-level courses for women opened in St. Petersburg, Kovalskaya went to the capital, where she soon made contact with like-minded young women, who met in groups to study and discuss. These women refused to wear the confining feminine clothing of the day—Kovalskaya remembered a young woman with "close-cropped hair" who first welcomed her; "she wore an outfit that seemed almost to have become the uniform for the advocates of the woman question: a Russian blouse, cinched with a leather belt, and a short, dark skirt."[63] Kovalskaya joined a number of these study groups. "In every case, the woman question was the center of debate, although other political and social issues were touched upon in passing," she recalled. "Women's meetings were so frequent that we barely had time to get from one place to another."[64]

Kovalskaya's study groups, like most others, transformed themselves into conspiratorial revolutionary organizations, admitting men and attempting to "go to the people" to spread education and subversion. As underground revolutionaries, women participated equally with men. Kovalskaya distributed illegal literature to workers in St. Petersburg in 1878; severely beaten by the police, she spent almost a year recuperating before she could resume her revolutionary activities. "The distant specter of revolution appeared, making me equal to a boy," recalled Vera Zasulich (1849–1919) of these years. "I, too, could dream of 'action,' or 'exploits,' of the 'great struggle.' "[65] Equal in the "great struggle," women suffered equal penalties when arrested. In 1881, Sophia Perovskaya (1853–1881), a member of Kovalskaya's study group, was publicly hanged with four men for their assassination of Czar Alexander II. "Really, Mamma dear, my fate isn't all that dismal," Perovskaya wrote a few days before her execution. "I have lived according to my convictions; I could not have acted otherwise; and so I await the future with a clear conscience."[66] In 1881, Kovalskaya was also arrested.

> In court, I declared that I did not recognize the government's tribunal and would not participate in the proceedings. I refused to have a lawyer or make a final statement to the court. They sentenced me to hard labor for life.[67]

Kovalskaya was sent to Siberia, where she spent the next twenty-three years in prison. Between 1880 and 1890, forty-three revolutionaries were sentenced to hard labor for life by the czarist government; twenty-one of them were women.[68] The Russian "Amazons," as they were called in their own day, created a heroic tradition for women. Socialist revolutionary women won positions of honor in the left-wing canon—their posters hung on walls, their writings reprinted, their lives invoked as examples. They made it easier for other women to participate in socialist movements. But they did not sponsor or support feminism.

They won equality with men in the movement, but at the price of feminist concerns.[69] Without exception, these Russian women abandoned "women's issues" for what they saw as the greater goal of popular revolution. In their memoirs, they stressed politics, and rarely mentioned private concerns. They insisted that their womanhood be overlooked and that they be treated as absolute equals to men. From the 1870s, strong women who followed this model created room for themselves in socialist circles. Rosa Luxemburg (1870–1919), for instance, carried this pattern on in the next generation. An active revolutionary socialist, she distanced herself from feminist issues.[70] Born to a Jewish family in Russian Poland, she studied at the University of Zurich before helping to found the Polish Social Democratic party. In 1899, she moved to Berlin, where she became a leader of the German Socialist party. Always a fighter, she actively engaged in party struggles, advocating revolution rather than reform and writing many important theoretical works before the outbreak of World War I. Imprisoned during the war for opposing German participation, she helped lead the ill-fated communist uprising in Germany in 1919. During this time, she received a letter from Clara Zetkin, her longtime ally within the German party.

> Oh Rosa, what days! I see before me so clearly the historic greatness and meanings of all your actions, but my knowledge of these things cannot still the urgent demands of my heart, I cannot overcome my terrible worry and fear for you personally.[71]

Zetkin's fears were well-founded: a few days later, Luxemburg was dead, murdered by right-wing soldiers after having been captured by the police.

Feminist socialists—like Clara Zetkin—formed a small minority within socialist parties. Sensitive to women's particular oppression, they had to balance feminism and socialism. They linked women's liberation to the success of socialism, arguing like Zetkin that "the involvement of the great mass of proletarian women in the emancipatory struggle of the proletariat is one of the pre-conditions for the victory of the socialist idea, for the construction of a socialist society."[72] Focusing on issues of women's work and women's welfare as well as the vote, Zetkin and other feminist socialists made new demands for women between 1875 and 1925, achieving their greatest success in Germany and Russia.

FEMINIST SOCIALISM IN GERMANY

In Imperial Germany, where women were barred from all political activity until 1908, feminist socialists created the largest working-class women's movement in Europe. In the process, they developed a strategy for feminist socialism which proved the most successful in gaining women rights within a party which, however radical, still remained largely opposed to women's independence and liberation. Women's low participation in socialism enabled feminist socialists to argue that recruiting working-class women necessitated the creation of institutions to make the movement more hospitable to women: a Women's Bureau, women's trade unions, a woman's newspaper. These institutions became bases of power within the Social Democratic party which feminists then used to make more demands for women. Treading the fine line between fighting for women and not threatening socialist unity, German feminist socialists established a position which was simultaneously impeccably Marxist and assertively feminist.

As a leading theoretician, political fighter, and socialist radical, Clara Eissner Zetkin (1857–1933) was responsible for developing the feminist socialist position which became accepted first by the German Socialist party and later by the Socialist Women's International, a loose association which she founded and led.[73] The daughter of a housewife and a teacher, Zetkin was one of the first German women to receive formal teacher training at the college established by Auguste Schmidt, herself a founder of a moderate German women's rights group in 1865. In the late 1870s, Zetkin encountered exiled Russian revolutionaries in Leipzig and became a socialist. She joined the party in 1881 and the following year fell in love and formed a *union libre* with Ossip Zetkin. In the 1880s, the couple lived in exile in Paris, part of the international socialist community. There, Clara Zetkin tutored to earn income and recruited women to socialism while raising two children and caring for her husband. "I am the court's tailor, cook, cleaning lady . . . in short, 'girl friday,' " she complained in a letter to a friend. "To that, add two sons who leave me not a minute's peace."[74]

Dramatically introduced to the difficult lives of working-class women by her own experiences, Zetkin argued—first as a representative of women workers at the 1889 meeting of the Second International, and later that year in her influential pamphlet *The Question of Women Workers and Women at the Present Time*—that socialism and feminism were intrinsically connected. Building on the thesis of Marx, Engels, and Bebel that earning income outside the home would ultimately liberate women, Zetkin added her feminist view that women also suffered from male dominance:

> Just as the male worker is subjugated by the capitalist, so is the woman by the man, and she will always remain in subjugation until she is economically independent. Work is the indispensible condition for economic independence.[75]

Forging the link between socialism and feminism, she asserted that the socialist party could not succeed without the support of women workers and that women workers would achieve their goals only through socialism. "By walking hand in hand with the Social Democratic Party, [women] are ready to share all burdens and sacrifices that this fight entails," she declared at the Second International. "But they are also fiercely determined to demand, after the achievement of victory, all of the rights which are rightfully theirs."[76]

In actuality, Zetkin and other feminist socialists had no intention of waiting until the revolution to press women's demands. In the 1890s, German feminist socialists formed a Women's Bureau and actively recruited tens of thousands of women to the Socialist party and its trade unions. Zetkin became the editor of the women's newspaper, which she renamed *Equality (Gleichheit)* instead of *The Working Woman*. In her first editorial, she stated the "party line" which would enable feminism to exist within socialism, linking women's oppression to capitalism:

Gleichheit proceeds from the conviction that the final cause for the thousand-year-old inferior social position of the female sex is not to be sought in the statutory legislation "made by men," but rather in the property relations determined by economic conditions.[77]

By linking women's subordination to capitalism rather than male dominance, Zetkin accomplished two important goals.

First, she identified feminist socialists primarily with socialism and only secondarily with feminism. This strategy was essential in a party where many favored subordinate women, at least in their own homes.[78] By claiming their rights as socialists, Zetkin and other feminist socialists used party doctrine to shame male socialists into supporting feminist positions. If female subordination was caused by capitalism and would only be ended by socialism, then socialists had no business subordinating women. Zetkin established the polite fiction that all male sexism was "petty bourgeois" in origin, a strategy which enabled her to criticize male comrades from an orthodox socialist position.

Zetkin also tied feminism to socialism by rejecting the charge that feminism was a bourgeois issue, of interest only to women of property. Zetkin consistently spoke and wrote vehemently against bourgeois women and their equal rights movements. She argued that any cross-class feminist alliance was impossible and that feminism could only succeed through socialism. "As far as the proletarian woman is concerned," she declared at the Socialist Party Congress in 1896,

it is capitalism's need to exploit and to search incessantly for a cheap labor force that has created the women's question. . . . Therefore, the liberation struggle of the proletarian woman cannot be similar to the struggle that the bourgeois woman wages against the male of her class. On the contrary, it must be a joint struggle with the male of her class against the entire class of capitalists.[79]

After refusing to link up with middle-class women's rights movements, Zetkin and other feminist socialists then insisted that socialism champion women's issues for working-class women. German feminist socialists created women's unions and educational "reading evenings"; they pressed for the improvement of women's working conditions and their political rights. In 1895, the German Socialist party supported women's suffrage; in 1907, Zetkin's Socialist Women's International resolved that all socialist parties commit to equal and universal women's suffrage. In 1910, the Socialist Women's International added demands for equal pay and maternity insurance to the suffrage and designated March 8 as International Women's Day, to be celebrated each year as a further means to bring women into socialism. By the eve of the First World War, advocating feminism seemed a successful strategy for socialist parties: hundreds of thousands of European women belonged to a movement which asserted that women's equality in work and society would come with the triumph of socialism.

But the very strategy which enabled feminist socialism to grow also limited its success. The tactic of subordinating women's issues to socialism when there was a conflict between them undermined issues important to many women. Zetkin spent a great deal of effort repudiating the feminist ideas of Lily Braun (1865–1916), who focused on communal housekeeping and day-care centers to make working women's lives easier. Stigmatizing Braun as a "bourgeois feminist," Zetkin disparaged efforts at the socialization of housework and child care before the socialist revolution.[80] When some feminist socialists raised women's right to birth control and abortion, Zetkin again took the position that only a socialist revolution could solve problems connected with motherhood. She saw contraception as "an easy out of all egotists who want to have as many and as convenient enjoyments in life as possible."[81] When some feminist socialists advocated a "birth strike" in the years just before the war, Zetkin joined with male party leaders to block them.[82]

Zetkin also limited the development of feminist socialism by her own example. A "superwoman" herself, she had successfully raised and supported two children alone while working and writing as a socialist. She consistently embraced the traditional view that women were happiest and most fulfilled as mothers and simply added on to it the socialist position that women should work full-time for pay outside the home as well. She assumed all women had her own extraordinary energy. "Women will develop their individuality as comrades advancing on a par with men with equal rights, an equal role in production and equal aspirations," she declared in 1896, "while at the same time they are able to fulfill their functions as wife and mother to the highest degree."[83] She resisted measures designed to ease women's "double burden," like communal housekeeping, and left unresolved the problem women faced trying to participate equally while also raising children.

Finally, the success of demands for women was limited because feminist socialists were associated with the minority radical wing of the German Socialist party. In 1908, when German women were allowed to participate legally in politics, the Women's Bureau merged with the party, and feminists found themselves isolated within an increasingly conservative institution.[84] Socialist women were now expected to confine themselves to issues of child care and social reform. When the German Social Democratic party voted to support Germany's entry into World War I, Zetkin and the other revolutionaries, like Rosa Luxemburg, who advocated pacifism, found themselves a

tiny minority. "The majority of organized Social-Democratic women sank
. . . to the position of defenders of the national 'fatherlands' of the imperialist
bourgeoisie," Zetkin wrote bitterly.

> They competed with the ladies of the bourgeoisie in their chauvinistic ideas and
> behavior. . . . They have given up their basic aim—the proletarian revolution—
> and have thus rendered themselves incapable also of representing the day-to-day
> demands of proletarian women.[85]

Disillusioned by socialism's "sellout," Zetkin welcomed the Russian Revolution and became a founder of the German Communist party. In the 1920s,
she spent much time in Moscow, returning to represent the communists in
the German Reichstag. Her last public appearance came in August 1932,
when she addressed the Reichstag in its last meeting before the Nazis took
control of the German government. As the oldest representative, Zetkin
became "honorary president" of the Reichstag, and in her opening address
she called for a "United Front of all workers to turn back fascism." She
concluded by hoping that "despite my current infirmities, I may yet have the
fortune to open as honorary president the first Soviet Congress of a Soviet
Germany."[86] From 1917 on, her hopes and those of many radical socialist
women lay with the Soviet Union. There, a new generation of Russian
women worked to incorporate feminism into the one European society where
socialism had triumphed by revolution.

FEMINIST SOCIALISM IN THE RUSSIAN REVOLUTION
AND THE SOVIET UNION

In Russia, feminism developed as in Germany, only later and slower. Until
the revolution of 1905, virtually all political and labor union activity was
outlawed in Russia. Thereafter, two women's movements emerged: an equal
rights movement and a socialist women's movement. The largely middle-class
Women' Union focused on winning women's rights equal to the newly
enfranchised men's. "Citizens!" went one statement,

> we the women of Russia, who chance to be living in this great epoch of Russia's
> renewal . . . appeal to your conscience and honor and demand—not request—
> recognition of civil and political rights equal to yours.[87]

Since trade unions had been illegal, the Russian socialist women's movement
drew its leaders not from the working class, but from more privileged women
who converted to socialism, continuing the tradition of the Amazons of the
1870s. Nadezhda Krupskaya (1869–1939), for instance, came from the impoverished aristocracy and trained as a teacher before becoming a socialist.

Focusing on the exploitation of women in the countryside and factories, Krupskaya published *The Woman Worker* (1900), a pamphlet which applied Zetkin's analysis of women's circumstances to Russia. Krupskaya, who married V. I. Lenin in 1896, was one of a number of women prominent in the leadership of the Marxist Russian Social Democratic party; women also comprised about 14 percent of the membership of the populist Socialist Revolutionary party.[88] The effort to organize working-class women into a political force was initiated after 1905 by Alexandra Kollantai (1872–1952), who became the leading feminist socialist in the Russian Social Democratic party. "In that period I realized for the first time how little our Party concerned itself with the fate of the women of the working class and how meager was its interest in women's liberation," she wrote in her memoirs. "Nevertheless in the years 1906–1908 I won a small group of women Party comrades over to my plans."[89]

After the Bolshevik Revolution of 1917, Kollontai became the only female commissar in Lenin's government. From 1920 to 1922, she headed Zhenodtel, the Soviet Women's Organization. Her accession to political power was unique for any feminist socialist in this era, and the successes and limits of her policies define the successes and limits of feminist socialism in Russia and the Soviet Union.

Like most Russian feminist socialists, Kollontai came from a relatively privileged background. Born into a liberal aristocratic family, Kollontai married at twenty and became a socialist three years later, in 1896. She left her husband and son to study in Zurich and joined the international socialist community there. From 1905, Kollontai concentrated on building a Russian socialist women's movement on the German model. Rosa Luxemburg had introduced her to Clara Zetkin in 1906, and Kollontai became Zetkin's disciple in these years.[90] In 1906, she attempted to set up a Women Worker's Bureau at Social Democratic party headquarters in St. Petersburg. The men reacted instantly, announcing: "The meeting for women only has been called off; tomorrow there will be a meeting for men only."[91] Despite such setbacks, Kollontai worked on two fronts: she made contact with the largely female Union of Textile Workers, which gave her political support, and in 1908 she published her *Social Bases of the Woman Question*. Applying Zetkin's theoretical analysis to the situation of Russian women, Kollontai went further in claiming not only women's right to fulfilling work, but also their rights to sexual freedom and control over their own fertility. She believed in socialism's power to liberate women from sexual oppression as well as political and economic exploitation. "In her difficult progression to the bright future," she wrote in 1908, the woman proletarian—this slave so recently oppressed, without rights, forgotten—at the same time learns to throw off all the virtues imposed on her by slavery; step by step she becomes an independent worker, an independent personality, a free lover. This person, in common struggle with the proletariat, wins the right to work for women; this person—"the little sister"—stubbornly, persistently, opens the way to the "free," "equal" woman of the future.[92]

In the years before the Russian Revolution, Kollontai spent much of her time in exile, and her strongest support came from Zetkin and her Socialist Women's International.

Following the Marxist argument that only paid labor outside the home could liberate women, Kollontai championed women's right to such labor. She also moved beyond Marxism and fought for women's rights to sexual freedom and to control of their own fertility. Harking back to claims made by the utopian socialists (whom she read as a young woman), she asserted that socialism would liberate women from "sexual slavery" and "maternal slavery," as well as wage slavery. Criticizing "ritual marriage," and "the compulsive isolated family" as characteristic of capitalism's oppression of women, she asserted that only socialism could make a truly "free" love possible for women:

Only a whole number of fundamental reforms in the sphere of social relations—reforms transposing obligations from the family to society and the state—could create a situation where the principle of "free love" might to some extent be fulfilled.[93]

Kollontai attempted to realize these reforms. At the Second Socialist Women's International Congress in 1910, she and some Finnish delegates pressed for two measures they believed essential for women: maternity benefits for single mothers and the funding of maternity benefits from taxes rather than individual insurance policies. Both met defeat: aid to women bearing children out of wedlock went too far beyond conventional morality even for socialist women. However, Kollontai remained a respected and vital presence in international socialism, speaking in Europe and the United States, and continuing to develop her vision of feminist socialism. Under socialism, she argued, motherhood would be transformed because of support services capitalism could not provide:

Society is there to help [the mother]. Children will grow up in the kindergarten, the children's colony, the crèche and the school under the care of experienced nurses. When the mother wants to be with her children, she only has to say the word; and when she has no time, she knows they are in good hands. Maternity is no longer a cross.[94]

Unlike Zetkin, Kollontai did not assume that socialist women would be "superwomen," able to be full-time wage earners and full-time mothers simultaneously. Kollontai envisioned the social support services to enable women to function as laborers and citizens as well as mothers and sexual partners. Her visions were partially realized as a result of the 1917 revolutions in Russia.

The February Revolution of 1917 overthrew the czar, enabled political exiles to return to Russia, and gave women and men equal and democratic political, legal, and voting rights. Kollontai, Lenin, and the Bolsheviks returned, and Kollontai became a popular speaker during the summer. After the Bolshevik Revolution of October 1917, Lenin appointed Kollontai commissar of social welfare and created Zhenodtel, the new communist Women's Bureau. From then until 1922, Kollontai implemented legislation and reforms designed to liberate women through socialism.

In late 1917, she promulgated a Decree on Marriage, which transformed a religious ceremony that gave the husband power over his wife into a civil arrangement where both were equal. Divorce became legal, easy to obtain, and based on equal grounds for women and men. Another decree committed the Soviet state to guaranteeing protection to mothers and children, and in early 1918, maternity hospital care became free. Religious education for girls ended, and religious institutions were converted into social service centers. The Labor Code of 1918 asserted the equal obligation of all to work, including women. In November 1918, Kollontai organized a conference of one thousand women industrial workers and peasants. Lenin addressed the women and proclaimed their importance; Kollontai linked feminism to the achievement of full communism:

> As long as prostitution is not destroyed, as long as the old forms of the family, home life, child-rearing are not abolished, it will be impossible to build socialism. If the emancipation of women is unthinkable without communism, then communism is unthinkable without the full emancipation of women.[95]

In 1920, the Soviet government legalized abortion and Kollontai became head of Zhenodtel, the communist Women's Bureau. Despite chronic underfunding, she worked to develop the social services she considered essential to liberate women. "We already have homes for very small babies," she wrote optimistically in 1920,

> crèches, kindergartens, children's colonies and homes, hospitals and health resorts for sick children, restaurants, free lunches at school and free distribution of text books, warm clothing and shoes to school children. All this goes to show that the responsibility for the child is passing from the family to the collective.[96]

Arguing that the disappearance of the family would "liberate women from domestic servitude, lighten the burden of motherhood and finally put an end to the terrible curse of prostitution," Kollontai urged the "worker-mother" "not to differentiate between yours and mine; she must remember that there are only our children, the children of Russia's communist workers."[97] In a fantasy written in 1922, Kollontai imagined life in a communist utopia, the Soviet Union of 1970:

> Life is organized so that people do not live in families but in groups, according to their ages. Children have their "palaces," the young people have their smaller houses; adults live communally in the various ways that suit them, and the old people live together in their "houses. . . ." The world is a federation of communes. The younger generation do not know what war is.[98]

Kollontai's vision of communal living included communal housekeeping. In 1920, she argued that under communism "housework ceases to be necessary" because "collective housekeeping" would replace unproductive individual labor. Cleaning would be done by "men and women whose job it is to go round in the morning cleaning rooms." Cooking would take place in communal kitchens and meals would be eaten in public restaurants. Clothes would be washed in "central laundries" and "special clothes-mending centers" would free the working woman. Under communism, both the family and housework would "wither away."[99]

Kollontai's visions failed to become reality. First, she attempted to implement massive social change during the chaos of the civil war years. As Russian society suffered war, famine, and disease, both women and men clung to traditional institutions—including the patriarchal family—rather than embrace new ones.[100] Second, Kollontai held power chiefly because of Lenin's support, not because of widespread approval of her policies. Many of her ideas echoed his: in 1919, for instance, Lenin published a pamphlet asserting that "public catering establishments, nurseries, kindergartens—here we have . . . the simple, everyday means . . . which can *really emancipate women*, really lessen and abolish their inequality with men."[101] But in 1921 Kollontai lost Lenin's political support. Joining the radical Workers' Opposition movement, she reunited with its leader and her former lover, Alexander Schliapnikov. Lenin disapproved of both her politics and her sexual behavior and stated in 1922 that he could no longer "vouch for the reliability or the endurance of a woman whose love affair is intertwined with politics."[102] When Kollontai persisted in supporting Workers' Opposition, Lenin removed her as head of Zhenodtel and sent her into virtual exile on minor diplomatic missions.

Kollontai also lost both Lenin's support and that of the other Russian feminist socialists because of her public statements on sexuality and love outside of marriage. As early as 1920, Lenin denounced sexual freedom, asserting that

> promiscuity in sexual matters is bourgeois. It is a sign of degeneration. The proletariat is a rising class. It does not need an intoxicant to stupify or stimulate it, neither the intoxicant of sexual laxity or of alcohol.[103]

He told Clara Zetkin that she had been wrong to try to attract women to communism by having discussions about sex and marriage; economic issues were more important.[104] Zetkin acquiesced. But Kollontai was a sexual radical, arguing like the Owenites and Saint-Simonians that women's liberation involved sexual as well as economic freedom. In her *New Morality and the Working Class,* reissued in 1918, she proclaimed women's right to sexual freedom:

> When the wave of passion sweeps over her [the woman worker of the future] does not hypocritically wrap herself up in a faded cloak of female virtue. No, she holds out her hand to her chosen one and goes away for several weeks to drink from the cup of love's joy, however deep it is, and to satisfy herself. When the cup is empty, she throws it away without regret or bitterness.[105]

Kollontai consistently asserted that feminist socialism meant women's liberation from "bourgeois morality" as well as from the duties of child care and housework.

In 1923, she published two collections of stories which dramatized her views, rejecting the "double standard" of sexual behavior and showing women asserting their rights to be sexually active and to control their own fertility. "We stay together as long as we get on with each other," declares Zhenya, a "new woman" of the revolution, to her mother,

> and when we no longer do, we just part company and nobody gets hurt. Of course, I'm going to lose two or three weeks' work because of this abortion, which is a pity, but that's my own fault and next time I'll take the proper precautions.[106]

By asserting women's rights to control their own sexuality and fertility, Kollontai pioneered a stage of feminism which achieved little political success in Europe until the 1970s. In the Soviet Union of the early 1920s, Kollontai's radical statements on sex met with severe criticism. "Now especially a profusion of articles were written about my 'horrid views' in relation to marriage and love," she wrote in her memoirs.[107] Articles denouncing her appeared in all the major communist journals, accusing Kollontai of the nineteenth-century sin of "George-Sandism" and calling her writings "feminist trash."[108] In 1923, the party congress passed a resolution warning of the danger of "feminist tendencies" which, "under the banner of improving the women's way of life, actually could lead to the female contingent of labor breaking away from the common class struggle."[109] Sofya Smidovich, who replaced Kollontai as the head of Zhenodtel, denounced women's sexual freedom and asserted that "it is best to love one person and stick to him."[110]

As a result of the Russian Revolution, women received equal political rights and the right—or obligation—to labor outside the home. By the late 1920s, however, Soviet economic policies like state-planned industrialization and the collectivization of agriculture shaped women's lives with little regard for feminist issues like women's working conditions or their right to equal pay. Women entered the labor force in the 1930s, and in those years, the numbers of state-supported crèches and kindergartens rose appreciably.[111] Women benefited from the literacy campaign and improved education. But the radical transformation of family and sexual life envisioned by Kollontai as essential for socialist women did not occur. When Kollontai suggested in the debate over the Marriage Law in 1926 that individual alimony should be replaced by a General Insurance Fund for wives, a voice shouted "Collective responsibility," and the overwhelmingly male audience erupted in laughter. Men's failure to pay alimony and child payments had made the easier divorce laws deprive women of economic support. Laughter also greeted her proposal that birth control be provided.[112] Kollontai gave up. In 1926, she rewrote her autobiography (which she published in Germany) and edited out all references to sexual freedom. Her first ending had declared,

> No matter what other tasks I shall be carrying out, it is perfectly clear to me that the complete liberation of the working woman and the creation of the foundation of a new sexual morality will always remain the highest aim of my activity, and of my life.

She replaced this with a politically correct ending: "Only the productive-working people is able to effect the complete equalization and liberation of women by building a new society."[113] She never wrote on sexual matters again and in 1948 praised Stalin's government for enabling woman "to fulfill her natural duty—to be a mother, the educator of her children and the mistress of her home."[114] By then, the Soviet government had abolished divorce, coeducation, and abortion, reviving traditional values, but now in the name of socialism.

In addition to limiting women's right to sexual freedom, the Soviet state also ended feminist socialism. Separate socialist women's movements had always existed on sufferance, trying to claim special rights for women without damaging party solidarity. In the early 1920s, Lenin had declared, "We want no separate organizations of communist women! She who is a Communist belongs as a member to the Party, just as he who is a Communist."[115] Admitting that the low participation of women in the party justified separate women's groups in the short run, Lenin looked to a future in which separate women's organizations would no longer be needed. After his death in 1924, they disappeared. The International Women's Secretariat (the successor to Zetkin's International Socialist Party Women's Group) was abolished in 1926; Zhenodtel in 1930.

Women's special needs and aspirations were assumed to have been met by communism. To disagree meant criticizing communism, a dangerous activity in the Stalinist era. After Stalin's accession to power in 1927, feminist socialists found little support within the Soviet Union. The "party line" on feminism—that Soviet socialism had solved all women's issues—stifled the development of feminist socialism for decades.

Related Causes, 1925–1945: Social Welfare, Pacificism, Anti-Fascism

SOCIAL WELFARE

By 1925, both feminist socialists and equal rights feminists believed they had won their battles for women's rights. In many nations, women had gained the vote, and in the interwar years, feminists generally turned to causes related to feminism rather than feminism itself. Instead of asserting women's claims to justice, joining with other women, and trying to overturn traditions subordinating women, those who had previously identified as feminists now looked in other directions. In 1919, in the first speech by a female deputy in the German Reichstag of the new Weimar Republic, Marie Juchacz (1880–1956) spoke for the socialist women:

> I should like to say now that the "Woman Question" in Germany no longer exists in the old sense of the term; it has been solved. It will no longer be necessary for us to campaign for our rights with meetings, resolutions and petitions. Political conflict, which will always exist, will from now on take place in another form. We women now have the opportunity to allow our influence to be exerted within the context of party groupings on the basis of ideology.[116]

English feminists now declared that "feminism is not enough."[117] Women's equal rights organizations and feminist socialist groups often turned to welfare work, and as socialist and communist parties eliminated their women's divisions, party women found they were expected to apply themselves to what had traditionally been considered "women's issues," causes associated with women's traditional roles as housekeeper and caretaker of the family. "The whole area of social politics, including maternal protection, infant and child welfare must become in the widest sense a special area for women," continued Juchacz in her opening Reichstag speech. "The housing question, preventative medicine, child care, unemployment insurance are areas in which the female sex has a special interest and is especially well suited."[118]

By identifying women with "social housekeeping"—a phrase used first by middle-class women's reform movements and signifying the application of women's skills as mother and housewife to society as a whole—Juchacz was harking back to an early feminist argument for women's involvement in government. No state could be well governed until "both sexes" are represented and influence "the enactment and administration of our laws," Anne Knight, the English Quaker feminist, had declared in 1851.

> The wise, virtuous, gentle mothers of a state or nation might contribute as much to the good order, the peace, the thrift of the body politic, as they severally do to the well being of their families.[119]

Men generally approved of this view: it maintained women's traditional role and functions within the new context of participatory politics. "With Marie Juchacz, we have a new and totally different kind of woman," declared the German Social Democratic party's newspaper in 1919. "Gone is the era when the activists of the women's movement believed that they had to prove their equality by taking on male characteristics."[120] Clara Zetkin had always insisted that women could do anything within socialism as well as men; Marie Juchacz, her successor, steered socialist women into more "womanly" concerns.

While a few feminists continued to assert that women should involve themselves in all areas of politics, they were a small and defensive minority. An analysis of Reichstag speeches by female and male deputies revealed that women spoke far more often on social welfare than men, and far less often on foreign policy.[121] Other female politicians complained that their parties only wanted their support services or ignored them entirely. The leaders of the English Labour party "only have any use for women so long as they have no opinions of their own but are willing to do the donkey work of the party," complained Dorothy Thurtle in 1930.[122] The French socialist Germaine Picard-Moch complained in the 1920s that socialist men acted as if socialist women did not exist.[123] This sort of exclusion or relegation to subjects considered appropriate to women was primarily a problem for women in left-wing parties.

Far fewer women held active positions in right-wing political parties. Women associated with right-wing parties saw themselves as men's supporters, as traditional women, and thought it appropriate that women deal only with social welfare issues. What was unusual about the interwar years in Europe was that women in left-wing parties turned to welfare work as well. The German socialist women's newspaper began to carry a regular supplement entitled "Woman and her House," and in 1924, *Gleichheit (Equality)* was renamed *Women's World,* and included fashions, household tips, and recipes. Juchacz devoted her energies to building Arbeiterwohlfahrt, a working-class welfare organization which focused on the special needs of working-class children. In England, the Women's Co-operative Guild, with sixty-seven thousand members in 1931, most from the working class, played a similar role.[124] Women who had been active as feminist socialists now worked to establish working-class welfare organizations. Middle-class women, with a longer tradition of social work, found established philanthropies eager for their services.

In devoting themselves to welfare, feminists previously active in either equal rights or socialist women's movements turned to causes related to feminism rather than feminism itself. Welfare did not involve rejecting traditions which limited women's lives so much as rejecting traditions which oppressed all people. And, by working in welfare, some feminists continued to press for women's issues: maternity benefits, reform of divorce laws, child allowances, even contraception and abortion.

Supporting the welfare of mothers and children could lead in two political directions. Conservative women—even those previously active in the struggles for the vote—fought against abortion, contraception (or at least advertisements for it), and pornography. The League of German Women's Associations *(BDF),* an equal rights group before World War I, opposed all efforts to gain women sexual freedom and control over their fertility in the 1920s.[125] In contrast, radical women demanded contraception, abortion, and sexual freedom as essential to improve the lives of working-class women and children. (By the 1920s abortion and contraception, although still illegal, were increasingly available to women with money.) The World League for Sexual Reform on a Scientific Basis linked advocates internationally. Their 1929 meeting in London, for instance, brought together 350 delegates, including Alexandra Kollontai for the Soviet Union; Helene Stöcker, head of the radical group Mutterschutz (Mothers' Protection), for Germany; Dora and Bertrand Russell, Marie Stopes, and George Bernard Shaw for England; and Margaret Sanger for the United States.[126]

While sexual radicals constituted an international community, they had little political influence in this era. The only nation where abortion was legal was the Soviet Union, and even there Stalin outlawed it in 1936, explaining,

We need men. Abortion which destroys life is not acceptable in our country. The Soviet woman has the same rights as the man, but that does not free her from a great and honorable duty which nature has given her: she is a mother, she gives life. And this is certainly not a private affair, but one of great social importance.[127]

Despite the efforts of socialist women, abortion remained illegal in Germany (although penalties were reduced between 1927 and 1933). France outlawed contraception in 1920 and raised the penalties for abortion in 1923 to try to increase its birth rate. Even in nations where contraception was available, like Denmark or England, many women found it difficult to obtain and of poor quality.[128] The triumph of fascist regimes in Italy, Portugal, Germany, and Spain made any reforms for women's control of their own fertility impossible in those nations.

The majority of politically active women in all nations, however, were not feminists and never had been. Many of them opposed feminism. In pre–World War I Italy, for instance, hundreds of thousands of women joined groups like the Catholic Association or the Women's Union of Catholic Action, while socialist or sexual reform groups attracted little female support.[129] But women did not have to be feminists to benefit from rights won for women by feminism. Intentionally or not, all these women solidified feminism's gains. Whether conservative or radical, communist or fascist, women now spoke in public, did welfare work, were active in all major political parties. From the 1920s on, no government, however repressive, denied women these rights, rights nineteenth-century feminists had had to fight for. Feminists in the interwar years who devoted their energies to causes like welfare, pacifism, and anti-fascism built on those gains.

PACIFISM AND ANTI-FASCISM

Women who worked for pacifist and anti-fascist movements in the interwar years believed that such causes were a natural continuation of their earlier feminist work. The association of feminists and pacifism stretched back to the 1840s. "Women reply to men who ask, 'What do you want, what are you trying to do?' " wrote the French socialist Jeanne Deroin in 1848,

"We want to construct a new world with you, where peace and truth will reign, we want justice in every spirit and love in every heart."[130]

During the Crimean War in 1854, Frederika Bremer (1801–1865), the Swedish equal rights feminist, appealed to women to form a peace league, arguing that "separately we are weak and can achieve only a little, but if we extend our hands around the whole world, we should be able to take the earth in our hands like a little child."[131] Bertha von Suttner (1843–1914), the Austrian "suffragette for peace," wrote *Lay Down Your Arms!* in 1889. The book's success prompted the industrialist Alfred Nobel to institute the Nobel Peace Prize, and Suttner was the first female recipient in 1905.[132]

This identification of men with war and women with peace was as old as European culture. Feminists often relied on the sentimental assumption that when women voted, war would end. A widely reproduced drawing published before World War I in the French Journal of the League of Women's Rights showed a man casting a ballot labeled "war" and a woman casting one marked "peace." The caption read, "World peace, social harmony and the well-being of humanity will only exist when women get the vote and are able to help men make the laws."[133] In 1910, the Socialist Women's International Conference made "the fight against war" a major topic, and urged that it be placed on the agenda of all socialist conferences.[134]

World War I split feminist organizations along pacifist and nationalist lines. Both equal rights and socialist women's movements put aside their feminist demands "for the duration" and supported the war, causing the minority of pacifist dissidents to leave groups they had long been members of. In England, both Pankhurst's Women's Social and Political Union and Fawcett's National Union of Women's Suffrage Societies dedicated their energies to war work: by spring 1915, all Fawcett's national officers except herself and the treasurer had resigned to form a peace group.[135] Emmeline Pethick-Lawrence (1867–1954), long active in the WSPU, left the organization and toured the United States in 1914 with the Hungarian feminist Rosika Schwimmer (1877–1948) speaking for peace.[136] In Germany, the suffragists Lida Gustava Heymann (1867–1943) and Anita Augsberg (1857–1943) broke with the *BDF*, which actively supported the war, and called for an international meeting of European women to work for peace. Clara Zetkin and Rosa Luxemburg, both socialist internationalists and therefore pacifists in World War I, found themselves at odds with the German Social Democratic party which supported the war.

Even this common opposition to the war did not heal the breach between equal rights feminists and feminist socialists. Two separate pacifist women's conferences met in Europe in the spring of 1915: the Conference of International Socialist Women, in Berne in March; and the equal rights International Women's Congress, in The Hague in April. Both called for an end to war and urged women to exert influence on governments to make peace. The equal rights International Women's Congress, under the leadership of Aletta Jacobs, the well-known Dutch doctor, suffragist, and pacifist, included 1,136 women, mostly from the Netherlands and other neutral nations like the United States, although there were also small delegations from most of the belligerent nations. Twenty-five delegates from Germany, England, Russia, France, Italy, the Netherlands, and Switzerland attended the rival Socialist Women's Conference under Clara Zetkin's leadership. Both groups issued manifestos calling for peace.[137]

During the war, all pacifists in the belligerent nations found themselves prosecuted and under police surveillance. In France, for instance, the feminist socialist Louise Saumoneau was jailed for distributing the Berne Manifesto. Hélène Brion (1882–1954), a feminist schoolteacher, was tried for treason and had her teacher's license revoked for distributing pacifist pamphlets in 1918. At her trial, she linked feminism and pacifism, declaring,

> I am an enemy of war because I am a feminist. War represents the triumph of brute strength, while feminism can only triumph through moral strength and intellectual values. Between the two there is total contradiction.[138]

After World War I, pacifism continued to cut across both gender and political lines. Some continued to believe that "womanly being, womanly essence are identical with pacifism," as Lida Gustava Heymann asserted in 1919, and large pacifist organizations, like the Women's International League for Peace and Freedom, continued to organize women exclusively.[139] Others became disillusioned about women's supposed innate pacifism. "I realized at once that my supporters were not the women," Emmeline Pethick-Lawrence, the English suffragist, observed of her campaign as a pacifist Labour candidate in 1919. "They were all for 'going over the top' to avenge their husbands and their sons. My supporters were the soldiers themselves."[140] Still others, like Clara Zetkin, insisted that peace would only come with a socialist revolution: "the struggle of the workers against imperialist wars is a life and death struggle against bloodsucking capitalism, for the salvation of socialism," she declared in her last written work, the 1933 pamphlet *Toilers Against War.*[141]

Many pacifist feminists and pacifists changed their philosophy in the 1930s, when the rise of fascism seemed to many to call for military action rather than peaceful solution. A number of groups called for the seemingly contradictory goals of peace and opposition to fascism. A Women's Congress Against War and Fascism met in Paris in July 1934, and opposition to both war and fascism was often voiced in this period.[142]

The most eloquent statement linking pacifism, feminism, and anti-fascism was Virginia Woolf's *Three Guineas*, published in 1938. By then Woolf (1882–1941) was one of the best-known English novelists of her generation. She had long supported and been active in feminist causes: working for the suffrage, hosting meetings of the Women's Co-operative Guild, and acting as secretary to a local Labour party branch.[143] In 1929, Woolf published her first avowedly feminist writing, *A Room of One's Own,* in which she examined how men had restricted women's lives in general and the life of the woman artist in particular. She urged women to independence—to "face the fact, for it is a fact, that there is no arm to cling to"—and to labor for future women artists. Using the imaginary figure of "Shakespeare's sister," the female writer of the past who died without ever having an opportunity to use her talent, Virginia Woolf argued that women should create a world in which female contributions would be valued:

> As for her [Shakespeare's sister] coming without that preparation, without that effort on our part, without that determination that when she is born again she shall find it possible to live and write her poetry, that we cannot expect, for that would be impossible. But I maintain that she would come if we worked for her, and that so to work, even in poverty and obscurity, is worth while.[144]

While valuing art, Woolf never lost sight of the economics which made art possible. In *A Room of One's Own* she admitted her financial legacy had been more important to her than the right to vote. The title came from her assertion that in order to be an artist, a woman needed "£500 a year and a room of one's own."

In 1930, Woolf contemplated a sequel about opening the professions to women, but the real impetus to write *Three Guineas,* her second feminist essay, came in 1935.[145] That year, Woolf was told by the novelist E. M. Forster that the London Library had voted to continue to exclude women from its administrative committee, maintaining the position established for the library by Woolf's long-dead tyrannical father, Leslie Stephen. Furious, Woolf thought of writing a feminist piece called *On Being Despised.* Instead, when Hitler's invasion of the Rhineland in 1936 convinced her war was imminent, she began to write an extended essay in answer to the question "How in your opinion are we to prevent war?"[146]

Woolf argued that the answer lay in the rejection of fascism and the traditional masculine, warlike values it glorified, and she created a feminist pacifist alternative to fascism. In *Three Guineas* Woolf asserted that if women had "no example of what we wish to be, we have, what is perhaps equally valuable, a daily and illuminating example of what we do not wish to be" in "the Fascist States."[147] To her, feminism was a means to peace, for it necessitated the rejection of fascism, of the warrior, of the institutionalized male oppression of women which Woolf called "the patriarchy."[148] In her own day, she saw this oppression embodied in Hitler and Mussolini:

> The figure of a man; some say, others deny, that he is Man himself. . . . His eyes are glazed; his eyes glare. His body, which is braced in an unnatural position, is

tightly cased in an uniform. Upon the breast of that uniform are sewn several medals and other mystic symbols. His hand is upon a sword. His is called in German and Italian Führer and Duce; in our own language Tyrant or Dictator.[149]

Having linked pacifism, feminism, and anti-fascism, Woolf noted that Hitler made the same connection when he had distinguished between "a nation of pacifists and a nation of men"; she herself urged women to action.[150] Woolf wrote that the "private" world of women's acquiescence to male aggression within the family was "inseparably connected" to the "public" world of fascist aggression in Europe: "the tyrannies and servilities of the one are the tyrannies and servilities of the other." Acquiescence to a tyrant in the home was like appeasement of a tyrant abroad. Woolf argued that both systems of aggression can—and must—be overthrown:

> We are not passive spectators doomed to unresisting obedience but by our thoughts and actions can ourselves change that figure [of the Führer]. A common interest unites us; it is one world, one life. How essential it is that we should realize that unity the dead bodies, the ruined houses prove.[151]

Three Guineas asserts that war is a problem of the male character and male dominance, and that women

> can best help [men] prevent war not by repeating your words and following your methods but by founding new words and creating new methods. We can best help you prevent war not by joining your society but by remaining outside your society but in co-operation with its aims.[152]

The heart of *Three Guineas* is this proposal that women, "anonymously and secretly," form "The Outsiders' Society," dedicated to subverting and transforming male-controlled society, war, and capitalism. Rejecting national identity for solidarity with other women, Woolf declared, "As a woman, I have no country. As a woman, I want no country. As a woman, my country is the whole world."[153] She listed the duties of the "outsiders." First, they must be militant pacifists, refusing to bear arms, to nurse the wounded, or to manufacture war matériel. They should take no part in patriotic demonstrations and should "maintain an attitude of complete indifference" to men's decisions about war.[154] Second, the outsiders must be militant feminists, supporting themselves, pressing for a living wage for all women, working for state wages for mothers.[155] Finally, outsiders should examine all facets of their culture and society, judging, rejecting, and transforming them as necessary. The essay concludes with Woolf addressing the male head of a pacifist organization and arguing that "we are both determined to do what we can to destroy the evil which that picture [of the male fascist warrior] represents, you by your methods, we by ours. And since we are different, our help must be different."[156]

Little time was left to form "The Outsiders' Society." The same year Virginia Woolf published *Three Guineas,* Helena Swanwick (1864–1939), a long-term English feminist pacifist, published her credo, *The Roots of War.* In it, Swanwick reasserted the belief that women were more pacifist than men: "Women do, I believe, hate war more fervently than men and this is not because they are better than men, or wiser, but because war hits them much harder and has very little to offer in return."[157] World War II began on September 3, 1939 in Europe. Through the winter of 1939, English women's and pacifist groups protested the war. A meeting of women's organizations agreed in the winter of 1939 that "universally women feel that men are responsible for the war . . . as every opportunity for co-operative effort has been wasted."[158] But active fighting ended such protests. In despair, Swanwick committed suicide in November 1939; by 1940, Virginia and Leonard Woolf had decided to gas or poison themselves in the event of a Nazi invasion.

For feminists generally, and Jewish, communist, socialist, and pacifist feminists in particular, the accession of fascists to power meant exile, imprisonment, and death. In the 1930s and 1940s, the worldwide depression, World War II, and the recovery from the war consumed the energies of all.

Purely women's issues were forgotten in the struggle for survival. For women, the daily effort needed to keep alive, to maintain their families, especially under conditions of extreme disruption, often made the feminist rejection of traditions seem dangerous, a peacetime luxury. The conditions of war ensured that women united with like-minded men, rather than with each other. In the resistance, in the postwar recoveries, women identified as Catholics, as communists, as members of one political, religious, or ethnic group or another, but not as feminists.

Restoring life to its usual patterns took precedence, and as after most wars, this meant woman in the home in her traditional role. European women generally accepted traditional roles and functions: in the late 1940s, birth rates rose and the age of marriage for women fell.[159] Throughout the 1950s and into the 1960s, European women ceased to defend their rights. Few voiced feminist concerns.

Women began to assert their rights again and to speak as feminists in the late 1960s. By then, the recoveries from the war had been completed. Across the Atlantic, in the United States, traditional values came under attack from the civil rights and anti-war movements. In Europe, 1968, like 1848, was a year of rebellion. Students and workers rioted in France, in Germany, in Italy, in Czechoslovakia. The "women's liberation movement," as feminism now called itself, dated from these upheavals of the late 1960s.

5

THE WOMEN'S LIBERATION MOVEMENT

❦

FEMINISM ORIGINATED in women's perception of the injustice of their situation and their refusal to accept it. The first European feminists pointed to the new ideals of individual achievement and education of Renaissance Humanism and asserted that they applied to women as well as men. Eighteenth- and nineteenth-century equal rights feminists claimed the new democratic and civil rights of their era for women, arguing that denying women political and legal equality was as unjust and oppressive as denying it to men. Feminist socialists often called attention to the gap between socialist ideals of economic liberation and equality and the actual lives of working-class women. The women's liberation movement of the 1960s continued this tradition. In Europe as in the United States, women participating in the anti-establishment and anti-government movements of the late 1960s realized that they still were not equal.[1] They noticed the disparity between the lofty ideals of their movements and the reality of their own lives, the gap between women's and men's roles in left-wing and progressive groups. "I was struck by the great words: liberation of peoples, liberation of women," wrote a Frenchwoman in a socialist journal in 1969.

> My liberation consists of serving him after my work while he reads or "thinks." While I peel [*épluche*] the vegetables, he can read [*éplucher*] at leisure—either *Le Monde* or works on marxist economy. Freedom only exists for the well-off ones and in the real world, the well-off one is the man.[2]

The new women's liberation movement, in contrast to earlier feminist movements, was founded on women's opposition to like-minded men. "Comrades, if you are not ready for this discussion [of women's liberation]," declared the German feminist Helke Sander to the Socialist German Students' (SDS) conference in 1968, "then we must draw the conclusion that the SDS is nothing more than an inflated mass of counter-revolutionary dough."[3]

Feminists insisted that the equality they were supposed to have achieved was not equality at all. "Women's Liberation Workshop believes that women in our society are oppressed," asserted an Englishwomen's group in 1969.

> We are economically oppressed: in jobs we do full work for half pay, in the home we do unpaid work full time. We are commercially exploited by advertisements, television and the press; legally we often have only the status of children. We are brought up to feel inadequate, educated to narrower horizons than men. This is our specific oppression as women. It is as women that we are, therefore, organizing.[4]

In 1405, Christine de Pizan began her feminist writing by trusting her own judgment and perceptions rather than the pronouncements of male writers; in the late 1960s, feminists of the women's liberation movement also asserted their own conclusions over received opinion, rejecting male standards and values and attempting to create a women-centered view of the world.

Western European feminists criticized the gap between ideals and practice, between promises and reality, between what they were told they had achieved and their private perceptions of their own situation. They questioned, as Helke Sander declared to the German SDS in 1968, both "the falsehood of the bourgeois way to emancipation" and the efficacy of socialism, "because a solely economic and political revolution does not suspend the repression of private life—as proven in all socialist countries."[5] This perception of oppression and injustice came from women who realized that the goals their mothers and grandmothers had fought for had not truly been achieved. They saw that the vast majority of women had neither political nor economic equality, despite assertions to the contrary.

The life of Simone de Beauvoir (1908–1986) shows the transformation of a privileged European woman into a feminist of the women's liberation movement. Born into the Parisian Catholic bourgeoisie, educated at the Sorbonne, de Beauvoir lived as "liberated" a life as was possible for a European woman in the twentieth century. A teacher who became a full-time writer in 1944, she supported herself financially by doing work she enjoyed. She established a lifelong liaison based neither on marriage nor sexual fidelity with the philosopher Jean-Paul Sartre. She used contraceptives and abortion—although they were illegal—to avoid having children. By the late 1940s, de Beauvoir was famous as the leading female intellectual of her day, a member of the influential circle of French existentialists. Active in left-wing causes in the 1950s and 1960s, she was admired for her writings, her philosophy, and her public defiance of tradition. Her analysis of women's condition, *The Second Sex*, became influential in the U.S. women's movement of the 1960s and a decade later in the new European women's movement as well.

The germ of *The Second Sex* came from Sartre's suggestion in 1946 that de Beauvoir write about what difference being a woman had made in her life. Initially resisting this proposal, de Beauvoir spent the next two and a half years producing her thousand-page study, published in 1949.[6] Starting from the premise that "women are not born but are made," de Beauvoir ranged through biology, mythology, history, and sociology to substantiate her view that culture far more than biology shaped women's nature and actions. Using existentialist philosophy and its emphasis on action rather than intention as her frame of reference, de Beauvoir argued that women's basic freedom had been limited by men, who took themselves as the measure, the standard, the model and could only view women as an inferior and deficient "other."[7] As the "other," women were the subject of men's projected fantasies and fears. In her lengthy and detailed analysis of women's lives and myths surrounding women, de Beauvoir portrayed woman as object, as victim, as oppressed by male values and a male perspective. Pioneering in her effort to analyze women's lives, de Beauvoir used male categories and language to express herself: the book ends with her call that "men and women unequivocally affirm their brotherhood," and the hope that a socialist revolution will liberate both sexes.[8]

Until the early 1970s, Simone de Beauvoir did not consider herself a feminist, but at most, a feminist socialist like Clara Zetkin or Alexandra Kollontai. In *The Second Sex,* she seemed to identify with men, calling women "they," and writing "objectively," as if she herself were not a woman. De Beauvoir later criticized the limitations of this viewpoint. "Because I more or less played the role of token woman," she wrote in 1974, "it seemed to me for a long time that certain inconveniences inherent in the 'feminine' condition should be either left alone or passed over and that there was no need to attack them."[9] She found her ending to *The Second Sex* inadequate: "I stopped on a note of vague confidence in the future, in the revolution, and in socialism," she stated in 1972. "Today I've changed my mind. I have become truly a feminist."[10]

This change, de Beauvoir explained, involved two connected steps which many feminists active in the women's liberation movement would also take. Becoming a feminist in Europe of 1970 meant first identifying with other women and then acknowledging that "socialism was not enough," insisting that women as a group still suffered special oppression in socialist as well as capitalist societies.

European feminists came to believe that women would remain oppressed until they themselves transformed the most basic conditions of their lives. "Abolishing capitalism will not mean abolishing the patriarchal tradition as long as the family is preserved," de Beauvoir told the German feminist Alice

Schwarzer in 1972. "I believe that not only must we change the ownership of the means of production, but that we must also change the family structure."[11] An early slogan and conviction of the women's liberation movement was that "the personal is political," that women's personal experiences were valid and had important political consequences for society and culture. Previously "unmentionable" topics, like abortion and rape, now became the subjects of political discourse and feminist action. For de Beauvoir, public acknowledgment that she had had an abortion marked her first political act as a feminist. Hers was the first signature on the Manifesto of 343 of 1971, in which 343 prominent Frenchwomen "confessed" to having had an abortion, which was then illegal in France. "A million women have abortions in France each year," read the manifesto.

> Because they are condemned to secrecy, they are aborted under dangerous conditions. If done under medical control, this operation is one of the simplest. These millions of women have been passed over in silence. I declare that I am one of them. Just as we demand free access to birth-control methods, we demand freedom to have abortions.[12]

By declaring, "I am one of them," de Beauvoir and other women became feminists: they identified as women and worked to improve the lives of all women. De Beauvoir abandoned her status as a "token woman," an "honorary man," and came to believe that her earlier "off-handedness involved complicity. In fact, if one accepts the slightest inequality between the two sexes one accepts Inequality."[13] Throughout the 1970s, de Beauvoir used her prestige to advance women's liberation. As president of the French League of Women's Rights and as editor of the journals *Nouvelles féministes* and *Questions féministes,* she called attention to problems like violence against women, sexual assault, sterilization, and lack of contraception. In analogies often used by other feminists in this era, she compared women to "natives" who had been "colonized" by male imperialists, to blacks who had been oppressed by whites. By becoming a feminist in her sixties, Simone de Beauvoir demonstrated that even for the most privileged of European women, equal rights and socialism had not brought liberation. Like de Beauvoir, increasing numbers of European women reasserted the unfilled demands of earlier women's movements. They also began to question anew all of the traditional assumptions about the relations between women and men.

From Consciousness-Raising to Political Action

The women's liberation movement of the late 1960s and the 1970s demanded more for women than earlier feminists had been able to. Now feminists challenged the ancient European tradition that man is the measure of all things: that male experience and perception are standard for all of humanity. Rejecting so deep and established a tradition demanded self-confidence and fortitude. Most women learned to trust their own experience and perceptions in small discussion groups with other women, groups designed to make women "conscious" of their gender, just as socialism had sought to make people conscious of their class. "Consciousness-raising" confirmed women's perceptions and validated them with other women's experiences. In the United States, consciousness-raising groups focused on psychology; in Europe, discussions were more political. These discussions confirmed that women's personal experiences had important political implications and gave many the confidence to demand more for women.[14] "We come together, having chosen to come," wrote the English poet Lilian Mohin of such a group,

> six women sit around a table, breakfast over,
> talking. . . .
>
> it is necessary to come together
>
> not faith, but the slow swelling
> of what we need; trust
> wrung from our distrust
> drop by globule
> spoonfuls, speech, touch, cupfuls
> a sea, our own tides.[15]

Like their counterparts in the United States, most European feminists in the women's liberation movement first made connections to other women in the late 1960s and early 1970s. "Since feminism doesn't have an official history, each woman initially rebels in isolation and alone," wrote the two French feminists who called themselves "Annie" and "Anne" in 1974.

> She has to find everything out by herself, and if she remains isolated her rebellion only leads to bitterness. If she encounters other women who have also thought through and written down what she experienced, then it is to her as if she discovered her origins. She gets back to herself and her life force.[16]

Feminists made contact with each other through women's newspapers and journals, in women's studies seminars, in the new women's bookstores, coffee shops, and shelters created by the growing network of women's liberation groups. "I look for women, hesitating at first, just because I need them," wrote Anja Meulenbelt, a Dutch teacher, of her life in this era. She rejected older equal rights and socialist women's organizations:

The Women's Association [the organization of the Dutch equal rights movement] is too much of a ladies' club, for that I am too left-wing, too much a political animal. . . . [But] the first [socialist] feminist organization drowns in party programmes dictated by men and slogans about the class struggle. Not much room to find myself.[17]

Instead Meulenbelt built her own group, finding congenial women with whom she could talk and share experiences. "Fay comes back from America," remembered Meulenbelt.

She says that in America small groups of women are coming together to talk about exactly those kinds of things [relationships with men] with each other. An idiotic idea. Ridiculous and attractive. Perhaps we should do it here too, we say, giggling and nervous. . . . We feel conspiratorial, a little ridiculous. We can always stop if nothing happens, I say. We condescendingly call it our ladies club.[18]

To her surprise, Meulenbelt found the meetings exhilarating: "Experiences tumble over each other, hardly patience to let each other talk. . . . Solidarity. I am not alone, I am not alone."[19] Confirmation of deep feelings and perceptions by other women proved liberating as well. "Together we can accomplish anything," wrote Meulenbelt,

Change women
Laugh over nothing
Make herstory
Swim naked in Mialet.[20]

In Portugal in these same years, the feminists who called themselves "the Three Marias," expressed similar feelings of power and support gained by talking with each other:

And again the three of us find ourselves together here, as on so many other occasions and at so many other times of decision, refusing to be shadows, a sedative, the warrior's repose. It is we who are warriors—women whose bodies are intact and whose hands are sure.[21]

When the Three Marias published their feminist writings in 1972, Salazar's government arrested them on charges of "abuse of the freedom of the press" and committing "an outrage to public decency." The book was banned and all copies confiscated until the authors' acquittal in 1974.

In more liberal societies, feminists had more success. In England, in France, in Germany and the Low Countries, in Scandinavia and in Italy, women's movements formed in the late 1960s and the 1970s. Feminists created new journals, new women's spaces, and new political networks very rapidly. Scores of feminist newspapers and magazines appeared in these years. While many were short-lived, others sold well enough to become established.

In West Germany, for instance, *Courage* reached a circulation of 70,000 by 1978; *Emma—Magazine by Women for Women* printed an initial run of 300,000 issues in 1977.[22] Celebrating the tenth anniversary of the English feminist magazine *Spare Rib* in 1981, the editorial collective explained the journal's success:

> *Spare Rib* aims to reflect women's lives in all their diverse situations so that they can recognize themselves in its pages. This is done by making the magazine a vehicle for their writing and their images. Most of all, *Spare Rib* aims to bring women together and support them in taking control of their lives. If there is one thing that sums up the common vision it is the letter that comes from Shropshire or Swansea or South London: "I thought I was the only woman in the world who felt as I did until I read *Spare Rib.*"[23]

Feminists all over Western Europe began to create not only separate women's movements, but also separate places where women could easily meet other women and feel comfortable. In cities and large towns, women established their own bookstores and coffeehouses, women's centers and shelters, courses and seminars. "What stands out is the *social* side of feminism," wrote Petra de Vries of the Netherlands.

> We put a lot of energy into creating good places for women to be, places where you can find out how much you will mix with sisters, how much you will agree with the ideas that are put forward in a course, where you can have *fun* and where it is *cozy*. . . . Why wait until after the revolution to have fun? Why not experience *now* that society can be different?[24]

For many feminists in these early years, solidarity with women meant hostility toward men. "More and more women from all over the world are getting together to rebel against what the world of men had made them," wrote the French feminists Annie and Anne in 1974.

> If we maintain that our sex unites us across all class differences because we are oppressed as women, regardless of class, race and age, then even those men who serve the revolution start to react violently and brutally against us. Since we conceive of ourselves as an oppressed sex, we naturally resist those who oppress us.[25]

"The time to scream is here," declared a poem published in *Donna é Bello* [*Woman Is Beautiful*], an Italian feminist journal in 1972, "at the destroyed man in his garbage."[26]

The confidence and solidarity generated by the new groups and institutions led European feminists to create political actions and symbols designed to shock people out of old traditions, to create a new political awareness of women's power. All over Europe, the recognized symbol of women's liberation was a vulva formed by aligning the raised fingers and lowered thumbs of both hands and turning the palms outward. In 1970, the Dutch feminist group *Dolle Mine* infiltrated a gynecologists' convention and raised their blouses to show the slogan "BOSS OF OUR OWN BELLY" written on their stomachs.[27] That same year a group of French feminists, including the writers Christine Rochfort and Monique Wittig, placed a wreath on the Tomb of the Unknown Soldier at the Arc de Triomphe in Paris. The wreath bore two ribbons: one dedicating it "to the unknown wife of the unknown soldier"; the other declaring, "One man out of every two is a woman."[28] By the early 1970s, such demonstrations and gestures attracted many to the cause of feminism. Women's liberation had become an important political movement, capable of mobilizing tens of thousands of European women and men to support its demands.

Asserting Women's Right to Control Their Own Fertility

The women's liberation movement successfully won Western European women a measure of control over their own bodies, especially their fertility and sexuality. Churches and governments had sought to regulate both throughout European history. Women's liberation fought successfully for divorce rights, for equality in marriage and parenting, for the end to legal disabilities for single mothers and their children. The movement concentrated its energies particularly on gaining women access to contraception and abortion and ending the laws which made both illegal. In addition, the movement initiated new attitudes about rape and made sexual violence against women an issue of national concern. It questioned basic sexual traditions of European culture: the "double standard" of sexual behavior between women and men, the outcaste status of prostitutes, even the assumption that humans were "naturally" heterosexual.

By insisting on adding the sexual dimension to politics, the women's liberation movement succeeded in an area where earlier reformers had at best achieved only partial victories. The history of public championing (as distinguished from private use) of contraception and abortion has numerous heroines and heroes, but no real success in changing attitudes or laws until the 1970s. Those who pioneered contraception and abortion faced almost universal opposition: from the Christian churches, many of which rejected all controls over fertility except sexual abstinence; from the medical establishment, which strenuously resisted making contraception and abortion easily available to women, especially poor women; and from public opinion, which often equated contraceptive information with obscenity and its use with promiscuity.[29]

Concern about control over fertility arose in the same circles which advocated equal political and legal rights for women. John Stuart Mill's first political act was to distribute contraceptive pamphlets in poor districts of London in the 1820s. The earliest public advocates of fertility control in both Europe and the United States were politically radical men who urged contraception as a way to make women's lives easier. From the 1820s in England, these men published contraceptive information in cheap, popular books which described the use of sponges, sheaths, and withdrawal to prevent pregnancy.[30] Risking prosecution for obscenity, they justified contraception as essential to liberate women from incessant childbearing. They advocated contraception as preferable to abstinence, which they considered inhuman, and abortion, which they condemned as dangerous. One of the first women to advocate contraception publicly was Annie Besant (1847–1933), an Englishwoman who deliberately courted arrest to make the issue public at her trial.

Unhappily married to a clergyman at twenty, Besant separated from her husband six years later and became a member of the radical Secular Society in 1875. She and Charles Bradlaugh, the head of the society and an avowed atheist, decided to make contraceptive information available to the poor by publishing a cheap edition of an earlier book on the subject. Besant and Bradlaugh were arrested, as they had expected, and they used their trial for obscene libel as a platform for their ideas. "I speak as counsel for hundreds of the poor," stated Besant,

> the poor woman who has only 6 pence to spare, and should be allowed to purchase with that 6 pence the knowledge which richer women can obtain for 2 shillings 6 pence, 5 shillings or 6 shillings at any of the railway bookstalls.[31]

Their verdict of guilty was later overturned on a technicality, allowing Besant to continue speaking and writing about contraception, although the notoriety caused by the trial lost her custody of her daughter. She helped to found the British Malthusian League to promulgate contraception information in 1877 and the following year published *The Law of Population*. The trial made contraception a public issue in England. Besant herself moved on to champion socialism and organize unions for working-class women, leaving the cause of contraception behind.[32]

Until after the First World War most European socialists remained at least publicly hostile to contraception, in large part because of its association with the population theory of Malthus, which they loathed. Thomas Malthus, the eighteenth-century English clergyman, had advocated sexual abstinence and a reduced birth rate as the only solution to working-class poverty. Socialists, who argued that poverty was not caused by the birth of "too many" workers, but by unfair distribution of society's wealth, viewed Malthusians as their enemies. This antagonism was intensified by the Malthusians' advocacy of eugenics: the improvement of the human race through selective breeding. Eugenicists, most active in northern Europe, deplored what they saw as the excessive fertility of "lower" types and urged restriction of births among the least-educated classes. By 1900, the British Malthusian League adopted "Non Quantitas Sed Qualitas" (Not Quantity But Quality) as its motto and displayed it prominently on its flag and stationery.[33] Despite its eugenicism, however, the Malthusian League remained the best source in the world for contraceptive information. When Emma Goldman (1869–1940), the Russian-born anarchist, finished her training as a midwife in Vienna in 1900, she went to a secret meeting of the league in Paris to educate herself about contraception.

Prior to the 1970s, the major sources of contraceptive information and devices for poor women were clinics opened by radical doctors and sex reformers in large northern European cities. In the Netherlands, for instance, Aletta Jacobs (1854–1929) pioneered making contraception available to poor women. Always an activist, and the first female doctor in the Netherlands, Jacobs's insistence on attending college had opened Dutch universities to women in 1875. She set up practice in Amsterdam in 1879, specializing in the diseases of women and children. She began dispensing advice and treatment to poor women through the Trades Unions Council. "From these women I learnt that they suffered a great deal from too many and too frequent pregnancies," she recollected;

> careful inquiry convinced me that if I could advise nothing but sexual abstinence as a means of avoiding pregnancy, my advice would be useless, for in the conditions under which they lived, abstinence was impracticable.[34]

In 1882, Jacobs opened the world's first birth-control clinic, where she fitted women with diaphragms. Faced with almost universal hostility from the medical community, she took careful notes on each case so she could "refute many false charges of harm resulting from the use of this method."[35] A tireless advocate, Jacobs fought for many feminist issues: the vote, the abolition of state-regulated prostitution, equal pay for equal work, and pacifism. By the 1920s, she was internationally respected as an authority on contraception. In 1927 she wrote that "actual experience has shown that forty-five years of Birth Control work has brought about the good results that I expected when I began my work."[36]

In the decades before World War I, a few European feminists worked for two causes they considered as important to women's well-being as contraception: the social acceptance of single mothers and the abolition of state-regulated prostitution. In Germany, for instance, Helene Stöcker (1869–1941) founded the League for the Protection of Mothers and Sexual Reform (known informally as Mutterschutz) in 1905. Drawing its membership from radical feminists, doctors, professors, and sex researchers, Mutterschutz argued, as Stöcker declared in 1903, that

> the women's movement has ignored all problems of love, marriage, and motherhood for decades. But today it is recognized that emancipation from economic subjection also involves emancipation from sexual subjection.[37]

Mutterschutz attacked the double standard of sexual morality, which permitted men free choice of sexual partners outside marriage, but condemned both single mothers and prostitutes. In the group's journal, *The New Generation*, and in public speeches, Stöcker and others advocated maternity insurance, homes and equal treatment for single mothers, reform of the marriage and divorce laws, sex education, the end of state-regulated prostitution, and the rights to contraception and abortion. Mutterschutz failed to win the support of the far larger League of German Women's Associations (the *Bund Deutscher Frauenverein*, or *BDF*) or the German Social Democratic party *(SPD)*. The *BDF* and the *SPD* refused to fight for legalized abortion and contraception—supported by the German Communist party—and members worked for these causes only on an individual basis. By the 1920s, those feminists and sex reformers who advocated contraception and abortion comprised a small, active, international community, but were not supported either by the public at large or any powerful social or political institution.

In the years between the two world wars, a few northern European nations allowed contraception to be dispensed in limited ways. Jacobs's birth-control clinics were never legal in the Netherlands, but the government rarely attempted to close them. The same was true of Germany until the Nazis came to power in 1933. In Denmark, women such as the novelist Thit Jensen and the socialist Marie Nielsen spoke in favor of contraception and sex education in the 1920s. Clinics opened for working-class women, but sex education remained untaught in many schools until after the Second World War.[38]

In England, Marie Stopes (1880–1958) opened a birth-control clinic in London in 1921; in 1923, however, the government prevented feminists from distributing a contraceptive pamphlet to poor women in East London on the grounds of obscenity. The next year, the first Labour government came into power and allowed the pamphlet to be printed. In 1924, the Labour Women's Conference voted 1,000 to 8 to recommend that state health authorities be allowed to give contraceptive information to those who asked for it. The Labour party defeated the resolution in 1925, declaring that "the subject of Birth Control is in its nature not one which should be made a political party issue."[39] Labour women formed the Workers' Birth Control Group (named to distinguish themselves from the elitist Malthusians) and pressed for contraception, using the slogan "It is four times as dangerous to bear a child as to work in a mine, and mining is man's most dangerous trade."[40] In 1930, the second Labour government ruled that contraceptive advice could be given if requested; the following year the Church of England allowed contraception in certain marital situations.

Gains were limited. Contraception was nominally available in a few northern European cities, but women often had difficulty obtaining it. In the 1950s, Catherine Barnes, an English nurse, tried to buy a diaphragm in the small city of Carlisle; no druggist had her size and, "very embarrassed," she ended up buying one a size too large, which did not fit. "That's the kind of thing women can be up against," she remembered. "Maybe you ought to be better prepared, maybe you ought to be better organized, but I think there is a heck of a lot of very poor service for a women in that way."[41] With contraception not easily available and often of poor quality, abortion became an important right for women: the English Women's Co-operative Guild, with sixty-seven thousand members, advocated legalized abortion in 1934. "Abortion must be the key to a new world for women," declared the English feminist socialist Stella Browne in 1935,

> not a bulwark for things as they are, economically nor biologically. Abortion should not be either a perquisite of a legal wife only, or merely a last remedy against illegitimacy. It should be available for any woman, without insolent inquisitions, nor ruinous financial charges, nor tangles of red tape. For our bodies are our own.[42]

These hopes would not be fulfilled until the women's liberation movement. The new feminism enabled women to question and challenge traditional policies. In the 1970s, hundreds of thousands of European women marched and protested, organized and lobbied to repeal the laws which prevented them from controlling their own fertility, whether by contraception or abortion.

By the late 1960s, contraception had become legal in some nations and surreptitiously available in others. Abortion remained illegal in Western Europe. The state socialist nations of Eastern Europe legalized abortion in the 1950s, but regulated its availability according to state population policy, not women's demands for it.[43] With no mass movement behind them, Eastern European feminists felt isolated and powerless, and women did not organize around women's issues. But in the West, the women's liberation movement focused on issues concerning women's control over their own fertility and made them the basis for its first mass campaigns.

In France a change in government policy on contraception emboldened feminists to oppose the laws forbidding abortion. In 1968, the French government legalized the sale of contraceptive devices and the dissemination of contraceptive information; in 1971, 343 Frenchwomen signed the manifesto admitting they had had abortions; in 1973, 345 French doctors signed a similar manifesto declaring they had performed abortions. (Under French laws of 1920 and 1939, both the abortionist and the woman aborted were liable to severe criminal penalties. The last execution for performing an abortion in France occurred in 1943, under the fascist Vichy government.)[44] In 1972, a trial in Bobigny provided a case feminists used to dramatize their cause. Michele Chevalier, then sixteen, claimed to have become pregnant by a schoolmate who raped her. She obtained an illegal abortion and was then denounced by the schoolmate. She and her abortionist were prosecuted. Gisèle Halimi (b. 1927), a radical attorney who had founded the association *Choisir* (To Choose) to aid the women who signed the Manifesto of 343 in 1971, made the Bobigny trial a *cause célèbre*. *Choisir* organized demonstrations and publicized the case. Simultaneously, another women's liberation group, the Movement for the Freedom of Abortion and Contraception *(MLAC)* defied the law and opened a number of abortion clinics in France. Michele Chevalier was acquitted, and *Choisir*'s membership rose from three hundred to two thousand in six months in 1973. The Bobigny trial changed public opinion and contributed to the French government's legalization of abortion in 1975.[45] That same year also saw the passage of laws equalizing rights of parents, of spouses, and of divorced people.

Passed by a narrow margin and with the support of left-wing parties, the new French abortion law allowed abortions up to the tenth week of pregnancy with a doctor's approval. The law expired in 1979, and that year feminists from many groups joined to work for its renewal. Over fifty thousand people demonstrated in a Paris march. While the law's limitations were criticized, its passage and renewal constituted a major political victory for the women's liberation movement.[46]

The pattern of political mobilization established in France proved equally successful in a number of other European nations. Achievements were most dramatic in Catholic societies, which had long forbidden abortion, contraception, and divorce. In Italy, the initial radicalizing issue was not control of fertility but divorce. In 1970, a new law allowed divorce under very controlled conditions. Between 1971 and 1974, the Roman Catholic church actively campaigned for the law's repeal, an effort which backfired in 1974 when 59.3

percent of the electorate reaffirmed the law. Women voted in favor of the law in greater numbers than men, and their outrage about the Church's role in the divorce issue led to action on related causes.

Like their French counterparts, Italian feminists used government prosecutions to publicize and dramatize their views. In 1974, the government charged 263 women in Trente with having had abortions; in response, over 2,500 Italian women signed a petition declaring they too had had illegal abortions. The feminist activist Gigliola Pierobon transformed her trial into a political case like Bobigny, as did Adele Faccio, arrested with forty other women for opening an abortion clinic in Florence. In response to these arrests, the women's liberation movement organized a series of large demonstrations in the major Italian cities. Ten thousand women marched in Florence in 1974; 20,000 in Rome in 1975; 50,000 in Rome in 1976. Over 500,000 Italian women and men signed petitions in the summer of 1975 to place the issue of abortion on the ballot as a referendum.

Arguing that the law against abortion was fascist, Italian feminists mobilized both women's liberation groups and left-wing political parties to work for its repeal. In the marches, women carried signs bearing the slogans of the international women's liberation movement: "My Body Belongs To Me!" "It's Women Who Decide," "My Womb Is My Own." Deaths from illegal abortions were emphasized: one march was led by women with their heads covered in black, each carrying a placard with the name and date of a woman who had died from an illegal abortion. After a law legalizing abortion passed the Italian assembly in 1977, the women's liberation movement demonstrated again with large marches in Rome to ensure its passage in the senate.[47] Given the opposition of the Roman Catholic church, which used its money, influence, and prestige to fight the women's campaigns, abortion would never have been legalized in Italy without the massive political action of these tens of thousands of Italian women. Similar campaigns succeeded in legalizing contraception and abortion in Spain (1978 and 1985) and other Western European nations. Recognition of abortion and contraception as women's rights came from feminists fighting to reject the European tradition that the male-dominated institutions of Church and state can control women's fertility. The women's liberation movement has gained Western European women control over their own fertility.

Asserting Women's Right to Determine Their Own Sexuality

The campaigns for women's contraceptive and abortion rights also led feminists to reject traditions which had sought to control women's sexuality as well as fertility. The refuge and support given by consciousness-raising groups and friendships, by women's groups and conferences, allowed women to question patterns of sexual behavior. They shared details of sexual experiences and explored and questioned their intimate behavior and attitudes. "At a women's conference we take a poll and then it appears that *three-quarters* of the women have at some time pretended to have orgasms," remembered the Dutch feminist Anja Meulenbelt of the early 1970s. "We think his [the man's] pleasure more important than ours. We have been crazy. We have been idiots."[48] Feminists discussed previous taboo subjects among themselves: masturbation, incest, homosexuality, rape, menstruation. Some learned to give themselves gynecological examinations; others questioned the validity of psychological theories about women's sexuality. They made three topics into issues of public policy and legal reform: the rights of prostitutes, sexual violence against women, and the rights of homosexuals, especially lesbians.

PROSTITUTES' RIGHTS

Nineteenth-century feminists usually saw the prostitute as victim. To equal rights feminists, she personified women's sexual oppression by the double standard of sexual morality which catered to male lust; to feminist socialists, she embodied capitalism's exploitation of working-class women. The connection between poverty and prostitution was glaring in the large new cities of the nineteenth century, where tens of thousands of women moved in and out of the profession. In an era in which prostitutes found themselves simultaneously despised and glamorized, feminists attempted to rescue them: to provide different opportunities to earn income, to build shelters for women, and to end state control of prostitutes through police regulation and mandatory medical inspections. The best-known feminist campaign for the abolition of state regulation of prostitution was that led by Josephine Butler in England. There, a coalition of middle-class women and working-class men achieved the repeal of state regulation in 1884. Butler founded an International Federation for Abolition (of state regulation) in 1875, framing the issue in moral terms:

> Injustice is immoral, oppression is immoral, the sacrifice of the weaker to the stronger is immoral, and all these immoralities are embodied in all the systems of legalized prostitution, in whatever part of the world or under whatever title they exist.[49]

But abolition campaigns succeeded only in England and Norway. German feminists fought for years to abolish state regulation in the port city of Hamburg without success. Many moderate feminists refused to join, believing that any connection with prostitutes would discredit their fight for the vote. Some nations ended state regulation in the 1920s, relying on better medical treatment to control venereal disease, and feminist campaigns on behalf of prostitutes ended.

In the 1970s, prostitutes themselves made it clear to women's liberation groups that they did not see themselves as victims nor the services they provided as shameful. They demanded an end to police and government harassment, not an end to their means of livelihood. "To my mind the ideal society would be one where you wouldn't need money to have what you wanted," declared the Lyons prostitute who identified herself as D.,

> so there wouldn't be a need for prostitutes either. Society will have really changed when no one uses "whore" as an insult any more, when the word "prostitute" itself no longer exists. Having said that, there's no need to wait for the ideal society. If the cops stopped chasing us, that would already make a big difference.[50]

French prostitutes in Lyons struck in 1975, declaring:

NO TO PROCURING
NO TO BROTHELS
NO TO POLICE REPRESSION
YES TO ALL THE ADVANTAGES AND RIGHTS TO BEING A WOMAN![51]

Some English prostitutes organized as the English Collective of Prostitutes and lobbied Parliament to pass the Protection of Prostitutes Bill in 1978, which decriminalized prostitution in Britain.[52] Some German prostitutes brought suit against their pimps in 1980, and the feminist press gave the case wide publicity.[53] Prostitutes and the women's liberation movement have tried to treat each other as equal allies in the struggle for women's rights, overcoming the nineteenth-century postures of rescuer and victim.

SEXUAL VIOLENCE AGAINST WOMEN

While nineteenth-century European feminists saw women's sexual oppression personified in the poor prostitute, twentieth-century feminists have focused on the female victim of rape, of incest, of molestation—all considered acts of male sexual violence against women. Feminists believed that all women, whether girls, "respectable" women, or prostitutes, sensed a male capacity for sexual violence, even if only some had experienced it firsthand. "Men's glances assault me, claw their way into the creases of my jeans between my legs as I descend the stairs to the subway," wrote German novelist and feminist Verena Stefan in 1975. "Whistles and clacking tongues cling to me. In the evenings all the bruisings of the day under the shower under the skin."[54] Becoming aware of the extent of male coercion, the women's liberation movement organized against sexual violence directed at women.

On March 8, 1976, International Women's Day, the International Tribunal of Crimes Against Women opened in Brussels. Its organizing slogan was "SISTERHOOD IS POWERFUL! INTERNATIONAL SISTERHOOD IS *MORE* POWERFUL!" Conceived in opposition to the United Nations Conference on the Decade for Women in Mexico City, 1975, the International Tribunal was modeled on both the Nuremberg trials and the international tribunals which had indicted the United States' involvement in the Vietnam War. "In contrast to Mexico where women, directed by their political parties, by their nations, were only seeking to integrate Woman into a male society," stated Simone de Beauvoir in her keynote speech, "you are gathered here to denounce the oppression to which women are subjected in this society." She concluded,

> Strengthened by your solidarity, you will develop defensive tactics, the first being precisely the one you will be using during these five days: talk to one another, talk to the world, bring to light the shameful truths that half of humanity is trying to cover up. The Tribunal is in itself a feat. It heralds more to come. I salute this Tribunal as being the start of a radical decolonization of women.[55]

Attended by over two thousand women from forty different nations, the tribunal discussed topics ranging from compulsory motherhood to compulsory heterosexuality, from clitoridectomies and child abuse to incest and rape. The tactic of the "speak-out," used at the tribunal, was especially effective with regard to rape, a topic traditionally mystified and masked by repressive and denigrating attitudes. Questioning the traditional views that women enjoy rape, that they bring it on themselves by their provocative behavior, that they deserve to be punished if they are raped, some European feminists analyzed rape as an individual act of male violence which enabled men to keep all women in a state of fear and subordination. "Rape emerges clearly as a terrorist tactic used by some men, but serving to perpetuate the power of all men over women," concluded the conference organizers, arguing that the fear of rape led many women to seek a male "protector" from the violence of other men.[56] Political action against such violence took four forms: mass demonstrations, further speak-outs, the creation of support institutions for victims of male violence, and legal efforts to change rape laws.

The tribunal sparked feminist action throughout Western Europe. In the autumn of 1976, an estimated 100,000 Italian women marched after dark in the major cities to "take back the night" and make the streets safe for women.[57] West German women held a similar march the following spring, and such demonstrations continue. Also in 1976, numerous small groups organized to provide support services to rape victims. Describing the formation of the British group Women Against Rape in 1976, the feminist writer Ruth Hall recalled how seven women, a number of them rape victims, began to meet to share their experiences. Taking public action in 1977, the group petitioned Parliament, demonstrated in courts, and held a public trial and speak-out in Trafalgar Square against the police and government. Such actions led to the formation of other rape counseling groups and major efforts to change rape laws, which forced a woman to prove her innocence rather than focusing on the guilt of the man.[58] In France, feminist publicity and agitation about a particularly brutal rape case in Aix-en-Provence in 1978 led in 1980 to a new law which defines rape as "any sexual act against a person's will."

Writing in 1980, the Council of Europe's Action Committee for Equality Between Women and Men argued that *"physical violence,* both *sexual* (rape, incest, indecent assault) and *domestic,"* should be the subject of legal action by the twenty Western European member states. "Despite its frequency and the gravity of its physical, psychological, and social consequences, which make it a genuine social scourge," wrote the committee,

> society at large has only recently become aware of this form of violence, the majority of whose victims are women. The laws pertaining to it are inappropriate in both substance and application, because the cultural attitudes they reflect have become obsolete.[59]

Women's liberation's writings, demonstrations, and political action have changed some European attitudes and traditions surrounding rape and other kinds of sexual violence.

LESBIANISM AND LESBIAN-FEMINISM

By questioning accepted traditions which controlled women's sexuality, the women's liberation movement began to question heterosexuality itself. If women had feigned sexual pleasure with men, where did their true sexual pleasure lie? Women studied their own sexual responses. In a widely reprinted pamphlet of 1969, the English feminist Anna Koedt wrote about *The Myth of the Vaginal Orgasm,* questioning the Freudian axiom that a "mature" woman moved beyond "clitoral" orgasms to "vaginal" ones. French feminists discussed the significance of women—and women only—possessing an organ "which is only for pleasure," the clitoris. They argued that much female sexual response and behavior had been warped by male dominance. In the open and trusting atmosphere created by feminist groups, women began to explore their sexual feelings for other women. Some openly acknowledged their attraction to and love of other women. Without feminism, wrote the French feminist author Monique Wittig in 1979, lesbian culture and society "would still be as secret as they have always been."[60] Women's liberation attempted to legitimize women's homosexuality and by the late 1970s had made the acceptance of lesbians and homosexuals by society in general a goal of the movement.

Lesbians had long been active in women's movements, but prior to the 1970s, lesbianism was either ignored or denied even by feminists. In the 1860s, for instance, the English women's rights movement was urged by John Stuart Mill to have only its more attractive, married women speak and to put the "mannish" Emily Faithful, the printer who ran the Victoria Press, in the back row at meetings.[61] "Considering the contributions made to the women's movement by homosexual women for decades," wrote the German lesbian feminist Anna Rueling in 1904,

> it is amazing that the large and influential organizations of the movement have never lifted a finger to improve the civil rights and social standings of their numerous Uranian [homosexual] members. . . . Without the active support of the Uranian women, the women's movement would not be where it is today—this is an indisputable fact.[62]

Christabel Pankhurst was a lesbian, and Emmeline Pankhurst, if not lesbian herself, inspired romantic attachments in lesbians like the English composer Ethel Smyth (1858–1944). Already famous as an outstanding musician by the 1890s, Smyth joined the suffragist WSPU. She composed the group's anthem, "The March of the Women," which she conducted with a toothbrush from her prison cell in Holloway when she was jailed for suffragette militancy. Dressing mannishly, she usually wore a tie in the WSPU colors of white, purple, and green. Late in life Smyth fell passionately in love with Virginia Woolf. In the German women's movement, Anita Augsberg and Lida Gustava Heymann had lived in a lesbian marriage for decades. Some women believed there to be an intrinsic connection between feminism and lesbianism. "I had realized, before Women's Liberation came into being, that feminism was inseparable from lesbianism," wrote Charlotte Wolff, a German physician. "The lesbian woman is a feminist by nature because she is free from emotional dependence on the male."[63]

The women's liberation movement explored this connection. If lesbians were "naturally" feminists, were feminists "naturally" lesbians? If the male dominance identified by the women's movement colored relations with men, what about relations with women? These questions were vividly portrayed in the 1975 novel *Shedding,* by the German Verena Stefan. Published by a small, feminist publishing house, *Shedding* had sold hundreds of thousands of copies in five European languages by 1980.[64] *Shedding* details the evolution of Veruschka, the narrator, from a victimized, helpless woman into a strong lesbian feminist. Becoming a feminist, she realizes that "sexism runs deeper than racism than class struggle," and turns from men to women:

> At first, I would leave Samuel to attend a woman's meeting only to come back again to the life with him. Gradually the emphasis shifted. I come back less and less.[65]

Veruschka comes to believe that relations with men are restricting. "When I am together with another woman I learn something about myself. With a man I learn only that I am different and that my body is supposed to be there for him. . . . it seems more and more unnatural, I really do mean *unnatural,* to have had access only to people of one sex."[66]

Drawn to women emotionally, and logically convinced that only with women can she truly be herself, Veruschka finds a more joyful sexuality with Fenna, a woman she has grown to love through their women's group. "I am beginning to see myself for what I really am," Veruschka sings,

> I assemble the separate parts to make one whole body.
> I have breasts and a pelvis.
> My legs run together to form curves, folds, lips.
> I glide and fall with Fenna.[67]

Veruschka transforms herself into "Cloe"—a new, stronger person. The novel ends with Cloe striding down the streets of Berlin, unafraid, "Cloe moves her hips. I am my own woman. People turn and stare."[68]

Shedding's popularity in feminist circles came from the heroine's radical "solution" of lesbianism. By the mid-1970s, lesbians were not only asserting their own sexual identity, but some also argued that only lesbians were true feminists. "Lesbian is the only concept that I know of which is beyond the categories of sex (woman and man), because lesbian societies are not based on women's oppression," declared Monique Wittig, the French lesbian feminist author, to a New York conference on de Beauvoir's *The Second Sex* in 1979;

> furthermore, what we aim at is not the disappearance of lesbianism, which provides the only social form that we can live in, but the destruction of heterosexuality—the political system based on women's oppression.[69]

Identifying heterosexuality itself as the enemy caused both individuals and movements to reevaluate their attitudes and views about sexuality. The most radical position, lesbian separatism, demanded that women relate sexually only to other women. "We don't see lesbianism as a purely personal issue," wrote Angela Stewart-Park and Jules Cassidy in 1977 in their introduction to an English anthology of lesbian writings.

> It's a political issue for us and for all women. We've made a choice. We've chosen to relate to women sexually and emotionally. We've chosen not to relate to men in these ways. We feel that the choice has made us strong.[70]

Equating lesbian separatism with women's liberation raised many questions for feminists. Were women who had sexual relations with men "consorting with the enemy"? Could women "choose" to become lesbians? Were all men the enemy? What about sons of lesbian feminists? What made a woman a feminist?

In the women's liberation movement, lesbian and heterosexual feminists found common ground. Feminism has been the only political movement to have championed homosexual rights, placing lesbians in the front ranks, taking pride in the contributions of lesbians. Moderation has replaced the separatist militancy of the mid-1970s. The tremendous asset lesbian-feminism brought women's liberation was a new perspective from which to view society and the world: a view in which woman was the center, not the "other," woman was the measure, not the measured, the standard, not the variant. The phrase "woman-centered" first designated lesbians, and then, by extension, the perspective from a feminist viewpoint. This viewpoint has proved crucial in analyzing and exploring all of European culture and society from a feminist perspective.

Toward a Woman-Centered World

By the late 1970s, the women's liberation movement turned increasingly to evaluating and exploring the consequences for women and men of a world defined exclusively from a male perspective. By taking little for granted, by seeing the world from a female viewpoint, by using women as the measure, feminists began to criticize systematically what they saw as the all-pervasive effects of male dominance. In her forward to a 1979 anthology entitled *Overcoming Speechlessness: Texts from the Women's Movement,* the German feminist Gabriele Dietze described the task ahead:

> No revolution in one sector alone—and certainly not an economic one only—can break through the complex system of the psychological, social and economic

conditioning of suppression. Everything has to be changed: the way of thinking, the way of governing, the whole way of life, the family, how to go about one's work—the whole technical/economic complex—yes, how we laugh, love and cry, and even how we dream—all this has to be changed.[71]

Feminists explored how they could bring about such vast changes. In the last decade, the women's liberation movement has focused on two goals: changing patterns of thought and cultural attitudes through the new field of "women's studies" and changing the natural and political environment by working with ecological and anti-nuclear movements. Their aim has been to transform not only the lives of women, but the lives of all.

WOMEN'S STUDIES

In literature and academia, feminists have created new perspectives to try to overcome the feeling of "speechlessness" so many women felt. "When I wanted to write about sensitivity, experiences, eroticism among women, I could not find the words," stated Verena Stefan in her introduction to *Shedding*. Writers created new modes of speech, new uses of language, new dictions and forms to articulate women's lives. In the field of literature, women's studies has produced writing and literary criticism from a feminist viewpoint. In her most famous novel, *Les Guérillères* (*The Women Guerrillas*, 1969), the French lesbian feminist Monique Wittig painted a feminist utopia, a future world where women dominate and create. Wittig attempted to inspire and empower other women through this creation. "There was a time when you were not a slave, remember that," runs one passage.

> You walked alone, full of laughter, you bathed bare-bellied. . . . You say there are no words to describe this time, you say it does not exist. But remember. Make an effort to remember. Or, failing that, invent.[72]

In Europe, where so many languages are gender-based, criticism has focused on the oppression of language itself, which places the male first. The masculine pronoun always precedes the feminine; feminine endings are classified as "weak" and are added on to a male root. The male article subsumes the female: children are still taught that if a group consists of one hundred women and one man, the masculine pronoun should be used. Some feminist writers, especially in France, are attempting to create a new woman-centered language. And feminists' efforts have brought some shifts. A new English edition of *Roget's Thesaurus* appeared in 1982, reflecting what its editor, Susan Lloyd, called "the enormous changes" of the last twenty years. She saw the most significant one as "making much more explicit the existence of women. Before, they were just assumed."[73] All these efforts have given women's writing new recognition and vitality.

As in literature, so in the arts generally. Women artists, women musicians, women sculptors and actors have consciously sought to create feminist productions and criticism. In 1977, for instance, the West German photographer Marianne Wex mounted an exhibit called " 'Let's Take Back Our Space': 'Female' and 'Male' Body Language as a Result of Patriarchal Structures." Using thousands of photographs of ordinary Germans, Wex demonstrated how men used space expansively, while women restricted and minimized themselves: sitting with their legs together, crossing their arms, bowing their heads. Wex also included a section analyzing classical European sculpture from a feminist perspective. Like much of women's studies, this exhibit introduced new standards by which to evaluate male dominance in European culture.

In the humanities, social sciences, and education, women's studies has had a profound impact. All the social sciences—anthropology, psychology, sociology, political science, and economics—have been forced to incorporate feminist scholarship. Theoretical models and assumptions previously unquestioned (like interviewing only the men of a group to be studied) have given way to theories and writings which attempt at least to incorporate women. More significantly, the humanities and social sciences have also been forced to rethink old modes of explanation and formulate new ones to take account of new feminist scholarship. For instance, the field of "women's history" did not exist before the 1970s. Writing in 1973, the German feminist Marielouise Janssen-Jurreit complained that in thirty-one history and social science textbooks used in German high schools, "there is only one sentence on the introduction of women's suffrage. . . . only one book mentions that there were female representatives in the Weimar Republic."[74] In the mid-1970s, the French women's liberation movement adopted the following hymn, which began,

> We who are without a past
> Without history, outcast
> Women lost in the dark of time
> Women whose continent is night.
> > Together slaves arise
> > To break our chains asunder
> > Arise![75]

Scholars have labored to provide women with histories of their own. In England, in Germany, in France, in Scandinavia, in Spain, in Italy, the study of women's past has become a formally acknowledged field, and its findings are changing the ways in which history is taught, organized, and understood. From individual disciplines, feminists have moved on to criticize much of society and culture. Once women challenged basic intellectual systems of Western culture, wrote the French feminist Hélène Cixous,

> then all the stories would have to be told differently, the future would be incalculable, the historical forces would, will, change the functioning of all of society. Well, we are living through this very period when the conceptual foundation of a millennial culture is in the process of being undermined by millions of a species of mole as yet not recognized.[76]

Historians have begun to realize that a history genuinely including women does not mean simply adding paragraphs on women's lives to chapters on traditional topics. The effort involves rethinking history itself and seeing how the entire narrative changes when women are considered an integral portion of humanity. Similarly, women's studies generally seeks to change not only the world of women, but to change the world itself.

RELATED CAUSES: ECOFEMINISM AND ANTI-NUCLEAR PROTESTS

In the 1970s and 1980s, European feminists have moved into new political activities which seek to change the world's future. By the mid-1970s, the women's liberation movement had created the theory of "ecofeminism." Ecofeminism connects male dominance and the destruction of the environment, and hopes that women can intervene to prevent destruction of the ecosystem, whether caused by pollution or nuclear war. "The feminist movement is not international, it is planetary," declared the Italian feminist Carla Lonzi.[77] In her 1974 work *Feminism or Death,* the French feminist Françoise d'Eaubonne connected male dominance to the future destruction of the world:

> Patriarchal man is therefore above all responsible for the demographic madness, just as he is for the destruction of the environment and for the accelerated pollution which accompanies this madness, bequeathing an uninhabitable planet to posterity.[78]

In Iceland, feminists formed "Women's Lists," slates of women candidates for political offices, reviving a tradition begun in the early twentieth century. In 1982, a Reykjavik Women's List Manifesto made clear the connections between feminism and the salvation of the world:

> Development of society is governed by the values of men—women's experience has not put any significant marks on this development. Now we face the near

destruction of nature and all living things. . . . In this world there often seems to be no room for human interaction, human feelings, creativeness and interaction with nature. Humankind is threatened by an arms race that will lead to total destruction. This development must be stopped—women must unite in order to stop it.[79]

As a result of such beliefs, feminists worked in ecological and anti-nuclear movements, seeking to reverse the disastrous consequences of men's destructive acts.

In the 1970s and 1980s, feminists have taken major roles in militant worldwide ecological groups like Greenpeace and in the Green political parties. The word *green* now carries the political meaning of being in favor of "ecopacifism" and is especially associated with women. *Donna è verde* (Woman is green) asserted Italian feminists at a Roman demonstration in 1983.[80] "I have hope for the world although it is 10 minutes before Doomsday," stated Petra Kelly in the early 1980s, when she was leader of the West German Greens, "because women all over the world are rising up, infusing the anti-nuclear, peace and alternative movements with a vitality and creativity never seen before."[81]

Women have been especially active in anti-nuclear protests. "WOMEN DECLARE WAR ON NUCLEAR POWER AND THE POLLUTION OF THE WORLD," read the banner of a group occupying a nuclear power plant in West Germany in 1975.[82] By the late 1970s, women all over northern Europe were organizing in specifically female anti-nuclear groups. They acknowledged their connection to female pacifists of previous generations, but based their convictions on different premises. "We don't think that women have a special role in the peace movement because we are 'naturally' more peaceful, more protective, or more vulnerable than men," wrote the Nottingham (England) women's group Women Oppose the Nuclear Threat (WONT) in the early 1980s, "nor do we look to women as the 'Earth Mother' who will save the planet from male aggression." Instead, they took a new view, valuing qualities traditionally associated with women equally with those associated with men.

> Rather, we believe that it is this very role division that makes the horrors of war possible. The so-called masculine, manly qualities of toughness, dominance, not showing emotion or admitting dependence can be seen as the driving force behind war; but they depend on women playing the opposite (but not equal) role, in which the caring qualities are associated with inferiority and powerlessness.[83]

The Nottingham women went on to argue that women's role in the anti-nuclear movement must be "assertive," "forcing men to accept women's ideas and organization, forcing them to do their own caring." European feminists' efforts in the anti-nuclear movements led to the Nordic Women's Peace March in 1982 and the creation of the English women's peace camp at the Greenham Common missile base in 1981.

Aware of their unique position between the Western powers and the Eastern bloc, Scandinavian women have attempted to cross cold war boundaries with letter-writing campaigns, mass petitions, and marches. By 1980, a gender gap of 15 percent existed between women and men voting on a nuclear referendum in Sweden.[84] That summer, at the opening of the United Nations Conference on Women in Copenhagen, eight Scandinavian women presented a petition that read: "We, half a million women from Northern Europe . . . ask you to stop the escalation of nuclear weapons."[85] This action led to the Nordic Peace March of 1982, when two groups of about three hundred women each marched westward from Oslo to Paris and eastward from Stockholm to Minsk. The women carried banners reading "No to Nuclear Weapons in Europe, East and West," and "Yes to Disarmament and Peace."

In 1981, Englishwomen invented a tactic which has since become part of the anti-nuclear strategy in many nations: the formation of a women's peace camp at the Greenham Common missile base. "As women we have been actively encouraged to stay at home and look up to men as our protectors," read an open letter encouraging other women to join, "but we reject this role. We cannot stand by while others are organizing to destroy life on our earth."[86] This appeal drew thirty thousand Englishwomen to a protest at Greenham in December 1982. In a novel political gesture, the women encircled the base, holding hands and forming a human chain to prevent nuclear missiles from entering the base. Peace camps have since been established in Italy, West Germany, the Netherlands, Switzerland, and Scotland, as well as in the United States, Canada, and Australia.[87]

Missiles entered Greenham in 1983; the following year, the camp was torn down, although women still remain there in tents and makeshift shelters. European feminists continue to assert that by changing traditional attitudes and behavior, women can transform contemporary life. "When we work together for our dreams they will come true," declared the Finnish feminist journal *We Women (Me Naiset)* in 1980.

> Women are not too weak to change the world. We have only to learn to trust ourselves. We have to take responsibility for our lives and our utopias. Even small deeds are important.[88]

Through "small deeds" and daring conceptions, through political actions and personal confrontations, feminists have gradually brought about the rejection of traditions which have restricted women's and men's lives for centuries.

From the fifteenth century on, they have changed the world, asserting women's humanity, women's rights, women's value, women's perspective. "Women witness the possibility that something may be accomplished," declared an English Quaker women's group in 1986,

> hand in hand, ordinary women declaring, "We know there is a healthy, sensible loving way to live and we intend to live that way." Nurses, midwives, teachers, lovers, nurturers, healers, prophets. . . . Now, once again, it is time.[89]

At the beginning of the fifteenth century, Christine de Pizan asserted that everything comes "in the right time" and that women's oppression, like "many other things which were tolerated for a long time," will be overturned.[90] The accomplishments of feminists in the past give hope for the power of feminists in the future.

Epilogue

In the 1990s, dramatic changes have transformed the lives of millions of European women. The "collapse of communism" in the former Eastern bloc—the replacement of single-party, state socialist governments and their centrally planned economies by multi-party democracies based on capitalist market economies—has had a disproportionate and often regressive effect on the lives and status of the female population. All state socialist governments, from the Soviet Union to the post-1945 satellite nations of the German Democratic Republic (East Germany), Poland, Czechoslovakia, Yugoslavia, and others, had imposed "emancipation from above": the implementation of laws and quotas designed to make women equal to men. Thirty percent of the seats in parliament were reserved for female representatives; child care and abortion were legal and routinely provided by the state; governments subsidized a wide range of maternity benefits, including job guarantees after a female worker returned from her birthing leave. No matter what their moral or political views of individual governments, women as a group received privileges under state socialism, which made the transition to capitalist democracy more difficult and more complicated for them than for men. "The day of [German] unification was for me a day of sad farewells . . . ," mused 27-year-old Angela Kunze of Leipzig in 1991, "I don't know if I shall find a home in the newly developing country."[1]

The end of state socialism meant the end of state-supported privileges for women. By 1990 it was clear, wrote Hanna Beate Schoepp-Schilling, first head of the West German Department of Women's Affairs, that East German policies of "affirmative action in the labor market, the extensive and inexpensive system of child care that covered children from birth until well into their teens, the unrestricted access to abortion in the first trimester, had little chance to be carried over into the policy framework of unified Germany."[2] In Germany, as in the rest of the former European communist world, such programs were rapidly dismantled. Debate soon focused on abortion. In Germany, a 1993 compromise bill—giving women far fewer rights than in the East but slightly extending their access to abortion in the West—was overturned by a Constitutional Court. Current law, passed in 1995, makes abortion "illegal and punishable in general." Exceptions can be made only during the first trimester, after the woman has received medical counseling designed to discourage the procedure, and only if she can pay the expenses herself.[3] Abortion rights are currently wider in Western Europe (with the exception of Ireland) than in Germany or in Poland (where, in 1997, abortion was made illegal under any circumstances).

The chief quantifiable results to date of the shift to capitalist democracy for women are economic. Women have borne a disproportionate share of the resulting unemployment. As of 1998, throughout Eastern and Central Europe women comprised about sixty percent of the unemployed.[4] Many turned to formerly illegal occupations. The unfettered market has given rise to thriving prostitution and pornography industries. At the height of the disruption, parents provided for their families with extreme measures. The sale of children for adoption became common in Romania and elsewhere. To many Western observers, the social and economic conditions of the transition period resembled those of the nineteenth century, especially with regard to shrinking subsidies, reduced social services, and wage opportunities for women.[5] In addition, women lacked political clout, and thus the ability to influence policies. Without electoral quotas, female parliamentary representation rapidly dropped to one-third of what it had been, to figures lower than those in Western European nations.[6]

The social, economic, and political changes associated with transition in Eastern Europe also affected the West. Decreased political commitment to social subsidies by all major political parties, and economic downturns, have eroded benefits for women there as well, prompting the historian Jane Jensen to characterize feminists' continuing support of current welfare states as "remaining loyal to a weakened friend."[7] In this, as in other key areas, the dramatic Cold War differences that divided Europe into East and West are no longer mirrored in women's lives. In their poverty, the work they do and the pay they receive, in their access to child care, their reproductive strategies, and their political representation, women across the continent have far more in common than in earlier decades of the twentieth century.

At the end of the 1990s, age-old inequalities in work continue to shape European women's lives, much as in earlier epochs. Occupational segregation persists. More than sixty percent of women workers cluster in their traditional low-paid occupations: child care worker, secretary, nurse, domestic servant, elementary school teacher. Increasing numbers of women continue to enter the labor force, chiefly as poorly paid, part-time workers with no benefits. As pay and status rise within a field, the number of women drastically declines: while seventy percent of Europe's elementary school teachers are female, women comprise only twenty-five percent of university professors.[8]

Women continue to retain primary responsibility for meal preparation, housework, family maintenance, and care of children, the sick, and the elderly, in addition to earning income. From Scandinavia to Italy, from Russia to Spain, women perform about seventy-five percent of the child care, laboring many more hours a week, both at home and at work, than men.[9] In addition, paralleling developments in the United States in the 1990s, increases in single-parent families headed by women, and women's disproportionate need to engage in part-time labor, has led to a "feminization of poverty" in Europe generally.[10]

Many European women have refused to accede to these multiple burdens. Throughout Europe, birth rates as of 1998 are the lowest ever recorded, well under the numbers needed to reproduce current populations in every nation except Iceland.[11] This demographic trend cannot be connected to the old East-West division, to Catholic-Protestant differences, or to the North-South axis: the lowest birth rates are in Bulgaria, Germany, Italy, and Spain, followed by Austria, Greece, Portugal, and Russia. Primarily using contraception rather than abortion, European women often limit themselves to a single child, citing problems in satisfying the demands of work and family. "I'm at the office all day and it is difficult to think about having a child," explained a 29-year-old married Italian lawyer, Francesca Casotti, in 1998. When called "selfish" for refusing to bear more children, women say the "tyranny of time" has determined their choices.[12]

Since the mid-1980s, increasing attention has been drawn to the many kinds of violence directed against women. Beginning with marches to "take back the night," to reclaim safe use of the streets, European women have pressed their governments to reform laws on domestic violence and rape, to provide shelters and refuges for victims of such crimes, and to begin antiviolence education in schools and the media.[13] The use of mass ethnic rape as an instrument of war in Serbia, Bosnia, and Croatia in the 1990s compelled Europeans to acknowledge how little progress had been achieved in controlling sexual violence. But in contrast to earlier eras, this time "rape warfare" was rapidly publicized and denounced. European and North American journalists, doctors, rape counselors, and social workers rushed to the afflicted areas. A worldwide coalition of feminist human rights activists lobbied successfully for international condemnation of all "gender-based violence" and for the designation of "rape in the conduct of armed conflict" as a war crime.[14]

Heightened feminist consciousness in Europe, as elsewhere, has brought about these developments in the last quarter of the twentieth century. Increased international contacts, through the European Community and its new parliament, as well as worldwide gatherings like the 1995 United Nations Women's Conference in Beijing, have focused attention on institutional structures that perpetuate Western traditions of male dominance and female subordination. However, problems remain. "Male domination is not exercised today as it was earlier in the century, through propaganda urging women to return to their homes," wrote the French sociologist Rose-Marie Lagrave in 1992. "Today it hides behind egalitarian laws, behind a minority of 'successful' women, behind rational explanations—so many ways of anesthetizing our awareness of the unequal likelihood of success in school and at work."[15]

As today's feminists explore the remaining structural and cultural inequalities of their societies, they build upon the gains of previous decades. In the 1970s, Europeans looked to the United States for leadership in politics and scholarship; at the end of the 1990s it is often the other way around.[16] Infused with the energy of recent accomplishments, younger European feminists turn to new issues of concern, like the social exploitation of women resulting from reproductive technology and genetic testing or the discriminatory effects of supposedly egalitarian legal systems and school curricula, which still privilege males.[17] The ongoing process of European unification has facilitated the development of new continent-wide feminist organizations, like WISE, Women's International Studies Europe. WISE publishes a newsletter and a scholarly journal, maintains an e-mail list, and hosts conferences. Feminist scholarship in Europe, as in other regions of the world, has grown exponentially in recent years, producing wide-ranging changes in how educated women and men view the world.

These intellectual and cultural changes are beginning to transform social and political existence as well. Despite the adverse consequences of the transition era, in their quality of life, their health, their life expectancy, their access to sex education and contraception, European women lead the world. In education at all levels they join the United States, Australia, and New Zealand in providing the most years of schooling for girls and women. New legislation, both in individual nations and the European Union, offers remedies for sexual harassment and discrimination for the first time in Europe's history. While these can sometimes work to women's disadvantage, as when a British Labour party goal of having half its elected representatives be female by 2000 was overturned for violating sex-discrimination laws, they represent real gains for most of Europe's women.[18]

At all levels of government, female participation is greater in Europe than in other regions of the world, including North America. Comprising twenty-five percent of the delegates to the new European Parliament, European women provide twenty percent or more of the representatives in nine northern national legislatures. Seventeen of the twenty-one states worldwide where women hold more than fifteen percent of their governments' ministerial and sub-ministerial positions are European.[19] Prior to 1979, when Margaret Thatcher became British prime minister, no European woman had ever been elected to head her government. Since then, Vigdis Finnsbogadóttir of Iceland, Milka Planinc of the former Yugoslavia, and Mary Robinson of Ireland have served as their nation's presidents; Gro Harlem Brundtland of Norway, Maria de Lourdes Pintasilgo of Portugal, Edith Cresson of France, and Hanna Suchocka of Poland, as well as Thatcher, have been prime ministers.[20]

Women democratically elected to hold the highest political offices overturn the ancient traditions of male political hegemony. The British Secretary for Northern Ireland, Mo (Marjorie) Mowlam (1950–   ), headed popularity polls in Britain in 1998, on the eve of the successful unification referendum, which her skillful diplomacy had helped produce. Prized by both colleagues and the public for her fearless irreverence and ability to cut through English formalities, she brings feminist principles to her public duties. When complaints arose about asking her Royal Ulster Constabulary bodyguards to buy tights, lipstick, and menstrual tampons, she replied, "It's my mission to civilize the Irish male." When a reporter criticized her hairdo at a press conference, she stripped off the wig she wore, revealing that she had lost her hair from radiation treatment.[21]

In contrast to female leaders like Margaret Thatcher who are hostile to women's rights issues, feminists like Mowlam strongly identify with other women and seek to improve women's lives. Some of these prime ministers and presidents have expressed feminist views and implemented feminist programs, further expanding the possibilities for all European women. Maria Lourdes de Pintasilgo, prime minister of Portugal from 1981–1985, wrote in 1995 that she always counters questions about her unique status by asserting that women have done far more in Portugal than they have been given credit for: "If statistics say that women are only 28% of the 'economically active population,' then what is wrong is the statistics—and the definition of being active!"

But Pintasilgo also argues that such questions belong to another era. "[T]he sheer quantity of the new legislation passed in recent years," she explained, "has drastically changed the identity of women, their image, and their status in society."[22] These basic shifts in both material reality and cultural attitudes give hope for the future, as feminism continues to develop and evolve. For the first time in European history, women as a group are beginning to assert values and to wield power on their own behalf, as well as for others. Pintasilgo wrote, "For my part, I believe deeply that women can change society; I feel that what we must say to one another is based on encouraging each of us to be true to herself."

Together, we can make it so.

Notes

PART VI WOMEN OF THE COURTS

PART VI: 1. The World of Absolute Monarchs from the Fifteenth to the Eighteenth Centuries

1. For the description of Catherine the Great's day, see Isabel de Madariaga, *Russia in the Age of Catherine the Great* (New Haven: Yale University Press, 1981), p. 573, and Edward Crankshaw, *Maria Theresa* (New York: Viking Press, 1969), pp. 199–202.
2. See Brigitte Hamann's biography *The Reluctant Empress*, trans. Ruth Hein (New York: Alfred A. Knopf, 1986).

PART VI: 2. The Life of the Courtier

1. An English study shows the change. Among the great ducal families from 1330–1479, 46 percent died a violent death. Between 1479 and 1679, 19 percent died this way, only 4 percent from 1730–1779. T. H. Hollingsworth, "A Demographic Study of the British Ducal Families," in D. V. Glass and D. E. Eversely, eds., *Population in History* (London: Edward Arnold, 1965), p. 359.
2. This devolution of independent power and changes in the role of the nobility have been studied extensively. See, for example: for France, Davis Bitton, *The French Nobility in Crisis 1560–1640* (Stanford, Calif.: Stanford University Press, 1969), pp. 28–33, 48ff., 69–72; for Germany, W. H. Bruford, *Germany in the Eighteenth Century: The Social Background of the Literary Revival* (New York: Cambridge University Press, 1952), pp. 51–53; for the Austrian Empire, H. G. Schenk selection on Austria in *The European Nobility in the Eighteenth Century*, ed. Albert Goodwin (New York: Harper & Row, Publishers, 1967), pp. 103–9; for Russia, Max Beloff, in Goodwin, ed., pp. 173–80, 186–88; for Great Britain, Lawrence Stone, *The Crisis of the Aristocracy, 1558–1641* (New York: Oxford University Press, 1967 ed.), pp. 13, 98–99.
3. For numbers of the elite in England see Stone, *Crisis*, table 50, figure 6; for France see Bitton, p. 40, and J. McManners, in Goodwin, ed., p. 23; for Austria see J. M. Roberts on Lombardy, in Goodwin, ed., p. 67, and Crankshaw, p. 153; for Hungary, see C. A. Macartney, in Goodwin, ed., pp. 128–29; for Russia, see Beloff, in Goodwin, ed., p. 181. For information on the selling of titles, see examples for France in Franklin Ford, *Robe and Sword: The Regrouping of the French Aristocracy After Louis XIV* (Cambridge: Harvard University Press, 1953); for the Stuarts and Hanovers in England see Stone, *Crisis*, pp. 46, 55; for the Hapsburgs in Austria, see Bruford, p. 61.
4. See Giovanni Boccaccio, *The Decameron*, trans. G. H. McWilliam (New York: Penguin Books, 1975), p. 130; see Baldassare Castiglione, *The Book of the Courtier*, trans. Charles S. Singleton (Garden City, N.Y.: Anchor Books, 1959), pp. 7, 31, 34, 36, 39, 97–98, 130. Castiglione was himself a noble in the service first of the Duke of Milan and then the Duke of Urbino. His guide, *The Book of the Courtier*, was translated and reprinted throughout the sixteenth century.

5. See, for the ideal of service, Castiglione, pp. 110–13, 117, 119, 289. The instructions and admonitions presented for courtiers in subsequent centuries also described qualities associated with women. In the seventeenth century Mme. de La Fayette, a member of the French court, in her novel *The Princess of Clèves*, made the hero noteworthy for his control. The French courtier, the *honnête homme*, was always gracious, never aggressive; all seemed effortless, and everyone felt at ease in his presence. See Ian Maclean, *Woman Triumphant: Feminism in French Literature, 1610–1652* (Oxford: Clarendon Press, 1977), p. 136. See also the maxims of the courtier François, Duke of La Rochefoucauld, in La Rochefoucauld, *Maxims*, trans. Leonard Tancock (New York: Penguin Books, 1981), pp. 57, 50, 128, 124.

6. Castiglione, pp. 206–7. Seventeenth-century French treatises emphasized adaptability, moderation, *la bien séance* of the *honnête femme* (the counterpart of the *honnetê homme*). See Maclean, *Woman Triumphant*, pp. 123–27.

7. Charles Perrault, *Perrault's Fairy Tales*, trans. A. E. Johnson (New York: Dover Publications, 1969), p. 4.

8. *Letters of Mme. de Sévigné to her Daughter and her Friends*, ed. Richard Aldington (London: Routledge & Kegan Paul, 1937), Vol. I, p. 208; for a description of the education given to Lady Diana, daughter of the seventeenth-century Earl of Bedford in England, see Gladys Scott Thomson, *Life in a Noble Household, 1641–1700* (Ann Arbor: University of Michigan Press, 1959), pp. 76, 110. For a general description of education see Phyllis Stock, *Better Than Rubies: A History of Women's Education* (New York: G. P. Putnam's Sons, 1978), p. 82.

9. Dorothy Gardiner, *English Girlhood at School: A Study of Women's Education through Twelve Centuries* (London: Oxford University Press, 1929), pp. 125, 123.

10. See Ruth Kelso, *Doctrine for the Lady of the Renaissance* (Urbana: University of Illinois Press, 1956), pp. 230–32.

11. See Kelso, pp. 222–30.

12. W. H. Lewis, *The Splendid Century: Life in the France of Louis XIV* (Garden City, N.Y.: Doubleday & Company, 1957), pp. 202–3, 208; Hans Zinsser, *Rats, Lice and History* (London: George Routledge & Sons, 1935), p. 186.

13. Samuel Edwards, *The Divine Mistress: A Biography of Emilie du Châtelet* (New York: David McKay Company, 1970), p. 4.

14. M. J. Tucker, "The Child As Beginning and End: Fifteenth and Sixteenth Century English Childhood," in Lloyd de Mause, ed., *The History of Childhood* (New York: Harper & Row, Publishers, 1974), p. 240.

15. Johan Huizinga, *The Waning of the Middle Ages* (Garden City, N.Y.: Doubleday & Company, 1954), p. 227.

16. Philippe Erlanger, *The Age of Courts and Kings: Manners and Morals, 1558–1717* (New York: Harper & Row, Publishers, 1967), p. 22; Marcelin Defourneaux, *Daily Life in Spain in the Golden Age*, trans. Newton Branch (London: George Allen and Unwin, 1970), p. 50.

17. Cited in Erlanger, p. 236.

18. Cited in G. P. Gooch, *Courts and Cabinets* (New York: Alfred A. Knopf, 1946), p. 238; see also pp. 224–38.

19. W. H. Lewis, *The Sunset of the Splendid Century: The Life and Times of Louis*

Auguste de Bourbon, Duc de Maine, 1670–1736 (Garden City, N.Y.: Doubleday & Company, 1963), p. 5.

20. Aside from the books cited in the text, on court rituals and service see also, for France, Cissie Fairchilds, *Domestic Enemies: Servants and Their Masters in Old Regime France* (Baltimore: Johns Hopkins University Press, 1984) and Lewis, *Splendid Century*; for England, J. E. Neale, *Queen Elizabeth I* (London: Jonathan Cape, 1952 ed.); and for Germany, G. P. Gooch, *Courts and Cabinets*.

21. Madame de La Fayette, *The Princesse de Clèves*, trans. Nancy Mitford, revised by Leonard Tancock (New York: Penguin Books, 1978), p. 41.

22. Cited in Gooch, *Courts and Cabinets*, pp. 32–33.

23. Gooch, *Courts and Cabinets*, pp. 1–2.

24. Cited in Edwards, p. 8.

25. Cited in Edwards, p. 8.

26. They had to know German, Italian, French, and Latin. See Crankshaw, p. 152.

27. Gooch, *Courts and Cabinets*, p. 257.

28. For a description of the treatise, see Edwards, pp. 223–29.

29. Enid McLeod, *The Order of the Rose: The Life and Ideas of Christine de Pizan* (Totowa, N.J.: Rowman and Littlefield, 1976), p. 23.

30. See, for example, *Great Palaces*, introduction by Sacheverell Sitwell (New York: Spring Books, 1964), p. 213.

31. Ruth Kleinman, *Anne of Austria: Queen of France* (Columbus: Ohio State University Press, 1985), pp. 258–59.

32. John F. Freeman, "Louise of Savoy: A Case of Maternal Opportunism," *Sixteenth Century Journal*, Vol. III, no. 2 (October 1972), p. 91.

33. See *Palaces*, p. 186.

34. See Joseph Calmette, *The Golden Age of Burgundy: The Magnificent Dukes and Their Courts*, trans. Doreen Weightman (New York: W. W. Norton & Company, 1963), p. 231, and Richard Vaughan, *Valois Burgundy* (Hamden, Conn.: Archon Books, 1975), p. 147.

35. See *Letters from Liselotte: Life in the Court of Louis XIV Observed by the Sister-in-Law of the King*, trans. and ed. Maria Kroll (New York: McCall Publishing Company, 1971), p. 83. There is a new edition of the letters edited by Elborg Forster, entitled *A Woman's Life in the Court of the Sun King: Letters of Liselotte von der Pfalz, 1652–1722* (Baltimore: Johns Hopkins University Press, 1986).

36. Cited in Erlanger, p. 233.

37. On embroidery, see Erica Wilson, *Crewel Embroidery* (New York: Charles Scribner's Sons, 1962), pp. 13–18. On accessories see, for example. Hannelore Sachs, *The Renaissance Woman*, (New York: McGraw-Hill Book Company, 1971), p. 35; *The Secular Spirit: Life and Art at the End of the Middle Ages* (New York: E. P. Dutton & Co., 1975), pp. 85–87.

38. Millia Davenport, *The Book of Costume*, Vol. II (New York: Crown Publishers, 1965), p. 511.

39. Maintenon had, however, a *robe de chambre* in "flame plush" and a violet satin skirt, perhaps to wear with the king in private. See Lewis, *Splendid Century*, p. 198.

40. Of all the queens and noblewomen, only Marie Antoinette left the name of her

dress designer, Rose Bertin. For the changes in women's fashions and the influence of individuals like Mme. de Montespan, see François Boucher, A History of Costume in the West, trans. John Ross (London: Thames and Hudson, 1967), p. 261. More dramatic, however, than the shifts in women's fashions are those that occurred for men. For the changes in men's dress, see Boucher, pp. 242, 261, 278; Davenport, Vol. I, p. 190, Vol. II, pp. 519, 534; James Laver, The Concise History of Costume and Fashion (New York: Harry N. Abrams, 1969), pp. 62–64, 114–18.

41. On cosmetics see, for example, Sachs, p. 31; Carroll Camden, The Elizabethan Woman (New York: Elsevier Press, 1952), pp. 179–84; Caterina Sforza's book is described in her biography by Ernst Breisach, Caterina Sforza: A Renaissance Virago (Chicago: University of Chicago Press, 1967).

42. Cited in Harriet Ray Allentuch, Mme. de Sévigné: a Portrait in Letters (Baltimore: Johns Hopkins University Press, 1963), p. 129.

43. Nancy Mitford, Madame de Pompadour (London: The Reprint Society, 1955), pp. 97–103.

44. For the Belvedere ball see Schenk, in Goodwin, ed., p. 115. Aside from the books cited in the text, descriptions for the full range of court activities may be found in the following: for Italy, Michael Mallet, The Borgias: The Rise and Fall of a Renaissance Dynasty (London: Bodley Head, 1969) and the popular illustrated work by J. H. Plumb, The Horizon Book of the Renaissance (New York: American Heritage Publishing Co., 1961); for Burgundy, see Calmette; for England, see Camden, Stone, Crisis, and Neville Williams, All the Queen's Men: Elizabeth I and Her Courtiers (New York: Macmillan Company, 1972); for Spain, see Defourneaux; for France, see G. P. Gooch, Courts and Cabinets, and Arno Schönberger, Halldor Soehner, and Theodor Müller, The Rococo Age: Art and Civilization of the Eighteenth Century, trans. Daphne Woodward (New York: McGraw-Hill Book Company, 1960), for Russia, see Vincent Cronin, Catherine, Empress of All the Russias (London: Collins, 1978).

45. For Burgundy, see Calmette, p. 223; for Elizabeth I, see Williams, p. 177.

46. Frances Steegmuller, The Grand Mademoiselle (New York: Farrar, Straus and Company, 1956), p. 185.

47. See Schenk, in Goodwin, ed., p. 115.

48. Historians are still uncertain about just how many of the courtly could read. Into the seventeenth century the estimates are low. See Rosemary Masek, "Women in an Age of Transition, 1485–1714," in The Women of England: From Anglo-Saxon Times to the Present, ed. Barbara Kanner (Hamden, Conn.: Archon Books, 1979), p. 152, for England; and Sara Nalle, "The Unknown Reader: Women and Literacy in Golden Age Spain," paper delivered at the Berkshire Conference on Women's History (Vassar College, Poughkeepsie, N.Y., June 1981), p. 2.

49. Sévigné, ed. Aldington, Vol. I, p. 57. In addition to the numerous editions of her letters, see also the most recent biography, Frances Mossiker, Madame de Sé'vigné. A Life and Letters (New York: Columbia University Press, 1985).

50. Fairchilds, Domestic Enemies, pp. 28–30.

51. The Memoirs of Mme. Elizabeth Louise Vigée-Le Brun, 1755–1789, trans. Gerard Shelley (London: John Hamilton Ltd., n.d.), p. 48.

52. See Lucien Febvre, *Life in Renaissance France*, ed. and trans. Marian Rothstein (Cambridge, Mass.: Harvard University Press, 1977), p. 18.
53. See Neale, *Elizabeth I*, p. 208.
54. Neale, *Elizabeth I*, pp. 210–11.
55. Madariaga, p. 564.
56. Cited in Steegmuller, p. 105. She was exiled for her part in the Fronde, the rebellion against the regent, Anne of Austria, and the young Louis XIV.
57. Cited in Gooch, *Courts and Cabinets*, p. 64.
58. Cited in Jeanne A. Ojala, "Mme. de Sévigné: Chronicler of an Age" in J. R. Brink, ed., *Female Scholars: A Tradition of Learned Women Before 1800* (Montreal: Eden Press Women's Publications, 1980), p. 108.
59. See Ojala, in Brink, ed., p. 103.
60. For accounts of her activities at "Les Rochers" see, for example, *Sévigné*, ed. Aldington, Vol. I, pp. 174–79, 265–68, and Allentuch, pp. 161–63.

PART VI: **3. The Traditional Life in a Grand Setting**

1. See, for example, Vives, in Kelso, p. 38.
2. Castiglione, p. 16; see also p. 15.
3. See William Amelia, "Castiglione and 'The Courtier,'" *History Today*, Vol. XXVIII (September 1978), pp. 580–82.
4. Cited in Gooch, *Courts and Cabinets*, p. 222.
5. See Kelso, pp. 110–14; Camden, p. 44.
6. Cited in Kelso, p. 47; cited in Paul N. Siegel, "Milton and the Humanist Attitude Toward Women," *Journal of the History of Ideas*, Vol. XI (1950), p. 49.
7. Cited in Gardiner, pp. 159–60.
8. Cited in Kelso, p. 84. For descriptions of the appropriate education for a wife, see chaps. 3, 4, and 5 in Kelso. Kelso's book includes two bibliographies: one of the didactic texts for women, one of the texts for men. See in particular the descriptions of the treatises by Orazio Lombardelli and Giovanni Michele Bruto. Also useful for similar prescriptions from the learned in England is Louis B. Wright, *Middle-Class Culture in Elizabethan England* (Chapel Hill: University of North Carolina Press, 1935), pp. 105–6, 131; for France see Donald M. Frame, *Montaigne: A Biography* (New York: Harcourt, Brace & World, 1965), chap. 6.
9. Cited in Siegel, p. 48.
10. Cited in David Hunt, *Parents and Children in History: The Psychology of Family Life in Early Modern France* (New York: Harper & Row, Publishers, 1972), p. 70; for discussion of the ideal wife as the subordinate, not the equal, see Kelso, chap. 5; see also Ian Maclean, *The Renaissance Notion of Woman: A Study of the Fortunes of Scholasticism and Medical Science in European Intellectual Life* (New York: Cambridge University Press, 1980), pp. 58–59; for a brief description of the sixteenth-century equivalent writing in Russia called *Domostroi*, which is even more restrictive, see Dorothy Atkinson, "Society and Sexes in the Russian Past," in *Women in Russia*, ed. Dorothy Atkinson, Alexander Dallin, and Gail Warshofsky Lapidus (Stanford, Calif.: Stanford University Press, 1977), p. 15 and fn. p. 15.

11. Cited in Gloria Kaufman, "Juan Luis Vives on the Education of Women," *Signs*, Vol. 3, no. 4 (Summer 1978), p. 894.

12. Castiglione, p. 188; see also Keith Thomas, "The Double Standard," *Journal of the History of Ideas*, Vol. XX, no. 2 (April 1959) for Vives' view.

13. Castiglione, p. 188.

14. Cited in Hunt, p. 74.

15. Castiglione, p. 210; see also p. 208; see Kelso, pp. 105–8, 217–18.

16. See W. Matthews, "The Wife of Bath and All Her Sect," *Viator: Medieval and Renaissance Studies*, Vol. 5 (1974), pp. 413–43; see also Kelso, pp. 128–31.

17. See Kelso, pp. 25, 36, 59, 95, 109; Gardiner, p. 160; Stock, pp. 12, 36; and Maclean, *Renaissance Notion*, p. 59, for differences in educational goals and for the view of the female nature. For the ideas of Vives in particular, see Siegel, p. 47, and for Fénelon, Mary W. Rowan, "Seventeenth-Century French Feminism: Two Opposing Attitudes," *International Journal of Women's Studies*, vol. 3, no. 3 (May–June 1980), p. 285.

18. Cited in Tucker, in de Mause, ed., p. 249; see also Kelso, chap. 3, especially pp. 39–41, 47–48, 53.

19. See Kelso, p. 267; Kaufman, p. 894.

20. Cited in Stock, p. 95.

21. In the seventeenth century to be presented at the French Bourbon court a woman or man needed to be able to show four generations of aristocracy (or quarterings) before 1400. Only about one thousand families qualified. See J. McManners, in Goodwin, ed., p. 23, and Lewis, *Splendid Century*, pp. 42–43. At the beginning of the seventeenth century the Swedish nobility, by the same reasoning, limited the number who could participate in the parliament, the Riksdag. See Georgina Masson, *Queen Christina* (New York: Farrar, Straus & Giroux, 1969), p. 45. To marry a baronet in seventeenth-century England a family had to provide a dowry of £5,000; for an earl or officer of the government they paid £10,000. See Maurice Ashley, *The Stuarts in Love: with some reflections on love and marriage in the sixteenth and seventeenth centuries* (New York: Macmillan, 1964), p. 27.

22. See Allentuch, fn. p. 203, and *Sévigné*, ed. Aldington, Vol. I, p. xvii.

23. Cited in Jean Heritier, *Catherine de' Medici*, trans. Charlotte Haldane (New York: St. Martin's Press, 1963), p. 67.

24. Cited in Neale, *Elizabeth I*, p. 103.

25. Heritier, p. 31.

26. Roland Bainton, *Women of the Reformation from Spain to Scandinavia* (Minneapolis: Augsburg Publishing House, 1977), p. 194.

27. See Stone, *Crisis*, p. 285; Jan DeVries, *The Economy of Europe in an Age of Crisis 1600–1750* (Cambridge: Cambridge University Press, 1958), pp. 102–3.

28. Carolly Erickson, *Bloody Mary* (Garden City, N.Y.: Doubleday & Company, 1978), pp. 54, 67–69.

29. Cited in George Edwin Fussell and K. R. Fussell, *The English Countrywoman: A farmhouse social history, 1500–1900* (London: A. Melrose, 1953), p. 78.

30. Boucalh cited in J. P. Cooper, "Patterns of Inheritance and Settlement by great

landowners from the fifteenth to the eighteenth centuries," in *Family and Inheritance: Rural Society in Western Europe, 1200–1800*, ed. Jack Goody, Joan Thirsk, and E. P. Thompson (New York: Cambridge University Press, 1978), p. 265. For a summary of jurists see Joan Thirsk, "The European Debate on Customs of Inheritance, 1500–1700," in Goody, ed., pp. 178–81. The descriptions that follow come from a variety of studies of the laws and customs of inheritance. The article by Cooper in Goody, ed., gives examples for France, Spain, Scotland, and England; P. C. Timbal, "L'esprit du droit privé au XVIIe siècle," *XVIIe Siècle*, nos. 58–59 (1963) describes the efforts of Louis XIV of France and gives examples from Spain; for France, see also Bitton, especially p. 93; see the collection edited by Goodwin—H. J. Habakkuk on England, Raymond Carr on Spain, and a. Goodwin on Prussia. For Denmark see Inga Dahlsgård, *Women in Denmark Yesterday and Today* (Copenhagen: Det Danske Selskab, 1980).

31. For example, Philip II of Spain specified that a lady in waiting should have a dowry of 1 million maravedas. See Cooper in Goody, ed., fn. p. 245; see Dahlsgård, p. 39.

32. For Catherine de' Medici, see De Lamar Jensen, "Catherine de' Medici and Her Florentine Friends," *Sixteenth Century Journal*, Vol. IX, no. 2 (1978), p. 63.

33. Meeting the terms of the widow's portion could be onerous. In the eighteenth century the wife of the third Duke of Leeds survived him by sixty-three years and collected a jointure of £190,000. H. J. Habakkuk, in Goodwin, ed., p. 8.

34. In sixteenth- and seventeenth-century England, circumventions of custom and law were arranged at the time of the marriage of the eldest son, or the daughter. It was called "strict settlement." Entailed properties were specified, and then provision for both daughters and younger sons was made, usually money for the females and disentailed income-producing properties for the males. English marriage settlements have been studied extensively; see appendixes I and II in Goody, ed., pp. 306–12, 313–27; see Stone, *Crisis*, for example, p. 289.

35. Cited in McLeod, p. 33.

36. Cited in McLeod, p. 34.

37. Timbal, p. 39. For examples of the laws see Cooper, in Goody, ed., p. 251; for France, see also Maclean, *Woman Triumphant*, p. 18, and Jean Brissaud, *A History of French Private Law* (Boston: Little, Brown and Company, 1912), pp. 116–17; for England, see Gooch, *Courts and Cabinets*, p. 215, and Thomas, "Double Standard"; for Russia, see Patrick P. Dunn, " 'That Enemy Is the Baby': Childhood in Imperial Russia," in de Mause, ed.; for Denmark, see Dahlsgård, p. 26.

38. Cited in Gooch, *Courts and Cabinets*, p. 31.

39. See Steegmuller, chap. 8.

40. See Roland Bainton, *Women of the Reformation in France and England* (Boston: Beacon Press, 1975), pp. 75ff.

41. Liselotte, p. 7.

42. Vaughan, p. 177.

43. Vaughan, p. 174.

44. *The Fugger News-Letters: Being a Selection of Unpublished Letters from the Correspondents of the House of Fugger Luring the Years 1568–1605*, ed. Victor von Karwill (New York: G. P. Putnam's Sons, 1925), p. 97.

45. *The Memoirs of Catherine the Great*, ed. Dominique Maroger and trans. Moura Budberg (New York: Collier Books, 1961), p. 76.

46. See Crankshaw, p. 327; Stanley Loomis, *Du Barry: A Biography* (New York: J. B. Lippincott Company, 1959), pp. 107–8.

47. For example, see Kelso, pp. 80–81, 92.

48. Cited in Stone, *Crisis*, p. 283.

49. Cited in Bainton, *France and England*, p. 122.

50. See Marie Louise Bruce, *Anne Boleyn* (New York: Coward, McCann & Geoghegan, 1972), pp. 233–34.

51. See Sven Stolpe, *Christina of Sweden*, ed. Sir Alec Randall and trans. Sir Alec Randall and Ruth Mary Bethell (New York: Macmillan Company, 1966), p. 38.

52. Cited in J. L. Flandrin, *Families in Former Times: Kinship, Household and Sexuality*, trans. Richard Southern (New York: Cambridge University Press, 1979), p. 217.

53. Bruce, p. 232.

54. The information comes from a variety of sources; see William Goodell, *A Sketch in the Life and Writings of Louyse Bourgeois* (Philadelphia: Collins, Printer, 1876), p. 21; Hunt, p. 84; Louise Bourgeois, *Les Six Couches de Marie de Medicis* (Paris: Leon Willem, 1875), pp. 108–9; Kleinman, p. 107.

55. G. P. Gooch, *Maria Theresa and Other Studies* (New York: Longmans, Green, 1952), p. 206; see Gooch, *Courts and Cabinets*, p. 259, for the account of the birth.

56. Liselotte, p. 30. See T. H. Hollingsworth, "A Demographic Study of British Ducal Families," in Glass and Eversley, eds., for statistics on children's deaths from 1500–1799 among England's elite. About one third died before the age of fifteen (see p. 363).

57. See Stone, *Crisis*, pp. 270, 283.

58. It is likely that privileged women used other traditional contraceptive techniques besides abstinence, such as herbal douches and cervical blocks. Their partners might use the condom (initially made of cloth) and withdrawal.

59. In his study of the British nobility from the fourteenth to the eighteenth centuries, the demographer Hollingsworth estimates that one sixth of the marriages remained childless. See Hollingsworth, in Glass and Eversley, eds., p. 377.

60. See her *Memoirs*, the epilogue, p. 250.

61. Cited in Erickson, *Mary*, p. 45.

62. See Bainton, *France and England*, p. 150; see the biography by Garrett Mattingly, *Catherine of Aragon* (Boston: Little, Brown and Company, 1941), and the Bruce biography of Anne Boleyn.

63. See, for example, Perrault, p. 20.

64. Cited in Ashley, p. 131.

65. Cited in Ashley, p. 131.

66. Cited in Crankshaw, p. 268.

67. Cited in Gooch, *Maria Theresa*, p. 18.

68. The writings and letters of English noblewomen of the sixteenth, seventeenth, and eighteenth centuries—for instance, Honor Lisle, a member of the Merchant Staple family; Mary Boyle, Countess of Warwick; and Sarah Churchill, Duchess of Marlborough—also show affection and love in their marriages. See, for example,

Ashley, pp. 30–33; Israel Shenker, "Familial History of the Tudor Age Is Brought to Life," *The Smithsonian*, Vol. 12, no. 1 (April 1981), pp. 110–12; *Viscountess Anne Finch Conway, the Conway Letters*, ed. Mariorie Hope Nicholson (New Haven: Yale University Press, 1930), p. 148; *Autobiography of Mary Rich, Countess of Warwick*, ed. Thomas Crofton Croker (Percy Society, 1848), Vol. XXII, pp. 6–13; Stenton, p. 229; and David Green, *Sarah, Duchess of Marlborough* (New York: Charles Scribner's Sons, 1967).

69. See Sachs, p. 45.
70. Cited in Kelso, p. 108.
71. Cited in Gooch, *Courts and Cabinets*, p. 177.
72. Lord John Hervey, *Memoirs of the Reign of George the Second*, ed. the Right Honorable John Wilson Croker (London: 1884), Vol. I, p. 300.
73. Cited in Gooch, *Maria Theresa*, pp. 142–43. See also *Maria Theresa*, ed. Karl A. Roider, Jr. (Englewood Cliffs, N.J.: Prentice-Hall, 1973), pp. 81–84, for the empress's views on the role of wife.
74. Cited in Gooch, *Maria Theresa*, p. 143.
75. Cited in Carolyn C. Lougee, *Le Paradis de Femmes: Women, Salons and Social Stratification in Seventeenth-Century France* (Princeton: Princeton University Press, 1976), p. 84.
76. Cited in Gooch, *Courts and Cabinets*, p. 8.

PART VI: **4. Women Rulers**

1. He did not approve of her success, and criticized, instead of praised her efforts on his return. See Maria Bellonci, "Beatrice and Isabella d'Este," in Plumb, p. 366.
2. For information on Margaret of Austria and Mary of Hungary, see Sachs, pp. 45–46, and Bainton, *Spain and Scandinavia*, pp. 213–14; for Margaret of Parma, see Pieter Geyl, *The Revolt of the Netherlands, 1555–1609* (New York: Barnes and Noble, 1958), especially pp. 75, 87, 100.
3. For the lives of the queen regents, see Freeman for Louise of Savoy (1476–1531); for Marie de' Medici (1573–1642), see Victor-L. Tapié, *France in the Age of Louis XIII and Richelieu*, trans. and ed. D. McN. Lockie (New York: Praeger Publishers, 1975); for Anne of Austria (1601–1666), see Gooch, *Courts and Cabinets*, and Kleinman. Catherine de' Medici (1519–1589) held power after Louise of Savoy.
4. Heritier, p. 210; see also J. E. Neale, *The Age of Catherine de'Medici*, (New York: Harper & Row, Publishers, 1962 ed.), p. 42.
5. See Neale, *Catherine de' Medici*, p. 51.
6. At one point she envisioned crowns for all of them: Charles as Holy Roman Emperor; Henry, King of Poland, her youngest boy, married to the Queen of England; her daughters ruling with kings of Spain and Navarre and with the Duke of Lorraine. Irene Mahoney, *Madame Catherine* (New York: Coward, McCann & Geoghegan, 1975), p. 339.
7. See Heritier, pp. 300, 340.
8. The Huguenots represented one sixth of France's population at the time (Heritier, p. 413).

9. On this interpretation of her policies and motives, see Heritier, pp. 142, 169–81.

10. On Catherine's role, see the views of Neale, *Catherine de' Medici*, pp. 76–79; Heritier, pp. 325–26; Mahoney, pp. 159–67; and especially Barbara Diefendorf, "Prologue to a Massacre: Popular Unrest in Paris, 1557–1572," *American Historical Review*, Vol. 90, no. 5 (December 1985), pp. 1,067–91.

11. For the life of Caterina Sforza, see the biography by Breisach.

12. For information about Catherine the Great, see the biography by Vincent Cronin, the collection of documents and articles about her in L. Jay Oliva, ed., *Catherine the Great* (Englewood Cliffs, N.J.: Prentice-Hall, 1971), and especially the study of the era by Isabel de Madariaga.

13. Others included Sophia (1682–1689), who ruled during Peter's own minority; Catherine (1725–1727); and Anna (1730–1740).

14. Cited in Madariaga, p. 7.

15. Peter probably did not consummate their marriage until 1752.

16. Cited in Madariaga, p. 12.

17. Catherine the Great took no direct part in their deaths, but showed her approval of what circumstances brought. When Peter III, her husband, died in a brawl, she rewarded the man in charge of his safety with lands, a promotion, and a medal. See Madariaga, p. 32.

18. See Madariaga, pp. 581–88.

19. See Madariaga, pp. 148–50.

20. Cited in Madariaga, p. 44.

21. Cited in G. P. Gooch, *Catherine the Great and Other Studies* (New York: Longmans, Green, 1954), p. 27.

22. See Catherine the Great, *Memoirs*, p. 308.

23. Cited in Oliva, ed., "Instruction," p. 53.

24. See Madariaga, p. 573, and Cronin, pp. 289–92, for descriptions of typical days.

25. Historians suggest that they may have married. See Madariaga, p. 357, and Gooch, *Catherine the Great*, p. 45.

26. Cited in Oliva, ed., p. 11.

27. Cited in Gooch, *Catherine the Great*, p. 53.

28. Cited in Madariaga, p. 580.

29. On the changes in warfare see DeVries, pp. 204–5, and Domenico Sella, "European Industries, 1500–1700," in *The Fontana Economic History of Europe: The Sixteenth and Seventeenth Centuries*, Carlo M. Cipolla, ed. (Glasgow: William Collins Sons & Co., 1976), pp. 384–86.

30. Cited in Carolyn Merchant, *The Death of Nature: Women, Ecology and the Scientific Revolution* (New York: Harper & Row, Publishers, 1980), pp. 145, 146.

31. Maclean, *Women Triumphant*, pp. 58–71. For a description of the arguments of scholars like Cornelius Agrippa, Torquato Tasso, and John Aylmer, see also pp. 19–20; and see Camden, p. 254.

32. See Kelso, p. 34, for a discussion of this attitude.

33. Giovanni Boccaccio, *Concerning Famous Women*, trans. Guido A. Guarino (New Brunswick, N.J.: Rutgers University Press, 1963), p. xxvii.

34. Cited in Maclean, *Renaissance Notion*, p. 61.

35. Biographies have been written for all of the queens highlighted in this and other sections. For Mary Tudor, see Erickson, *Mary*; for Elizabeth I, see especially Neale, *Elizabeth I*; for Mary Stuart, see Antonia Fraser, *Mary, Queen of Scots* (New York: Dell Publishing Co., 1971); for Queen Anne, see Edward Gregg, *Queen Anne* (Boston: Routledge and Kegan Paul, 1980); for Queen Christina of Sweden, see Masson, *Queen Christina*, and the more psychologically oriented biography by Stolpe; for Maria Theresa of Austria, see especially Crankshaw and the collection of documents and views of historians in *Maria Theresa*, Roider, ed.

36. Cited in Doris Mary Stenton, *The English Woman in History* (New York: Macmillan Company, 1957), p. 129.

37. Cited in Neale, *Elizabeth I*, p. 26.

38. Cited in Masson, *Queen Christina*, p. 58.

39. On her education see M. L. Clarke, "The Making of a Queen, the Education of Christina of Sweden," *History Today*, Vol. 28, no. 4 (April 1978), especially pp. 230–33.

40. In the sixteenth century neither Mary Tudor in England nor Mary Stuart in Scotland succeeded in this. Both were welcomed as rulers. Neither, however, retained her popularity, nor convinced contemporaries that she was governing wisely. Mary Tudor married Philip II of Spain as part of her reinstitution of the Catholic religion in England, but he gave her little affection and involved her in a war that cost England territory. The marriage was unsuccessful in another way. Mary never became pregnant and died with no heir to carry on her Catholic rule. For Mary Stuart, marriage to Henry Stewart, Earl of Darnley, gave her a son but no support or guidance as a monarch. She never established an image of herself as a conscientious ruler. Instead her subjects assumed she had a frivolous, immoral nature, believed she had been involved in her husband's murder, and rebelled against her and her lover James Hepburn, Earl of Bothwell. She was forced to abdicate and flee her kingdom.

41. Cited in Neale, *Elizabeth I*, p. 215. See Williams for descriptions of her handling of men and of how she gained their service.

42. John Foxe, *Acts and Monuments*, ed. G. A. Williamson (Boston: Little, Brown and Company, 1965), p. 429; see also pp. xxiv, xxxi.

43. Cited in Neale, *Elizabeth I*, p. 298.

44. All aspects of Elizabeth I's reign have been reexamined by historians. New interpretations have arisen out of study of the periods before and after her reign. They call into question the extent to which she created unity and maintained peace. They suggest that the problems were not so great on her accession and that circumstances, not her actions, ameliorated others. For a summary of the new perspectives and a list of the relevant books, see the review article by Christopher Haigh, "The Reign of Elizabeth I," *History Today*, Vol. 35 (August 1985), pp. 53–55, or see *The Reign of Elizabeth I*, ed. Christopher Haigh (Athens: University of Georgia Press, 1985) for the latest scholarship.

45. Cited in Williams, p. 175. Of all the royal candidates, only Catherine de' Medici's youngest son, François, Duke of Alençon, seems seriously to have engaged her fancy. On his visit to England in August of 1579 she made a public pledge of

marriage and exchanged rings with him. But by the spring after he left she had retreated to her old stance. She gave him loans and an alliance for his war in the Netherlands, but no marriage.

46. Cited in Alison Heisch, "Queen Elizabeth I and the Persistence of Patriarchy," *Feminist Review*, no. 4 (1980), p. 50.
47. Williams, p. 177.
48. Cited in Masson, *Queen Christina*, p. 27.
49. Cited in M. L. Clarke, p. 235.
50. Cited in Stolpe, p. 40.
51. From a description of her after her abdication, when she was in Italy. Cited in Stolpe, p. 194.
52. Cited in Masson, *Queen Christina*, p. 389.
53. Historians speculate on her sexual preferences, suggesting from her correspondence with her favorite, Ebba Brahe, that she was a lesbian. Subsequent events after her abdication, in particular her attachment to the Italian Cardinal Azzolino, suggest that she was probably reluctant to engage sexually with women or men. On Brahe, see Masson, *Queen Christina*, p. 86, and Stolpe, pp. 60–61; on Azzolino, see Stolpe, p. 254.
54. Cited in Masson, *Queen Christina*, p. 217; see also Stolpe, p. 170.
55. Cited in C. A. Macartney, ed., *The Hapsburg and Hohenzollern Dynasties in the Seventeenth and Eighteenth Centuries* (New York: Walker and Company, 1970), p. 100.
56. Cited in Crankshaw, p. 92.
57. Cited in Macartney, ed., pp. 99–100.
58. Cited in Macartney, ed., p. 115.
59. Cited in Roider, ed., pp. 100, 101.
60. Cited in Roider, ed., p. 116.
61. Cited in Macartney, ed., p. 105; see Oliva, ed., pp. 70–72, for an example of the way in which she manipulated her principal minister, Kaunitz.
62. Cited in Macartney, ed., p. 114.
63. See Crankshaw, p. 192.
64. Cited in Gooch, *Maria Theresa*, p. 33.
65. Cited in Gooch, *Maria Theresa*, p. 52.
66. Cited in Roider, ed., p. 25.

PART VI: **5. New Opportunities**

1. See Geoffrey Parker, "Philip II of Spain: A Reappraisal," *History Today*, Vol. XXIX (December 1979), p. 804. Information on jesters comes from a variety of sources: for Burgundy, see Calmette, p. 233, and Huizinga, p. 26; for England, Erickson, *Mary*, p. 193; and for France, Natalie Zemon Davis, *Society and Culture in Early Modern France* (Stanford, Calif.: Stanford University Press, 1975), p. 141.
2. Cited in Carol Neuls-Bates, ed., *Women in Music: An Anthology of Source Readings from the Middle Ages to the Present* (New York: Harper & Row, Publishers, 1982), p. 38.

3. Anthony Newcomb, "Courtesans, Muses, or Musicians? Professional Women Musicians in Sixteenth Century Italy," in *Women Making Music: The Western Art Tradition, 1150–1950*, ed. Jane Bowers and Judith Tick (Urbana: University of Illinois Press, 1986), pp. 8–9.

4. For example: madrigals by the sixteenth-century Venetian Maddalena Casulana, and sacred music for the mass and work for instruments by the seventeenth-century composer Marieta Prioli. See Newcomb, in Bowers and Tick, eds., pp. 22–24, and Jane Bowers, "The Emergence of Women Composers in Italy, 1566–1700," in Bowers and Tick, eds., pp. 1–4.

5. Bowers, in Bowers and Tick, eds., p. 10.

6. Cited in Neuls-Bates, ed., p. 42. Only a book of songs and accompaniments, and one opera, survive of her works (Neuls-Bates, ed., p. 41).

7. There is a possibility that she also acted as courtesan to them. Courtesans like Anna Maria Sardelli who gained reputations as singers were a part of seventeenth-century Venetian culture. See Ellen Rosand, "The Voice of Barbara Strozzi," in Bowers and Tick, eds., pp. 7–8.

8. See Bowers and Tick, eds., pp. 12, 14.

9. See Julie Ann Sadie, "*Musiciennes* of the Ancien Régime," in Bowers and Tick, eds.

10. See Sadie, in Bowers and Tick, eds., p. 27.

11. Cited in Neuls-Bates, ed., p. 48.

12. Neuls-Bates, ed., p. 48.

13. For material on these musicians and composers, in addition to Bowers and Tick and Neuls-Bates, see the bibliographical essay by Elizabeth Wood, "Women in Music," *Signs*, Vol. 6, no. 2 (Winter 1980), pp. 283–97.

14. Only in the nineteenth century did ballet positions, leaps, and combinations evolve into dances with a story to tell, and the costumes for women and men allow them easier movement. See Mary Clarke and Clement Crisp, *Ballet: An Illustrated History* (New York: Universe Books, 1973).

15. See Ann Sutherland Harris and Linda Nochlin, *Women Artists, 1550–1950* (New York: Alfred A. Knopf, 1978), pp. 36–37.

16. In addition to the works cited, for information about the artists mentioned see Pearl Hogrefe, *Tudor Women: Commoners and Queens* (Ames: Iowa State University Press, 1975); Germaine Greer, *The Obstacle Race: The Fortunes of Women Painters and Their Work* (New York: Farrar, Straus & Giroux, 1979); Karen Petersen and J. J. Wilson, *Women Artists: Recognition and Reappraisal from the Early Middle Ages to the Twentieth Century* (New York: Harper & Row, Publishers, 1976).

17. Vigée-Lebrun, p. 46.

18. Vigée-Lebrun, p. 47.

19. Vigée-Lebrun, p. 38.

20. Vigée-Lebrun, pp. 67–71.

21. Vigée-Lebrun, p. 42.

22. Vigée-Lebrun, pp. 52, 53, 56.

23. Crankshaw, p. 139.

24. More than 150 years later the English writer Virginia Woolf used the fictional figure

of "Shakespeare's sister" to make this same point in her 1929 essay A *Room of One's Own.*

25. Historians have noted that the feminine equivalent to the masculine term *courtier* in Italian is *cortigiana*, the Italian word for courtesan. In fact, the courtesan acted the sycophant much like the ideal male courtier described by Baldassare Castiglione. See Georgina Masson, *Courtesans of the Italian Renaissance* (New York: St. Martin's Press, 1976), p. 5, for the connection between the two terms, and see also Tita Rosenthal, "The Venetian Courtesan and the Inquisition: Was Veronica Franco a Witch?" unpublished paper given in New York, Fordham University, March 16, 1986.

26. Cited in Kenneth Clark, *The Nude, a Study in Ideal Form* (New York: Pantheon Books, 1956), p. 77. For the evolution of taste in the representations of the nude form in Europe, see also pp. 27, 91ff., 114, 127ff., 295–96.

27. For example: "What love is?" "What the effect of love is?" "How it may be acquired, retained, increased, decreased and ended?" and the fact of sexual intercourse, "her solaces," as Cappellanus called it, especially between lovers who were not married. See Andreas Cappellanus, *The Art of Courtly Love,* trans. John Jay Parry (New York: W. W. Norton & Company, 1969), pp. 28, 153; see also p. 154. For discussion of the interpretations of the text, see Sidney Painter, *Medieval Chivalry* (Ithaca: Cornell University Press, 1957), p. 159, and especially John F. Benton, "Clio and Venus: An Historical View of Courtly Love," in F. X. Newman, ed., *The Meaning of Courtly Love* (Albany: State University of New York Press, 1968). From the seventeenth century on courtiers read books like *The School for Women (L'école des femmes)* and Choderlos de Laclos's *Dangerous Affairs (Les liaisons dangereuses)* that were openly erotic. *Fanny Hill,* by John Cleland, is the eighteenth-century English equivalent.

28. Edwards, p. 36.

29. See Masson, *Courtesans,* p. 66.

30. See Masson, *Courtesans,* p. 153.

31. Cited in Masson, *Courtesans,* p. 11.

32. Masson, *Courtesans,* p. 153.

33. Cited in Masson, *Courtesans,* p. 157.

34. Cited in Masson, *Courtesans,* p. 164.

35. *The Penguin Book of Women Poets,* ed. Carol Cosman, Joan Keefe, and Kathleen Weaver (New York: Penguin Books, 1979), p. 117.

36. Cited in Masson, *Courtesans,* p. 69.

37. Cited in Evelyne Sullerot, *Women on Love: Eight Centuries of Feminine Writing,* trans. Helen R. Lane (Garden City, N.Y.: Doubleday & Company, 1979), p. 134.

38. For example, Agnes Sorel (1420?—1450) and Diane de Poitiers (1499–1566).

39. Cited in Lewis, *Sunset,* p. 81.

40. Charlotte Haldane, *Mme. de Maintenon: Uncrowned Queen of France* (New York: Bobbs-Merrill Company, 1970), pp. 77–78.

41. Contemporaries and subsequent generations have speculated on whether or not Louis XIV married Mme. de Maintenon. When students at St. Cyr (the school she founded) suggested it, she was quick to answer, "Who told you that?" (cited

in Haldane, p. 271). Historians assume that the king and Mme. de Maintenon went through some sort of ceremony in the fall of 1683 or the summer of 1684. See Haldane, pp. 150–55, and Lewis, *Sunset*, p. 90.

42. Cited in Haldane, pp. 265, 165.

43. Cited in Haldane, p. 252.

44. For the life of Mme. de Pompadour, see the biography by Nancy Mitford.

45. See Mitford, p. 159.

46. See Mitford, p. 88.

47. Cited in Mitford, p. 251.

48. Haldane, p. 125.

49. Sévigné, ed. Aldington, Vol. I, p. 170.

50. Masson, *Courtesans*, pp. 128–29.

51. See John Harold Wilson, *Nell Gwyn, Royal Mistress* (New York: Pellegrini & Cudahy, 1952), especially pp. 152–53, 192.

52. Cited in Masson, *Courtesans*, p. 165; see also the translation in Brian Pullan, *Rich and Poor in Renaissance Venice* (Cambridge: Harvard University Press, 1971), p. 392.

53. For information on these women see, for Marguerite of Navarre, C. J. Blaisdell, "Marguerite de Navarre and Her Circle," in Brink, ed.; for Zayas, Sandra M. Foa, "María de Zayas y Sotomayor: Sibyl of Madrid," in Brink, ed. For Gournay, see Rowan, pp. 273–91; the biography by Marjorie Ilsley, *A Daughter of the Renaissance, Marie le Jars de Gournay* (The Hague: Mouton, 1963); and see Frame, pp. 277–80, on the most recent speculation about her relationship with Montaigne. For Burney, see, for example, the short biography in Gooch, *Courts and Cabinets*, on her life at court.

54. Note that there are disagreements over dates given for the principal events of Christine de Pizan's life. In each instance those of Charity Cannon Willard, her most recent biographer, have been used. See Charity Cannon Willard, *Christine de Pizan: Her Life and Works* (New York: Persea Books, 1984), pp. 39–40.

55. See Mcleod, pp. 124–25, and Willard, *Christine de Pizan*, pp. 180, 184–85, for summaries of her works. Willard's bibliography lists modern editions and translations.

56. Willard, *Christine de Pizan*, p. 164.

57. Cited in Willard, *Christine de Pizan*, p. 163. Records of the court note only one payment from 1408–1410 from her patrons (Willard, *Christine de Pizan*, p. 177).

58. See Leslie Altman, "Christine de Pizan: First Professional Woman of Letters," in Brink, ed., p. 12.

59. See Mcleod, pp. 135–36.

60. See Hindman.

61. Cited in Nadia Margolis, "Christine de Pizan: The Poetess as Historian," unpublished manuscript (1981), p. 9.

62. See, for example, Willard, *Christine de Pizan*, pp. 118–19, 200.

63. See Erica Harth, *Ideology and Culture in Seventeenth Century France* (Ithaca: Cornell University Press, 1983), pp. 206–13, on the context of this novel and its publishing history. For her life, see Dorothy Anne Liot Backer, *Precious Women: A Feminist Phenomenon in the Age of Louis XIV* (New York: Basic Books, 1974),

pp. 281–87, and Natalie Zemon Davis, "Gender and Genre: Women as Historical Writers, 1400–1820," in Patricia H. Labalme, ed., *Beyond Their Sex: Learned Women of the European Past* (New York: New York University Press, 1980).

64. See for example Foa, in Brink, ed., pp. 63–65, and Margaret, Queen of Navarre, *The Heptameron*, trans. Walter K. Kelly (London, 1857), pp. 109, 142.

65. La Fayette, p. 191.

66. See the description of her playing at thirteen in Neuls-Bates, ed., p. 46.

PART VI: **6. The Legacies of Renaissance Humanism and the Scientific Revolution**

1. For information on elite women and their books from the ninth to the sixteenth centuries, see Susan Groag Bell, "Medieval Book Owners: Arbiters of Lay Piety and Ambassadors of Culture," *Signs*, Vol. 7, no. 4 (Summer 1982), pp. 742–68.

2. There are over two thousand extant letters (see Stock, p. 39).

3. "Prima donna del mondo," cited in Sachs, p. 15. The court of Mantua already had a reputation for educated women when Isabella d'Este joined it. The first Margrave of Mantua, Gianfrancesco Gonzaga, had his daughter Cecilia taught with his sons. See Margaret Leah King, "Thwarted Ambitions: Six Learned Women of the Italian Renaissance," *Soundings*, Vol. LIX, no. 3 (Fall 1976), pp. 291–92.

4. Felipe Fernández-Armesto, *Ferdinand and Isabella* (New York: Taplinger Publishing Company, 1975), pp. 110–11.

5. On the education of Isabella's children, see Irene L. Plunket, *Isabel of Castile and the Making of the Spanish Nation, 1451–1504* (New York: G. P. Putnam's Sons, 1978), pp. 331–35. For a life of Makin, see J. R. Brink, "Bathsua Makin: Educator and Linguist," in Brink, ed. For selections from her writings see Moira Ferguson, ed. *First Feminists: British Women Writers, 1578–1799* (Bloomington: Indiana University Press, 1985), pp. 128–42.

6. Blaisdell, in Brink, ed., p. 38; see Blaisdell for short biographies of mother and daughter, and Bainton, *France and England*.

7. See M. A. Screech, *The Rabelaisian Marriage: Aspects of Rabelais's Religion, Ethics and Philosophy* (London: Edward Arnold Publishers, 1958), p. 26.

8. In connection with her religious studies, she corresponded with, intervened for, and gave shelter to those like Lefèvre d'Etaples, whose writings were condemned by the religious authorities in France. She filled her poetry and her prayers with ideas associated with Luther and Calvin. Her *Mirror of the Sinful Soul* of 1531 brought her into conflict with the theologians at the Sorbonne in Paris. Only intervention by her brother, King Francis I, removed the formal condemnation of her reflections. See Geoffrey Brereton, A *Short History of French Literature* (Baltimore: Penguin Books, 1968), pp. 176, 109, on this and on her secular writings.

9. Margaret Leah King estimates that between 1350 and 1530 the lives of twelve such women are known; another twenty are named in the sources. See Margaret Leah King, "Book-lined Cells: Women and Humanism in the Early Italian Renaissance," in Labalme, ed., *Beyond Their Sex*, p. 67. For their lives, see King, "Six Learned Women."

10. Cited in King, "Book-lined Cells," in Labalme, ed., *Beyond Their Sex*, p. 68.

11. For her life, see Roland Bainton, *Women of the Reformation in Germany and Italy* (Boston: Beacon Press, 1971), pp. 253–65.

12. Cited in Bainton, *France and England*, p. 182; see pp. 181–83 for a brief biography.

13. See Camden, p. 58, and Minna F. Weinstein, "Reconstructing Our Past: Reflections on Tudor Women," *International Journal of Women's Studies*, Vol. I, no. 2 (March–April 1978), p. 135.

14. See *The Conway Letters*, ed. Nicholson.

15. For her life, see Joyce L. Irwin, "Anna Maria van Schurman: The Star of Utrecht," in Brink, ed.

16. It was for "dialectics"; the University of Padua refused her father's suggestion that it be in "theology." For her life see Patricia H. Labalme, "Elena Lucrezia Cornaro Piscopia: Piety and Pride in Seventeenth-Century Venice," unpublished manuscript, and Maria Remiddi, "A Woman of High Degree," *UNESCO Courier*, July 1978, pp. 12–13. This was not the first instance of a woman being associated with a university. They had studied and taught in Italy, Germany, and Spain. For example, Dorothea Bocchi and Maddalena Buonsignori in the fifteenth century at the University of Bologna; Olivia Fulvia Morata at the University of Heidelberg, and Lucia de Medrano and Caterina Ribira at the University of Salamanca. In the eighteenth century the University of Bologna granted two doctorates to women. See Paul Oskar Kristeller, "Learned Women of Modern Italy: Humanists and University Scholars," in Labalme, ed., *Beyond Their Sex*.

17. See Kate Campbell Hurd-Mead, *A History of Women in Medicine: From the Earliest Times to the Beginning of the Nineteenth Century* (Haddam, Conn.: Haddam Press, 1938), pp. 424–32.

18. Cited in "A Surinam Portfolio," *Natural History*, Vol. LXXI, no. 10 (December 1962), p. 30.

19. For her life and descriptions of her studies, in addition to the work cited, see Harris and Nochlin, eds., pp. 153–54.

20. For her life and work see the biography by Edwards. For Châtelet and other scientists see also Margaret Alic, *Hypatia's Heritage: A History of Women in Science from Antiquity through the Nineteenth Century* (Boston: Beacon Press, 1986).

21. Cited in Edwards, p. 11.

22. Cited in Edwards, p. 12.

23. Edwards, p. 83.

24. See Edwards, pp. 110–15.

25. For her life see King, "Six Learned Women."

26. For the arguments which, like that of the seventh-century Church Father St. Ambrose, defend Eve as the weaker, less perfect creature and so not responsible for her sin in contrast to the stronger, more perfect Adam, see Margaret L. King, "The Religious Retreat of Isotta Nogarola (1418–1466): Sexism and Its Consequences in the Fifteenth Century, *Signs*, Vol. 3, no. 4 (Summer 1978), pp. 819–20.

27. Cited in King, "Religious Retreat," p. 816.

28. Cited in King, "Six Learned Women," p. 285.

29. Cited in King, "Book-lined Cells," in Labalme, ed., p. 76. Tiraqueau quoted on Marguerite of Navarre in M. A. Screech, p. 26.

30. Cited in King, "Book-lined Cells," in Labalme, ed., fn. p. 88

31. Cited in Altman, in Brink, ed., p. 11.

32. Ruth Kelso and Joan Kelly emphasize the uniqueness of this development, women arguing on their own behalf. Kelly sees it as the beginning of "feminism." See Kelso, p. 28; Joan Kelly, "Early Feminist Theory and the *Querelles des Femmes*," in Joan Kelly, *Women, History and Theory* (Chicago: University of Chicago Press, 1984), pp. 65–109.

33. They also accepted that the majority of women would live according to prevailing customs. For discussion of this aspect of Christine de Pizan's writing, see Sylvia Huot, "Seduction and Sublimation: Christine de Pizan, Jean de Meun, and Dante," *Romance Notes*, Vol. XXV, no. 3 (Spring 1985) and Susan Groag Bell, "Christine de Pizan (1364–1430): Humanism and the Problem of a Studious Woman," *Feminist Studies*, Vol. III, no. 3–4 (Spring–Summer 1976), p. 179; in *The Book of the City of Ladies*, trans. Earl Jeffrey Richards (New York: Persea Books, 1982), pp. 204, 206, and 212ff, Christine de Pizan identifies contemporary exemplary women by their relationship to a man, not by their own names (pp. 255–56); her *Book of the Three Virtues*, trans. Sarah Lawson (New York: Penguin Books, 1985) describes the life of women in many social strata and sets forth the traditional models of virtuous behavior for them to emulate.

34. Cited in Marianne S. Meijer, "Women's Lib in Sixteenth-Century France," unpublished manuscript, p. 3; for a summary of the qualities attributed to the female, see Kelso, pp. 11–12; for specific examples see Cappellanus, pp. 128, 201, 202, 208; Guillaume de Lorris and Jean de Meun, *The Romance of the Rose*, Charles W. Dunn, ed., trans. Harry W. Robbins (New York: E. P. Dutton, 1962), pp. 349, 351ff; Castiglione, pp. 132, 208; La Rochefoucauld, pp. 89, 111, 124; and Maclean, *Renvissance Notion*, which includes the sixteenth-century Italian ideas (see p. 52, for example). Sixteenth-century Italian jurists in their tracts on the law, with their rediscovery of the Eastern Roman Emperor Justinian's *Digest*, used this adverse image of women's nature to justify women's exclusion from the normal acts of citizenship. Their "*levitas, fragilitas, imbecillitas, infirmatas*" meant they could not inherit property. They could not fight, be with men and remain chaste, keep a secret or a promise. Their other supposed vices, like greed and ambition, made them even more untrustworthy as witnesses, or holders of civic office. Their weakness made them "simple-minded" in the eyes of the law, acting out of fear or self-interest. The sixteenth-century jurist Jacques Cujas took the reasoning to its logical conclusion in his comments on Roman law. As women were not included within the designation of humanity in one section of the *Digest*, perhaps they were not human. Then to murder a woman could not be a crime. For these arguments see Maclean, *Renaissance Notion*, pp. 70, 74, 77–79.

35. Cappellanus, p. 206; this is but an early version of a commonplace saying.

36. Cappellanus, p. 148.

37. For the English pamphlet by Joseph Swetnam, see Camden, pp. 255–56; for other examples see summaries in Blanche H. Dow, *The Varying Attitudes Towards Women*

in French Literature of the Fifteenth Century (New York: French Studies Institute, 1936), pp. 111, 107, and Joan Ferrante, *Women as Image in Medieval Literature from the Twelfth Century to Dante* (New York: Columbia University Press, 1975), pp. 108–12; see also Cappellanus, pp. 148–49, 207–9; the speech of the Duenna in *Romance of the Rose*, especially pp. 270, 272, 277, 286–88, 299; for a Humanist's reasoning see Sachs, p. 28; also La Rochefoucauld, p. 124.

38. See Pizan, *City of Ladies* p. 185, 253–54.

39. Pizan, *City of Ladies*, pp. 4, 119, 209; cited in Charity Cannon Willard, "The Franco-Italian Professional Writer: Christine de Pizan," in Katharina M. Wilson, ed., *Medieval Women Writers* (Athens, Ga.: University of Georgia Press, 1984), pp. 342, 343.

40. See, for example, Pizan, *City of Ladies*, p. 343.

41. Cited in Mcleod, p. 24. See, in addition to *City of Ladies*, citations from Charity Cannon Willard, *Christine de Pizan*, p. 63; Mcleod, pp. 131, 24. Selections from the correspondence on *The Romance of the Rose* and from *The Book of Three Virtues* can be found in Willard in Wilson, ed. Although *The City of Ladies* leans heavily for stories on Boccaccio's *Concerning Famous Women* and *The Decameron*, Christine de Pizan made original use of the material. For example, Medea and Dido became examples of constancy. See Christine de Pizan, *City of Ladies*, pp. xxvii, xxxviii, 69–70, 188–90. Her writings in defense of women from her "Tale of the Rose" to *The Book of Three Virtues* knew great popularity in the courts of her own time. *The Letter to the God of Love* is considered the first salvo in her battle with contemporary learned males, followed by her correspondence over Jean de Meun's *Romance of the Rose*. *The City of Ladies* went into numerous manuscript copies—twenty-seven are still extant—and was translated into English and Dutch. Scenes from it formed the designs for tapestries given to Marguerite of Austria, the Holy Roman Emperor Charles V's sister. See Mcleod, p. 133, and Stock, p. 44.

42. For these arguments of Zayas, see Foa, in Brink, ed., p. 57.

43. See Pizan, *City of Ladies*, pp. 118–19, 164, 166.

44. Cited in Patricia Labalme, "Three Feminists of Early Modern Venice: Modesta de Pozzo, Lucrezia Marinelli, Arcangela Tarabotti," unpublished manuscript p. 12.

45. See Zayas, cited in Nalle, p. 10; see Brachart, cited in Kelso, p. 64; *Women's Sharpe Revenge* cited in Margaret George, "From Goodwife to Mistress: The Transformation of the Female in Bourgeois Culture," *Science and Society*, Vol. XXXVII, no. 2 (Summer 1973), p. 171, and for the whole argument, pp. 171–73.

46. Cited in Mcleod, p. 77.

47. Cited in Judith Thurman, "Louise Labé: Still Scandalous After 400 Years," *Ms.*, Vol. 8 (March 1980), p. 92; Davis, *Society and Culture*, p. 74.

48. Cited in Foa, in Brink, ed., p. 59.

49. Cited in Irwin, in Brink, ed., p. 74.

50. Cited in Bell, "Christine de Pizan," p. 175.

51. Cited in Foa, in Brink, ed., p. 65.

52. Cited in Nalle, p. 10.

53. Cited in Foa, in Brink, ed., p. 60; see also p. 65.

54. Cited in Labalme, "Three Feminists," pp. 12, 15.

55. See Irwin, in Brink, ed. p. 76. Christine de Pizan made a similar point, citing Minerva and the Virgin Mary for their outstanding contributions to human history (*City of Ladies*, p. 63); see, for exemplars, *City of Ladies*, pp. 81, 91–92, 187, and Altman, in Brink, ed., p. 16. The English pamphleteer of *Women's Sharpe Revenge* insisted women were more moral than men. See George, p. 174. As part of the *querelles des femmes*, men made the same point. See especially Poullain de la Barre, cited in Rowan, p. 282. In the seventeenth century a group of male French writers also spoke of women as the superiors of men because of their superior virtue. A *femme forte*, as such a woman was called, showed the greater strength, for she by sheer force of will had overcome all that her nature decreed about her potential behavior. "Honor" came to such a woman as the perfect wife. "Honor" combined with chastity, one writer argued, "is their true liberty"; for then the woman could assume male roles and functions (Du Bosc, cited in Maclean, *Woman Triumphant*, p. 84); see for the whole argument especially pp. 58–71, 74–79, 86–87. The contemporary queen regents like Marie de' Medici and Anne of Austria, as well as a whole list of classical and biblical heroines, could be exemplars. Even Joan of Arc became a symbol of this model female.
56. Cited in Stock, p. 16.
57. See Kelso, pp. 77, 11, and Hilda Smith, "Gynecology and Ideology in Seventeenth-Century England," in Berenice A. Carroll, ed., *Liberating Women's History: Theoretical and Critical Essays* (Urbana: University of Illinois Press, 1976), p. 107. Gournay cited in Mäité Albistur and Daniel Armogathe, eds., *Histoire du féminisme français* (Paris: des femmes, 1977), Vol. I, p. 185.
58. See Maclean, *Renaissance Notion*, p. 29, for popularity of *Gynaecea*; question cited in Lewis, *Splendid Century*, p. 186.
59. See Stolpe, p. 44; Maclean, *Renaissance Notion*, pp. 34–35, 42–43.
60. See Smith, in Carroll, ed., p. 100; Maclean, *Renaissance Notion*, p. 43.
61. Cited in Merchant, p. 159.
62. Cited in Merchant, p. 160.
63. Cited in Julia O'Faolain and Lauro Martines, eds., *Not in God's Image: Women in History from the Greeks to the Victorians* (New York: Harper & Row, Publishers, 1973), pp. 121–22.
64. R. V. Schnucker, "The English Puritans and Pregnancy, Delivery and Breast-Feeding," *History of Childhood Quarterly*, Vol. I, no. 4 (Spring 1974), footnote, p. 653.
65. This idea of the uterus's influence emerged in classical and Christian learned treatises as the female-physiological equivalent to the idea of the penis as the uncontrollable part of the male anatomy. See, for example, Maclean, *Woman Triumphant*, pp. 9–11, 46–47, and Davis, *Culture and Society*, pp. 124–25.
66. See Maclean, *Renaissance Notion* p. 40; for a description of Rabelais, see Screech, pp. 92–93, 95; for the surgeon Ambrose Paré and others on hysteria, see Ilza Veith, *Hysteria: The History of a Disease* (Chicago: University of Chicago Press, 1965), pp. 29–39, 113–18.
67. Historians have described the impact of such dramatic changes in information gathering and dissemination for Europe. They cite the Protestant Reformation and the Scientific Revolution as impossible without printing presses. See Elizabeth L.

Eisenstein, *The Printing Press as an Agent of Change: Communications and Cultural Transformations in Early-Modern Europe* (New York: Cambridge University Press, 1979), Vol. I, p. 303.
68. Pizan, *City of Ladies*, p. 185.

PART VII WOMEN OF THE SALONS AND PARLORS

PART VII: **1. Women in the Salons**

1. For the Marquise de Rambouillet, see Dorothy Anne Liot Backer, *Precious Women: A Feminist Phenomenon in the Age of Louis XIV* (New York: Basic Books, 1974), section I.
2. Backer, p. 36.
3. Description in W. H. Lewis, *The Splendid Century* (New York: William Sloane Associates, 1954), p. 27. For photographs of the *chambre bleue* and a drawing of the Hôtel de Rambouillet, see *Esistere come donna*, catalogue of a 1982 exhibit produced by the Comune di Milano (Milan: Mazzota, 1983), p. 66.
4. Gloria Orenstein, "Natalie Barney's Parisian Salon: The Savoir Faire and Joie de Vivre of a Life of Love and Letters," *Thirteenth Moon*, Vols. I–II, combined issue (1980), pp. 76–94.
5. Cited in Evelyne Sullerot, *Women on Love: Eight Centuries of Feminine Writing* (London: Jill Norman, 1979), p. 121.
6. *Une ruelle*, by Abraham Bosse, reproduced in Backer, p. 5.
7. Cited in Lillian Day, *Ninon: A Courtesan of Quality* (Garden City, N.Y.: Doubleday, 1957), p. 197.
8. On this point see Carolyn C. Lougee, *Le Paradis des Femmes: Women, Salons, and Social Stratification in Seventeenth-Century France* (Princeton, N.J.: Princeton University Press, 1976), p. 122ff.
9. For Suzanne Necker, see J. Christopher Herold, *Mistress to an Age: A Life of Madame de Staël* (Indianapolis: Bobbs-Merrill Company, 1958), chap. 31; for Germaine de Staël, see Harold and Madelyn Gutwirth, *Madame de Staël, Novelist: The Emergence of the Artist as Woman* (Urbana: University of Illinois Press, 1978).
10. J. Christopher Herold, *Love in Five Temperaments* (New York: Atheneum. 1961), p. 54.
11. Herold, *Five Temperaments*, p. 201.
12. Cited in Anny Latour, *Uncrowned Queens*, trans. A. A. Dent (London: J. M. Dent and Sons, 1970), p. 68.
13. Cited in Latour, p. 73.
14. Cited in Latour, pp. 72 and 74.
15. Cited in Herold, *Five Temperaments*, p. 215. Guests and room descriptions from Herold, *Five Temperaments*, pp. 211–15.
16. G. Lemonnier's "A Reading at the Salon of Mme. Geoffrin," reproduced in *Esistere come donna*, p. 67.
17. Herold, *Five Temperaments*, p. 55.

18. Cited in Katharine M. Rogers, "The View from England," in Samia I. Spencer, ed., *French Women and the Age of Enlightenment* (Bloomington: Indiana University Press, 1984), p. 358.
19. Cited in Barbara Corrado Pope, "Revolution and Retreat: Upper-Class French Women After 1789," in Carol R. Berkin and Clara M. Lovett, eds., *Women, War, and Revolution* (New York: Holmes and Meier, 1980), p. 233, footnote 10.
20. Cited in Katharine M. Rogers, ed., *Before Their Time: Six Women Writers of the Eighteenth Century* (New York: Frederick Ungar, 1979), p. 109.
21. Many fanciful etymologies have been concocted to explain the origin of *Bluestocking*. This explanation (and supporting quotations) is from the *Oxford English Dictionary*.
22. Mary Wollstonecraft, *Letters Written during a Short Residence in Sweden, Norway, and Denmark* (Lincoln: University of Nebraska Press, 1976 [1796]), p. 181.
23. Deborah Hertz, *Jewish High Society in Old Regime Berlin* (New Haven, Conn.: Yale University Press, 1987), chap. 1.
24. Cited in Latour, p. 101.
25. Herz called her league a *Tugendbund*. Cited in Geneviève Bianquis, *Love in Germany*, trans. James Cleugh (Frederick Muller, 1964), p. 182.
26. For these salons, see Prudence Hannay, "Lady Blessington" and "Lady Holland," in Peter Quennell, ed., *Affairs of the Mind: The Salon in Europe and America from the 18th to the 20th Century* (Washington, D.C.: New Republic Books, 1980), pp. 23–46.
27. Cited in Latour, p. 101.
28. Mary Hargrave, *Some German Women and Their Salons* (London: T. Werner Laurie, 1912?), p. 151.
29. Cited in Latour, p. 116.
30. Cited in Rogers, *Before Their Time*, p. 66.
31. For Elstob, see Katharine M. Rogers, *Feminism in Eighteenth-Century England* (Urbana: University of Illinois Press, 1982), p. 261.
32. Quotations in Walter S. Scott, *The Bluestocking Ladies* (London: John Green & Co., 1947), pp. 47, 51.
33. Rogers, *Eighteenth-Century Feminism*, p. 255.
34. Cited in Scott, p. 203.
35. The first phrase is from the *Second Treatise on Government*; the second from the *First Treatise on Government* Cited in Lorenne M. G. Clark, "Women and Locke: Who owns the apples in the Garden of Eden," in Lorenne M. G. Clark and Lynda Lange, eds., *The Sexism of Social and Political Theory: Women and Reproduction from Plato to Nietzsche* (Toronto: University of Toronto Press, 1979), pp. 18–19.
36. David Hume, "My Own Life," in *An Inquiry Concerning Human Understanding* (Indianapolis: Bobbs-Merrill Company, 1955), p. 9.
37. Cited in Rogers, in Spencer, ed., pp. 358–59.
38. On this point, see Abby R. Kleinbaum, "Women in the Age of Light," in Renate Bridenthal and Claudia Koonz, eds., *Becoming Visible: Women in European History* (Boston: Houghton Mifflin, 1977), p. 221.

39. Terry Smiley Dock, "Women in the *Encyclopédie*" (1979), cited in Sara Ellen Procious Malueg, "Women and the *Encyclopédie*," in Spencer, ed., pp. 265–66.

40. From *Inquiries Concerning Human Understanding*, cited in Clark and Lange, eds., p. 60.

41. *The Spectator*, edited and with an introduction by Donald F. Bond, 5 vols. (Oxford: Clarendon Press, 1965), no. 435, July 19, 1712, p. 1.

42. Pauline Kra, "Montesquieu and Women," in Spencer, ed., pp. 281–82. Kra argues Montesquieu was quite favorable to women.

43. Cited in Lougee, p. 84.

44. Cited in Alain Decaux, *Histoire des françaises*, 2 vols. (Paris: Librarie Académique Perrin, 1972), vol. II, p. 619.

45. From *The Philosophy of Right*, cited in Rosemary Agonito, ed., *History of Ideas on Women: A Source Book* (New York: G. P. Putnam's Sons, 1977), p. 167.

46. Cited in Samuel Edwards, *The Divine Mistress* (New York: David McKay Co., 1970), p. 86. For more on Châtelet, see part VI, chaps. 1 and 5.

47. Edwards, frontispiece.

48. Jean-Jacques Rousseau, *Émile*, trans. Barbara Foxley (London: Everyman's Library, 1974), p. 370.

49. From *Observation on the Feeling of the Beautiful and Sublime*, in Agonito, p. 131.

50. "Si tu veux être heureux / N'épousez pas un bas-bleu," cited in Margaret Crosland, *Women of Iron and Velvet: French women writers after George Sand* (New York: Taplinger, 1976), p. 39.

51. Cynthia L. White, *Women's Magazine: 1693–1968* (London: Michael Joseph, 1970), p. 39.

52. Cited in Scott, p. 45.

53. *Spectator*, no. 342, April 2, 1712.

54. Addison, *Spectator*, no. 15, March 17, 1711.

55. Elinor Barber, *The Bourgeoisie in Eighteenth-Century France* (Princeton, N.J.: Princeton University Press, 1955), p. 80.

56. Reproduced in Carol Duncan, "Happy Mothers and Other New Ideas in Eighteenth-Century French Art," in Norma Broude and Mary D. Garrard, eds., *Feminism and Art History: Questioning the Litany* (New York: Harper & Row, Publishers, 1982), p. 202.

57. Cited in Duncan, in Groude and Garrard, eds., p. 202.

58. Rousseau, *Émile*, p. 13.

59. Cited in Ute Gerhard, *Verhältnisse und Verhinderungen: Frauenarbeit, Familie und Rechte der Frauen im 19. Jahrhundert. Mit Dokumenten* (Frankfurt am Main: Suhrkamp Verlag, 1978), p. 146.

60. Cited in Gisela Brinker-Gabler, ed., *Deutsche Dichterinnen vom. 16. Jahrhundert bis zur Gegenwart* (Frankfurt am Main: Fischer, 1978), p. 48.

61. Cited in Hilde Spiel, "Rahel Varnhagen," in Quennell, ed., p. 17.

62. From *A Treatise on Human Nature*, in Agonito, p. 126.

63. Jean-Jacques Rousseau, *Julie, or the New Heloise*, p. 325.

64. Samuel Richardson, *Pamela, or Virtue Rewarded*, intro. William M. Sale, Jr. (New York: Norton, 1958), p. ix.

65. So much so that a study of heroines in English novels from 1750 to 1900 was aptly titled *Pamela's Daughters*.
66. Rousseau, *Julie*, Vol. IV, p. 137.
67. Rousseau, *Émile*, p. 356.
68. Among them were Germaine de Staël, Manon Roland, George Sand, and George Eliot. See Gita May, "Rousseau's Anti-Feminism Reconsidered," in Spencer, ed., p. 310.
69. *Spectator*, no. 155, August 28, 1711.
70. Cited in Kleinbaum, in Bridenthal and Koonz, eds., p. 222.
71. Cited in Germaine Greer, *The Obstacle Race* (New York: Farrar, Straus & Giroux, 1979), p. 259.
72. For this point, see Pope, in Berkin and Lovatt, eds., pp. 216–17.
73. Cited in Elisabeth Badinter, *Mother Love: Myth and Reality—Motherhood in Modern History* (New York: Macmillan, 1981), p. 142.
74. Herold, *De Staël*, p. 179.
75. For this argument, see Pope, in Berkin and Lovett, eds.
76. Margaret H. Darrow, "French Noblewomen and the New Domesticity, 1750–1850," in *Feminist Studies*, Vol. 5, no. 1 (Spring 1979), pp. 41–65.
77. Cited in Adeline Daumard, *Les Bourgeois de Paris au XIXe Siècle* (Paris: Flammarion, 1970), p. 186.
78. Cited in Spiel, in Quennell, ed., p. 17.
79. For this point, see Richard Stites, *The Women's Liberation Movement in Russia: Feminism, Nihilism and Bolshevism, 1860–1930* (Princeton, N.J.: Princeton University Press, 1978), p. 15.
80. Johann Wolfgang von Goethe, *The Sorrows of Young Werther*, trans. Catherine Hutter (New York: New American Library, 1962), p. 35.
81. Rousseau, *Émile*, p. 350.
82. Mary Wollstonecraft, A *Vindication of the Rights of Woman* (New York: W. W. Norton and Company, 1977), ed. Carol H. Poston, pp. 9–10.
83. Wollstonecraft, *Vindication*, pp. 148–49.
84. Eleanor Flexner, *Mary Wollstonecraft* (Baltimore: Penguin, 1973), p. 90; for Wollstonecraft's life, also see Clare Tomalin, *The Life and Death of Mary Wollstonecraft* (London: Weidenfeld and Nicholson, 1974).
85. Wollstonecraft, *Letters*, p. 26.
86. Wollstonecraft, *Letters*, p. 160.
87. Mary Wollstonecraft, *Maria, or the Wrongs of Woman* (New York: Norton, 1975 [1797]), p. 104.
88. Wollstonecraft, *Maria*, p. 131.
89. Wollstonecraft, *Maria*, p. 65.
90. Cited in Scott, p. 207.
91. M. G. Jones, *Hannah More* (Cambridge: Cambridge University Press, 1952), p. 81.
92. Cited in Flexner, pp. 164–65.
93. Cited in Vineta Colby, *Yesterday's Women: Domestic Realism in the English Novel* (Princeton, N.J.: Princeton University Press, 1974), p. 121.
94. For this odd episode in More's life, see Jones, p. 16ff.

95. Cited in Jones, p. 115.
96. Cited in Flexner, p. 165.
97. Hannah More, *Coelebs in Search of a Wife*, Vol. II of *The Complete Works of Hannah More* (New York: Harper & Bros., 1847), p. 328.
98. More, *Coelebs*, p. 60.
99. More, *Coelebs*, p. 78.
100. More named her cats "Passive Obedience" and "Non-Resistance" (Jones, p. 219).
101. Cited in Jones, p. 265, footnote 25.

PART VII: **2. Women in the Parlors**

1. Cited in Winifred Gérin, *Elizabeth Gaskell: A Biography* (Oxford: Oxford University Press, 1980), p. 56.
2. Cited in Robert Palfrey Utter and Gwendolyn Bridges Needham, *Pamela's Daughters* (New York: Macmillan Company, 1936), p. 26.
3. For this argument, see Lousie A. Tilly and Joan W. Scott, *Women, Work and Family* (New York: Holt, Rinehart and Winston, 1978).
4. Cited in Bonnie G. Smith, *Ladies of the Leisure Class: The Bourgeoises of Northern France in the Nineteenth Century* (Princeton, N.J.: Princeton University Press, 1981), p. 41.
5. Smith, passim.
6. Cited in G. E. and K. R. Fussell, *The English Countrywoman: A Farmhouse Social History* (London: Andrew Melrose, 1953), p. 106.
7. Cited in Utter and Needham, p. 26.
8. From the thirteenth century on, according to the *Oxford English Dictionary*.
9. Cited in Ivy Pinchbeck, *Women Workers and the Industrial Revolution, 1750–1850* (New York: F. S. Crofts & Co., 1930), p. 31.
10. For this argument, see Barbara Corrado Pope, "Angels in the Devil's Workshop: Leisured and Charitable Women in Nineteenth-Century England and France," in Bridenthal and Koonz, eds., pp. 296–324.
11. Cited in Rita McWilliams-Tullberg, "Women and Degrees at Cambridge University, 1862–1897," in Martha Vicinus, ed., A *Widening Sphere: Changing Roles of Victorian Women* (Bloomington: Indiana University Press, 1977), p. 121.
12. Cited in Ute Gerhard, *Verhältnisse und Verhinderungen. Frauenarbeit, Familie und Rechte der Frauen im 19, Jahrhundert. Mit Dokumenten* (Frankfurt am Main: Suhrkamp, 1978), pp. 282–83. Authors' translation
13. Gerhard, pp. 290–91. Authors' translation.
14. *Journals of Dorothy Wordsworth*, 2nd ed., edited by Mary Moorman (New York: Oxford University Press, 1983), p. 33.
15. For this argument, see Olwen H. Hufton, *The Poor of Eighteenth-Century France: 1750–1789* (Oxford: Clarendon Press, 1974), p. 26.
16. Theresa M. McBride, *The Domestic Revolution: The Modernisation of Household Service in England and France, 1820–1920* (New York: Holmes and Meier, 1976), p. 45; Patricia Branca, *Silent Sisterhood: Middle-Class Women in the Victorian Home* (Pittsburgh: Carnegie-Mellon University Press, 1975), p. 36.

17. Cited in J . A. and Olive Banks, *Feminism and Family Planning in Victorian England* (New York: Schocken, 1964), p. 74.

18. T. H. Hollingsworth, "The Demography of the British Peerage," supplement to *Population Studies* XVIII (1964); Smith, p. 225.

19. Cited in Yvonne Knibiehler and Catherine Fouquet, *L'histoire des mères du moyenâge à nos jours* (Paris: Edition Montalba, 1980), p. 150.

20. Cited in Edward Shorter, *A History of Women's Bodies* (New York: Basic Books, 1982), p. 142.

21. Cited in Elizabeth Longford, *Queen Victoria: Born to Succeed* (New York: Pyramid Books, 1966), p. 234.

22. Shorter, pp. 148–49.

23. Shorter, chap. 6.

24. Margaret H. Darrow, "French Noblewomen and the New Domesticity, 1750–1850," *Feminist Studies*, Vol. 5, no. 1 (Spring 1979), p. 60.

25. For this argument, see Randolph Trumbach, *The Rise of the Egalitarian Family: Aristocratic Kinship and Domestic Relations in Eighteenth-Century England* (New York: Academic Press, 1978).

26. Scott and Tilly, p. 91; B. R. Mitchell, *European Historical Statistics, 1750–1950* (New York: Columbia University Press, 1976), pp. 127–33.

27. Cited in Gérin, p. 72.

28. Cited in Gérin, p. 74.

29. By W. Pickett, reproduced in R. J. White, *Life in Regency England* (London: B. T. Batsford, 1963), p. 46.

30. Cited in Gerhard, p. 295.

31. *The Poems of Mary Collier, The Washerwoman of Petersfield* (Petersfield, Eng.: W. Minchin, 1739).

32. Branca, p. 108.

33. Jan DeVries, *The Economy of Europe in an Age of Crisis: 1600–1750* (Cambridge: Cambridge University Press, 1976), p. 199.

34. For Germany, see Hugh Wiley Puckett, *Germany's Women Go Forward* (New York: AMS Press, 1967 [1929]), pp. 22–24; for France, Nina Gelbart, "The *Journal des Dames*," paper presented at Fifth Berkshire Conference on the Histon, of Women, Vassar College, Poughkeepsie, N.Y., 1981; for England, Alison Adburgham, *Women in Print: Writing Women and Women's Magazines from the Restoration to the Accession of Victoria* (London: George Allan & Unwin, 1972).

35. Adburgham, p. 68.

36. Phyllis Stock, *Better Than Rubies: A History of Women's Education* (New York: G. P. Putnam's Sons, 1978), p. 121; J.M.S. Tompkins, *The Popular Novel in England 1770–1800* (Lincoln: University of Nebraska Press, 1961), p. 120.

37. *The Lady's Museum* (London, 1760), Vol. I, pp. 11–12.

38. Rachel M. Brownstein, *Becoming a Heroine: Reading About Women in Novels* (New York: Penguin, 1984), p. 240.

39. Cited in Utter and Needham, p. 24.

40. *Oxford English Dictionary*, entry on *work*.

41. More precise figures about numbers of women in the middle or upper classes are

currently unavailable. For these figures and definitions, see Sarah Freeman, *Isabella and Sam: The Story of Mrs. Beeton* (New York: Coward, McCann and Geoghegan, 1978), p. 192.

42. Cited in C. Willett Cunnington, *Feminine Attitudes in the Nineteenth Century* (New York: Macmillan Company, 1936), p. 159.

43. Freeman, p. 203.

44. Gwen Raverat, *Period Piece: A Cambridge Childhood* (London: Faber and Faber, 1968 [1952]), p. 79.

45. Marguerite Perrot, *Le mode de vie des familles bourgeoises, 1873–1953* (Paris: Librarie Armand Colin, 1961), p. 89.

46. Leonore Davidoff, *The Best Circles: Women and Society in Victorian England* (Totowa, N.J.: Rowman and Littlefield, 1973), passim.

47. Raverat, p. 264.

48. Mrs. Alfred Sidgewick, *Home Life in Germany* (New York: Macmillan Company, 1909), pp. 200–201.

49. For the French *Encyclopédie*, see Malueg, in Spencer; and Kleinbaum, in Bridenthal and Koonz, eds., p. 223; for Germany, see Karin Hausen, "Family and Role-Division: The Polarisation of Sexual Stereotypes in the Nineteenth Century—An Aspect of the Disassociation of Work and Family Life," in Richard J. Evans and W. R. Lee, eds., *The German Family: Essays on the Social History of the Family in Nineteenth- and Twentieth-Century Germany* (London: Croom Helm, 1981), pp. 53–56.

50. Cited in Hausen, in Evans and Lee, eds., p. 64.

51. "Princess Ida." These lines are said by the Prince's father, who is the spokesman for conservatism.

52. Cited in Ruby V. Redinger, *George Eliot: The Emergent Self* (New York: Knopf, 1975), p. 61.

53. Cited in Janet Horowitz Murray, ed., *Strong-Minded Women and Other Lost Voices from Nineteenth-Century England* (New York: Pantheon, 1982), p. 102.

54. On this subject, see Ellen Moers, *The Dandy: From Brummell to Beerbohm* (Garden City, N.Y.: Doubleday & Company, 1965).

55. Only the short-lived Empire style of the early nineteenth century avoided emphasizing the waist and featured short hair until the fashion revolutions of the World War I era.

56. For earlier cases of women who cross-dressed, see Lillian Faderman, *Surpassing the Love of Men: Romantic Friendship and Love Between Women from the Renaissance to the Present* (New York: William Morrow, 1981), pp. 47–54.

57. George Sand, *My Life*, translated and adapted by Dan Hofstadter (New York: Harper & Row, Publishers, 1979 [1854–1855]), p. 203.

58. Sand, *My Life*, p. 204.

59. Brinker-Gabler, ed., p. 197.

60. The permit is reproduced in Karen Petersen and J. J. Wilson, *Women Artists: Recognition and Reappraisal from the Early Middle Ages to the Twentieth Century* (New York: Harper & Row, Publishers, 1976), p. 77. The printed part of the form says "dress like a man"; there was no equivalent form for men who wished to dress like women.

61. Richard J. Evans, *The Feminists: Women's Emancipation Movements in Europe, America and Australasia, 1840–1920* (London: Croom Helm, 1977), p. 94.

62. Such laws were passed in France in 1850; in Prussia in 1851; and in Austria in 1867.

63. Cited in Darlene Gay Levy, Harriet Branson Applewhite, and Mary Durham Johnson, eds., *Women in Revolutionary Paris: 1789–1795* (Urbana: University of Illinois Press, 1979), p. 219.

64. Cited in Evans, *Feminists*. The woman is not identified.

65. Thanks to Dorothy Helly for help with this point.

66. Inga Dahlsgård, *Women in Denmark Yesterday and Today* (Copenhagen: Det Danske Selskab, 1980), p. 39.

67. Barbara Beuys, *Familienleben in Deutschland: Neue Bilder aus der deutschen Vergangenheit* (Reinbeck bei Hamburg: Rowohlt, 1980), p. 339.

68. Bernard Schwartz, ed., *The Code Napoleon and the Common Law World* (New York: New York University Press, 1956), p. 146.

69. Daumard, p. 186; Theodore Zeldin, *France, 1848–1945: Ambition and Love* (New York: Oxford University Press, 1979), p. 18.

70. Cited in Stites, pp. 6–7; this code also gave a married woman the most liberal property rights in Europe, but retained her traditional obedience to her husband.

71. Cited in Julia O'Faolain and Laura Martines, eds., *Not In God's Image: Women in History from the Greeks to the Victorians* (New York: Harper & Row, Publishers, 1973), p. 318.

72. Patrick Kay Bidelman, *Pariahs Stand Up! The Founding of the Liberal Feminist Movement in France, 1858–1889* (Westport, Conn.: Greenwood Press, 1982), p. 5.

73. Cited in O'Faolain and Martines, eds., p. 321.

74. Catholic nations, except for France from 1792–1816 and after 1884, did not allow divorces. Some Protestant nations allowed divorce, but varied on whether it was easy or difficult to obtain. Relatively easy in Prussia and Scandinavia, it was much harder in Austria and Russia. The determining factor was the idea of guilt and punishment: nations which thought divorce should be punished made it harder to obtain. See Robert Chester, ed., *Divorce in Europe* (Leiden, Netherlands: Martinus Nijhoff, 1977), pp. 288–89.

75. Cited in Beuys, p. 339.

76. Theodore Stanton, ed., *The Woman Question in Europe: A Series of Original Essays* (New York: G. P. Putnam, 1970 [1884]), passim.

77. In Agonito, ed., p. 252.

78. Cited in Agonito, ed., pp. 260, 262–63.

79. From *The Study of Sociology*, p. 373, cited in Judith Jeffrey Howard, *The Woman Question in Italy, 1861–1880* (unpublished Ph.D. dissertation, 1977). On Spencer, also see T. S. Gray, "Herbert Spencer on Women: A Study in Personal and Political Disillusion," *International Journal of Women's Studies*, Vol. 7, no. 3 (1984), pp. 218–20.

80. Cited in Susan Groag Bell and Karen M. Offen, eds., *Women, the Family and Freedom: The Debate in Documents*, Vol. I, *1750–1880* (Stanford, Calif.: Stanford University Press, 1983), p. 221.

81. Cited in Marielouise Janssen-Jurreit, *Sexism: The Male Monopoly on History and Thought*, trans. Verne Moberg (New York: Farrar, Straus, & Giroux, 1982), p. 187.

82. Cited in Erna Olafson Hellerstein, Leslie Parker Hume, and Karen M. Offen, eds., *Victorian Women: A Documentary Account of Women's Lives in Nineteenth-Century England, France and the United States* (Stanford, Calif.: Stanford University Press, 1981), p. 90.

83. Hellerstein et al., eds., p. 94; Elaine and English Showalter, "Victorian Women and Menstruation," in Martha Vicinus, ed., *Suffer and Be Still: Women in the Victorian Age* (Bloomington, Ind.: Indiana University Press, 1972), pp. 38–44.

84. Cited in Bell and Offen, eds., Vol. I, p. 338.

85. Hellerstein et al., eds., p. 176ff.

86. Cited in Angus McLaren, *Birth Control in Nineteenth-Century England* (New York: Holmes and Meier, 1978), p. 83.

87. Priscilla Robertson, *An Experience of Women: Pattern and Change in Nineteenth-Century Europe* (Philadelphia: Temple University Press, 1982), pp. 32–33, Badinter, p. 231.

88. Cited in Bonnie S. Anderson, "The Writings of Catherine Gore," in the *Journal of Popular Culture*, Vol. X, no. 2 (Fall 1976), p. 102.

89. Cited in Anderson, p. 82.

90. Cited in Hellerstein et al., eds., p. 61.

91. Cited in Knibiehler, p. 199. Authors' translation.

92. Cited in Margaret Fuller, *Woman in the Nineteenth Century* (New York: W. W. Norton & Company, 1971 [18553]), p. 160.

93. Cited in Elaine Partnow, ed., *The Quotable Woman*, Vol. I, *1800–1899* (Los Angeles: Pinnacle Books, 1980), p. 24.

94. Cited in Maïté Albistur and Daniel Armogathe, eds., *Le grief des femmes: anthologie du textes féministes du IIe empire à nos jours* (Poitiers: Editions Hier et Demain, 1978), p. 131. Authors' translation.

95. All Marjory's selections are in her original spelling. Cited in Mary Jane Moffat and Charlotte Painter, eds., *Revelations: Diaries of Women* (New York: Random House, 1974), pp. 26, 22–23.

96. Cited in Moffat and Painter, eds., p. 23.

97. Cited in Ellen Moers, *Literary Women* (Garden City, N.Y.: Doubleday & Company, 1977), p. 27.

98. Cited in Hellerstein et al., eds., p. 63.

99. Cited in Mary S. Hartman, *Victorian Murderesses: A True History of Thirteen Respectable French and English Women Accused of Unspeakable Crimes* (New York: Schocken, 1977), p. 20.

100. Malwida von Meysenbug, *Memoirs: Rebel in Bombazine*, trans. Elsa von Meysenbug Lyons, ed. Mildred Adams (New York: W. W. Norton & Company, 1936 [1875]), p. 87.

101. Cited in Puckett, pp. 149–50.

102. Sand, *My Life*, p. 145.

103. Cited in Hartman, pp. 21, 23.

104. Cited in Hellerstein et al., eds., p. 146.

105. Cited in Ray Strachey, *The Cause: A Short History of the Women's Movement in Great Britain* (London: Virago, 1978 [1928]), p. 92.

106. Cited in Beuys, p. 357. Authors' translation.

107. Cited in Leonore Davidoff, *The Best Circles: Women and Society in Victorian England* (Totowa, N.J.: Rowman and Littlefield, 1973), p. 66

108. Cited in Hellerstein et al., eds., p. 148.

109. Daumard, p. 170.

110. Gore, *The Diary of a Disennuyée*, p. 204.

111. Cited in Robert Haven Schauffler, *Florestan: The Life and Work of Robert Schumann* (New York: Dover Publications, Inc., 1963 [1945]), p. 155.

112. Cited in Carl Storck, ed., *The Letters of Robert Schumann*, trans. Hannah Bryant (New York: Benjamin Blom, Inc., 1907), p. 230. For Clara Schumann, see Nancy B. Reich, *Clara Schumann: The Artist and the Woman* (Ithaca, N.Y.: Cornell University Press, 1985).

113. Cited in Bianquis, p. 162.

114. Cited in Hellerstein et al., eds., p. 181.

115. Cited in André Maurois, *Lélia: The Life of George Sand*, trans. Gerard Hopkins (New York: Harper & Bros., 1953), p. 74. Italics are in the original.

116. See, for instance, Lanier, in Laurel Holliday, ed., *Heart Songs. The Intimate Diaries of Young Girls* (Guerneville, Calif.: Bluestocking Books, 1978), passim; Dora Russell, *The Tamarisk Tree: My Quest for Liberty and Love* (London: Virago, 1978), p. 24, Vicki Baum, *It Was All Quite Different: The Memoirs of Vicki Baum* (New York: Funk and Wagnalls, 1964), pp. 123–26.

117. Cited in Hartman, p. 25.

118. Cited in Gérin, p. 53.

119. Cited in Hellerstein et al., eds., p. 209.

120. Cited in Moffat and Painter, eds., p. 142.

121. Cited in Hellerstein et al., eds., p. 200.

122. Cited in Schauffler, p. 165.

123. Cited in Laura S. Strumingher, "L'Ange de la Maison: Mothers and Daughters in Nineteenth-Century France," in *International Journal of Women's Studies*, Vol. 2, no. 1 (January–February 1979), p. 53.

124. Cited in Amy Kathleen Hackett, "The Politics of Feminism in Wilhelmine Germany, 1890–1918" (Ph.D. dissertation, Columbia University, 1976), p. 292.

125. Richard Wortman, "The Russian Empress as Mother," in David L. Ransel, ed., *The Family in Imperial Russia* (Urbana: University of Illinois Press, 1978), pp. 61, 63.

126. She was christened Marie Josèphe Rose Tascher de la Pagerie; her first husband was Alexandre de Beauharnais.

127. David Stacton, *The Bonapartes* (New York: Simon and Schuster, 1966), p. 18.

128. Cited in Hargrave, p. 236.

129. Cited in Hargrave, p. 217.

130. Cited in Longford, pp. 184, 174.

131. Longford, p. 500.

PART VII: **3. Leaving the Parlors**

1. Cited by Myra Stark in the introduction to *Cassandra*, by Florence Nightingale (Old Westbury, N.Y.: Feminist Press, 1979), p. 8.
2. Nightingale, pp. 11, 17.
3. Cited in Marghanita Laski, *George Eliot and Her World* (New York: Charles Scribner's Sons, 1973), p. 91.
4. Cited in Gérin, p. 254.
5. Cited in Gérin, pp. 138–39.
6. Cited in Gordon Rattray Taylor, *The Angel Makers: A Study in the Psychological Origins of Historical Change, 1750–1850* (New York: E. P. Dutton & Co., 1974 [1958]), p. 97.
7. Cited in Herold, *Mistress to an Age*, pp. 68–69, 233.
8. Cited in Cecil Woodham-Smith, *Florence Nightingale* (New York: McGraw-Hill, 1951), p. 20.
9. Simone de Beauvoir, *Memoirs of a Dutiful Daughter*, trans. James Kirkup (New York: Harper & Row, Publishers, 1959), pp. 295–96.
10. Cited in Maurois, p. 109.
11. Cited in Partnow, Vol. I, p. 20.
12. Cited in Maurois, p. 324. Also see Edith Thomas, *Les femmes en 1848* (Paris: Presses Universitaires de France, 1948), chap. 5.
13. "Dem Reich der Freiheit werb' ich Bürgerinnen" *Die Frauen-Zeitung von Louise Otto*, ed. Ute Gerhard, Elizabeth Hannover-Drück, and Romina Schmitter (Frankfurt am Main: Syndikat, 1979), p. 85. Authors' translation.
14. Both poems are in Elaine Hedges and Ingrid Wendt, *In Her Own Images: Women Working in the Arts* (Old Westbury, N.Y.: Feminist Press, 1980), p. 226.
15. Cited in Brinker-Gabler, p. 183. Authors' translation.
16. Cited in Partnow, Vol. I, p. 17.
17. Elaine Showalter, in *A Literature of Their Own: British Women Novelists from Brontë to Lessing* (Princeton, N.U.: Princeton University Press, 1977), p. 21, calls these novelists "feminine" novelists.
18. From "Am Turme," in Brinker-Gabler, ed., p. 170. Author's translation.
19. Cited in Françoise Basch, *Relative Creatures: Victorian Women in Society and the Novel*, trans. Anthony Rudolf (New York: Schocken, 1974), p. 106.
20. Far more has currently been written on women writers than painters or musicians. For painters, the chief works are Ann Sutherland Harris and Linda Nochlin, *Women Artists, 1550–1950* (New York: Knopf, 1982); Karen Petersen and J. J. Wilson, *Women Artists: Recognition and Reappraisal from the Early Middle Ages to the Twentieth Century* (New York: Harper & Row, Publishers, 1976); and Germaine Greer, *The Obstacle Race* (New York: Farrar, Straus & Giroux, 1979). For musicians, see Jane Bowers and Judith Tick, eds., *Women Making Music: The Western Art Tradition, 1150–1950* (Urbana: University of Illinois Press, 1986) and Carol Neuls-Bates, ed., *Women in Music: An Anthology of Source Readings from the Middle Ages to the Present* (New York: Harper & Row, Publishers, 1982).
21. Harris and Sutherland say *no* woman artist has since achieved such success, p. 190.

22. Even the very popular Lady Elizabeth Butler, who specialized in battle scenes, failed of election by two votes in 1879 (Harris and Sutherland, pp. 174, 249).

23. Greer, p. 315.

24. Harris and Sutherland, p. 223.

25. Reproduced in Greer, p. 311; see Greer, p. 230, for Bonheur's influence.

26. Paula Harper, "The Women of the Impressionist Movement: Berthe Morisot, Mary Cassatt, Marie Bracquemond," (unpublished manuscript, 1982), p. 3.

27. For Kollwitz's work, see Mina C. Klein and H. Arthur Klein, *Käthe Kollwitz: Life in Art* (New York: Schocken, 1976).

28. *Paula Modersohn-Becker: The Letters and Journals,* ed. Günter Busch and Liselotte von Reinken, ed. and trans. Arthur S. Wensinger and Carole Clew Hoey (New York: Taplinger, 1983), pp. 408–9.

29. The best source for Modersohn-Becker is her diaries and letters; for her paintings, see Gillian Perry, *Paula Modersohn-Becker: Her Life and Work* (New York: Harper & Row, Publishers, 1979).

30. Cited in Neuls-Bates, ed., p. 144.

31. Cited in Neuls-Bates, ed., pp. 147–48.

32. Cited in Neuls-Bates, ed., pp. 154–55.

33. Cited in Schauffler, p. 155; Neuls-Bates, ed., p. 154.

34. Cited in Jones, p. 19.

35. Nightingale, p. 37.

36. F. K. Prochaska, *Women and Philanthropy in Nineteenth-Century England* (Oxford: Clarendon Press, 1980), p. 141.

37. Smith, pp. 137–40.

38. Cited in Stanton, ed., p. 167.

39. Thomas Fowell Buxton, cited in Mervyn Williams, ed., *Revolutions, 1775–1830* (Baltimore: Penguin, 1971), p. 531. For Fry also see June Rose, *Elizabeth Fry, 1780–1845* (London: Macmillan, 1980).

40. Buxton in Williams, ed., p. 531.

41. Cited in Caroline Emelia Stephen, *The Service of the Poor: Being an Inquiry into the Reasons for and against the Establishment of Religious Sisterhoods for Charitable Purposes* (London: Macmillan, 1871), p. 100.

42. Cited in Hellerstein et al., eds., p. 431.

43. Cited in Prochaska, p. 114.

44. Claude Langlois, "Les Effectifs des congrégations féminines au XIXe siècle. De l'enquête statistique à l'histoire quantitatife," *Revue d'Histoire de l'Èglise en France,* Vol. 60 (January–June 1974), pp. 39–64.

45. Stanton, pp. 215–16.

46. Smith, p. 138; Prochaska, p. 30.

47. From *Woman's Mission* by Louis-Aimé Martin, translated in 1839 by Sarah Lewis, cited in Murray, p. 24.

48. Josephine G. Butler, *Autobiography,* ed. G.A.W. (London, 1911), p. 82.

49. Butler p. 31.

50. Cited in Introduction, by Gardner B. Taplin, to Elizabeth Barrett Browning, *Aurora Leigh* (Chicago: Academy Chicago, 1979), p. xx.

51. Butler, p. 58.
52. Butler pp. 60, 281.
53. Cited in Elizabeth Longford, *Eminent Victorian Women* (New York: Knopf, 1981), p. 115.
54. Cited in Prochaska, p. 220.
55. Cited in Eleanor S. Riemer and John C. Fout, eds., *European Women: A Documentary History, 1789–1945* (New York: Schocken, 1980), p. 10. For similar attitudes among French Catholic women, see Smith, p. 155.
56. Cited in Prochaska, p. 154.
57. For the French group, see Evans, *Feminists*, p. 152; for the Germans, Renate Bridenthal, "Class Struggle Around the Hearth: Women and Domestic Service in the Weimar Republic," in Michael Dubkowski and Isidor Walliman, eds., *Towards the Holocaust: Anti-Semitism and Fascism in Weimar Germany* (Westport, Conn.: Greenwood Press, 1983), pp. 243–64.
58. Cited in Stanton, ed., p. 163.
59. Cited in Margaret Bryant, *The Unexpected Revolution: A Study in the History of the Education of Women and Girls in the Nineteenth Century* (Windsor, Eng.: NFER Press, 1979).
60. On this subject, see Stock, p. 126, and part VIII, chap. 3.
61. Cited in Bryant, p. 93.
62. Cited in Stock, p. 164.
63. Catherine M. Prelinger, "Religious Dissent, Women's Rights, and the Hamburger Hochschule für das weibliche Geschlecht in mid-nineteenth-century Germany," *Church History*, Vol. 45 (March 1976), p. 55.
64. George Bernstein and Lottelore Bernstein, "The Curriculum for German Girls' Schools, 1870–1914," *Paedagogica Historica: International Journal of the History of Education*, Vol. XVIII, no. 2 (1978), p. 281.
65. Cited in Hackett, p. 80.
66. George Bernstein and Lottelore Bernstein, "Attitudes Toward Women's Education in Germany, 1870–1914," *International Journal of Women's Studies*, Vol. II, no. 5 (September–October 1979), pp. 473–81.
67. Cited in Richard J. Evans, *The Feminist Movement in Germany, 1894–1933* (London: Sage, 1976), p. 21.
68. Cited in Bernstein and Bernstein, "Attitudes," p. 485.
69. Cited in Hackett, p. 211, footnote 1.
70. Evans, *German Feminism*, p. 20.
71. Zeldin, p. 344.
72. For European women's earlier role in medicine, and their ouster from the profession in the fifteenth through seventeenth centuries, see Vol. I, part V, chap. 3.
73. Cited in Strachey, pp. 168–69.
74. Stock, p. 208.
75. Cited in Strachey, p. 181.
76. Stock, p. 208; Stanton, ed., p. 351.
77. Stites, pp. 85–86, 175.

78. James F. McMillan, *Housewife or Harlot: The Place of Women in French Society, 1870–1940* (New York: St. Martin's Press, 1981), p. 56.

79. Cited in Eve Curie, *Madame Curie: A Biography*, trans. Vincent Sheean (New York: Doubleday, Doran, 1939), p. 60.

80. Cited in Curie, p. 72.

81. Cited in Curie, p. 227.

82. Cited in Curie, p. 252.

83. Cited in Curie, p. 306.

84. Cited in Curie, p. 345.

85. Cited in Curie, p. 357.

86. Karen Horney, *The Adolescent Diaries of Karen Horney* (New York: Basic Books, 1980), p. 25.

87. Horney, *Diaries*, p. 110.

88. Horney, *Diaries*, pp. 147, 86.

89. See "The Flight from Womanhood" (1926) and "The Dread of Woman" (1932), in Karen Horney, *Feminine Psychology* (New York: W. W. Norton & Company, 1973)

90. Cited in McWilliams-Tullberg, in Vicinus, ed., *Suffer*, p. 122.

91. Cited in Dåhlstrom, p. 80.

92. Cited in Murray, ed., p. 157.

93. Cited in Murray, ed., p. 278.

94. Cited in Lee Holcombe, *Victorian Ladies at Work: Middle-Class Working Women in England and Wales, 1850–1914* (Hamden, Conn.: Archon Books, 1973), p. 27. Bedford College, founded a year after Queen's and also in London, evolved into a true college. A group of women who went on to found the English women's movement met there. See part IX, chap. 3. See also Joan N. Burstyn, *Victorian Education and the Ideal of Womanhood* (New Brunswick, N.J.: Rutgers University Press, 1984), pp. 23–24.

95. Cited in Murray, ed., p. 279; *The Red Virgin: Memoirs of Louise Michel*, ed. and trans. Bullitt Lowry and Elizabeth Ellington Gunter (University, Ala.: University of Alabama Press, 1981), pp. 39–40.

96. M. V. Hughes, *A London Home of the 1890s* (Oxford: Oxford University Press, 1979 [1934]) pp. 20–21.

97. Cited in David Thomson, ed., *France: Empire and Republic, 1850–1940: Historical Documents* (New York: Harper & Row, Publishers, 1968), p. 241.

98. Cited in Holcombe, p. 203.

99. Holcombe, pp. 165, 212; McMillan, p. 56. Also see Susan Bachrach, *Dames Employées: The Feminization of Postal Work in Nineteenth-Century France* (New York: Haworth, 1985).

100. Cited in Stanton, ed., p. 233.

PART VII: **4. Opportunities and Limits: Change and Tradition in the Twentieth Century**

1. Cited in Susan Groag Bell and Karen M. Offen, eds., *Women, the Family, and*

Freedom: The Debate in Documents, Vol. II: *1880–1950* (Stanford, Calif.: Stanford University Press, 1983), p. 277.

2. The number of female white-collar employees in Germany rose from 493,000 in 1907 to 1,446,000 in 1925. Renate Bridenthal and Claudia Koonz, "Beyond *Kinder, Küche, Kirche:* Weimar Women in Politics and Work," in Renate Bridenthal, Atina Grossmann, and Marion Kaplan, eds., *When Biology Became Destiny: Women in Weimar and Nazi Germany* (New York: Monthly Review Press, 1984), p. 64; in Great Britain, there were 505,000 women in commerce in 1914, 934,000 in 1918. In 1914, 262,000 women were in government employ (including education); by 1918, there were 460,000. Arthur Marwick, *The Deluge: British Society and the First World War* (New York: W. W. Norton & Company, 1965), p. 92. In France, 293,000 women were in the liberal professions and public service in 1906; by 1925, there were 491,000 of them. McMillan, p. 11.

3. Cited in Marwick, p. 88.

4. Sylvia Pankhurst, *The Suffragette Movement* (London: Virago, 1977 [1931]), p. 593; Janssen-Jurreit, p. 112; Stites, pp. 280–81.

5. Cited in Janssen-Jurreit, p. 130.

6. Cited in Karen Payne, ed., *Between Ourselves: Letters Between Mothers and Daughters, 1750–1982* (Boston: Houghton Mifflin, 1983).

7. Cited in Friedrich G. Kürbisch and Richard Klucsarits, eds., *Arbeiterinnen Kämpfen um ihr Recht: Autobiographische Texte zum Kampf rechtloser und entrechteter 'Frauenpersonen' in Deutschland, Osterreich, und der Schweiz des 19. und 20 Jahrhunderts* (Wuppertal: Peter Hammer Verlag, n.d.), p. 165. Authors' translation.

8. Cited in Marwick, p. 89; also Strachey, pp. 337, 348.

9. Cited in Payne, ed., p. 160.

10. Cited in McMillan, p. 110.

11. Cited in Stites, p. 300.

12. Cited in Strachey, p. 354; also see Marwick, pp. 99–105.

13. The document is reproduced in Renate Pore, *A Conflict of Interest: Women in German Social Democracy, 1919–1933* (Westport, Conn.: Greenwood Press, 1981), p. 46, footnote 36.

14. Cited in Jean McCrindle and Sheila Rowbotham, eds., *Dutiful Daughters: Women Talk About Their Lives* (Austin: University of Texas Press, 1977), p. 161.

15. Marwick, p. 89.

16. Strachey, pp. 386–87.

17. Cited in Marwick, p. 112.

18. Cited in Murray, ed., p. 73.

19. Horney, *Diaries*, pp. 61–62.

20. Cited in Ruth Hall, ed., *Dear Dr. Stopes: Sex in the 1920s* (Harmondsworth, Eng.: Penguin, 1981), p. 51.

21. Roseanna Ledbetter, *A History of the Malthusian League: 1877–1927* (Columbus: Ohio State University Press, 1976), p. 237.

22. Cited in Hall, ed., p. 180.

23. McMillan, p. 166.

24. de Beauvoir, *Dutiful Daughter*, p. 178.

25. de Beauvoir, p. 106.
26. de Beauvoir, pp. 151–52.
27. de Beauvoir, p. 174.
28. de Beauvoir, pp. 225, 240.
29. de Beauvoir, pp. 251–52.
30. de Beauvoir, p. 271.
31. de Beauvoir, p. 241.
32. Simone de Beauvoir, *The Second Sex*, trans. H. M. Parshley (New York: Bantam, 1961), p. xxviii.
33. McMillan, p. 159.
34. Cited in Werner Thönessen, *The Emancipation of Women: The Rise and Decline of the Women's Movement in German Social Democracy, 1863–1933*, trans. Joris des Bres (Glasgow: Pluto Press, 1976), p. 91.
35. Claudia Koonz, "Conflicting Allegiances: Political Ideology and Women Legislators in Weimar Germany," *Signs*, Vol. I, no. 3, pt. 1 (Spring 1976), p. 680; Pore, p. 110, footnote 46.
36. Sheila Lewenhak, *Women and Trade Unions. An Outline History of Women in the British Trade Union Movement* (New York: St. Martin's Press, 1977), p. 183.
37. Edith Summerskill, *A Woman's World* (London: Heinemann, 1967), pp. 21–22, Carol Dyhouse, "Towards a 'Feminine' Curriculum for English Schoolgirls: The Demands of Ideology, 1870–1963," *Women's Studies International Quarterly*, Vol. I, no. 4 (1978), p. 307.
38. Dahlsgård, pp. 147–48; Bridenthal and Koonz, in Bridenthal et al., eds., p. 53.
39. Cited in Bell and Offen, eds., Vol. II, p. 314.
40. Cited in Bell and Offen, eds., Vol. II, pp. 369–70.
41. Cited in Bell and Offen, eds., Vol. II, p. 307.
42. Knibiehler and Fouquet, p. 319.
43. *Esistere come donna*, p. 202; Gisela Bock, "Racism and Sexism in Nazi Germany: Motherhood, Compulsory Sterilization and the State," in Bridenthal et al., eds.; Jill Stephenson, *Women in Nazi Society* (New York: Harper & Row, Publishers, 1975), p. 49.
44. Claudia Koonz, *Mothers in the Fatherland: Women, the Family, and Nazi Politics* (New York: St. Martin's Press, 1987), chap. 6.
45. Cited in Malcolm Potts, Peter Diggory, and John Peel, *Abortion* (Cambridge: Cambridge University Press, 1977), p. 382.
46. Gail Warshofsky Lapidus, *Women in Soviet Society: Equality, Development and Social Change* (Berkeley: University of California Press, 1978), pp. 113, 117.
47. Bell and Offen, eds., Vol. II, pp. 410–11; Carola Hansson and Karen Lidén, *Moscow Women: Thirteen Interviews*, trans. Gerry Bothmer, George Blecher, and Lone Blecher (New York: Pantheon, 1983), p. 125.
48. For more on maternity legislation, see part VIII, chap. 3.
49. Gisbert H. Flanz, *Comparative Women's Rights and Political Participation in Europe* (Dobbs Ferry, N.Y.: Transnational Publishers, 1983), p. 30ff.
50. Cited in Strachey, p. 384.
51. Cited in Stephenson, pp. 15–16; cited in Hackett, p. 1052.

52. Evans, *Feminists*, p. 213.
53. Lapidus, pp. 64, 210.
54. Koonz "Conflicting Allegiances," pp. 672–73.
55. Cited in Charles Sowerwine, *Sisters or Citizens? Women and Socialism in France since 1876* (Cambridge: Cambridge University Press, 1982), p. 168.
56. Summerskill pp. 40–41.
57. Summerskill pp. 179–80.
58. Bridenthal and Koonz, in Bridenthal et al., eds., p. 36.
59. See, for example, Hannah Mitchell, *The Hard Way Up: The Autobiography of Hannah Mitchell, Suffragette and Rebel*, ed. Geoffry Mitchell (London: Virago, 1977), p. 141.
60. A Nazi song, cited in Annette Kuhn and Gerhard Schneider, eds., *Frauen in der Geschichte* (Düsseldorf: Schwann, 1979), p. 210. Authors' translation.
61. Koonz, *Mothers*, p. 206. Thanks to the author for letting us read this important work in manuscript.
62. Koonz, *Mothers*, p. 59; Leila J. Rupp, " 'I Don't Call That *Volksgemeinschaft!*': Women, Class, and War in Nazi Germany," in Berkin and Lovett, eds., pp. 37–54.
63. World War II and the experiences of the majority of European women are analyzed in part VIII, chap. 4.
64. David Schoenbrun, *Soldiers of the Night: The Story of the French Resistance* (New York: E. P. Dutton & Co., 1980), p. 91.
65. Schoenbrun, pp. 98–99. For Fourcade, also see Marie Madeleine Fourcade, *Noah's Ark*, trans. Kenneth Morgan (New York: E. P. Dutton & Co., 1974) and Margaret L. Rossiter, *Women in the Resistance* (New York: Praeger, 1986), pp. 125–29.
66. Cited in Vera Laska, ed., *Women in the Resistance and in the Holocaust: The Voices of Eyewitnesses* (Westport, Conn.: Greenwood Press, 1983), p. 149.
67. Françoise Giroud, *I Give You My Word*, trans. Richard Seaver (Boston: Houghton Mifflin, 1974), pp. 80–81.
68. Giroud, p. 108; Cited in Elizabeth Wilson, *Only Halfway to Paradise: Women in Postwar Britain, 1945–1968* (London: Tavistock, 1980), p. 83.
69. Giroud, p. 115.
70. On Pappenheim, see Marion A. Kaplan, *The Jewish Feminist Movement in Germany: The Campaigns of the Jüdischer Frauenbund, 1904–1933* (Westport, Conn.: Greenwood Press, 1979), chap. 2, Horney, *Diaries*, p. 270.
71. Sigmund Freud, *An Outline of Psychoanalysis*, trans. James Strachey (New York: W. W. Norton & Company, 1963), p. 107.
72. Horney, "Flight from Womanhood," in *Feminine Psychology*, p. 54.
73. For a detailed analysis of these attitudes, see Vol. I, part I, "Traditions Inherited."
74. Sigmund Freud, *Three Contributions to the Theory of Sex*, Book III in *The Basic Writings of Sigmund Freud*, trans. A. A. Brill (New York: Random House, 1938), p. 612, footnote 3; Sigmund Freud, *Analysis Terminable and Interminable*, cited in Karen Horney, *New Ways in Psychoanalysis* (New York: W. W. Norton & Company, 1939), p. 110.
75. Freud, "Femininity," p. 314.

76. Freud, "Femininity," pp. 315, 321.
77. Cited in Bell and Offen, eds., Vol. II, p. 357.
78. Karen Horney, "Maternal Conflicts," in *Feminine Psychology*, p. 175.
79. Atina Grossmann, lecture, Barnard Conference on the Scholar and the Feminist, Barnard College, New York, 1981.
80. On this see Elizabeth Mavor, *The Ladies of Langollen: A Study in Romantic Friendship* (Baltimore: Penguin, 1971).
81. Faderman, p. 149.
82. Charlotte Wolff, *Hindsight* (London: Quartet Books, 1980), p. 76.
83. Cited in Badinter, p. 317.
84. Shirley Conran, *Superwoman* (New York: Bantam, 1978), p. 298.
85. Conran, p. 298.
86. Helge Pross, cited in Harry G. Shaffer, *Women in the Two Germanies: A Comparative Study of a Socialist and a Non-Socialist Society* (New York: Pergamon Press, 1981), p. 137.
87. Viola Klein, cited in Wilson, p. 55; the French study is Catherine Bodard Silver, "Salon, Foyer, Bureau: Women and the Professions in France," in Mary Hartman and Lois W. Banner, eds., *Clio's Consciousness Raised: New Perspectives on the History of Women* (New York: Harper & Row, Publishers, 1974), pp. 72–85.
88. See Norton T. Dodge, "Women in the Professions," in Dorothy Atkinson, Alexander Dallin, and Gail Warshofsky Lapidus, eds., *Women in Russia* (Stanford, Calif.: Stanford University Press, 1977), pp. 205–24; Evelyne Sullerot, *Women, Society and Change*, trans. Margaret Scotford Archer (New York: McGraw-Hill, 1976), pp. 148, 155; Joni Lovenduski and Jill Hills, eds., *The Politics of the Second Electorate: Women and Public Participation* (London: Routledge and Kegan Paul, 1981), p. 27.
89. Cited in Payne, ed., pp. 185–86.
90. Cited in Robertson, p. 508.
91. Cited in Ann Oakley, *The Sociology of Housework* (New York: Random House, 1974), p. 191.
92. Cited in Shaffer, p. 115.
93. Cited in Albistur and Armogathe, Vol. II, p. 649. Authors' translation.

PART VIII   WOMEN OF THE CITIES

PART VIII: **1. Family Life**

1. The sources for their lives are Lucy Luck, "A Little of My Life," *The London Mercury* XIII (November 1925–April 1926), pp. 354–73, and May Hobbs, *Born To Struggle* (Plainfield, Vt.: Daughters, 1975).
2. Luck, p. 354.
3. Luck, p. 355.
4. Luck, p. 356.
5. Luck, p. 365.
6. Luck, p. 366.
7. Luck, p. 369.

8. Luck, p. 370.
9. Luck, p. 371.
10. Luck, p. 372.
11. Luck, p. 373
12. Luck, p. 373.
13. Luck, p. 373.
14. Luck, p. 354.
15. Hobbs, p. 16.
16. Hobbs, p. 28.
17. Hobbs, p. ii.
18. Hobbs, p. 44.
19. Hobbs, p. 5.
20. Hobbs, p. 62.
21. Cited in Hobbs, back jacket.
22. Hobbs, p. 86.
23. Luck, p. 355.
24. Hobbs, p. i.
25. The argument and figures are from Fernand Braudel, *Capitalism and Material Life: 1400–1800*, trans. Miriam Kochan (New York: Harper & Row, Publishers, 1973). Also see Jan DeVries, *European Urbanization: 1500–1800* (Cambridge, Mass.: Harvard University Press, 1984), pp. 43–48.
26. Jan DeVries, *The Economy of Europe in an Age of Crisis: 1600–1750* (Cambridge: Cambridge University Press, 1976), p. 154; De Vries, *Urbanization*, table 3.8, p. 46.
27. Adna Ferrin Weber, *The Growth of Cities in the Nineteenth Century: A Study in Statistics* (Ithaca, N.Y.: Cornell University Press, 1967, [1899]), p. 450.
28. Jürgen Kuczynski, *The Rise of the Working Class*, trans. C. T. A. Ray (New York: McGraw-Hill, 1971), table 6, p. 154; DeVries, *Urbanization*, p. 43.
29. DeVries, *Urbanization*, table 3.8, p. 45.
30. Weber, p. 449; DeVries, *Urbanization*, table 3.8, p. 46.
31. *The London Journal of Flora Tristan, 1842, or Aristocracy and the Working Class of England*, trans. Jean Hawkes (London: Virago, 1982), p. 156.
32. Tristan, p. 157.
33. Weber, p. 276.
34. Janet Roebuck, *The Shaping of Urban Society: A History of City Forms and Functions* (New York: Charles Scribner's Sons, 1974), p. 157.
35. Roebuck, p. 145.
36. Roebuck, p. 145.
37. E. Royston Pike, *"Golden Times": Human Documents of the Victorian Age* (New York: Schocken, 1972), p. 279.
38. Cited in Erna Olafson Hellerstein, Leslie Parker Hume, and Karen M. Offen, eds., *Victorian Women: A Documentary Account of Women's Lives in Nineteenth-Century England, France, and the United States* (Stanford, Calif.: Stanford University Press, 1981), p. 321.
39. Cited in Barbara Taylor, "'The Men Are As Bad As Their Masters . . .': Socialism,

Feminism, and Sexual Antagonism in the London Tailoring Trade in the Early 1830s," *Feminist Studies*, Vol. 5, no. 1 (Spring 1979), p. 20.

40. Henry Mayhew, *London Labour and the London Poor*, 4 vols. (New York: Dover, 1968 [1861–1862]), Vol. I, p. 88.

41. Cited in Louise A. Tilly and Joan W. Scott, *Women, Work, and Family* (New York: Holt, Rinehart and Winston, 1978), pp. 121–22; also see Barbara Franzoi, *At the Very Least She Pays the Rent: Women and German Industrialization, 1871–1914* (Westport, Conn.: Greenwood Press, 1985), chap. 2.

42. Grace Foakes, *My Part of the River* (London: Futura Books, 1976), p. 116.

43. Cited in Ivy Pinchbeck, *Women Workers and the Industrial Revolution, 1750–1850* (New York: F. S. Crofts, 1930), pp. 1–2. See also the London court case, c. 1880, in which a wife justified throwing a cup at her husband because he sent her out to look for work and stayed home himself. Cited in Ellen Ross, " 'Fierce Questions and Taunts': Married Life in Working-Class London, 1870–1914," *Feminist Studies*, Vol. 8, no. 3 (Fall 1982), pp. 581–82.

44. Cited in Maud Pember Reeves, *Round About a Pound a Week* (London: Virago, 1979 [1913]), p. 16.

45. Foakes, p. 145.

46. Robert Roberts, *The Classic Slum: Salford Life in the First Quarter of the Century* (New York: Penguin, 1973), p. 44.

47. Enid Gauldie, *Cruel Habitations: A History of Working-Class Housing, 1780–1918* (London: George Allen and Unwin, 1974), p. 22.

48. Roger Thabault, *Education and Change in a Village Community: Mazières-en-Gâtine, 1848–1914*, trans. Peter Tregear (New York: Schocken, 1971 [1945]), pp. 100–101.

49. Roberts, p. 54.

50. Cited in Friedrich G. Kürbisch and Richard Klucsarits, eds., *Arbeiterinnen Kämpfen um ihr Recht: Autobiographische Texte zum Kampf rechtloser und entrechteter 'Frauenpersonen' in Deutschland, Österreich, und der Schweiz des 19. und 20. Jahrhunderts* (Wuppertal: Peter Hammer, n.d.), p. 55. See also Nancy A. Tomes, "A 'Torrent of Abuse': Crimes of Violence between Working-class Men and Women in London, 1840–1875," in *The Journal of Social History*, Vol. 4, no. 1 (Spring 1978), p. 337.

51. Pember Reeves, p. 64.

52. *My Song Is My Own: 100 Women's Songs*, ed. Kathy Henderson with Frankie Armstrong and Sandra Kerr (London: Pluto Press, 1979), pp. 63–64.

53. Pember Reeves, p. 133.

54. Cited in Jane Lewis, *Women in England, 1820–1950: Sexual Divisions and Social Change* (Bloomington: Indiana University Press, 1984), p. 48.

55. Peter N. Stearns, "Working-Class Women in Britain, 1890–1914," in Martha Vicinus, ed., *Suffer and Be Still: Women in the Victorian Age* (Bloomington: Indiana University Press, 1979), p. 116.

56. Foakes, p. 100.

57. Tomes, p. 338.

58. Cited in Taylor, p. 22.

59. Tomes, p. 336.

60. Mayhew, Vol. I, p. 24.

61. Foakes, p. 145.

62. See M. Dorothy George, *London Life in the Eighteenth Century* (Harmondsworth, Eng.: Penguin, 1978), p. 118.

63. Carlo M. Cipolla, *The Economic History of World Population*, 7th ed. (Harmondsworth, Eng.: Penguin, 1978), p. 118.

64. E. A. Wrigley, *Population and History* (New York: McGraw-Hill, 1969), p. 152.

65. Thomas McKeown, *The Modern Rise of Population* (New York: Harcourt Brace Jovanovich, 1976), p. 34.

66. See Rudolf Braun, *Industrialisierung und Volksleben* (Zurich: Erlenbach, 1960); David Levine, *Family Formation in the Age of Nascent Capitalism* (New York: Harcourt Brace Jovanovich, 1977); and DeVries, *Age of Crisis*, p. 11ff.

67. Eda Sagarra, *A Social History of Germany: 1648–1914* (New York: Holmes and Meier, 1977), p. 10.

68. McKeown, passim.

69. Tilly and Scott, pp. 172–73.

70. Wrigley, pp. 185, 197.

71. Antonina Martynova, "Life of the Pre-Revolutionary Village As Reflected in Popular Lullabies," in David Ransel, ed., *The Family in Imperial Russia: New Lines of Historical Research* (Urbana: University of Illinois Press, 1978), p. 172.

72. Margaret Llewelyn Davies, ed., *Life As We Have Known It, by Co-operative Working Women* (New York: W. W. Norton, 1975 [1931]), p. 63.

73. Joan W. Scott and Louise A. Tilly, "Women's Work and the Family in Nineteenth-Century Europe," in Charles E. Rosenberg, ed., *The Family in History* (Philadelphia: University of Pennsylvania Press, 1975), p. 159, footnote 44.

74. Sheila Ryan Johansson, "Sex and Death in Victorian England: An Examination of Age and Sex Specific Death Rates, 1840–1910," in Martha Vicinus, ed., *A Widening Sphere: Changing Roles of Victorian Women* (Bloomington: Indiana University Press, 1977), p. 163.

75. Arthur E. Imhof, "Women, Family and Death: Excess Mortality of Women of Child-Bearing Age in Four Communities in Nineteenth-Century Germany," in Richard J. Evans and W. R. Lee, eds., *The German Family: Essays on the Social History of the Family in Nineteenth- and Twentieth-Century Germany* (London: Croom Helm, 1981), p. 149.

76. Imhof, P. 161.

77. McKeown, p. 105.

78. Although the Viennese doctor Ignaz Semmelweiss analyzed the causes of puerperal fever in the 1850s, his ideas did not gain wide acceptance until the 1890s.

79. Robert Lee, "Family and 'Modernization': The Peasant Family and Social Change in Nineteenth-Century Bavaria," in Evans and Lee, eds., p. 95

80. Johansson, in Vicinus, ed., p. 169.

81. Wrigley, p. 188, and William L. Langer, "The Origins of the Birth Control Movement in England in the Early Nineteenth Century," in Robert I. Rotberg and

Theodore K. Rabb, eds., *Marriage and Fertility: Studies in Interdisciplinary History* (Princeton, N.U.. Princeton University Press, 1980), p. 283, footnote 39.

82. McKeown, p. 147.

83. On this subject, see Jacques Depauw, "Illicit Sexual Activity and Society in Eighteenth-Century Nantes," in *Family and Society: Selections from the Annals*, ed. Robert Forster and Ornest Ranum, trans. Elborg Forster and Patricia M. Ranum (Baltimore: Johns Hopkins University Press, 1976); John R. Gillis, "Servants, Sexual Relations, and the Risks of Illegitimacy in London, 1801–1900," *Feminist Studies* Vol. 5, no. 1 (Spring 1979), pp. 142–73; P.E.H. Hair, "Bridal Pregnancy in Rural England in Earlier Centuries," *Population Studies*, Vol. XX, no. 2 (November 1966), pp. 233–43; and J. Michael Phayer, "Lower Class Morality: The Case of Bavaria," *Journal of Social History*, Vol. 8 (Fall 1974), pp. 79–95.

84. Law, in Ute Gerhard, *Verhältnisse und Verhinderungen, Frauenarbeit, Familie und Rechte der Frauen im 19. Jahrhundert. Mit Dokumenten* (Frankfurt am Main: Suhrkamp, 1978), pp. 457–58. The unwed mother received no payment if she was of "ill repute" (*bescholtenes*), i.e., if she had had intercourse with other men, if she had taken money from other men, if she had been pregnant before, if she had been divorced, or if the man were under twenty.

85. Judith Jeffrey Howard, "The Woman Question in Italy, 1861–1880," unpublished Ph.D. dissertation, 1977, passim.

86. K. H. Connell, *Irish Peasant Society* (Oxford: Clarendon Press, 1968), p. 58.

87. On this subject, see Olwen H. Hufton, *The Poor of Eighteenth-Century France: 1750–1789* (Oxford: Clarendon Press, 1974), p. 231; Gillis, pp. 162–63; Cissie Fairchilds, "Female Sexual Attitudes and the Rise of Illegitimacy: A Case Study," in Rotberg and Rabb, pp. 163–204; and Cissie Fairchilds, *Domestic Enemies: Servants and Their Masters in Old Regime France* (Baltimore: Johns Hopkins University Press, 1984), pp. 86–87.

88. Cited in *Arberterinnen*, p. 178. Authors' translation.

89. This point has become controversial, primarily because of Edward Shorter's assertion in his influential book *The Making of the Modern Family* (1975) that the rise in illegitimacy rates signals a "sexual revolution" in which young women took the lead in initiating sex for pleasure. He connected this assertion to his belief that romantic love and "affective" sexual relations arose only in the late eighteenth century, with the growth of capitalism. The history of European women provides myriad examples of true love and sexual pleasure before the eighteenth century; it supplies no evidence that women of the working class suddenly discovered sexual pleasure and acted on this discovery around 1800.

90. Foakes, p. 78.

91. Davies, ed., p. 1. For similar testimonies, also see Margaret Llewellyn Davies, ed., *Maternity: Letters from Working Women* (New York: W. W. Norton & Company, 1978 [1915]).

92. Angus McLaren, "Abortion in France: Women and the Regulation of Family Size, 1800–1914," in *French Historical Studies*, Vol. X, no. 3 (Spring 1978), p. 462.

93. McLaren, "Abortion in France," p. 472, and Angus McLaren, *Birth Control in Nineteenth-Century England* (New York: Holmes and Meier, 1978), chap. 13.

94. Mary Wollstonecraft, *Letters Written During a Short Residence in Sweden, Norway, and Denmark* (Lincoln: University of Nebraska Press, 1976 [1796]), p. 34.

95. For reasons which still remain unclear, Frenchwomen of all classes used wet nurses in greater numbers than did other Europeans. Olwen Hufton, "Women and the Family Economy in Eighteenth-Century France," *French Historical Studies*, Vol. 9 (Spring 1975), p. 12, and George D. Sussman, *Selling Mother's Milk: The Wet-Nursing Business in France, 1715–1914* (Urbana: University of Illinois Press, 1982).

96. Fanny Fäy-Sallois, *Les Nourrices à Paris au XIX siècle* (Paris: Payot, 1980), p. 243.

97. David L. Ransel, "Abandonment and Fosterage of Unwanted Children: The Women of the Foundling System," in Ransel, ed., p. 206.

98. Hufton, *The Poor of Eighteenth-Century France*, p. 326.

99. Ann Oakley, "Wisewomen and Medicine Man: Changes in the Management of Childbirth," in Juliet Mitchell and Ann Oakley, eds., *The Rights and Wrongs of Women* (Harmondsworth, Eng.: Penguin, 1979), pp. 42–43.

100. William L. Langer, "Infanticide: A Historical Survey," *History of Childhood Quarterly*, Vol. I, no. 3 (Winter 1974), p. 358.

101. Hufton, *The Poor of Eighteenth-Century France*, p. 319; Langer, p. 359; Ransel, in Ransel, ed., p. 193.

102. Langer, pp. 358–59.

103. Hufton, *The Poor of Eighteenth-Century France*, p. 332.

104. Langer, "Infanticide," p. 356.

105. Pierre Guiral and Guy Thuillier, *La Vie Quotidienne des Domestiques en France au XIXe siècle* (Paris: Hachette, 1978), p. 141.

106. Figures from Nottingham, England, 1856–1861, cited in Hellerstein et al., p. 204.

107. Louise Otto, *"Dem Reich der Freiheit werb' ich Burgerinnen": Die Frauen-Zeitung von Louise Otto*, ed. Ute Gerhard, Elizabeth Hannover-Drück, and Romina Schmitter (Frankfurt am Main: Syndikat, 1980), p. 182. Authors' translation.

108. McKeown, p. 147.

PART VIII: **2. Earning Income**

1. Theresa M. McBride, *The Domestic Revolution: The Modernisation of Household Service in England and France, 1820–1920* (New York: Holmes and Meier, 1976), p. 111.

2. Figures from Inga Dahlsgård, *Women in Denmark: Yesterday and Today* (Copenhagen: Det Danske Selskab, 1980), p. 104.

3. Cited in Barbara Taylor, *Eve and the New Jerusalem: Socialism and Feminism in the Nineteenth Century* (New York: Pantheon, 1983), p. 96.

4. For this argument, see Jean H. Quataert, "The Shaping of Women's Work in Manufacturing: Guilds, Households, and the State in Central Europe, 1648–1870," in the *American Historical Review*, Vol. 90, no. 4 (December 1985), pp. 1, 122–48.

5. Sally Alexander, "Women's Work in Nineteenth-Century London: A Study of the Years 1820–1850," in Mitchell and Oakley, pp. 74, 82–83.

6. Cited in Jill Liddington and Jill Norris, *One Hand Tied Behind Us: The Rise of the Women's Suffrage Movement* (London: Virago, 1978), p. 36.

7. Cited in Margaret Hewitt, *Wives and Mothers in Victorian Industry* (Westport, Conn.: Greenwood Press, 1975 [1958]), p. 22.

8. Cited in Pinchbeck, p. 194.

9. Cited in Pinchbeck, pp. 304–5.

10. *Arbeiterinnen*, pp. 66–67. Authors' translation.

11. *Arbeiterinnen*, p. 81. Thanks to Diana Ellis for help with the translation.

12. Cited in Janet Horowitz Murray, ed., *Strong-Minded Women and Other Lost Voices from Nineteenth-Century England* (New York: Pantheon, 1982), pp. 378–79.

13. Murray, ed., pp. 344–45.

14. Cited in Hellerstein et al., p. 46.

15. Phyllis Stock, *Better Than Rubies: A History of Women's Education* (New York: G. P. Putnam's Sons, 1978), passim. Girls were educated quite equally on the primary-school level in Germany and Scandinavia, where literacy was required of both sexes by the turn of the nineteenth century.

16. Emma Goldman, *Living My Life* (New York: Alfred A. Knopf, 1931), p. 12.

17. Mayhew, Vol. I, p. 24.

18. *The Poems of Mary Collier, The Washerwoman of Petersfield* (Petersfield, Eng.: W. Minchin, 1739), p. 1.

19. Adelheid Popp, *A Working-Class Childhood*, cited in Eleanor S. Riemer and John C. Fout, eds., *European Women: A Documentary History* (New York: Schocken, 1980), pp. 130–31.

20. Liddington and Norris, p. 34.

21. See McBride, p. 14, for England and France; see Ransel, in Ransel, ed., p. 195, table 2, for Russia.

22. Sarah C. Maza, *Servants and Masters in Eighteenth-Century France: The Uses of Loyalty* (Princeton, N.J.: Princeton University Press, 1983), p. 62.

23. McBride, p. 45, table 2.7; for a similar feminization of domestic service in Germany, see Sagarra, p. 385.

24. Tilly and Scott, *Women, Work, and Family*, pp. 82–83; Liddington and Norris, p. 101.

25. The Servants Code is reproduced in Gerhard, pp. 261–77; the law about whipping is from the Prussian Civil Code, in Gerhard, p. 241.

26. Guiral and Thuillier, pp. 224, 114.

27. Cited in Frank Dawes, *Not in Front of the Servants: Domestic Service in England, 1850–1939* (London: Wayland, 1973), p. 31.

28. Gerhard, p. 51.

29. Cited in Katharine M. Rogers, *Feminism in Eighteenth-Century England* (Urbana: University of Illinois Press, 1982), p. 19.

30. Davidoff, "Class and Gender," p. 140, footnote 106.

31. Fairchilds, *Domestic Enemies*, pp. 86–87.

32. See Ivy Pinchbeck and Margaret Hewitt, *Children in English Society* (1969), Vol. II, p. 534.

33. Maza, p. 63; male servants showed a much wider age distribution.

34. Maza, p. 62.
35. Cited in Leonore Davidoff, "Class and Gender in Victorian England: The Diaries of Arthur J. Munby and Hannah Cullwick," *Feminist Studies*, Vol. 5, no. 1 (Spring 1979), p. 108.
36. Davidoff, "Class and Gender," p. 98.
37. Cited in Guiral and Thuillier, pp. 48–49.
38. Cited in Guiral and Thuillier, p. 79. Authors' translation
39. Collier, p. 10.
40. Dawes, p. 16.
41. Dawes, p. 83.
42. Frank E. Huggett, *Life Below Stairs: Domestic Servants in England from Victorian Times* (London: Book Club Associates, 1977), p. 50; Bonnie G. Smith, *Ladies of the Leisure Class: The Bourgeoises of Northern France in the Nineteenth Century* (Princeton, N.J.: Princeton University Press, 1981), p. 75.
43. Renate Bridenthal, "Class Struggle around the Hearth: Women and Domestic Service in the Weimar Republic," in Michael Dobkowski and Isidor Walliman, eds., *Towards the Holocaust: The Social and Economic Collapse of the Weimar Republic* (Westport, Conn.: Greenwood Press, 1983), pp. 247–50.
44. Cited in Guiral and Thuillier, p. 39.
45. Theresa M. McBride, "The Modernization of 'Women's Work,'" in the *Journal of Modern History*, Vol. 49, no. 2 (June 1977), p. 242.
46. Guiral and Thuillier, p. 124.
47. Cited in *Arbeiterinnen*, p. 89. Authors' translation.
48. Gwen Raverat, *Period Piece: A Cambridge Childhood* (London: Faber and Faber, 1968 [1952]), p. 68.
49. Hufton, *The Poor of Eighteenth-Century France*, p. 25.
50. Pike, p. 162.
51. Cited in *Arbeiterinnen*, pp. 87–88. Authors' translation.
52. Yvonne Kapp, *Eleanor Marx: Volume One* (New York: Pantheon, 1972), p. 26.
53. Cited in Kapp, p. 194
54. Kapp, p. 26 and Appendix I.
55. Cited in *Arbeiterinnen*, p. 97. Authors' translation.
56. McBride, *Domestic Revolution*, p. 105; Guiral and Thuillier, p. 132.
57. Mayhew, Vol. IV, p. 223.
58. Tilly and Scott, *Women, Work and Family*, pp. 82–83.
59. Percentages of married women in the female industrial labor force varied from nation to nation, although they were always a minority. They were 32 percent of the women employed by the French textile industry in 1896; 42 percent of the female factory workers in the Moscow area in the late 1880s. McMillan, p. 38, and Rose L. Glickman, "The Russian Factory Woman, 1880–1914," in Dorothy Atkinson, Alexander Dallin, and Gail Worshofsky Lapidus, eds., *Women in Russia* (Stanford, Calif.: Stanford University Press, 1977), p. 76. On women's attitudes to factory work, see Betty Messenger, *Picking Up the Linen Threads: A Study in Industrial Folklore* (Austin: University of Texas Press, 1980), p. 207.
60. Cited in Liddington and Norris, p. 39.

61. Cited in Liddington and Norris, p. 94.
62. Cited in *Arbeiterinnen*, pp. 188–89.
63. Both cited in Messenger, pp. 27, xv.
64. Cited in Liddington and Norris, p. 86.
65. Madeleine Guilbert, *Les fonctions des femmes dans l'industrie* (Paris: Mouton, 1966), p. 36. For similar findings in Russia, see Rose L. Glickman, *Russian Factory Women: Workplace and Society, 1880–1914* (Berkeley, Calif.: University of California Press, 1984).
66. Pinchbeck, p. 193.
67. Liddington and Norris, pp. 84–90.
68. For the 1830s, see Pinchbeck, p. 186; for the 1890s, Liddington and Norris, pp. 84–90.
69. The mule-spinner, like its namesake, was a hybrid—in this case, of machines which spun both warp and woof threads. Liddington and Norris, p. 90.
70. Liddington and Norris, p. 85.
71. Cited in Riemer and Fout, eds., p. 15.
72. Cited in *Arbeiterinnen*, p. 72.
73. Cited in Pike, p. 192.
74. Cited in Liddington and Norris, p. 39.
75. Cited in Taylor, "The Men Are As Bad," p. 23.
76. For the French delegates, see Marilyn J. Boxer, "Foyer or Factory: Working-Class Women in Nineteenth-Century France," in *Western Society for French Historical Proceedings*, Vol. II (November 1974), pp. 195–96; for the Germans, Werner Thönnessen, *The Emancipation of Women: The Rise and Decline of the Women's Movement in German Social Democracy. 1863–1933*, trans. Joris de Bres (Glasgow: Pluto Press, 1976), p. 20.
77. Cited in Messenger, p. 135.
78. *Arbeiterinnen*, pp. 77–78.
79. Cited in Sheila Rowbotham, *Women, Resistance and Revolution: A History of Women and Revolution in the Modern World* (New York: Vintage, 1974), p. 112.
80. Cited in Louis James, ed., *English Popular Literature, 1819–1851* (New York: Columbia University Press, 1976), p. 330.
81. There were 45,000 laundresses counted. They were outnumbered by 730,700 domestic servants, 115,400 women in cotton manufacture, and 89,000 seamstresses. Pinchbeck, pp. 317–19.
82. McMillan, p. 69.
83. George, p. 207.
84. Sheila Lewenhak, *Women and Work* (1980), p. 106; Eunice Lipton, "The Violence of Ideological Distortion: The Imagery of Laundresses in 19th Century French Culture," *Heresies* (Summer 1978), p. 77.
85. Collier, p. 6.
86. Census figures from *The Unknown Mayhew: Selections from the Morning Chronicle, 1849–1851*, ed. E. P. Thompson and Eileen Yeo (Baltimore: Penguin, 1973), pp. 195, 274–75, 500.
87. George, p. 172.

88. Mayhew, Vol. I, p. 159.
89. Judith R. Walkowitz, *Prostitution and Victorian Society: Women, Class and the State* (Cambridge, Eng.: Cambridge University Press, 1980), p. 195.
90. Richard J. Evans, "Prostitution, State and Society in Imperial Germany," in *Past and Present*, Vol. 70 (February 1976), p. 117.
91. E. M. Sigsworth and T. J. Wyke, "A Study of Victorian Prostitution and Venereal Disease," in Martha Vicinus, ed., *Suffer and Be Still*, p. 79; Evans, "Prostitution," p. 108.
92. Daubié was the first Frenchwoman to receive a doctoral degree; her dissertation was a study of poor women. Cited in Edith Thomas, *The Women Incendiaries* (London: Secker and Warburg, 1967), p. 5.
93. Walkowitz, passim, and Alain Corbin, *Les Filles de Noce: misère sexuelle et prostitution (19e and 20e siècles)* (Paris: Aubier Montaigne, 1978), passim.
94. Walkowitz, p. 197; Corbin, pp. 59–61; Evans, p. 116.
95. Corbin, pp. 72–73; Walkowitz, p. 15; Laura S. Strumingher, *Women and the Making of the Working Class: Lyon, 1830–1870* (St. Albans, Vt.: Eden Press Women's Publications, 1979), pp. 34–35.
96. Cited in Johanna Richardson, *The Courtesans: The Demi-Monde in Nineteenth-Century France* (Cleveland, Ohio: World Publishing Co., 1967), pp. 196–98.
97. Corbin, pp. 55–56.
98. Corbin, pp. 143–46, 163–65.
99. Vern and Bonnie Bullough, *Prostitution: An Illustrated Social History* (New York: Crown, 1978), pp. 178–80.
100. Cited in Constance Rover, *Love, Morals and the Feminists* (London: Routledge and Kegan Paul, 1970), p. 75.
101. Corbin, p. 143; Walkowitz, p. 202.
102. Corbin, p. 134, footnote 254.
103. Cited in Walkowitz, p. 128.
104. Corbin, p. 15; Walkowitz, pp. 209–10.
105. The Dumas *fils* 1878 play *La Femme de Claude* used the career of La Païva for Claude's wife, denounced as "the Beast which undermines society, dissolves the family, profanes love, dismembers the country" (Richardson, p. 96).
106. Material on La Païva drawn from Richardson, chap. 5.
107. Cited in Susan Raven and Alison Weir, *Women of Achievement* (New York: Harmony Books, 1981), p. 182.
108. "Where? When? How much?" "Your place. Tonight. For free." Cited in Raven and Weir, p. 182.
109. For Elssler, see Mary Clarke and Clement Crisp, *Ballet: An Illustrated History* (New York: Universe Books, 1973), pp. 66–67.
110. Clarke and Crisp, p. 89.
111. Liddington and Norris, p. 47.
112. Mayhew, Vol. I, p. 72.
113. Liddington and Norris, p. 47.
114. Leonore Davidoff, "The Separation of Home and Work? Landladies and Lodgers

in Nineteenth- and Twentieth-Century England," in Sandra Burman, ed., *Fit Work for Women* (New York: St. Martin's Press, 1979), p. 85.

115. Robyn Dasey, "Women's Work and the Family: Women Garment Workers in Berlin and Hamburg Before the First World War," in Evans and Lee, eds., p. 232.

116. Davidoff, in Berman, p. 82.

117. Landes, pp. 187–90; E. P. Thompson, *The Making of the English Working Class* (New York: Vintage, 1963), p. 260.

118. Luck, p. 372.

119. Pinchbeck, p. 210.

120. Dasey, in Evans and Lee, eds., p. 228. Emphasis in the original.

121. Cited in George, pp. 195–96.

122. Cited in Jean H. Quataert, "The Shaping of Women's Work," p. 1, 134; also see Joan Scott, "Parisian Tailors and Women Seamstresses," in *Essays in Honor of Eric Hobsbawm* (Cambridge: Cambridge University Press, 1984).

123. *Oxford English Dictionary*.

124. Jean H. Quataert, *Reluctant Feminists in German Social Democracy: 1885–1917* (Princeton, N.J.: Princeton University Press, 1979), p. 35.

125. Cited in Darlene Gay Levy, Harriet Branson Applewhite, and Mary Durham Johnson, eds., *Women in Revolutionary Paris: 1789–1795* (Urbana: University of Illinois Press, 1979), p. 19.

126. The ballad is by Thomas Hood and is reproduced in Miriam Schneir, ed., *Feminism: The Essential Historical Writings* (New York: Random House, 1972), pp. 58–61.

127. *The Unknown Mayhew*, pp. 137–216.

128. Marie-Hélène Zylberberg-Hocquard, *Femmes et féminisme dans le movement ouvrier français* (Paris: Éditions ouvrières, 1981), p. 33.

129. *The Unknown Mayhew*, p. 523.

130. McMillan, p. 65.

131. *The Unknown Mayhew*, p. 197.

132. Cited in Pike, p. 205.

133. *The Unknown Mayhew*, p. 148.

134. McMillan, p. 67; Thomas, p. 11.

135. Dasey, in Evans and Lee, eds., p. 242.

136. "Die Heimatarbeiterin," in Gisela Brinker-Gabler, ed., *Deutsche Dichterinnen vom 16. Jahrhundert bis zur Gegenwart* (Frankfurt am Main: Fischer, 1978), p. 267. Authors' translation.

137. Collier, p. 9.

138. Evelyne Sullerot, *Histoire et sociologie du travail féminin* (Paris: Editions Gonthier, 1968), p. 156; Hufton, *Poor of Eighteenth-Century France*, p. 213ff.; George, p. 174; Pinchbeck, p. 2, footnote 2.

139. Lutz K. Berkner, "The Stem Family and the Developmental Cycle of the Peasant Household: An Eighteenth-Century Austrian Example," *American Historical Review*, Vol. 77, no. 2 (April 1972), p. 404; Michael Drake, *Population and Society in Norway: 1735–1865* (Cambridge, Eng.: Cambridge University Press, 1969), p. 116; Strumingher, p. 88.

140. *The Unknown Mayhew*, p. 168.

141. Mayhew, Vol. IV, p. 232.
142. *The Unknown Mayhew*, p. 311. For examples of widows living with others, see Mayhew, Vol. I, pp. 48, 94, 393–94.
143. Mayhew, Vol. I, pp. 393–94.
144. Mayhew, Vol. I, p. 462.
145. Mayhew, Vol. IV, p. 245.
146. See J. M. Beattie, "The Criminality of Women in Eighteenth-Century England," *Journal of Social History*, Vol. 8 (Summer 1975), p. 109. Poor women's participation in crime in nineteenth-century Europe is a virtually unstudied topic.
147. Mayhew, Vol. I, p. 47.
148. Cited in J. Michael Phayer, *Sexual Liberation and Religion in Nineteenth Century Europe* (London: Croom Helm, 1977), p. 96.
149. Phayer, p. 95; also see Thabault, p. 101; Alain Decaux, *Histoire des françaises*, 2 vols. (Paris: Librairie Académique Perrin, 1972), Vol. II, p. 702; for England, see Ian Bradley, *The Call to Seriousness: The Evangelical Impact on the Victorians* (1976), p. 41.
150. Hewitt, p. 156.
151. Linda L. Clark, "The Molding of the *Citoyenne*: The Image of the Female in French Educational Literature," *Third Republic*, Vol. I, nos. 3–4 (Spring-Fall 1977), pp. 77–78.
152. Cited in Strumingher, p. 52.
153. Mayhew, Vol. I, p. 47.

PART VIII: **3. Revolutions and Reforms**

1. The German artist Käthe Kollwitz portrayed Black Anna in the fifth drawing of her series on the Peasants' War. Jeanne Laisne—"Hachette"—led women of Beauvais, armed with hachets, against the Burgundians in 1470. "Mad Meg" and "Long Meg" appear repeatedly in English ballads and folktales. See Patricia Gartenberg, "An Elizabethan Wonder Woman: The Life and Fortunes of Long Meg of Westminster," *Journal of Popular Culture*, Vol. 17 (Winter 1983), pp. 49–59.
2. For an analysis of these bread riots, see Michelle Perrot, "La femme populaire rebelle," in Christine Dufrancatel et al., *L'histoire sans qualitiés* (Paris: Editions Galiliee, 1979), pp. 134–35.
3. Lois G. Schwoerer, "Women in Revolutionary Politics: 1680–1690," paper presented at Fifth Berkshire Conference, Smith College, Northampton, Mass., 1984, p. 4.
4. Cited in Perrot, pp. 138–39.
5. For a petition of "Women of the Third Estate to the King," see Darline Gay Levy, Harriet Branson Applewhite, and Mary Durham Johnson, eds., *Women in Revolutionary Paris, 1789–1795* (Urbana: University of Illinois Press, 1979), pp. 18–21.
6. The Third Estate was a rank which included all citizens except the clergy (First Estate) and the aristocracy (Second Estate). Quotation cited in Julia O'Faolain and Lauro Martines, eds., *Not in God's Image: Women in History from the Greeks to the Victorians* (New York: Harper & Row, Publishers. 1973), p. 304.

7. Cited in Levy et al., p. 28.
8. Cited in Levy et al., p. 30.
9. Cited in Levy et al., p. 38.
10. For contemporary Italian drawings, see *Esistere come donna* (Milan: Mazzotta, 1983), p. 85.
11. Cited in Louise Otto, *Dem reich der Freiheit*, p. 70. Authors' translation.
12. Edith Thomas, *Les femmes en 1848* (Paris: Presses Universitaires de France, 1948), pp. 34, 71; Marie Collins and Sylvie Weil Sayre, eds., *Les Femmes en France* (New York: Charles Scribner's Sons, 1974), p. 127.
13. Cited in Edith Thomas, *The Women Incendiaries* (London: Secker and Warburg, 1967), p. 45.
14. Thomas, *Women Incendiaries*, p. 128.
15. Thomas, *Women Incendiaries*, p. 83.
16. The professions of sixty members are known and also included waistcoat-makers, linen-drapers, makers of men's clothes, boot-stitchers, hat-makers, cardboard-makers, and "one embroiderer of military decorations, one braid-maker, one tie-maker, one school-teacher, one book-stitcher and one book-binder" (Thomas, *Women Incendiaries*, p. 62).
17. Levy et al., p. 310.
18. Cited in Levy et al., p. 72.
19. Cited in Levy et al., p. 132.
20. Cited in Levy et al., pp. 160–65. On the society, also see Margaret George, "The 'World Historical Defeat' of the Republicaines-Révolutionnaires," *Science and Society*, Vol. 40, no. 4 (Winter 1976–1977), pp. 410–37.
21. Maïte Albistur and Daniel Armogathe, *Histoire du féminisme français*, 2 vols. (Paris: Editions des femmes, 1977), Vol. II, p. 453
22. Thomas, *Les femmes en 1848*, p. 47.
23. Thomas, *Les femmes en 1848*, p. 47, and Albistur and Armogathe, Vol. II, p. 454.
24. Cited in Louise Otto, *Dem Reich der Freiheit*, p. 55. Authors' translation.
25. For the 1790 workshops, see Levy et al., p. 308; for 1848, see Thomas, *Les femmes en 1848*, pp. 52–53.
26. Reproduced in *Esistere come donna*, p. 98.
27. Thomas, *Women Incendiaries*, p. 100.
28. *The Red Virgin: Memoirs of Louise Michel*, ed. and trans. Bullitt Lowry and Elizabeth Ellington Gunter (University, Ala.: University of Alabama Press, 1981), p. 67.
29. Male Communards worked in more elite working-class jobs. Louise A. Tilly, "Women's Collective Action and Feminism in France, 1870–1914," in Louise A. Tilly and Charles Tilly, eds., *Class Conflict and Collective Action* (Beverly Hills. Calif.: Sage Press, 1981), p. 219.
30. Cited in Thomas, *Women Incendiaries*, p. 157.
31. Cited in Thomas, *Women Incendiaries*, p. 123.
32. Cited in Levy et al., p. 215.
33. Cited in Levy et al., pp. 219–20.
34. Thomas, *Les femmes en 1848*, p. 41.

35. Thomas, *Les femmes en 1848*, p. 74, and Edith Thomas, *Pauline Roland: Socialisme et Feminisme au XIXe Siècle* (Paris: Librairie Marcel Riviere, 1956), p. 151.
36. Thomas, *Women Incendiaries*, p. 53.
37. McKeown, p. 153.
38. Evelyne Sullerot, *Women, Society and Change*, trans. Margaret Scotford Archer (New York: McGraw-Hill, 1976), p. 48, table 2.2.
39. Edward Shorter, *A History of Women's Bodies* (New York: Basic Books, 1982), p. 133.
40. B. R. Mitchell, *European Historical Statistics, 1750–1970* (New York: Columbia University Press, 1976), pp. 130–31.
41. E. A. Wrigley, *Population and History* (New York: McGraw-Hill, 1969), p. 185.
42. Wrigley, pp. 186–88.
43. Mitchell, pp. 114–20.
44. Mitchell, pp. 108–20.
45. Angus McLaren, "Abortion in France: Women and the Regulation of Family Size, 1800–1914," *French Historical Studies*, Vol. X, no. 3 (Spring 1978), pp. 461–85; Angus McLaren, "Abortion in England, 1890–1914," chap. 13 of *Birth Control in Nineteenth-Century England* (New York: Holmes and Meier, 1978).
46. For a discussion of these statistics and their interpretation, see Shorter, Appendix A.
47. On this subject, see Karen Offen, "Depopulation, Nationalism, and Feminism in Fin-de-Siècle France," *American Historical Review*, Vol. 89, no. 3 (June 1984), pp. 648–76, for France; Jane Lewis, *The Politics of Motherhood* (London: Croom Helm, 1980) for England; Anneliese Bergmann, "Frauen, Männer, Sexualität und Geburtenkontrolle. Zur 'Gebärstreikdebatte' der SPD 1913," in Karin Hausen, ed., *Frauen Suchen Ihre Geschichte: Historische Studien zum 19. und 20. Jahrhundert* (Munich: C. H. Beck, 1982), pp. 81–109, for Germany.
48. Cited in *Dear Dr. Stopes: Sex in the 1920s*, ed. Ruth Hall (New York: Penguin, 1981), p. 37.
49. Cited in Mary Chamberlain, *Fenwomen: A Portrait of Women in an English Village* (London: Virago, 1977), pp. 77, 74.
50. Cited in Ellen Ross, " 'Fierce Questions and Taunts': Married Life in Working-Class London, 1870–1914," *Feminist Studies*, Vol. 8, no. 3 (Fall 1982), p. 594.
51. Henry J. Harris, *Maternity Benefit Systems in Certain Foreign Countries* (Washington, D.C.: General Printing Office, 1919), p. 11.
52. Harris, p. 67.
53. Cited in Janet Horowitz Murray, *Strong-Minded Women and Other Lost Voices from Nineteenth-Century England* (New York: Pantheon, 1982), pp. 373–74.
54. Cited in Pinchbeck, p. 248.
55. Russia forbade night work for women in the textile industry in 1885, but the law was unenforceable; France forbade women from working between 9 P.M. and 5 A.M. in 1892; Germany forbade women from working at night in 1910.
56. Jean Quataert, *Reluctant Feminists in German Social Democracy, 1885–1917* (Princeton, N.1.: Princeton University Press, 1979), p. 42.
57. Cited in Marie-Hélène Zylberberg-Hocquard, *Femmes et Féminisme dans le Mouvement Ouvrier Français* (Paris: Éditions ouvrières, 1981), p. 187. Authors' translation.

58. Cited in Karen Offen, "Depopulation, Nationalism, and Feminism," p. 672.
59. Cited in Charles Sowerwine, *Sisters or Citizens? Women and socialism in France since 1876* (Cambridge: Cambridge University Press, 1982), p. 196.
60. Cited in Hal Draper and Anne C. Lipow, "Marxist Women versus Bourgeois Feminism," *The Socialist Register* (1976), p. 224.
61. Sowerwine, p. 196.
62. August Bebel, *Women and Socialism*, trans. Daniel De Leon (New York: Schocken, 1971 [1879]), p. 180.
63. Sheila Lewenhak, *Women and Trade Unions: An Outline History of Women in the British Trade Union Movement* (New York: St. Martin's Press, 1977), p. 20.
64. Barbara Taylor, *Eve and the New Jerusalem: Socialism and Feminism in the Nineteenth Century* (New York: Pantheon, 1983), p. 18.
65. Cited in Pinchbeck, pp. 213–14.
66. Dorothy Thompson, "Women and Nineteenth-Century Radical Politics: A Lost Dimension," in Oakley and Mitchell, eds., p. 133.
67. Cited in *Arbeiterinnen*, p. 188. Authors' translation.
68. Cited in Norbert C. Soldon, *Women in British Trade Unions, 1874–1976* (Dublin: Gill and Macmillan, 1978), p. 55.
69. Figures from Barbara Drake, *Women in Trade Unions* (London: Virago, 1984 [1920]), table 1.
70. Cited in Drake, p. 41.
71. Quataert, *Reluctant Feminists*, p. 185.
72. The lower French figures represent the lower rate of industrialization in France. German figures from Quataert, *Reluctant Feminists*, p. 185; French figures from Madeleine Guilbert, *Les femmes et l'organisation syndicale avant 1914* (Paris: Centre National de la Recherche Scientifique, 1966), p. 29; English figures from 1913, in Lewenhak, *Women and Trade Unions*, p. 97.
73. Boxer, "Foyer or Factory," p. 196.
74. Inga Dahlsgård, *Women in Denmark: Yesterday and Today* (Copenhagen: Det Danske Selskab, 1980), p. 94.
75. Cited in Richard Stites, *The Women's Liberation Movement in Russia: Feminism, Nihilism, and Bolshevism, 1860–1930* (Princeton, N.J.: Princeton University Press, 1978), p. 163.
76. Information on Sillanpää drawn from John Wuorinen, *A History of Finland* (New York: Columbia University Press, 1965), p. 324, footnote 2, and Richard J. Evans, *The Feminists: Women's Emancipation Movements in Europe, America and Australasia, 1840–1920* (London: Croom Helm, 1977), p. 168.
77. Russian information from Stites, p. 334.
78. Tilly and Scott, *Women, Work, and Family*, p. 213.
79. Cited in Gail Braybon, *Women Workers in the First World War: The British Experience* (London: Croom Helm, 1981), pp. 182–83.
80. McBride, p. 112.
81. Tilly and Scott, *Women, Work, and Family*, p. 184.
82. Renate Bridenthal and Claudia Koonz, "Beyond *Kinder, Kuche, Kirche*: Weimar Women in Politics and Work," in Renate Bridenthal, Atina Grossmann, and Mar-

ion Kaplan, eds., *When Biology Became Destiny: Women in Weimar and Nazi Germany* (New York: Monthly Review Press, 1984), p. 64, footnote 98.

83. Bridenthal and Koonz, in Bridenthal et al., p. 52.

84. Françoise Giroud, *I Give You My Word*, trans. Richard Seaver (Boston: Houghton Mifflin, 1974), pp. 12–13. Giroud became France's first minister for women in the 1970s.

85. Renate Bridenthal, "Something Old, Something New: Women Between the Two World Wars," in Renate Bridenthal and Claudia Koonz, eds., *Becoming Visible: Women in European History* (Boston: Houghton Mifflin, 1977), p. 435, and McMillan, p. 157.

86. For this argument, see Braybon for England; McMillan for France; Bridenthal and Koonz, in Bridenthal et al., for Germany.

87. For England and Germany, see Renate Pore, *A Conflict of Interest Women in German Social Democracy, 1919–1933* (Westport, Conn.: Greenwood Press, 1981), p. 90; for France, McMillan, pp. 136, 138.

88. For France, McMillan, p. 132; for England, Arthur Marwick, *The Deluge: British Society and the First World War* (New York: W. W. Norton, 1965), pp. 91–93.

89. Cited in Braybon, p. 79; for men's opposition to women in factories, see Braybon, p. 72.

90. Braybon, p. 76; McMillan, p. 140.

91. McMillan, p. 140.

92. Pore, p. 90.

93. Cited in Catherine W. Reilly, ed., *Scars Upon My Heart: Women's Poetry and Verse of the First World War* (London: Virago, 1981), p. 90.

94. For England, Braybon, p. 229; for France, McMillan, p. 155; for Germany, Thonnessen, p. 91.

95. Mitchell, pp. 155, 163.

96. Tilly and Scott, *Women, Work, and Family*, p. 208.

97. Corbin, pp. 479–80.

98. McMillan, pp. 140, 150–51.

99. Temma Kaplan, "Women's Networks and Social Change in Twentieth-Century Petrograd, Turin, and Barcelona" (1981), unpublished paper, pp. 5–6.

100. Sheila Rowbotham, *Hidden from History: Rediscovering Women in History from the 17th Century to the Present* (New York: Pantheon, 1974), pp. 112–13.

101. Cited in Temma Kaplan, "Female Consciousness and Collective Action: The Case of Barcelona, 1910–1918," in *Signs*, Vol. 7, no. 3 (Spring 1982), p. 561.

102. Kaplan, "Women's Networks," p. 17; McMillan, p. 151.

103. The city's traditional name, St. Petersburg, was changed to the more Russian Petrograd as a patriotic gesture in 1914. After 1924, the city was called Leningrad.

104. Cited in Nancy Frieden, "The Bolshevik Party and the Organization and Emancipation of Women, 1914–1921" (1975), unpublished paper.

105. Cited in William M. Mandel, *Soviet Women* (Garden City, N.Y.: Doubleday & Company, 1975), p. 43.

106. On March 8, 1908, hundreds of women clothing workers held a large demonstration in New York City, demanding the vote and a stronger union. In 1910, Clara

Zetkin, the German head of the Socialist Women's International, persuaded the International to declare March 8 International Women's Day. It was sporadically celebrated in Europe before 1914, but the war led governments to cancel such demonstrations.

107. Frieden, p. 30; Mandel, p. 44.
108. Cited in Barbara Evans Clements, "Working-Class and Peasant Women in the Russian Revolution, 1917–1923," *Signs*, Vol. 8, no. 2 (Winter 1982), p. 226. Additional information on Rodionova from Mandel, p. 42.
109. Cited in William Henry Chamberlin, *The Russian Revolution: 1917–1921* (New York: Grosset and Dunlop, 1963 [1935]), Vol. I, p. 73.
110. Cited in Kaplan, "Women's Networks," p. 14.
111. Stites, p. 292.
112. Mandel, pp. 47–48.
113. Stites, p. 306.
114. Mandel, p. 49.
115. Cited in Bernice Glatzer Rosenthal, "Love on the Tractor: Women in the Russian Revolution and After," in Bridenthal and Koonz, eds., *Becoming Visible*, p. 380.
116. Cited in Mandel, p. 44.
117. V. I. Lenin, *The Emancipation of Women* (New York: International Publishers, 3: 1966 [1934]), pp. 63–64.
118. For more on Kollontai, see part IX, ch. 4.
119. Clements, passim.
120. Stites, p. 328.
121. Cited in Gail Warshofsky Lapidus, *Women in Soviet Society: Equality, Development, and Social Change* (Berkeley: University of California Press, 1978), p. 137.
122. Rosenthal, in Bridenthal and Koonz, eds., p. 380.
123. Stites, p. 395.
124. Cited in *The Island: The Life and Death of an East London Community, 1870–1970* (London: Centerprise Trust, 1979), p. 32.
125. Lewenhak, *Women and Work*, pp. 183–85.
126. Soldon, p. 31.
127. Hannah Mitchell, *The Hard Way Up: The Autobiography of Hannah Mitchell, Suffragette and Rebel* (London: Virago, 1977), p. 217.
128. *The Island*, p. 40.
129. M. Jahoda, P. F. Lazarsfeld, and H. Zeisel, *Marienthal: The Sociography of an Unemployed Community* (London: 1972 [1933]), pp.74–75. Thanks to Atina Grossmann for this reference.
130. Margery Spring Rice, *Working-Class Wives: Their Health and Conditions* (Harmondsworth: Penguin, 1939), pp. 91, 214.
131. At this date, far more research has been done on women in Nazi Germany than women in fascist Italy. For women in Italy during the fascist era, see Alexander de Grand, "Women Under Italian Fascism," *The Historical Journal*, Vol. 19, no. 4 (1976), pp. 947–68; Victoria di Grazia, *The Culture of Consent* (New York: Cambridge University Press, 1981); Maria-Antoinette Macciocchi, *La donna 'nera: consenso femminile e faschismo* (Rome: Feltrinelli, 1977); and Laura Mariana, "Moglie

e madri per la patria" and "Nel fascismo all'opposizione," in *Esistere come donna*. Thanks to Claudia Koonz for help with these references.

132. Speech in Susan Groag Bell and Karen M. Offen, eds., *Women, the Family, and Freedom: The Debate in Documents*, Vol. II, *1880–1950* (Stanford, Calif.: Stanford University Press, 1983), pp. 375–77.

133. The first phrase is from Hitler's 1934 speech; the second from Rosenberg's *Mythus*. Cited in Claudia Koonz, *Mothers in the Fatherland: Women, the Family and Nazi Politics* (New York: St. Martin's Press, 1987), p. 105.

134. Cited in Koonz, *Mothers*, p. 75.

135. Richard J. Evans, "German Women and the Triumph of Hitler," *Journal of Modern History*, Supplement (March 1976), p. 157.

136. Cited in Koonz, *Mothers*, p. 136.

137. Karin Hausen, "Mother's Day in the Weimar Republic," in Bridenthal et al., pp. 131–52.

138. Jill Stephenson, *Women in Nazi Society* (New York: Harper & Row, Publishers, 1975), p. 49.

139. Koonz, *Mothers*, in manuscript draft, p. 233.

140. Gisela Bock, "Racism and Sexism in Nazi Germany: Motherhood, Compulsory Sterilization, and the State," in Bridenthal et al., p. 276.

141. Bock, in Bridenthal et al., pp. 279–84.

142. B. R. Mitchell, p. 116. Explanations for the rise in the birth rate include the improvement in the German economy and the desire of women to avoid war work by having children.

143. The League of German Women's Associations (*Bund Deutscher Frauen*) disbanded rather than nazify, but its largest member organizations—the housewives, the Protestant and the Catholic women's groups, the teachers, and the office workers—all separately nazified.

144. Susanna Dammer, "Kinder, Küche, Kriegsarbeit—Die Schulung der Frauen durch die NS-Frauenschaft," in Frauengruppe Faschismusforschung, eds., *Mutterkreuz und Arbeitsbuch: Zur Geschichte der Frauen in der Weimarer Republik und im Nationalsozialismus* (Frankfurt am Main: Fischer, 1981), p. 236.

145. Leila J. Rupp, *Mobilizing Women for War German and American Propaganda, 1939–1945* (Princeton, N.J.: Princeton University Press, 1978), p. 38.

146. Dammer, in Frauengruppe Faschismusforschung, eds., p. 224.

147. Bridenthal et al., p. 25.

148. Lore Kleiber, " 'Wo ihr seid, da soll die Sonne scheinen!'—Der Frauenarbeitsdienst am Ende der Weimarer Republik und im Nationalsozialismus," in Frauengruppe Faschismusforschung, eds., p. 211.

149. Leila J. Rupp, " 'I Don't Call that *Volksgemeinschaft!*': Women, Class, and War in Nazi Germany," in Carol R. Berkin and Clara M. Lovett, eds., *Women, War and Revolution* (New York: Holmes and Meier, 1980), p. 38.

150. Rupp, *Mobilizing Women*, p. 77.

151. Hiltgunt Zassenhaus, *Walls: Resisting the Third Reich—One Woman's Story* (Boston: Beacon Press, 1976), p. 14.

152. Sybil Milton, in Esther Katz and Joan Miriam Ringelheim, eds. *Women Surviving*

the Holocaust: Proceedings of the Conference (New York: Occasional Papers for the Institute for Research in History, 1983), p. 13.

153. Cited in Ingrun Lafleur, "Five Socialist Women: Traditional Conflicts and Socialist Visions in Austria, 1893–1934," in Marilyn J. Boxer and Jean H. Quataert, eds., *Socialist Women: European Socialist Feminism in the Nineteenth and Early Twentieth Centuries* (New York: Elsevier, 1978), p. 230.

154. Cited in *Arbeiterinnen*, p. 236. Thanks to Joan Reuterschan for help with the translation.

155. *Arbeiterinnen*, pp. 236–37. Leichter's son, Franz Leichter, survived to become a New York State assemblyman.

PART VIII: **4. Continuity and Change: Women in World War II and After**

1. In 1911, there were 1,503,000 men between twenty and twenty-four in the United Kingdom and 1,673,000 young women of the same age. By 1921, there were 1,703,000 women between twenty and twenty-four, but only 1,448,000 young men. German and French populations show a similar skewing. B. R. Mitchell *European Historical Statistics, 1750–1970* (New York: Columbia University Press 1976), pp. 52, 36–37.

2. Alexander Werth, *Russia at War, 1941–1945* (New York: E. P. Dutton & Co., 1964), pp. 195–96.

3. Cited in Angus Calder, *The People's War Britain, 1939–1945* (New York: Pantheon, 1969), p. 190.

4. Calder, p. 331.

5. Werth, p. 1,004.

6. Cited in Werth, p. 239.

7. Calder, pp. 268, 331.

8. Cited in Calder, p. 306.

9. Calder, p. 400.

10. Cited in Calder, p. 402.

11. On this subject, see Raynes Minns, *Bombers and Mash: The Domestic Front, 1939–1945* (London: Virago, 1980), pp. 38–39.

12. Poster in University of Connecticut Art Gallery, Storrs, Conn.

13. Sheila B. Kamerman, "Work and Family in Industrialized Societies," *Signs*, Vol. IV, no. 4 (Summer 1979), pp. 638–46.

14. Cited in Minns, p. 91. Government rationing actually improved the health of the general population through more equal food distribution.

15. Cited in Minns, p. 147.

16. Mary Lee Settle, *All the Brave Promises: Memories of Aircraft Woman 2nd Class 2146391* (New York: Ballantine, 1980 [1960]), p. 22.

17. Settle, p. 40.

18. Cited in N. Vishneva-Sarafanova, *Soviet Women: A Portrait*, trans. Sheena Wakefield (U.S.S.R.: Progress Publishers, 1981), p. 11.

19. Jack Cassin-Scott, *Women at War, 1939–1945* (London: Osprey Publishing, 1980), p. 34.

20. Cited in Bruce Myles, *Night Witches: The Amazing Story of Russia's Women Pilots in World War Two* (London: Panther, 1983), p. 59.

21. Cited in Myles, p. 232.

22. David Schoenbrun, *Soldiers of the Night: The Story of the French Resistance* (New York: E. P. Dutton & Co., 1980), p. 36.

23. Schoenbrun, p. 378.

24. Cited in Jean McCrindle and Sheila Rowbotham, eds., *Dutiful Daughters: Women Talk about Their Lives* (Austin: University of Texas Press, 1977), p. 89.

25. Harrison E. Salisbury, *The 900 Days: The Siege of Leningrad* (New York: Harper & Row, Publishers, 1969), p. 513.

26. Cited in Salisbury, p. 122.

27. Cited in Salisbury, p. 389.

28. Cited in Salisbury, p. 484.

29. On this subject, see Margaret L. Rossiter, *Women in the [French] Resistance* (New York: Praeger, 1986), passim, and Vera Laska, ed., *Women in the Resistance and in the Holocaust: The Voices of Eyewitnesses* (Westport, Conn.: Greenwood Press, 1983), pp. 4–9.

30. For Fourcade, see part VII, chap. 4.

31. Testimony of Reuven Dafne in *Hannah Senesh: Her Life and Divry*, trans. Marta Cohn (New York: Schocken, 1973), p. 177.

32. Vladka Meed, *On Both Sides of the Wall: Memoirs from the Warsaw Ghetto*, trans. Steven Meed (New York: Holocaust Library, 1979 [1948]), p. 137.

33. For women in German resistance movements, including the 1944 attempt to assassinate Hitler, see Sybil Milton, "Women and the Holocaust: The Case of German and German-Jewish Women," in Bridenthal et al., p. 317; John Cammett, "Women and the Resistance," in Jane Slaughter and Robert Kern, eds., *European Women on the Left: Socialism, Feminism and the Problem Faced by Political Women, 1880 to the Present* (Westport, Conn.: Greenwood Press, 1981).

34. For the servant, see Schoenbrun, p. 6; for the fishmonger, Laska, ed., pp. 61–62; for the partisans, Meed, p. 221.

35. Senesh, p. 32.

36. Catherine Senesh, Hannah's mother, cited in Senesh, p. 7

37. Senesh, p. 63.

38. Senesh, p. 84.

39. Senesh, p. 125.

40. Cited in Senesh, p. 174.

41. Senesh, p. 256.

42. Senesh, p. 257.

43. Except in Denmark, where the king appeared wearing a Jewish star.

44. Ka-Tzetnik 135633, *House of Dolls*, trans. Moshe M. Kohn (London: Cranada, 1981 [1956]), p. 54.

45. Milton, in Bridenthal, et. al., p. 312.

46. Meed, p. 20.

47. Ellipses in original. Isabella Leitner, *Fragments of Isabella* (New York: Dell, 1978), p. 28.

48. Charlotte Delbo, *None of Us Will Return*, trans. John Githens (Boston: Beacon, 1968), p. 35.
49. Katz and Ringelheim, eds., *Women Surviving the Holocaust*, pp. 171–72. Simone St. Clair entitled her memoir *Ravensbruck, The Women's Hell*.
50. Fania Fénelon, *Playing for Time*, trans. Judith Landry (New York: Atheneum, 1979), pp. 19–20.
51. Leitner, p. 14.
52. Cited in Laska, ed., p. 257.
53. Cited in Laska, ed., p. 259.
54. Ka-Tzetnik, pp. 165–89.
55. Milton, in Bridenthal et al., pp. 311–16.
56. Katz and Ringelheim, eds., *Women Surviving the Holocaust*, pp. 17–20; Laska, ed., p. 206.
57. Cited in Eleni Fourtouni, ed. and trans., *Contemporary Greek Women Poets* (New Haven, Conn.: Thelphini Press, 1978), p. 10.
58. Fénelon, p. 263.
59. *Anne Frank: The Diary of a Young Girl*, trans. B. M. Mooyaart-Doubleday (New York: Doubleday & Company, 1953), p. 177.
60. Frank, p. 237; "march of death" phrase, p. 48.
61. Cited in Aliki and Willis Barnstone, eds., *A Book of Women Poets from Antiquity to Now* (New York: Schocken, 1980), p. 345.
62. Cited in Michael Sissons and Philip French, eds., *Age of Austerity 1945–51* (Baltimore: Penguin, 1964), p. 139.
63. Hildegard Knef, *The Gift Horse: Report on a Life*, trans. David Anthony Palastanga (New York: Dell, 1971), p. 138.
64. Sullerot, *Women, Society and Change*, p. 75; Tilly and Scott, *Women, Work, and Family*, p. 225.
65. Sullerot, *Women, Society and Change*, p. 75.
66. Bernard Berelson, ed., *Population Policy in Developed Countries* (New York: McGraw-Hill, 1974), table 1, p. 3, and B. R. Mitchell, pp. 121–24.
67. Cited in Mary Cornelia Porter and Corey Venning, "Catholicism and Women's Role in Italy and Ireland," in Lynne B. Iglitzin and Ruth Ross, eds., *Women in the World: A Comparative Study* (Santa Barbara, Calif.: Clio Press, 1976), p. 85.
68. B. R. Mitchell, p. 123.
69. For the Soviet Union, see Lapidus, p. 299, footnote 25; for Eastern Europe, Hilda Scott, *Women and Socialism: Experiences from Eastern Europe* (London: Alison and Busby, 1976), chap. 7.
70. Cited in Carola Hansson and Karin Lidén, *Moscow Women: Thirteen Interviews*, trans. Gerry Bothmer, George Blecher, and Lone Blecher (New York: Pantheon, 1983), p. 21.
71. Harry G. Shaffer, *Women in the Two Germanies: A Comparative Study of a Socialist and a Non-Socialist Society* (New York: Pergamon Press, 1981), p. 16.
72. Giroud, p. 244.
73. For 1960 figures, Tilly and Scott, *Women, Work and Family*, p. 70; Shaffer, p. 57; H. Scott, p. 83; Sullerot, *Women, Society and Change*, p. 139; for 1980, Periodic

Reports to the United Nations World Conference on the Decade of Women; and Joni Lovenduski and Jill Hills, *The Politics of the Second Electorate: Women and Public Participation* (London: Routledge and Kegan Paul, 1981), passim.

74. Cited in Hilda Scott, *Sweden's 'Right To Be Human'—Sex-Role Equality: The Goal and the Reality* (London: Alison and Busby, 1982), p. 125.

75. *Report of the Seminar on the Participation of Women in the Economic Evolution of the EEC Region and the Economic Role of Women in the EEC Region*, CONF. 94/14 (Copenhagen, 1980), p. 12; for 1971 figures for Great Britain, Sweden, Ireland, and the United States, see Virginia Novarra, *Women's Work, Men's Work: The Ambivalence of Equality* (London: Marion Boyers, 1980), p. 94.

76. Lapidus, passim; Atkinson, Dallin, and Lapidus, eds., section 2.

77. Cited in McCrindle and Rowbotham, p. 389.

78. Lindsay Mackie and Polly Pattullo, *Women at Work* (London: Tavistock, 1977), p. 40.

79. Sally Alexander, Introduction to *From Hand to Mouth: Women and Piecework*, by Marianne Herzog, trans. Stanley Mitchell (New York: Penguin, 1980), p. 19.

80. Cited in McCrindle and Rowbotham, p. 151.

81. Male workers do not cluster in this way. Mackie and Pattullo, pp. 40–41.

82. Madeleine Guilbert, *Les fonctions des femmes dans l'industrie* (Paris: Mouton, 1966), pp. 133–35.

83. Mackie and Pattullo, p. 40.

84. Herzog, p. 43.

85. Shaffer, p. 99.

86. Guilbert, *Fonctions*, p. 137.

87. Cited in Herzog, p. 139.

88. Clothing Economic Development Report, 1972, cited in Mackie and Pattullo, p. 48.

89. Vern Bullough and Bonnie Bullough, *Prostitution: An Illustrated Social History* (New York: Crown, 1978), p. 265.

90. Bullough and Bullough, p. 278.

91. Cited in Claude Jaget, ed., *Prostitutes: Our Life*, trans. Anna Furse, Suzie Fleming, and Ruth Hall (Bristol, Eng.: Falling Wall Press, 1980), pp. 163, 172.

92. Cited in Jaget, ed., p. 69.

93. Jaget, ed., pp. 158–59.

94. Jaget, ed., pp. 78, 111.

95. Cited in Hilda Scott, *Women and Socialism*, p. 164.

96. For a comparison of the policies of France, the Federal Republic of Germany, the German Democratic Republic, Hungary, and Sweden, see Sheila B. Kamerman, "Work and Family in Industrialized Societies," *Signs*, Vol. 4, no. 4 (Summer 1979), pp. 632–50. For the Soviet Union as of 1975, Lapidus, p. 132.

97. Ann Oakley, *Subject Women* (New York: Pantheon, 1981), tables 11.1 and 11.2, pp. 250–51.

98. Shaffer, p. 109.

99. Cited by Adele Pesce, "Work Representation and Time in Women Workers' Memories," speech at International Conference on Oral History and Women's History, Columbia University, 1983; also see Zylberberg-Hocquard, p. 155.

100. Natalia Baranovskaya, "A Week Like Any Other," in George St. George, ed., *Our Soviet Sister* (Washington, D.C.: Robert B. Luce, 1973), p. 237.

101. Baranovskaya, in St. George, ed., p. 235.

102. Baranovskaya, in St. George, ed., p. 256.

103. Cited in Herzog, p. 31.

104. Ursula Scheu, *Wir werden nicht als Mädchen geboren—wir werden dazu gemacht: Zur fruhkindlichen Erziehung in unserer Gesellschaft* (Frankfurt am Main: Fischer, 1977), p. 109.

105. Cited in Mary Chamberlain, *Fenwomen* (London: Virago, 1975), pp. 113–14.

106. For the Soviet Union, see Ethel Dunn, "Russian Rural Women," in Atkinson, Dallin, and Lapidus, eds., p. 180; for the Federal Republic of Germany (West Germany), Ministry of Health Report, 1977, p. 12; for France, Albistur and Armogathe, Vol. II, p. 644; for Sweden, Hilda Scott, *Sweden's Right*, p. 24.

107. Cited in Vera St. Erlich, *Family in Transition: A Study of 300 Yugoslav Villages* (Princeton, N.J.: Princeton University Press, 1966), p. 258.

PART IX TRADITIONS REJECTED

PART IX: **1. Feminism in Europe**

1. Christine de Pizan, *The Book of the City of Ladies*, trans. Earl Jeffrey Richards (New York: Persea Books, 1982), p. 187. Although the attitudes comprising feminism were first expressed in the early fifteenth century, the word "feminism" was not used until the late nineteenth century. The word did not come into common French usage until the feminist Hubertine Auclert (1848–1914) used it in 1882 to mean the extension of rights to women. The historian Jane Rendell has traced the first recorded use of "feminism" in English to 1894. In English, "feminism" originally meant "the possession of womanly qualities"—in German it retained this meaning, which evolved to also denote effeminacy. By 1911, "feminism" and "feminist" were being used in Russian in their modern meanings. The women's liberation movement of the 1960s and 1970s revived or introduced the word to most European languages and also used "women's liberation" to signify the attitudes comprising feminism. See Jane Rendell, *The Origins of Modern Feminism: Women in Britain, France, and the United States* (New York: Schocken, 1984), p. 1; Patrick Kay Bidelman, *Pariahs Stand Up! The Founding of the Liberal Feminist Movement in France, 1858–1889* (Westport, Conn.: Greenwood Press, 1982), p. 215, footnote 3; Richard J. Evans, *The Feminist Movement in Germany, 1894–1933* (London: Sage, 1976), p. 202, footnote 27; Richard J. Evans, *The Feminists: Women's Emancipation Movements in Europe, America and Australasia, 1840–1920*, rev. ed. (London: Croom Helm, 1979), p. 39, footnote 1; Richard Stites, *The Women's Liberation Movement in Russia: Feminism, Nihilism, and Bolshevism, 1860–1930* (Princeton, N.J.: Princeton University Press, 1978), p. 191.

2. Hannah Mitchell, *The Hard Way Up: The Autobiography of Hannah Mitchell, Suffragette and Rebel*, ed. Geoffrey Mitchell (London: Virago, 1977), p. 43.

3. Cited in Marielouise Janssen-Jurreit, *Sexism: The Male Monopoly on History and Thought*, trans. Verne Moberg (New York: Farrar, Straus & Giroux, 1982), p. 8.

4. Feminism and Nonviolence Study Group, *Piecing It Together Feminism and Nonviolence* (Devon, Eng.: Feminism and Nonviolence Study Group, 1983), p. 53.

5. Italics in the original. Cited in Claire Goldberg Moses, *French Feminism in the Nineteenth Century* (Albany, N.Y.: SUNY Press, 1984), p. 168.

6. Simone de Beauvoir, *Memoirs of a Dutiful Daughter*, trans. James Kirkup (New York: Harper & Row Publishers, 1959), pp. 295–96.

7. Simone de Beauvoir, "From an Interview," in Elaine Marks and Isabelle de Courtivron, eds., *New French Feminisms: An Anthology* (New York: Schocken, 1981), p. 143.

8. Cited in Janet Horowitz Murray, *Strong-Minded Women and Other Lost Voices from Nineteenth-Century England* (New York: Pantheon, 1982), p. 437.

9. For Glückl, see her memoirs (New York: Schocken, 1977), trans. Marvin Lowenthal, and Vol. I, part V; Collier's poetry is used in Vol. II, part VIII; for Charpentier, see Darlene Gay Levy, Harriet Branson Applewhite, and Mary Durham Johnson, eds., *Women in Revolutionary Paris, 1789–1795* (Urbana: University of Illinois Press, 1979), p. 29.

10. Hélène Brion, *La voie féministe* (Paris: Editions Syros, 1978 [1916]), p. 63. Authors' translation. Since this is a history of women, throughout this discussion male feminists have been somewhat slighted in favor of their female contemporaries. For instance, Marie de Gournay, the seventeenth-century French feminist, has been used instead of her male contemporary Poullain de la Barre, author of *On the Equality of the Two Sexes* (1673). Mary Wollstonecraft, the eighteenth-century English feminist, is examined instead of contemporary male feminists like the French Marquis de Condorcet (1743–1794) or the German Theodor von Hippel (1741–1796). The nineteenth-century male feminists like the Englishman John Stuart Mill (1806–1873) or the Frenchman Leon Richer (1824–1911) receive less discussion than contemporary female feminists. A number of these men have been the subject of biographies; to date there is no systematic history of male European feminists.

11. Cited in Les Garner, *Stepping Stones to Women's Liberty: Feminist ideas in the [English] women's suffrage movement, 1900–1918* (London: Heinemann Educational Books, 1984), p. 65.

12. Cited in Mary Lefkowitz and Maureen B. Fant, *Women's Life in Greece and Rome: A source book in translation* (Baltimore: Johns Hopkins University Press, 1982), p. 64.

13. *Genesis* 3:16. This is part of God's punishment of Eve for having been disobedient.

14. St. Paul, *Letter to the Corinthians* I, 11:4–9.

15. "Abuse of Women," cited in Katharine M. Rogers, *The Troublesome Helpmate: A History of Misogyny in Literature* (Seattle: University of Washington Press, 1966), p. 73.

16. Joan Kelly, "Did Women Have a Renaissance?" in *Women, History, and Theory: The Essays of Joan Kelly* (Chicago: University of Chicago Press, 1984), p. 19.

17. Baldasarre Castiglione, *The Book of the Courtier* (Garden City, N.Y.: Doubleday, 1959), trans. Charles Singleton, p. 213.

18. François Fénelon, *Fénelon on Education*, trans. H. C. Barnard (Cambridge, Eng.: Cambridge University Press, 1966), p. 6.

19. For these arguments, see Joan Kelly, "Early Feminist Theory and the *Querelles des Femmes*," in Kelly, *Women, History, and Theory*, pp. 69–71.

20. For feminism's origins as polemic, written in opposition to denigrating male views of women, see Kelly, p. 78, and Ruth Kelso, *Doctrine for the Lady of the Renaissance* (Urbana: University of Illinois Press, 1956), p. 28.

21. Cited in Kelly, p. 84.

22. For an example of this genre, see [Judith Drake?] *An Essay in Defense of the Female Sex* (New York: Source Book Press, 1970 [London, 1696]), passim.

23. European works expressing this view include Suzanne Brøgger, *Deliver Us From Love*, trans. Thomas Teal (New York: Dell, 1973); Françoise d'Eaubonne, *La féminisme ou la mort* (Paris: Horay, 1974), and Carla Lonzi, *Spit on Hegel* (1972).

24. Louise Michel, *The Red Virgin: Memoirs of Louise Michel*, ed. and trans. Bullitt Lowry and Elizabeth Ellington Gunther (University, Ala.: University of Alabama Press, 1981 [1886]), p. 142.

PART IX: **2. Asserting Women's Humanity: Early European Feminists**

1. For this point see Joan Kelly, "Early Feminist Theory and the *Querelle des Femmes*," in *Women, History, and Theory: The Essays of Joan Kelly* (Chicago: University of Chicago Press, 1984), p. 79.

2. Christine de Pizan, *The Book of the City of Ladies*, trans. Earl Jeffrey Richards (New York: Persea Books, 1982), p. 3.

3. Pizan, p. 4.

4. Pizan, p. 4.

5. Pizan, p. 5.

6. Pizan, p. 6.

7. Pizan, p. 10.

8. Pizan, p. 16.

9. Pizan, p. 118.

10. Pizan, p. 63.

11. Pizan, p. 119

12. Pizan, p. 161.

13. Cited in Maïté Albistur and Daniel Armogathe, *Histoire du féminisme français*, 2 vols. (Paris: Editions des Femmes, 1977), Vol. I, p. 185. Authors' translation.

14. Pizan, p. 254.

15. Moira Ferguson, ed., *First Feminists: British Women Writers, 1578–1799* (Bloomington: Indiana University Press, 1985), p. 188. For Astell, see Ferguson, pp. 180–200; Hilda L. Smith, *Reason's Disciples: Seventeenth-Century English Feminists* (Urbana: University of Illinois Press, 1982), chap. 4; and Ruth Perry, *The Celebrated Mary Astell: An Early English Feminist* (Chicago: University of Chicago Press, 1986).

16. Mary Wollstonecraft, *A Vindication of the Rights of Woman*, ed. Carol H. Poston (New York: W. W. Norton & Company, 1977 [1792]), p. 25.

17. Albistur and Armogathe, Vol. I, p. 183.
18. Albistur and Armogathe, Vol. I, p. 184, Kelly, p. 92.
19. Kelly, p. 92; Ferguson, p. 191.
20. Pizan, pp. 18–20.
21. Ferguson, p. 129.
22. From *Letters on Education* (1790), Ferguson, p. 402.
23. On van Schurman, see Joyce Irwin, "Anna Maria van Schurman: The Star of Utrecht," in J. R. Brink, ed., *Female Scholars: A tradition of learned women before 1800* (Montreal: Edwin Press, 1980), pp. 68–85.
24. Ferguson, p. 136.
25. Ferguson, p. 135.
26. Ferguson, p. 188.
27. Ferguson, pp. 137, 195.
28. Cited in Sara Nalle, "The Unknown Reader: Women and Literacy in Golden Age Spain," paper delivered at Fifth Berkshire Conference on the History of Women, Vassar College, Poughkeepsie, New York, 1981, p. 10.
29. Ferguson, p. 230.
30. For more on Wollstonecraft's life, see part VII, chap. 1.
31. Wollstonecraft, *Vindication*, p. 168.
32. Wollstonecraft, *Vindication*, p. 149.
33. Wollstonecraft, *Vindication*, p. 19.
34. Wollstonecraft, *Vindication*, pp. 53, 36.
35. Wollstonecraft, *Vindication*, p. 81.
36. Wollstonecraft, *Vindication*, p. 19.
37. Wollstonecraft, *Vindication*, p. 34.
38. Ferguson, p. 237.
39. Ferguson, pp. 192–93.
40. Wollstonecraft, *Vindication*, p. 167.
41. Wollstonecraft, *Vindication*, p. 167.

PART IX **3. Asserting Women's Legal and Political Equality: Equal Rights Movements in Europe**

1. Cited in Gail Warshofsky Lapidus, *Women in Soviet Society: Equality, Development, and Social Change* (Berkeley: University of California Press, 1978), p. 20.
2. On this subject, see Lois G. Schwoerer, "Women in Revolutionary Politics: 1680–1690," paper presented at Sixth Berkshire Conference on the History of Women, Smith College, Northampton, Mass., 1984; and Jane Abray, "Feminism in the French Revolution," *American Historical Review*, Vol. 80, no. 1 (February 1975), pp. 43–62.
3. For a fuller account of women's actions during the French Revolution, see part VIII, chap. 3.
4. Cited in Jane Rendall, *The Origins of Modern Feminism: Women in Britain, France and the United States, 1780–1860* (New York: Schocken, 1984), p. 46. The journal's

title in French was *Étrennes Nationales des Dames*; *étrennes* is an obsolete word for gifts given on New Year's Day.

5. D'Aelders was active in French revolutionary politics from 1789 to 1793. Little is known of her life. Levy, Applewhite, and Johnson, eds., pp. 62, 75.

6. Levy, Applewhite, and Johnson, eds., p. 92.

7. Levy, Applewhite, and Johnson, eds., p. 93.

8. From the speech which persuaded the National Convention to outlaw "women's societies and popular clubs." Women's deputations were barred from the Paris Commune a few weeks later. Levy, Applewhite, and Johnson, eds., p. 216.

9. Mary Wollstonecraft, *Maria, or The Wrongs of Woman* (New York: W. W. Norton & Company, 1975 [1798]), pp. 104, 108.

10. Ferguson, p. 405.

11. Wollstonecraft, *Vindication*, p. 43; Goldman cited in Alice Wexler, "Emma Goldman on Mary Wollstonecraft," *Feminist Studies*, Vol. 7, no. 1 (Spring 1981), p. 114.

12. John Stuart Mill and Harriet Taylor Mill, *Essays on Sex Equality*, ed. Alice S. Rossi (Chicago: University of Chicago Press, 1970), p. 147.

13. Cited in Richard S. Evans, *The Feminists: Women's Emancipation Movements in Europe, America and Australasia*, 1840–1920 (London: Croom Helm, 1977), p. 86.

14. de Beauvoir, *Dutiful Daughter*, p. 57.

15. Maria Isabel Barreno, Maria Terese Horta, Maria Velho da Costo, *The Three Marias: New Portuguese Letters*, trans. Helen R. Lane (New York: Doubleday & Company, 1976), p. 17.

16. By 1920, women could vote in the Scandinavian nations, Great Britain, Germany, Austria, the U.S.S.R., and Czechoslovakia. Women in France, Italy, Portugal, Belgium, and Switzerland did not receive the vote until after the Second World War. Non-European nations where women first won the vote were also predominantly Protestant: New Zealand, Australia, and the United States of America.

17. For these arguments, see Evans, *Feminists*, pp. 28–31.

18. Cited in Amy Kathleen Hackett, "The Politics of Feminism in Wilhelmine Germany, 1890–1918" (Ph.D. dissertation, Columbia University, 1976), p. 673.

19. Cited in Marian Ramelson, *The Petticoat Rebellion: A Century of Struggle for Women's Rights* (London: Lawrence and Wishart, 1967), p. 87.

20. Cited in Ray Strachey, *The Cause: A Short History of the Women's Movement in Great Britain* (London: Virago, 1978 [1928]), p. 32. The petition was introduced to the House of Commons by the radical M.P., Henry "Orator" Hunt.

21. Cited in Barbara Taylor, *Eve and the New Jerusalem: Socialism and Feminism in the Nineteenth Century* [in England] (New York: Pantheon, 1983), p. 277.

22. Hamelson, p. 72.

23. Mill and Mill, p. 120. Little is known of Harriet Mill's early life.

24. Mill and Mill, p. 114.

25. Mill and Mill, p. 107.

26. Mill and Mill, pp. 112–13.

27. Mill and Mill, p. 95.

28. Mill and Mill, p. 102.

29. Mill and Mill, p. 57.

30. Cited in Murray, p. 119. Bodichon was quoting a contemporary English judge.
31. The committee also included Howitt's daughter, Anna Mary Howitt, an artist; Eliza Fox, also an artist; and Elizabeth Reid, the founder of Bedford College. Information drawn from Lee Holcombe, *Wives and Property: Reform of the Married Women's Property Law in Nineteenth-Century England* (Toronto: University of Toronto Press, 1983), pp. 58–62.
32. Petition and list of chief petitioners in Holcombe, Appendix I.
33. Holcombe, p. 109.
34. Oxford and Cambridge did not grant women degrees, even if they had attended courses and passed all examinations, until the early 1920s. Oxford then allowed the women degree-holders to vote in university elections; Cambridge did not until 1948.
35. Cited in F. K. Prochaska, *Women and Philanthropy in Nineteenth-Century England* (New York: Oxford University Press, 1980), p. 228.
36. Cited in Marilyn J. Boxer, " 'First-Wave' Feminism in Nineteenth-Century France: Class, Family and Religion," *Women's Studies International Forum*, Vol. V, no. 6 (1982), p. 553.
37. Cited in Friedrich G. Kürbisch and Richard Klucsarits, eds., *Arbeiterinnen Kämpfen um ihr Recht: Autobiographische Texte Zum Kampf rechtloser und entrechteter "Frauenpersonen" in Deutschland, Osterreich und der Schweiz des 19. und 20. Jahrhunderts* (Wuppertal: Peter Hammer, n.d.), p. 272. Authors' translation.
38. Cited in Sheila Rowbotham, *A New World for Women: Stella Browne, Socialist Feminist* (London: Pluto Press, 1977), p. 37.
39. Cited in Jill Liddington and Jill Norris, *One Hand Tied Behind Us: The Rise of the Women's Suffrage Movement* (London: Virago,1978), p. 145. For the NUWSS and working-class women, see pp. 224–25.
40. Strachey, p. 301.
41. Cited in Strachey, p. 338.
42. Cited in Liddington and Norris, p. 248.
43. Cited in Jilly Cooper and Tom Hartman, *Violets and Vinegar: Beyond Bartlett's, Quotations by and About Women* (New York: Stein and Day, 1982), p. 177.
44. For working-class repudiation of the WSPU, see Liddington and Norris, chap. 11.
45. Cited in Cooper and Hartman, p. 177.
46. Hannah Mitchell, *The Hard Way Up: The Autobiography of Hannah Mitchell. Suffragette and Rebel*, ed. Geoffrey Mitchell (London: Virago, 1977), p. 150.
47. Evans, p. 191.
48. Cited in Midge Mackenzie, *Shoulder to Shoulder A Documenhty* (New York: Alfred A. Knopf, 1975), pp. 128–29.
49. Mitchell, p. 184.
50. Cited in Mackenzie, p. 277.
51. Cited in Mackenzie, p. 261.
52. Phrases used by Lord Cromer, a leading anti-suffragist, in 1910. Cited in Brian Harrison, *Separate Spheres: The Opposition to Women's Suffrage in Britain* (New York: Holmes and Meier, 1978), p. 34. Anti-suffrage groups in England had many more members than suffrage groups. On this subject, see Harrison generally.

53. The exact role of World War I in hastening women's suffrage is still under debate by historians. For the view that the war was crucial, see Arthur Marwick, *The Deluge: British Society and the First World War* (New York: W. W. Norton & Company, 1965), pp. 95–104.

54. Cited in Strachey, p. 348.

55. Garner, pp. 109–10. In Germany, women did vote more for conservative and Catholic parties than men.

56. Cited in Garner, p. 104.

57. On women's rights movements in Scandinavia, see Evans, *The Feminists*, pp. 69–91, and Inga Dahlsgård, *Women in Denmark Yesterday and Today* (Copenhagen: Det Danske Selskab, 1980), pp. 98–140.

58. Cited in Ann Oakley, *Subject Women* (New York: Pantheon, 1981), p. 11.

59. For a similar argument about the U.S. women's rights movement, see Barbara Berg, *The Remembered Gate: Origins of American Feminism, 1800–1860* (1978), passim.

60. Mitchell, p. 141.

61. Cited in Evans, *German Feminism*, p. 77.

62. Evans, *Feminists*, p. 212.

63. Evans, *Feminists*, p. 213.

64. Cited in Albistur and Armogathe, Vol. II, p. 567. Authors' translation.

65. The one exception was Republican Spain, where women voted from 1931 to Franco's accession to power in 1939.

66. Cited in Hal Draper and Anna C. Lipow, "Marxist Women versus Bourgeois Feminism," in *The Socialist Register* (1976), p. 224.

67. Cited in Hackett, p. 430.

68. Evans, *German Feminism*, p. 90.

69. Virginia Woolf, *A Room of One's Own* (New York: Harcourt, Brace, 1957 [1929]), p. 37.

PART IX: **4. Feminist Socialism in Europe**

1. Italics in original. From Fourier's *Théorie des quartre movements et des destinées générales*, cited in Susan Groag Bell and Karen M. Offen, eds., *Women, the Family, and Freedom: The Debate in Documents*, vol. I, *1750–1880*, p. 41. Prior to the nineteenth century, even authors of utopian fantasies who urged communal ownership of property rarely advocated feminism. Plato's *Republic*, written in the fifth century B.C., was a rare exception.

2. Louise Michel, *The Red Virgin: Memoirs of Louise Michel*, ed. and trans. Bullitt Lowry and Elizabeth Ellington Gunter (University, Ala.: University of Alabama Press, 1981), pp. 197, 142.

3. Michel, p. 59.

4. Cited in Alfred G. Meyer, *The Feminism and Socialism of Lily Braun* (Bloomington: Indiana University Press, 1985), p. 52.

5. Cited in Renate Pore, *A Conflict of Interest: Women in German Social Democracy*,

1919–1933 (Westport, Conn.: Greenwood Press, 1981), pp. xvii–xviii, footnote 5. Thanks to Diana Ellis for help with the translation.

6. On this subject, see Moses, chap. 1, and Patrick Kay Bidelman, *Pariahs Stand Up! The Founding of the Liberal Feminist Movement in France, 1858–1889* (Westport, Conn.: Greenwood Press, 1982), p. 26.

7. Friedrich Engels, *The Origin of the Family, Private Property and the State*, ed. Eleanor Burke Leacock (New York: International Publishers, 1973 [1884]), pp. 137–38.

8. In *The Woman Question: Selections from the Writings of Karl Marx, Friedrich Engels, V. I. Lenin, Joseph Stalin* (New York: International Publishers, 1951), p. 30.

9. For law, see Taylor, pp. 283–84.

10. Marx, *The Woman Question*, p. 30; Engels, in *The Condition of the Working Class in England* (1844), cited in Bell and Offen, eds., Vol. I, p. 217.

11. Cited in Werner Thönnessen, *The Emancipation of Women: The Rise and Decline of the Women's Movement in German Social Democracy, 1863–1933*, trans. Joris des Bres (Glasgow: Pluto Press, 1976), p. 20; for similar sentiments among the Prudhonnist French socialists, see Marilyn J. Boxer, "Foyer or Factory: Working Class Women in Nineteenth-Century France," *Western Society for French History Proceedings*, Vol. II (November 1974), p. 196.

12. *Clara Zetkin: Selected Writings*, ed. Philip S. Foner (New York: International Publishers, 1984), p. 21.

13. August Bebel, *Women Under Socialism*, trans. Daniel De Leon (New York: Schocken, 1971 [1879]), p. 343.

14. Cited in Jean H. Quataert, "Unequal Partners in an Uneasy Alliance: Women and the Working Class in Imperial Germany," in Marilyn J. Boxer and Jean H. Quataert, eds., *Socialist Women: European Socialist Feminism in the Nineteenth and Early Twentieth Centuries* (New York: Elsevier, 1978), p. 120.

15. For this argument, see Richard Stites, "Women and the Russian Intelligentsia: Three Perspectives," in Dorothy Atkinson, Alexander Dallin, and Gail Warshofsky Lapidus, eds., *Women in Russia* (Stanford, Calif.: Stanford University Press, 1977), p. 46.

16. Richard Stites, *The Women's Liberation Movement in Russia: Feminism, Nihilism, and Bolshevism, 1860–1930* (Princeton, N.J.: Princeton University Press, 1978), p. 149.

17. Richard Evans, *The Feminists: Women's Emancipation Movements in Europe, America and Australasia, 1840–1920*, rev. ed. (London: Croom Helm, 1979), p. 162.

18. Italics in the original. Cited in Charles Sowerwine, *Sisters or Citizens? Women and Socialism in France since 1876* (Cambridge: Cambridge University Press, 1982), p. 64.

19. Michel, p. 142.

20. The majority of early female feminist socialists came from the working class. A sizable minority were lower-middle-class in origin. Taylor, pp. 88, 73; Moses, pp. 65–67.

21. Taylor estimates that the Owenist *Pioneer* was the second-largest working-class paper in England in these years, and that its "Women's Page" was "the singlemost

important platform for working-class feminist ideas in the early 1830s" (Taylor, p. 97, footnote 1).

22. Cited in Taylor, p. 96.

23. Cited in Barbara Taylor, " 'The Men Are As Bad As Their Masters . . .': Socialism, Feminism, and Sexual Antagonism in the London Tailoring Trade in the Early 1830s," *Feminist Studies*, Vol. 5, no. 1 (Spring 1979), p. 23.

24. Cited in Taylor, *Eve*, p. 96.

25. Cited in Taylor, "The Men Are As Bad," p. 11.

26. Cited in Taylor, *Eve*, p. 150.

27. Père Enfantin, cited in Albistur and Armogathe, Vol. II, p. 409.

28. Moses, pp. 52–53.

29. Cited in Moses, p. 63.

30. Moses, p. 65.

31. Cited in Moses, pp. 68, 67.

32. Cited in Albistur and Armogathe, Vol. II, p. 421. Authors' translation.

33. Suzanne Voilquin, *Mémoires d'une saint-simonienne en Russie*, ed. Maïté Albistur and Daniel Armogathe (Paris: Edition des femmes, 1977), p. 15. Authors' translation.

34. Cited in Moses, p. 128.

35. Cited in Moses, pp. 145–46. For Deroin's campaign, also see Edith Thomas, *Les femmes en 1848* (Paris: Presses Universitaires de France, 1948), chap. 9.

36. Letter in Miriam Schneir, ed., *Feminism: The Essential Historical Writings* (New York: Random House, 1972), pp. 91–92. Deroin and Roland were congratulating the Americans on the same convention in Worcester, Mass., which had inspired Harriet Mill to write her essay on women's enfranchisement.

37. Cited in Domenique Desanti, *A Woman in Revolt: A Biography of Flora Tristan*, trans. Elizabeth Zelvin (New York: Crown, 1976), pp. 270–71.

38. Cited in Lillian Faderman, *Surpassing the Love of Men: Romantic Friendship and Love Between Women from the Renvissance to the Present* (New York: William Morrow, 1981), p. 216.

39. Cited in Desanti, pp. 271–72.

40. "Free love" was used in all European languages to criticize untraditional sexual views of marriage. Originally used by critics, it gradually became adopted by adherents, although many pointed out that without contraception, "love" was not "free" for women. See Stites, *Women's Liberation*, p. 97.

41. Cited in Taylor, *Eve*, pp. 187, 184.

42. Cited in Taylor, *Eve*, p. 46.

43. Moses, p. 71.

44. Cited in Moses, p. 73.

45. Cited in Edith Thomas, *Pauline Roland: Socialisme et Féminisme au XIXe Siècle* (Paris: Marcel Rivière, 1956), p. 64. Authors' translation.

46. Moses, pp. 81–82.

47. Cited in Thomas, *Roland*, p. 116.

48. Cited in Moses, pp. 147–48.

49. Cited in Thomas, *Roland*, pp. 155–56.

50. Cited in Thomas, *Roland*, p. 172.
51. Cited in *The Woman Question*, p. 35.
52. Cited in Stites, *Women's Liberation*, p. 136.
53. Claudia Koonz, "Conflicting Allegiances: Political Ideology and Women Legislators in Weimar Germany," *Signs*, Vol. I, no. 3, part 1 (Spring 1976), table 1, p. 667.
54. Adelheid Popp, *The Autobiography of a Working Woman*, trans. E. C. Harvey (Westport, Conn.: Hyperion Press, 1983 [1909]), p. 107. For Popp's life, also see Ingrun Lafleur, "Five Socialist Women: Traditionalist Conflicts and Socialist Visions in Austria, 1893–1934," in Boxer and Quataert, eds., *Socialist Women*, pp. 223–25.
55. Popp, p. 107.
56. Popp, p. 120.
57. The University of Zurich accepted women from 1864 and granted women medical degrees from 1867. By 1870, a sizable community of Russian women studied there.
58. On this point, see, for Germany, Jean H. Quataert, *Reluctant Feminists in German Social Democracy, 1885–1917* (Princeton, N.J.: Princeton University Press, 1979), p. 153ff.; for Russia, Rose L. Glickman, *Russian Factory Women: Workplace and Society, 1880–1914* (Berkeley: University of California Press, 1984); for France, Sowerwine, passim.
59. Cited in Quataert, *Reluctant Feminists*, p. 139.
60. On this subject, see Stites, *Women's Liberstion*, chap. 5, and Barbara Alpern Engel, *Mothers and Daughters: Women of the Intelligentsia in Nineteenth-Century Russia* (Cambridge, Eng.: Cambridge University Press, 1983).
61. *Five Sisters: Women Against The Tsar*, ed. and trans. Barbara Alpern Engel and Clifford N. Rosenthal (New York: Schocken, 1977), p. 210.
62. *Five Sisters*, p. 211.
63. *Five Sisters*, p. 212.
64. *Five Sisters*, p. 216.
65. *Five Sisters*, p. 69.
66. *Five Sisters*, p. xxxiv.
67. *Five Sisters*, p. 240.
68. Gail Warshofsky Lapidus, *Women in Soviet Society: Equality, Development, and Social Change* (Berkeley: University of California Press, 1978), p. 38.
69. On this point, see Engel, *Mothers and Daughters*, p. 197.
70. Peter Nettl, *Rosa Luxemburg*, abridged ed. (New York: Oxford University Press, 1969), pp. 38, 88, 415.
71. Cited in Nettl, p. 476, footnote 1.
72. Cited in Hal Draper and Anne G. Lipow, "Marxist Women versus Bourgeois Feminism," *The Socialist Register* (1976), p. 201.
73. On this point, see Quataert, *Reluctant Feminists*, pp. 15, 65, 108.
74. Cited in Quataert, *Reluctant Feminists*, p. 66.
75. Cited in Thönnessen, p. 39.
76. Zetkin, p. 50.
77. Zetkin, p. 25.

78. On this point, see Quataert, *Reluctant Feminists*, pp. 152–53.
79. Zetkin, pp. 76–77.
80. For Braun and Zetkin, see Quataert, *Reluctant Feminists*, p. 230ff., and Alfred G. Meyer, *The Feminism and Socialism of Lily Braun* (Bloomington: Indiana University Press, 1985).
81. Cited in Karen Honeycutt, "Clara Zetkin: A Socialist Approach to the Problem of Woman's Oppression," *Feminist Studies*, Vol. 4, nos. 1 and 2 (Spring–Summer 1976), p. 136.
82. On this see Quataert, *Reluctant Feminists*, p. 231 and Anneliese Bergmann, "Frauen, Männer, Sexualität und Geburtenkontrolle: Zur 'Gebärstreikdebatte' der SPD 1913," in Karin Hausen, ed., *Frauen Suchen Ihre Geschichte: Historische Studien zum 19. und 20. Jahrhunderts* (Munich: C. H. Beck, 1983), pp. 81–109.
83. Cited in Draper and Lipow, p. 201.
84. Quataert, *Reluctant Feminists*, p. 237.
85. Cited in Hilda Scott, *Women and Socialism: Experiences from Eastern Europe* (London: Alison and Busby, 1976), p. 63.
86. Zetkin, p. 175.
87. Cited in Stites, *Women's Liberation*, p. 207. For the Russian equal rights movement, see Stites, chap. 7, and Linda Harriet Edmonson, *Feminism in Russia, 1900–1917* (Stanford, Calif.: Stanford University Press, 1984).
88. About 6 percent of the Social Democratic delegates to the Party Congress in the summer of 1917 were female. More detailed figures are not available. Lapidus, pp. 38–39.
89. Alexandra Kollontai, *The Autobiography of a Sexually Emancipated Communist Woman*, trans. Salvator Attanasio (London: Herder and Herder, 1971 [1926]), pp. 13–14.
90. Stites, *Women's Liberation*, p. 250.
91. Cited in *Selected Writings of Alexandra Kollontai*, ed. and trans. Alix Holt (New York: W. W. Norton & Company, 1977), p. 33.
92. Cited in Barbara Evans Clements, *Bolshevik Feminist: The Life of Alexandra Kollontai* (Bloomington: Indiana University Press, 1979), p. 61.
93. Kollontai, pp. 66–67.
94. From an essay of 1914. Kollontai, p. 134.
95. Cited in Clements, p. 155.
96. Kollontai, *Selected Writings*, p. 256.
97. Kollontai, *Selected Writings*, p. 259.
98. Kollontai, *Selected Writings*, p. 233.
99. Kollontai, *Selected Writings*, pp. 253–56.
100. For this point, see Barbara Evans Clements, "Working-Class and Peasant Women in the Russian Revolution, 1917–1923," *Signs*, Vol. 8, no. 2 (Winter 1982), pp. 234–35.
101. V. I. Lenin, *The Emancipation of Women* (New York: International Publishers, 1978), p. 64.
102. Cited in Stites, *Women's Liberation*, p. 325.
103. Lenin, pp. 107–8.

104. Lenin, p. 102.
105. Cited in Clements, *Bolshevik Feminist*, p. 73.
106. From the story "Three Generations," in Alexandra Kollontai, *Love of Worker Bees*, trans. Cathy Porter (Chicago: Cassandra Editions, 1978), p. 203.
107. Kollontai, *Autobiography*, p. 45.
108. Clements, *Bolshevik Feminist*, pp. 235, 233.
109. Cited in Lapidus, p. 71.
110. Cited in Stites, *Women's Liberation*, pp. 383–84; also see p. 334 on Smidovich.
111. For figures, see Lapidus, table 11, p. 166, and table 2, p. 131.
112. Kollontai, *Selected Writings*, pp. 308–9. On this subject, also see Beatrice Brodsky Farnsworth, "Bolshevik Alternatives and the Soviet Family: The 1926 Marriage Law Debate," in Atkinson et al., eds., pp. 139–66.
113. Kollontai, *Autobiography*, pp. 47–48.
114. Kollontai, *Selected Writings*, p. 315.
115. Lenin, p. 110.
116. Juchacz had replaced Zetkin as head of the German Socialist Party Women's Group. Cited in Jill Stephenson, *Women in Nazi Society* (New York: Harper & Row, Publishers, 1975), pp. 15–16.
117. Sheila Rowbotham, *Hidden from History: Rediscovering Women in History from the Seventeenth Century to the Present* (New York: Pantheon, 1974), p. 160.
118. Cited in Pore, p. 41.
119. Cited in Garner, p. 3.
120. Cited in Pore, p. 34.
121. Koonz, "Conflicting Allegiances," p. 673.
122. Cited in Sheila Rowbotham, *A New World for Women: Stella Browne, Socialist Feminist* (London: Pluto Press, 1977), p. 35.
123. Cited in Sowerwine, p. 168.
124. Margaret Llewelyn Davies, ed., *Life As We Have Known It* (New York: W. W Norton & Company, 1975 [1931]), p. xi.
125. Richard J. Evans, *The Feminist Movement in Germany, 1894–1933* (London: Sage, 1976), p. 237.
126. Dora Russell, *The Tamarisk Tree: My Quest for Liberty and Love* (London: Virago, 1978), pp. 217–18.
127. Cited in Stites, *Women's Liberation*, p. 386.
128. On this subject, see Ruth Hall, ed., *Dear Dr. Stopes: Sex in the 1920s* (New York: Penguin, 1978) for England; for Germany, Atina Grossmann, "Abortion and Economic Crisis: The 1931 Campaign Against Paragraph 218," in Renate Bridenthal, Atina Grossmann, and Marion Kaplan, eds., *When Biology Became Destiny: Women in Weimar and Nazi Germany* (New York: Monthly Review Press, 1984); for Denmark, Inga Dahlsgård, *Women in Denmark Yesterday and Today* (Copenhagen: Det Danske Selskab, 1980); and for the Soviet Union, Gail Warshofsky Lapidus, *Women in Soviet Society: Equality, Development, and Social Change* (Berkeley: University of California Press, 1978), pp. 60, 299.
129. Carla Bielli, "Some Aspects of the Condition of Women in Italy," in Lynne B.

Iglitzin and Ruth Ross, eds., *Women in the World: A Comparative Study* (Santa Barbara, Calif.: Clio Press, 1976), p. 111.

130. Cited in Lynne Jones, ed., *Keeping the Peace: A Woman's Peace Handbook* (London: The Women's Press, 1983), p. xiii.

131. Cited in Jones, p. 273.

132. Suttner had worked as a secretary for Nobel in the late 1860s. *Suffragette for Peace* is the title of her biography by B. Kempf (1972).

133. Reproduced in Cambridge Women's Peace Collective, *My Country Is the Whole World: An Anthology of Women's Work on Peace and War* (London: Pandora Press, 1984), p. 67.

134. Gisela Brinker-Gabler, ed., *Frauen gegen den Krieg* (Frankfurt am Main: Fischer, 1980), p. 24.

135. Liddington and Norris, p. 253.

136. Gertrude Bussey and Margaret Tims, *Pioneers for Peace: Women's International League for Peace, 1915–1965* (London: WILPF British Section, 1980), pp. 17–18.

137. Brinker-Gabler, pp. 24–33.

138. Cited in Bell and Offen, eds., Vol. II, p. 274; also see Albistur and Armogathe, Vol. II, p. 538.

139. Brinker-Gabler, p. 14.

140. Cambridge Women's Peace Collective, p. 113.

141. Zetkin, p. 178.

142. Bussey and Tims, pp. 126–30.

143. Bernice A. Carroll, "'To Crush Him in Our Own Country': The Political Thought of Virginia Woolf," paper read at Third Berkshire Conference on the History of Women, Bryn Mawr College, Bryn Mawr, Pa., 1976, p. 34.

144. Virginia Woolf, *A Room of One's Own* (New York: Harcourt, Brace and World, 1957 [1929]), p. 118.

145. Quentin Bell, *Virginia Woolf: A Biography*, 2 vols. (St. Albans, Eng.: Triad/Paladin, 1976), Vol. II, p. 156.

146. Bell, Vol. II, p. 193; Carroll, p. 38. Woolf's title comes from her decision that to help peace she must donate "three guineas": one to a women's college, one to a society to promote women's employment in the professions, and one to a pacifist organization. She concludes that "the causes are the same and inseparable." Virginia Woolf, *Three Guineas* (New York: Harcourt, Brace and World, 1966 [1938]), p. 144.

147. Woolf, p. 114.

148. For Woolf's disinclination to use the word *feminism* and her use of alternate terms, see Carroll, pp. 36–37.

149. Woolf, *Three Guineas*, p. 124.

150. Woolf, *Three Guineas*, p. 186, footnote 48.

151. Woolf, *Three Guineas*, p. 142.

152. Woolf, *Three Guineas*, p. 143.

153. Woolf, *Three Guineas*, p. 109.

154. Woolf, *Three Guineas*, pp. 106ff.

155. Woolf, *Three Guineas*, p. 110.

156. Woolf, *Three Guineas*, p. 143.
157. Cambridge Women's Peace Collective, p. 123.
158. Cited in Raynes Minns, *Bombers and Mash: The Domestic Front, 1939–1945* (London: Virago, 1980), p. 3.
159. Evelyne Sullerot, *Women, Society, and Change*, trans. Margaret Scotford Archer (New York: McGraw-Hill, 1971), pp. 74–75.

PART IX: 5. The Women's Liberation Movement

1. On this subject, see Joni Lovenduski, *Women and European Politics: Contemporary Feminism and Public Policy* (Amherst: University of Massachusetts Press, 1986), pp. 110–11; David Bouchier, *The Feminist Challenge: The Movement for Women's Liberation in Britain and the United States* (New York: Schocken, 1983), chap. 2; Maïté Albistur and Daniel Armogathe, *Histoire du féminisme français*, 2 vols. (Paris: Editions des femmes, 1977), Vol. II, chap. 5; Elaine Marks and Isabelle de Courtivron, eds., *New French Feminisms* (New York: Schocken, 1981), pp. 30–31; Lucia Chiavola Birnbaum, *liberazione della donna: feminism in Italy* (Middletown. Conn.: Wesleyan University Press, 1986), chap. 7; Eleonore Eckmann Pisciotta, "The strength and the powerlessness of the new Italian women's movement: the case of abortion," in Drude Dahlerup, ed., *The New Women's Movement: Feminism and Political Power in Europe and the USA* (London: Sage, 1986), p. 28; Edith Hoshino Altbach, "The New German Women's Movement," in Edith Hoshino Altbach, Jeanette Clausen, Dagmar Schultz, and Naomi Stephan, eds., *German Feminism: Readings in Politics and Literature* (Albany, N.Y.: SUNY Press, 1984), pp. 3–26; Petra De Vries, "Feminism in the Netherlands," in Jan Bradshaw, ed., *The Women's Liberation Movement: Europe and North America* (Oxford, Eng.: Pergamon Press, 1982; originally Vol. 4, no. 4 of the *Women's Studies International Forum*), p. 391.
2. Cited in Marie Collins and Sylvie Weil Sayre, eds., *Les Femmes en France* (New York: Simon and Schuster, 1974), p. 302. Authors' translation.
3. Altbach et al., p. 310.
4. Marsha Rowe, ed., *Spare Rib Reader* (Harmondsworth, Eng.: Penguin, 1982), p. 574.
5. Altbach et al., pp. 308–9. Such criticism was not allowed in state socialist nations: the Soviet Union, the German Democratic Republic, etc. As a result, women's liberation movements did not appear there. Those few women who asserted feminism found themselves isolated, and occasionally expelled. On this subject, see Lovenduski, p. 114; Barbara Einhorn, "Socialist Emancipation: The Women's Movement in the German Democratic Republic," in Bradshaw, ed., pp. 435–52; Alix Holt, "The First Soviet Feminists" [the dissidents who published *Women and Russia* in 1979], in Barbara Holland, ed., *Soviet Sisterhood* (Bloomington: Indiana University Press, 1985), pp. 237–65; and Tatyana Mamonova, ed., *Women and Russia*, trans. Rebecca Park and Catherine A. Fitzpatrick (Boston: Beacon Press, 1984).
6. Carol Ascher, *Simone de Beauvoir: A Life of Freedom* (Boston: Beacon Press, 1981), p. 128.

7. Simone de Beauvoir, *The Second Sex*, trans. H. M. Parshley (New York: Bantam, 1961), p. xxviii.
8. de Beauvoir, p. 689.
9. Marks and Courtivron, eds., p. 192.
10. Marks and Courtivron, eds., pp. 149–50.
11. Marks and Courtivron, eds., p. 146.
12. Marks and Courtivron, eds., p. 190.
13. Cited in Ascher, p. 128.
14. Thanks to Claudia Koonz for help with this point. Consciousness-raising was more pervasive in the United States than in Europe, but a recent analysis calls it the first step for the women's liberation movement in Europe as well. Drude Dahlerup, Introduction, in Dahlerup, ed., p. 8.
15. Lilian Mohin, ed., *One Foot on the Mountain: An Anthology of British Feminist Poetry, 1969–1979* (London: Onlywomen Press, 1979), p. 147.
16. Cited in Theresia Sauter-Bailliet, "The Feminist Movement in France," in Bradshaw, ed., p. 409. Some feminists, like the women of the Saint-Simonians in the 1830s, discarded their last names to signify their liberation from male dominance. Annie and Anne occasionally assumed the surnames "de Pizan" and "Tristan" in honor of the earlier feminists Christine de Pizan and Flora Tristan.
17. Anja Meulenbelt, *The Shame Is Over: A Political Life Story*, trans. Ann Oosthuizen (London: Women's Press, 1980), p. 137.
18. Meulenbelt, p. 138.
19. Meulenbelt, p. 139.
20. Meulenbelt, p. 11.
21. Maria Isabel Barreno, Maria Teresa Horta, Maria Velho da Costa, *The Three Marias: New Portuguese Letters*, trans. Helen R. Lane (New York: Doubleday & Company, 1976), pp. 32–33.
22. Altbach, et al., eds., p. 48.
23. Rowe, ed., p. 607.
24. Bradshaw, ed., p. 400.
25. Cited in Sauter-Bailliet, in Bradshaw, ed., pp. 409–10.
26. *Esistere come donna* (Milan: Mazzotta, 1983), p. 276.
27. Diane E. H. Russell and Nicole Van de Ven, eds., *Crimes Against Women: Proceedings of the International Tribunal* (Millbrae, Calif.: Les Femmes, 1976), p. 14.
28. Marks and Courtivron, eds., p. 31; Anne Batiot, "Radical democracy and feminist discourse: the case of France," in Dahlerup, ed., p. 93.
29. On these subjects, see John T. Noonan, *Contraception: A History of Its Treatment by the Catholic Theologians and Canonists* (Cambridge, Mass.: Belknap Press, 1966); Thomas McKeown, *The Modern Rise of Population* (New York: Harcourt Brace Jovanovich, 1976); Angus McLaren, *Birth Control in Nineteenth-Century England* (New York: Holmes and Meier, 1978); Angus McLaren, *Sexuality and the Social Order: The Debate over the Fertility of Women and Workers in France, 1770–1920* (New York: Holmes and Meier, 1983); Peter Fryer, *The Birth Controllers* (New York: Stein and Day, 1965); Roseanna Ledbetter, *A History of the Malthusian League: 1877–1927* (Columbus, Ohio: Ohio State University Press, 1976); Margaret

Sanger and Hannah Stone, eds., *The Practice of Contraception. An International Symposium and Survey* (Baltimore, Md.: Williams and Wilkins, 1931); Malcolm Potts, Peter Diggory, and John Peel, *Abortion* (Cambridge: Cambridge University Press, 1977).

30. For publication figures and book titles, see F. Barry Smith, "Sexuality in Britain, 1800–1900: Some Suggested Revisions," in Martha Vicinus, ed., *A Widening Sphere: Changing Roles of Victorian Women* (Bloomington: Indiana University Press, 1977), pp. 188–90. The history of contraception and abortion advocacy in England has been most thoroughly studied.

31. Cited in Elizabeth Longford, *Eminent Victorian Women* (New York: Alfred A. Knopf, 1981), p. 139.

32. Her life was characterized by a series of disparate enthusiasms, culminating in Indian religion and politics. A. H. Nethercott, her biographer, entitled his volumes, *The First Five Lives of Annie Besant* and *The Last Four Lives of Annie Besant*.

33. Ledbetter, p. 206; McLaren, *England*, p. 109.

34. Eleanor S. Riemer and John C. Fout, eds., *European Women: A Documentary History, 1789–1945* (New York: Schocken, 1980), p. 215.

35. Riemer and Fout, eds., p. 217.

36. Riemer and Fout, eds., p. 216.

37. Cited in Evans, *Feminist Movement in Germany*, p. 118. For Stöcker and *Mutterschutz*, see this work, part IX, chap. 4, and Amy Hackett, "Helene Stöcker: Left-Wing Intellectual and Sex Reformer," in Bridenthal et al., eds., pp. 109–30.

38. Dahlsgård, pp. 177–79.

39. Cited in Rowbotham, p. 37.

40. Dora Russell, *The Tamarisk Tree: My Quest for Liberty and Love* (London: Virago, 1978), p. 175.

41. Jean McCrindle and Sheila Rowbotham, eds., *Dutiful Daughters: Women Talk About Their Lives* (Austin: University of Texas Press, 1977), pp. 104–5.

42. Cited in Rowbotham, *New World*, p. 114.

43. In Rumania, for instance, an extremely low birth rate led to severe restrictions on abortion; in 1984 President Ceausescu declared bearing four children to be women's "patriotic duty," and women who abort without permission face a year in prison. On this subject, see "For the Sake of Reproduction," in *Connexions: Women in Eastern Europe* (Summer 1982); *National NOW Times* (January–February 1985); Alena Heitlinger, *Women and State Socialism: Sex Inequality in the Soviet Union and Czechoslovakia* (Montreal: McGill-Queens University Press, 1979), chaps. 12 and 17; and Robert J. McIntyre, "Demographic Policy and Sexual Equality: Value Conflicts and Policy Reappraisal in Hungary and Romania," in Sharon J. Wolchik and Alfred G. Meyer, eds., *Women, State, and Party in Eastern Europe* (Durham, N.C.: Duke University Press, 1985), pp. 270–85.

44. Potts et al., p. 383.

45. Gisèle Halimi, *La cause des femmes* (Paris: Grasset, 1973), p. 98; Joni Lovenduski and Jill Hills, eds., *The Politics of the Second Electorate: Women and Public Participation* (London: Routledge and Kegan Paul, 1981), p. 121.

46. In France, the name "Women's Liberation Movement" (*Mouvement de la Libéra-*

tion des Femmes) was used by one group, but copyrighted in a bitter lawsuit by another. In this text, it is used to refer to the feminist movement of the late 1960s and 1970s as a whole.

47. Louise Vandelac, *L'italie au féminisme* (Paris: Editions tierce, 1978), pp. 16–19, 46.
48. Meulenbelt, p. 158.
49. Cited in Longford, p. 123. For a fuller account of Butler's campaign, see part VII.
50. Claude Jaget, ed., *Prostitutes: Our Life* (Bristol, Eng.: Falling Wall Press, 1980), p. 131.
51. Cited in Marks and Courtivron, eds., p. 196.
52. Jaget, ed., p. 29.
53. Altbach et al., pp. 184–86.
54. Verena Stefan, *Shedding*, trans. Johanna Moore and Beth Weckmueller (New York: Daughters Publishing Co., 1978), p. 31.
55. Cited in Russell and Van der Ven, p. xiii.
56. Russell and Van der Ven, p. 127.
57. Vandelac, p. 17.
58. Ruth Hall, *Women at WAR* [Women Against Rape] (London, 1978), pp. 20, 26.
59. Italics in the original. Council of Europe Information Document submitted to the World Conference of the United Nations Decade for Women (1980), p. 27.
60. Cited in Lillian Faderman, *Surpassing the Love of Men: Romantic Friendship and Love Between Women from the Renaissance to the Present* (New York: William Morrow, 1981), p. 388.
61. Cited in Lillian Faderman and Brigitte Eriksson, eds., *Lesbian-Feminism in Turn-of-the-Century Germany* (Weatherby Lake, Mo.: Naiad Press, 1981), pp. 88, 91.
62. Barbara Caine, "Feminism, Suffrage and the Nineteenth-Century English Women's Movement," *Women's Studies International Forum*, Vol. 5, no. 6 (1982), pp. 537–50.
63. Charlotte Wolff, *Hindsight* (London: Quartet Books, 1980), p. 216.
64. Heinz Puknus, ed., *Neue Literatur der Frauen: Deutschsprachige Autorinnen der Cegenwart* (Munich: C. H. Beck, 1980), p. 208.
65. Stefan, pp. 28, 45.
66. Stefan, pp. 77–78.
67. Stefan, p. 91.
68. Stefan, p. 118.
69. "The Second Sex—Thirty Years Later," Conference papers delivered at the New York Institute for the Humanities, 1979, pp. 74–75.
70. Cited in Faderman, *Surpassing*, p. 390.
71. Cited in Wolff, p. 231.
72. Monique Wittig, *Les Guérillères*, trans. David Le Vay (New York: Avon, 1971), p. 89.
73. *New York Times*, April 18, 1982, p. 4.
74. Marielouise Janssen-Jurreit, *Sexism: The Male Monopoly on History and Thought*, trans. Verne Moberg (New York: Farrar, Straus & Giroux, 1982), p. 17.
75. Marks and Courtivron, eds., p. 189.

76. Cited in Elaine Marks, "Women and Literature in France," *Signs*, Vol. 3, no. 4 (Summer 1978), p. 841.
77. Marks and Courtivron, eds., p. 64.
78. Marks and Courtivron, eds., p. 64.
79. Cited by Audur Styrkársdóttir, "From social movement to political party: the new women's movement in Iceland," in Dahlerup, ed., p. 151.
80. Cited in Birnbaum, p. 223.
81. Cambridge Women's Peace Collective, p. 259.
82. Lynne Jones, ed., *Keeping the Peace: A Woman's Peace Handbook* (London: Women's Press, 1983), p. 8.
83. Jones, p. 23.
84. Lovenduski and Hills, p. 222.
85. Cambridge Women's Peace Collective, p. 211.
86. Cambridge Women's Peace Collective, p. 263.
87. *Ms.* magazine, Vol. XIII, no. 7 (January 1985), p. 23.
88. Cited in Elina Juusola-Halonen, "The Women's Liberation Movement in Finland," in Bradshaw, ed., p. 453.
89. Quakers Women's Group, *Bring the Invisible into the Light: Some Quaker Feminists Speak of Their Experience* (London: Quaker Home Service, 1986), pp. 95–96.
90. Pizan, p. 185.

Epilogue

1. Cited in Barbara Einhorn, "The Great Divide? Women's Rights in Eastern and Central Europe Since 1945," in Renate Bridenthal, Susan Mosher Stuard, and Merry E. Weisner, eds., *Becoming Visible: Women in European History*, 3rd ed. (Boston: Houghton Mifflin Company, 1998), p. 517. Also see Barbara Einhorn, *Cinderella Goes to Market: Citizenship, Gender and Women's Movements in East Central Europe* (London: Verso, 1993) and Renate Siemienska, "Women in the Period of Systemic Changes in Poland," *Journal of Women's History*, vol. 5, no. 3 (Winter 1994), pp. 70–90.
2. Hanna Beate Schoepp-Schilling, "The Impact of German Unification of Women: Losses and Gains," in Alida Brill, ed., *A Rising Public Voice: Women in Politics Worldwide* (New York: The Feminist Press at the City University of New York, 1995), p. 33.
3. On this subject, see Schoepp-Schilling and Nanette Funk, "Abortion and German Unification," in Nanette Funk and Magda Mueller, eds., *Gender Politics and Post-Communism: Reflections from Eastern Europe and the Former Soviet Union* (New York: Routledge, 1993), pp. 194–200. Also see Malgorzata Fuszara, "Abortion and the Formation of the Public Sphere in Poland," in Funk and Mueller, pp. 241–252.
4. Einhorn in Bridenthal et. al., p. 529.
5. Editors' Note, "The More Things Change The Worse They Become for Women," *Journal of Women's History*, Special Issue on the Current Situation of European Women, vol. 5, no. 3 (Winter 1994), pp. 6–9; Barbara Einhorn and Eileen James

Yeo, eds., *Women and Market Societies: Crisis and Opportunity* (Aldershot, U.K.: Edward Elgar, 1995); Wilma Rule and Norma C. Noonan, eds., *Russian Women in Politics and Society* (Westport, Conn.: Greenwood Press, 1996).

6. *The World's Women 1995: Trends and Statistics* (New York: United Nations, 1995), p. 154.

7. Jane Jensen, "Friend or Foe? Women and State Welfare in Western Europe," in Bridenthal et. al., p. 505.

8. *The World's Women 1995*, pp. 126–127, 96.

9. *The World's Women 1995*, pp. 105, 108.

10. Liisa Rantalaiho and Raija Julkunen, "Women in Western Europe: Socioeconomic Restructuring and Crisis in Gender Contracts," *Journal of Women's History*, vol. 5, no. 3 (Winter 1994), pp. 19–22.

11. *The New York Times*, 10 July 1998, pp. A1, A6.

12. *New York Times*, 10 July 1998, p. A6.

13. *The World's Women 1995*, p. 162.

14. Beverly Allen, *Rape Warfare: The Hidden Genocide in Bosnia-Herzegovina and Croatia* (Minneapolis: University of Minnesota Press, 1996). For a description of the international campaign, see Charlotte Bunch and Niamh Reilly, *Demanding Accountability: The Global Campaign and Vienna Tribunal for Women's Human Rights* (New York: UNIFEM, 1994).

15. Rose-Marie Lagrave, "A Supervised Emancipation," in Françoise Thébaud, ed., *Toward a Cultural Identity in the Twentieth Century*, vol. V of Georges Duby and Michelle Perrot, gen. eds., *A History of Women in the West* (Cambridge, Mass.: Harvard University Press, 1994), p. 487.

16. The International Federation for Research in Women's History was founded in 1987, its goal "to encourage and co-ordinate research in all aspects of women's history at the international level, by promoting exchange of information and publications, and by organising large-scale international conferences as well as more specialised meetings." See the Federation's first collaborative publication, Karen Offen, Ruth Roach Pierson, and Janet Rendall, eds., *Writing Women's History: International Perspectives* (Bloomington: Indiana University Press, 1991).

17. Seminar, University of the Balearic Islands, April, 1995.

18. Sheila Rowbotham, *A Century of Women: The History of Women in Britain and the United States* (New York: Viking, 1997), p. 556.

19. Renate Bridenthal, "Women in the New Europe," in Bridenthal, et. al., p. 568; *The World's Women 1995*, p. 153.

20. *The World's Women 1995*, updated summer 1998, by Jennifer M. Morris.

21. *New York Times*, 22 May 1998, p. A3.

22. This and the subsequent quotation, Maria de Lourdes Pintasilgo, "Portugal: Daring to Be Different," in Brill, pp. 127–132.

Bibliography

This bibliography is selective and is designed to give the reader the names of the most useful and most accessible primary and secondary works. It is divided as follows:

1. General categories of works useful to all parts: general works, essay collections, and national histories;
2. Works about women essential to the analysis and narrative arranged topically in each part.

References to very specialized primary sources, articles, and monographs may be found in the Notes.

General Works and Essay Collections about Women

Abrams, Lynn, and Elizabeth Harvey, eds. *Gender Relations in German History: Power, Agency and Experience from the Sixteenth to the Twentieth Century.* Durham, N.C.: Duke University Press, 1997.

Baranski, Zygmunt G., and Shirley W. Vinall, eds. *Women and Italy: Essays on Gender, Culture and History.* New York: St. Martin's Press, 1991.

Benjamin, Marina. *A Question of Identity: Women, Science and Literature.* [women and the history of science] New Brunswick, N.J.: Rutgers University Press, 1993.

Berkin, Carol R., and Clara M. Lovett, eds. *Women, War and Revolution.* New York: Holmes and Meier, 1980.

Bremer, Jan, and Lourens van den Bosch, eds. *Between Poverty and the Pyre: Moments in the History of Widowhood.* New York: Routledge, 1995.

Bridenthal, Renate, Susan Mosher Stuard, and Merry Wiesner, eds. *Becoming Visible: Women in European History.* 3rd ed. Boston: Houghton Mifflin, 1998.

Carroll, Berenice A., ed. *Liberating Women's History: Theoretical and Critical Essays.* Urbana: University of Illinois Press, 1976.

Clark, Lorenne M. G., and Lynda Lange, eds. *The Sexism of Social and Political Theory: Women and Reproduction from Plato to Nietzsche.* Toronto: University of Toronto Press, 1979.

Clements, Barbara Evans, Barbara Alpern Engel, and Christine D. Worobec, eds. *Russia's Women: Accommodation, Resistance, Transformation.* Berkeley: University of California Press, 1991.

Cohen, Sherrill. *The Evolution of Women's Asylums since 1500: From Refuges for Ex-Prostitutes to Shelters for Battered Women.* New York: Oxford University Press, 1992.

Duby, Georges, and Michelle Perrot, eds. *A History of Women in the West,* 5 vols. Cambridge, Mass.: Harvard University Press, 1994.

Faderman, Lillian. *Surpassing the Love of Men: Romantic Friendship and Love Between Women from the Renaissance to the Present.* New York: William Morrow, 1981.

Fildes, Valerie. *Wet Nursing: A History from Antiquity to the Present.* New York: Basil Blackwell, 1988.

Gillis, John R., Louise A. Tilly, and David Levine, eds. *The European Experience of Declining Fertility: The Quiet Revolution.* Oxford: Basil Blackwell, 1992.

Goscilo, Helena, and Beth Holmgren, eds. *Russia, Women, Culture.* Bloomington: University of Indiana Press, 1996.

Hartman, Mary S., and Lois W. Banner, eds. *Clio's Consciousness Raised: New Perspectives on the History of Women.* New York: Harper & Row, 1974.

Hyman, Paula. *Gender and Assimilation in Modern Jewish History: The Roles and Representations of Women.* Seattle: University of Washington Press, 1995.

Kelly, Joan. *Women, History and Theory: The Essays of Joan Kelly.* Chicago: University of Chicago Press, 1984.

Knibiehler, Yvonne, and Catherine Fouquet. *L'histoire des Mères du moyen-âge à nos jours.* Paris: Editions Montalba, 1980.

London Feminist History Group. *The Sexual Dynamics of History: Men's Power, Women's Resistance.* London: Pluto Press, 1983.

Maynes, Mary Jo, ed. *Gender, Kinship, Power: A Comparative and Interdisciplinary History.* New York: Routledge, 1996.

Mitchell, Juliet, and Ann Oakley, eds. *The Rights and Wrongs of Women.* Harmondsworth, U.K.: Penguin, 1979.

Offen, Karen, Ruth Roach Pierson, and Jane Rendall, eds. *Writing Women's History: International Perspectives.* Bloomington: University of Indiana Press, 1991.

Okin, Susan Moller. *Women in Western Political Thought.* Princeton, N.J.: Princeton University Press, 1979.

Pushkareva, Natalia. *Women in Russian History: From the Tenth to the Twentieth Century.* Translated and edited by Eve Levin. Armonk, N.Y.: M.E. Sharpe, 1997.

Reynolds, Siân, ed. *Women, State and Revolution: Essays on Power and Europe since 1789.* Amherst: University of Massachusetts Press, 1987.

Rotberg, Robert I., and Theodore K. Rabb, eds. *Marriage and Fertility: Studies in Interdisciplinary History.* Princeton, N.J.: Princeton University Press, 1980.

Scott, Joan Wallach. *Gender and the Politics of History.* New York: Columbia University Press, 1988.

Shapiro, Ann-Louise, ed. *Feminists Revision History.* New Brunswick, N.J.: Rutgers University Press, 1994.

Smart, Carol, ed. *Regulating Womanhood: Historical Essays on Marriage, Motherhood and Sexuality.* London: Routledge, 1992.

Smith, Bonnie G. *Changing Lives: Women in European History Since 1700.* Lexington, Mass.: D.C. Heath & Co., 1989.

Stanley, Mary Lyndon, and Carole Pateman, eds. *Feminist Interpretations and Political Theory*. University Park: Pennsylvania State University Press, 1991.

Stock, Phyllis. *Better Than Rubies: A History of Women's Education*. New York: G. P. Putnam's Sons, 1978.

Tilly, Louise A., and Joan W. Scott. *Women, Work, and Family*. New York: Holt, Rinehart and Winston, 1978.

The following journals have useful articles and reviews: *Feminist Studies, Gender and History, Journal of Women's History, Signs: Journal of Women in Culture and Society, Women's History Review, Women's Studies International Forum*.

The following web sites are useful resources:

Dutch Women's History (Vereniging voor Vrouwengeschiedenis, Amsterdam)
http://www.let.ruu.nl/hist/info/VVG

EUROPA—Official web site for the European Community (EU)
http://europa.eu.int

GABRIEL—Gateway to Europe's National Libraries
http://portico.bl.uk/gabriel/

International Federation for Research in Women's History—Access to the newsletter of the international organization of women historians' groups
http://www.arts.unimelb.edu.au/Dept/History/ifrwh

Italian Women's History (Societa italiana dele storiche)
http://www.idg.fi.cnr.it/wwwdonna/storiche.htm

Modern British Women's History
http://info.ox.ac.uk/^shilinfo/women3.html

Modern German Women's History
http://www.kgw.tu-berlin.de/ZIFG   [University of Berlin]
http://www.uni-bonn.de/Frauengesschichte/in   [University of Bonn]

United Nations web site for women
http://www.un.org/womenwatch/

United Nations web site for International Organizations
http://www.library.nwu.edu/govpub/idtf/igo.html

ViVa—a bibliography of articles in women's history from sixty journals from 1995 to the present.
http://www.iisg.nl/~womhist

See also WISTAT—a CD-ROM of United Nations statistics on women.

National Histories of Women

Allen, Anne Taylor. *Feminism and Motherhood in Germany, 1800–1914*. New Brunswick, N.J.: Rutgers University Press, 1991.

Atkinson, Dorothy, Alexander Dallin, and Gail Warshofsky Lapidus, eds. *Women in Russia*. Stanford, Calif.: Stanford University Press, 1977.

Beuys, Barbara. *Familienleben in Deutschland: Neue Bilder der Deutschen Vergangenheit*. Hamburg, Germany: Rowohlt, 1980.

Campbell, Beatrix. *Wigan Pier Revisited: Poverty and Politics in the '80s*. London: Virago, 1984.

Charnon-Deutsch, Lon, and Jane Labanyi, eds. *Culture and Gender in Nineteenth-Century Spain*. Oxford: Clarendon Press, 1995.

Clark, Linda L. "The Molding of the *Citoyenne*: The Image of the Female in French Educational Literature," *Third Republic/Troisième République*, vol. I, nos. 3–4 (Spring–Fall 1977), pp. 74–103.

Collins, Marie, and Sylvie Weil Sayre, eds. *Les Femmes en France*. New York: Charles Scribner's Sons, 1974.

Dahlsgård, Inga. *Women in Denmark: Yesterday and Today*. Copenhagen: Det Danske Selskab, 1980.

Davidoff, Leonore. *The Best Circles: Women and Society in Victorian England*. Totowa, N.J.: Rowman and Littlefield, 1973.

Decaux, Alain. *Histoire des Françaises*. 2 vols. Paris: Librairie Académique Perrin, 1972.

Edmondson, Linda. *Women and Society in Russia and the Soviet Union*. Cambridge, U.K.: Cambridge University Press, 1992.

Fout, John R., ed. *German Women in the Nineteenth Century: A Social History*. New York: Holmes and Meier, 1984.

Frevert, Ute. *Women in German History: From Bourgeois Emancipation to Sexual Liberation*. Oxford: Berg, 1989.

Fritz, Paul, and Richard Morton, eds. *Women in the Eighteenth Century and Other Essays*. Toronto: Samuel Stevens, 1976.

Gardiner, Dorothy. *English Girlhood at School: A Study of Women's Education through Twelve Centuries*. London: Oxford University Press, 1929.

Good, David F., Margarete Gradner, and Mary Jo Maynes, eds. *Austrian Women in the Nineteenth and Twentieth Centuries: Cross-Disciplinary Perspectives*. Providence, R.I.: Berghahn, 1996.

Hausen, Karin, ed. *Frauen Suchen Ihre Geschichte: Historische Studien zum 19. und 20. Jahrhundert*. Munich: C. H. Beck, 1982.

Holland, Barbara, ed. *Soviet Sisterhood*. Bloomington: Indiana University Press, 1985.

Holmes, Janice, and Diane Urquhart, eds. *Coming Into the Light: The Work, Politics, and Religion of Women in Ulster 1840–1940*. Belfast: Queens University, Institute of Irish Studies, 1994.

Hudson, Kenneth. *The Place of Women in Society*. London: Ginn and Co., 1970.

Hufton, Olwen. *The Poor of Eighteenth-Century France*. Oxford: Clarendon Press, 1974.

Jancar, Barbara Wolfe. *Women Under Communism*. Baltimore: Johns Hopkins University Press, 1978.

Kanner, Barbara, ed. *The Women of England: From Anglo-Saxon Times to the Present: Interpretive Bibliographical Essays*. Hamden, Conn.: Archon Books, 1979.

Kuhn, Annette, and Gerhard Schneider, eds. *Frauen in der Geschichte*. Düsseldorf: Schwann, 1976.

Lapidus, Gail Warshofsky. *Women in Soviet Society: Equality, Development and Social Change*. Berkeley: University of California Press, 1978.

Lewis, Jane. *Women in England, 1870–1950: Sexual Divisions and Social Change*. Bloomington: Indiana University Press, 1984.

———. *Women in Britain Since 1945*. Oxford: Blackwell, 1992.

Luddy, Maria, and Cliona Murphy, eds. *Women Surviving: Studies in Irish Women's History in the Nineteenth and Twentieth Centuries*. Dublin: Poolbeg Publishers, 1990.

MacCurtain, Margaret, and Donncha O'Corráin, eds. *Women in Irish Society: The Historical Dimension*. Westport, Conn.: Greenwood Press, 1979.

Mandel, William. *Soviet Women*. Garden City, N.Y.: Doubleday & Company, 1975.

Marshall, Rosalind K. *Virgins and Viragos: A History of Women in Scotland from 1080 to 1980*. Chicago: Academy Chicago, 1983.

McMillan, James F. *Housewife or Harlot: The Place of Women in French Society, 1870–1940*. New York: St. Martin's Press, 1981.

Offen, Karen M. "Aspects of the Woman Question during the Third Republic," *Third Republic/Troisième Republique*, vol. I, nos. 3–4 (Spring–Fall 1977), pp. 1–19.

Puckett, Hugh Wiley. *Germany's Women Go Forward*. New York: AMS Press, 1967 [1929].

Rowbotham, Sheila. *A Century of Women: The History of Women in Britain and the United States*. [twentieth century] New York: Viking, 1997.

Shaffer, Harry G. *Women in the Two Germanies: A Comparative Study of a Socialist and a Non-Socialist Society*. New York: Pergamon Press, 1981.

Thompson, Roger. *Women in Stuart England and America: A Comparative Study*. London: Routledge and Kegan Paul, 1974.

Vicinus, Martha, ed. *Suffer and Be Still: Women in the Victorian Age*. Bloomington: Indiana University Press, 1972.

———. ed. *A Widening Sphere: Changing Roles of Victorian Women*. Bloomington: Indiana University Press, 1977.

Vishneva-Sarafanov, N. *Soviet Women: A Portrait*. U.S.S.R.: Progress Publishers, 1981.

Wilson, Elizabeth. *Women and the Welfare State*. London: Tavistock, 1977.

———. *Only Halfway to Paradise: Women in Postwar Britain 1945–1968*. London: Tavistock, 1980.

PART VI WOMEN OF THE COURTS

Court Life, Courtiers, and Courtesans

Aldington, Richard, ed. *Letters of Mme. de Sévigné to her Daughter and her Friends*. London: Routledge and Kegan Paul, 1937.

Allentuch, Harriet Ray. *Mme. de Sévigné: A Portrait in Letters*. Baltimore: Johns Hopkins University Press, 1963.

Backer, Dorothy Anne Liot. *Precious Women: A Feminist Phenomenon in the Age of Louis XIV*. New York: Basic Books, 1974.

Bucholz, R. O. *The Augustan Court: Queen Anne and the Decline of Court Culture*. Stanford, Calif.: Stanford University Press, 1993.

Calmette, Joseph. *The Golden Age of Burgundy: The Magnificent Dukes and Their Courts*. Translated by Doreen Weightman. New York: W. W. Norton, 1963 [1949].

Chaussinand-Nogaret, Guy. *The French Nobility in the Eighteenth Century From Feudalism to Enlightenment*. Translated by William Doyle. New York: Cambridge University Press, 1989.

The Memoirs of Princess Dashkova. Translated and edited by Kiril Fitzlyon. Durham, N.C.: Duke University Press, 1995.

Erlanger, Philippe. *The Age of Courts and Kings: Manners and Morals, 1558–1715*. New York: Harper & Row, 1967.

Fairchilds, Cissie. *Domestic Enemies: Servants & Their Masters in Old Regime France.* Baltimore: Johns Hopkins University Press, 1984.

Gibson, Wendy. *Women in Seventeenth-Century France.* Basingstoke, U.K.: Macmillan, 1989.

Gooch, G.P. *Courts and Cabinets.* New York: Alfred A. Knopf, 1946.

———. *Maria Theresa and Other Studies.* New York: Longmans, Green & Co., 1952.

———. *Catherine the Great and Other Studies.* New York: Longmans, Green & Co., 1954.

Goodwin, Albert, ed. *The European Nobility in the Eighteenth Century.* New York: Harper & Row, 1967.

Haldane, Charlotte. *Mme. de Maintenon: Uncrowned Queen of France.* New York: Bobbs-Merrill, 1970.

Hardwick, Julie. "Seeking Separations: Gender, Marriages, and Household Economies in Early Modern France," *Journal of French Historical Studies,* vol. 21, no. 1 (Winter 1998), pp. 157–80.

Harris, Barbara J. "Marriage and Politics in Early Tudor England," *Historical Journal,* vol. 33 (1990), pp. 259–81.

Harris, Frances. *A Passion for Government: The Life of Sarah, Duchess of Marlborough.* New York: Clarendon Press, 1991.

Kettering, Sharon. "The Household Service of Early Modern French Noblewomen," *French Historical Studies,* vol. 20, no. 1 (1997), pp. 55–85.

Lewis, W. H. *The Splendid Century: Life in the France of Louis XIV.* Garden City, N.Y.: Doubleday & Company, 1957.

———. *The Sunset of the Splendid Century: The Life and Times of Louis Auguste de Bourbon, Duc du Maine, 1670–1736.* Garden City, N.Y.: Doubleday & Company, 1963.

Loomis, Stanley. *Du Barry: A Biography.* New York: J. B. Lippincott Company, 1959.

Maguire, Nancy Klein. "The Duchess of Portsmouth: English Royal Consort and French Politician, 1670–85," in R. Malcolm Smuts, ed. *The Stuart Court and Europe: Essays in Politics and Political Culture.* New York: Cambridge University Press, 1996.

Mitford, Nancy. *Madame de Pompadour.* London, U.K.: The Reprint Society of London, 1955.

Renaissance Studies (Special Issue). "Women Patrons of Renaissance Art, 1300–1600," vol. 10, no. 2 (June 1996).

Ribeiro, Aileen. *The Art of Dress: Fashion in England and France 1750–1820.* New Haven: Yale University Press, 1995.

Roberts, David. *The Ladies: Female Patronage of Restoration Drama, 1660–1700.* New York: Oxford University Press, 1989.

Roche, Daniel. *The Culture of Clothing: Dress and Fashion in the Ancien Regime.* New York: Cambridge University Press, 1994.

Sánchez, Magdalena S. *The Empress, the Queen, and the Nun: Women and Power at the Court of Philip III of Spain.* Baltimore: Johns Hopkins University Press, 1998.

Steegmuller, Frances. *The Grand Mademoiselle.* New York: Farrar, Straus and Company, 1956.

von der Pfalz, Liselotte. *A Woman's Life in the Court of the Sun King: Letters of Liselotte von der Pfalz, 1652–1722.* [Elisabeth-Charlotte, Duchesse d'Orléans]. Translated by Elborg Forster. Baltimore: Johns Hopkins University Press, 1986.

Wilson, John Harold. *Nell Gwyn: Royal Mistress.* New York: Pellegrini & Cudahy, 1952.

Traditional Life: Wife and Queen Consort

Ashley, Maurice. *The Stuarts in Love, with some reflections on love and marriage in the sixteenth and seventeenth centuries.* New York: Macmillan, 1964.

Bainton, Roland H. *Women of the Reformation in Germany and Italy.* Boston: Beacon Press, 1971.

———. *Women of the Reformation in France and England.* Boston: Beacon Press, 1975.

———. *Women of the Reformation from Spain to Scandinavia.* Minneapolis: Augsburg Publishing House, 1977.

Bourgeois, Louise. *Les six couches de Marie de Médicis.* Paris: Leon Willem, 1875.

Erickson, Amy Louise. *Women and Property in Early Modern England: 1580–1720.* London: Routledge, 1993.

Goody, Jack, Joan Thirsk, and E. P. Thompson, eds. *Family and Inheritance: Rural Society in Western Europe, 1200–1800.* New York: Cambridge University Press, 1978.

Hollingsworth, T. H. "A Demographic Study of the British Ducal Families," in D. V. Glass and D. E. Eversley, eds. *Population in History.* London: E. Arnold, 1965.

Hunt, David. *Parents and Children in History: The Psychology of Family Life in Early Modern France.* New York: Harper & Row, 1972.

Kaufman, Gloria. "Juan Luis Vives on the Education of Women," *Signs*, vol. 3, no. 4 (Summer 1978), pp. 891–96.

Kelso, Ruth. *Doctrine for the Lady of the Renaissance.* Urbana: University of Illinois Press, 1956.

Lindsey, Karen. *Divorced, Beheaded, Survived: A Feminist Reinterpretation of the Wives of Henry VIII.* Reading, Mass.: Addison-Wesley, 1995.

de Mause, Lloyd, ed. *The History of Childhood.* New York: Harper & Row, 1974.

Phillips, Roderick. *Putting Asunder: A History of Divorce in Western Society.* New York: Cambridge University Press, 1988.

Rich, Mary. *The Autobiography of Mary Rich, Countess of Warwick.* Vol. XXII. Edited by Thomas Crofton Croker. London: Percy Society, 1848.

Sachs, Hannelore. *The Renaissance Woman.* Translated by Marianne Hertzfeld. New York: McGraw-Hill, 1971.

Sánchez, Magdalena S., and Alain Saint-Saens, eds. *Spanish Women in the Golden Age: Images and Realities.* Westport, Conn.: Greenwood Press, 1996.

Thomson, Gladys Scott. *Life in a Noble Household, 1641–1700.* Ann Arbor: University of Michigan Press, 1959.

Tillyard, Stella. *Aristocrats: Caroline, Louisa, and Sarah Lennox, 1740–1832.* New York: Farrar, Straus & Giroux, 1994.

Watt, Jeffrey R. *The Making of Modern Marriage: Matrimonial Control and the Rise of Sentiment in Neuchâtel, 1550–1800.* Ithaca, N.Y.: Cornell University Press, 1992.

Women Rulers: Regents and Monarchs

Alexander, John T. *Catherine the Great: Life and Legend*. New York: Oxford University Press, 1989.

Breisach, Ernst. *Caterina Sforza: A Renaissance Virago*. Chicago: University of Chicago Press, 1967.

The Memoirs of Catherine the Great. Edited by Dominique Maroger, translated by Moura Budberg. New York: Collier Books, 1961.

Clarke, M. L. "The Making of a Queen: The Education of Christina of Sweden," *History Today*, vol. 28, no. 4 (April 1978), pp. 228–34.

Crankshaw, Edward. *Maria Theresa*. New York: Viking Press, 1969.

Erickson, Carolly. *Bloody Mary*. Garden City, N.Y.: Doubleday & Company, 1978.

Fraser, Antonia. *Mary, Queen of Scots*. New York: Dell Publishing, 1971.

Freeman, John F. "Louise of Savoy: A Case of Maternal Opportunism," *Sixteenth Century Journal*, vol. III, no. 2 (October 1972), pp. 77–98.

Frye, Susan. *Elizabeth I: The Competition for Representation*. New York: Oxford University Press, 1993.

Gregg, Edward. *Queen Anne*. Boston: Routledge and Kegan Paul, 1980.

Haigh, Christopher, ed. *The Reign of Elizabeth I*. Athens: University of Georgia Press, 1985.

Hanley, Sarah. "The Monarchic State in Early Modern France: Marital Regime Politics and Male Right, 1500–1800," in Adrianna E. Bakos, ed. *Politics, Ideology and the Law in Early Modern Europe*. Rochester, N.Y.: University of Rochester Press, 1994.

Heisch, Alison. "Queen Elizabeth I and the Persistence of Patriarchy," *Feminist Review*, no. 4 (1980), pp. 45–56.

Heritier, Jean. *Catherine de'Medici*. Translated by Charlotte Haldane. New York: St. Martin's Press, 1963.

Hopkins, Lisa. *Women Who Would be Kings: Female Rulers of the Sixteenth Century*. New York: St. Martin's Press, 1991.

Hughes, Lindsey. *Sophia, Regent of Russia, 1657–1704*. New Haven: Yale University Press, 1990.

Kleinman, Ruth. *Anne of Austria: Queen of France*. Columbus: Ohio State University Press, 1985.

Levin, Carole. *The Heart and Stomach of a King: Elizabeth I and the Politics of Sex and Power*. Philadelphia: University of Pennsylvania Press, 1994.

de Madariaga, Isabel. *Russia in the Age of Catherine the Great*. New Haven: Yale University Press, 1981.

Masson, Georgina. *Queen Christina*. New York: Farrar, Straus & Giroux, 1969.

Neale, J. E. *Queen Elizabeth I*. London: Jonathan Cape, 1952.

Oliva, L. Jay, ed. *Catherine the Great*. Englewood Cliffs, N.J.: Prentice-Hall, 1971.

Richards, Judith M. "Mary Tudor as 'Sole Quene'?: Gendering Tudor Monarchy," *Historical Journal*, vol. 40, no. 4 (1997), pp. 895–924.

Roberts, Michael. "Queen Christina and the General Crisis of the Seventeenth Century," in Trevor Aston, ed. *Crisis in Europe, 1560–1660*. New York: Basic Books, 1965.

Roider, Karl A., Jr., ed. *Maria Theresa*. Englewood Cliffs, N.J.: Prentice-Hall, 1973.

Stolpe, Sven. *Christina of Sweden*. Edited by Sir Alec Randall, translated by Sir Alec Randall and Ruth Mary Bethell. New York: Macmillan Company, 1966.

Williams, Neville. *All the Queen's Men: Elizabeth I and Her Courtiers*. New York: Macmillan Company, 1972.

New Opportunities: Performers, Composers, and Painters

Bowers, Jane, and Judith Tick, eds. *Women Making Music: The Western Art Tradition, 1150–1950*. Chicago: University of Illinois Press, 1986.

Citron, Marcia J. *Gender and the Musical Canon*. New York: Cambridge University Press, 1993.

Clarke, Mary, and Clement Crisp. *Ballet: An Illustrated History*. New York: Universe Books, 1973.

Greer, Germaine. *The Obstacle Race: The Fortunes of Women Painters and Their Work*. New York: Farrar, Straus & Giroux, 1979.

Halliwell, Ruth. *The Mozart Family: Four Lives in a Social Context*. New York: Oxford University Press, 1997.

Harris, Ann Sutherland, and Linda Nochlin. *Women Artists, 1550–1950*. New York: Alfred A. Knopf, 1978.

Jacobs, Fredrika H. *Defining the Renaissance Virtuosa: Women Artists and the Language of Art History and Criticism*. New York: Cambridge University Press, 1997.

Keener, Frederick M., and Susan E. Lorsch, eds. *Eighteenth Century Women and the Arts*. New York: Greenwood Press, 1988.

Lawrence, Cynthia. *Women and Art in Early Modern Europe: Patrons, Collectors, and Connoisseurs*. University Park: Pennsylvania State University Press, 1997.

Neuls-Bates, Carol, ed. *Women in Music: An Anthology of Source Readings from the Middle Ages to the Present*. New York: Harper & Row, 1982.

Perlingieri, I.S. *Sofonisba Anguissola: The First Great Woman Artist of the Renaissance*. New York: Rizzoli, 1992.

Petersen, Karen, and J. J. Wilson. *Women Artists: Recognition and Reappraisal from the Early Middle Ages to the Twentieth Century*. New York: Harper & Row, 1976.

Roworth, Wendy Wassyng. "Painting for Profit and Pleasure: Angelica Kauffman and the Art Business in Rome," *Eighteenth-Century Studies*, vol. 29, no. 2 (Winter 1995–96), pp. 225–28.

Sheriff, Mary D. *The Exceptional Woman: Elisabeth Vigée-Lebrun and the Cultural Politics of Art*. Chicago: University of Chicago Press.

Vigée-le-Brun, Elisabeth Louise. *The Memoirs of Mme. Elisabeth Louise Vigée-le-Brun, 1755–1789*. Translated by Gerard Shelley. London: John Hamilton, n.d.

New Opportunities: Writers and Scientists

Alic, Margaret. *Hypatia's Heritage: A History of Women in Science from Antiquity through the Nineteenth Century*. Boston: Beacon Press, 1986.

Beasley, Faith E. *Revising Memory: Women's Fiction and Memoirs in Seventeenth-Century France*. New Brunswick, N.J.: Rutgers University Press, 1990.

Beilin, Elaine V. *Redeeming Eve: Women Writers of the English Renaissance.* Princeton, N.J.: Princeton University Press, 1987.

Brabant, Margaret, ed. *Politics, Gender, and Genre: The Political Thought of Christine de Pizan.* Boulder, Colo.: Westview, 1992.

Cereta, Laura. *Collected Letters of a Renaissance Feminist.* Edited and translated by Diana Robin. Chicago: University of Chicago Press, 1997.

Conway, Anne. *The Principles of the Most Ancient and Modern Philosophy.* Edited and translated by Allison P. Coudert and Taylor Corse. New York: Cambridge University Press, 1996.

Davis, Natalie Zemon. *Women on the Margins: Three Seventeenth-Century Lives.* Cambridge, Mass.: Harvard University Press, 1995.

Ehrman, Esther. *Mme. Du Châtelet: Scientist, Philosopher and Feminist of the Enlightenment.* Leamington Spa, U.K.: Berg, 1986.

Grundy, Isobel, and Susan Wiseman, eds. *Women, Writing, History 1640–1740.* London: Batsford, 1992.

Harth, Erica. *Cartesian Women: Versions and Subversions of Rational Discourse in the Old Regime.* Ithaca, N.Y.: Cornell University Press, 1992.

Haselkorn, Anne M., and Betty S. Travitsky, eds. *The Renaissance Englishwoman in Print: Counterbalancing the Canon.* Amherst: University of Massachusetts Press, 1990.

Krontiris, Tina. *Oppositional Voices: Women as Writers and Translators of Literature in the English Renaissance.* New York: Routledge, 1992.

Lewalski, Barbara Kiefer. *Writing Women in Jacobean England.* Cambridge, Mass.: Harvard University Press, 1993.

Logan, Gabriella Berti. "The Desire to Contribute: An Eighteenth-Century Italian Woman of Science," *American Historical Review,* vol. 99, no. 3 (June 1994), pp. 785–812.

Phillips, Patricia. *The Scientific Lady: A Social History of Women's Scientific Interests, 1520–1918.* New York: St. Martin's Press, 1990.

The Selected Writings of Christine de Pizan. Translated by Renate Blumenfeld-Kosinski and Kevin Brownlee and edited by Renate Blumenfeld-Kosinski. New York: W. W. Norton & Company, 1997.

The Writings of Christine de Pizan. Edited by Charity Cannon Willard. New York: Persea Books, 1994.

Schiebinger, Londa. "The History and Philosophy of Women in Science: A Review Essay," *Signs,* vol. 12, no. 2 (1987), pp. 305–32.

———. *The Mind Has No Sex? Women in The Origins of Modern Science.* Cambridge, Mass.: Harvard University Press, 1989.

"A Surinam Portfolio" (Maria Sibylla Merian), *Natural History,* vol. LXXI, no. 10 (December 1962), pp. 28–41.

Warner, Marina. *From the Beast to the Blonde: On Fairy Tales and Their Tellers.* New York: Farrar, Straus & Giroux, 1995.

Willard, Charity Cannon. *Christine de Pizan: Her Life and Works.* New York: Persea Books, 1984.

Wilson, Katharina M., ed. *Medieval Women Writers.* Athens: University of Georgia Press, 1984.

————, ed. *Women Writers of the Renaissance and Reformation*. Athens: University of Georgia Press, 1987.

Wilson, Katharina M., and Frank J. Warnke, eds. *Women Writers of the Seventeenth Century*. Athens: University of Georgia Press, 1989.

Learned Women and the Querelles des Femmes

Bell, Susan Groag. "Christine de Pizan (1364–1430): Humanism and the Problem of a Studious Woman," *Feminist Studies*, vol. III, nos. 3–4 (Spring–Summer 1976), pp. 173–84.

————. "Medieval Women Book Owners: Arbiters of Lay Piety and Ambassadors of Culture," *Signs*, vol. 7, no. 4 (Summer 1982), pp. 742–68.

Benson, Pamela Joseph. *The Invention of the Renaissance Woman: The Challenge of Female Independence in the Literature and Thought of Italy and England*. University Park: Pennsylvania State University Press, 1992.

Bordo, Susan. "The Cartesian Masculinization of Thought," *Signs*, vol. 11, no. 3 (Spring 1986), pp. 439–56.

Dow, Blanche H. *The Varying Attitude towards Women in French Literature of the Fifteenth Century*. New York: Institute of French Studies, 1936.

Ferguson, Moira, ed. *First Feminists: British Women Writers, 1578–1799*. Bloomington: Indiana University Press, 1985.

————. "Feminist Polemic: British Women's Writings in English from the Late Renaissance to the French Revolution," *Women's Studies International Forum*, vol. 9, nos. 5–6 (1986), pp. 451–64.

Hannay, Margaret Patterson, ed. *Silent but for the Word: Tudor Women as Patrons, Translators, and Writers of Religious Works*. Kent, Ohio: Kent State University Press, 1985.

Henderson, Katherine Usher, and Barbara F. McManus. *Half Humankind: Contexts and Texts of the Controversy About Women in England 1540–1640*. Urbana: University of Illinois Press, 1985.

Horowitz, Maryanne Cline. "The 'Science' of Embryology before the Discovery of the Ovum," in Marilyn J. Boxer and Jean H. Quataert, eds. *Connecting Spheres: Women in the Western World, 1500 to the Present*. New York: Oxford University Press, 1987.

Hull, Suzanne. *Chaste, Silent and Obedient: English Books for Women 1475–1640*. San Marino, Calif.: Huntington Library, 1982.

Huot, Sylvia. "Seduction and Sublimation: Christine de Pizan, Jean de Meun, and Dante," *Romance Notes*, vol. XXV, no. 3 (Spring 1985), pp. 361–73.

Ilsley, Marjorie. *A Daughter of the Renaissance: Marie le Jars de Gournay, her Life and Works*. The Hague: Mouton, 1963.

Jordan, Constance. *Renaissance Feminism: Literary Texts and Political Models*. Ithaca, N.Y.: Cornell University Press, 1990.

King, Margaret L. *Women of the Renaissance*. Chicago: University of Chicago Press, 1991.

King, Margaret L., and Albert Rabil, Jr., eds. *Her Immaculate Hand: Selected Works by and about the Women Humanists of Quattrocento Italy*. Binghamton, N.Y.: State University of New York at Binghamton, 1981–1983.

Kitts, Sally-Ann. *The Debate on the Nature, Role, and Influence of Woman in Eighteenth-Century Spain.* Lewiston, N.Y.: E. Mellen Press, 1995.

Labalme, Patricia H., ed. *Beyond Their Sex: Learned Women of the European Past.* New York: New York University Press, 1980.

Levin, Carole, and Patricia A. Sullivan, eds. *Political Rhetoric, Power, and Renaissance Women.* Albany: State University of New York Press, 1995.

Maclean, Ian. *Woman Triumphant: Feminism in French Literature 1610–1652.* Oxford: Clarendon Press, 1977.

———. *The Renaissance Notion of Woman: A Study of the Fortunes of Scholasticism and Medical Science in European Intellectual Life.* New York: Cambridge University Press, 1980.

Merchant, Carolyn. *The Death of Nature: Women, Ecology and the Scientific Revolution.* New York: Harper & Row, 1980.

Perry, Ruth. *The Celebrated Mary Astell: An Early English Feminist.* Chicago: University of Chicago Press, 1986.

Schiebinger, Londa. *Nature as Body: Gender in the Making of Modern Science.* Boston: Beacon Press, 1993.

Smith, Hilda L., ed. *Women Writers and the Early Modern British Political Tradition.* New York: Cambridge University Press, 1998.

Veith, Ilza. *Hysteria: The History of a Disease.* Chicago: University of Chicago Press, 1965.

Wilson, Katharina M. *Women Writers of the Renaissance and Reformation.* Athens: University of Georgia Press, 1986.

Wilson, Katharina M., and Frank J. Warnke, eds. *Women Writers of the Seventeenth Century.* Athens: University of Georgia Press, 1989.

Zimmermann, Margaret, and Dina De Rentiis, eds. *City of Scholars: New Approaches to Christine de Pizan.* New York: Walter de Gruyter, 1994.

PART VII   WOMEN OF THE SALONS AND PARLORS

Autobiographies, Biographies, Diaries, and Memoirs of Women

Baum, Vicki. *It Was All Quite Different: The Memoirs of Vicki Baum.* New York: Funk and Wagnalls, 1964.

de Beauvoir, Simone. *Memoirs of a Dutiful Daughter.* Translated by James Kirkup. New York: Harper & Row, 1959.

Butler, Josephine G. *Autobiography.* Edited by G. A. W. London, 1911.

Curie, Eve. *Madame Curie: A Biography.* Translated by Vincent Sheean. New York: Doubleday, Doran and Co., 1939.

Day, Lillian. *Ninon: A Courtesan of Quality.* [Ninon de L'Enclos] Garden City, N.Y.: Doubleday & Company, 1957.

Edwards, Samuel. *The Divine Mistress.* [Emilie du Châtelet] New York: David McKay, 1970.

Flexner, Eleanor. *Mary Wollstonecraft.* Baltimore: Penguin, 1973.

Freeman, Sarah. *Isabella and Sam: The Story of Mrs. Beeton.* New York: Coward, McCann and Geoghegan, 1978.

Gérin, Winifred. *Elizabeth Gaskell: A Biography.* Oxford: Oxford University Press, 1980.

Giroud, Françoise. *I Give You My Word.* Translated by Richard Seaver. Boston: Houghton Mifflin, 1974.

Gray, Francine du Plessix. *Rage and Fire: A Life of Louise Colet, Pioneer Feminist, Literary Star, Flaubert's Muse.* New York: Simon and Schuster, 1994.

Halsband, Robert. *The Life of Lady Mary Wortley Montague.* Oxford: Clarendon Press, 1956.

Herold, J. Christopher. *Mistress to an Age: A Life of Madame de Staël.* Indianapolis: Bobbs-Merrill Company, 1958.

Holliday, Laurel, ed. *Heart Songs: The Intimate Diaries of Young Girls.* Guerneville, Calif.: Bluestocking Books, 1978.

Horney, Karen. *The Adolescent Diaries of Karen Horney.* New York: Basic Books, 1980.

Hughes, M. V. *A London Family: 1870–1900.* 3 vols. Oxford: Oxford University Press, 1979 [1934].

Jacobs, Aletta. *Memories: My Life as an International Leader in Health, Suffrage, and Peace.* Edited by Harriet Feinberg. Translated by Annie Wright. New York: The Feminist Press, 1996.

Jones, M. G. *Hannah More.* Cambridge: Cambridge University Press, 1952.

Klein, Mina C., and H. Arthur Klein. *Käthe Kollwitz: Life in Art.* New York: Schocken, 1976.

Laski, Marghanita. *George Eliot and Her World.* New York: Charles Scribner's Sons, 1973.

Latour, Anny. *Uncrowned Queens.* [Salonières from Isabella D'Este to Gertrude Stein] Translated by A. A. Dent. London: J. M. Dent and Sons, 1970.

Linder, Doris H. *Crusader for Sex Education: Elise Ottesen-Jansen (1886–1973) in Scandinavia and on the International Scene.* Lanham, Md.: University Press of America, 1996.

Longford, Elizabeth. *Queen Victoria: Born to Succeed.* New York: Pyramid Books, 1966.

———. *Eminent Victorian Women.* New York: Alfred A. Knopf, 1981.

Maurois, André. *Lélia: The Life of George Sand.* Translated by Gerard Hopkins. New York: Harper & Row, 1953.

Mavor, Elizabeth. *The Ladies of Llangollen: A Study in Romantic Friendship.* [Lady Eleanor Butler and Sarah Ponsonby] Baltimore: Penguin, 1971.

Meysenbug, Malwida von. *Memoirs: Rebel in Bombazine.* Translated by Elsa von Meysenbug Lyons. Edited by Mildred Adams. New York: W. W. Norton & Company, 1936.

Modersohn-Becker, Paula. *Paula Modersohn-Becker: The Letters and Journals.* Edited by Günter Busch and Liselotte von Reinken. Edited and translated by Arthur S. Wensinger and Carole Clew Hoey. New York: Taplinger, 1983.

Perry, Gillian. *Paula Modersohn-Becker: Her Life and Work.* New York: Harper & Row, Publishers, 1979.

Peters, H. F. *Zarathustra's Sister: The Case of Elisabeth and Friedrich Nietzsche.* New York: Crown, 1977.

Quennell, Peter, ed. *Affairs of the Mind: The Salon in Europe and America from the 18th to the 20th Century.* [Biographies of salonières from Rahel to Lady Cunard] Washington, D.C.: New Republic Books, 1980.

Raverat, Gwen. *Period Piece: A Cambridge Childhood.* London: Faber and Faber, 1968.
    [1952]

Redinger, Ruby V. *George Eliot: The Emergent Self.* New York: Alfred A. Knopf, 1975.

Reich, Nancy B. *Clara Schumann: The Artist and the Woman.* Ithaca, N.Y.: Cornell
    University Press, 1985.

Richardson, Johanna. *Princess Mathilde.* New York: Charles Scribner's Sons, 1969.

Rose, June. *Elizabeth Fry.* London: Macmillan, 1980.

Sand, George. *My Life.* Translated and adapted by Dan Hofstadter. New York: Harper
    & Row, 1979.

Summerskill, Edith. *A Woman's World.* London: Heinemann, 1967.

Autobiography of St. Thérèse of Lisieux. Translated by Ronald Knox. New York: P. J.
    Kennedy and Sons, 1958.

Tomalin, Clare. *The Life and Death of Mary Wollstonecraft.* London: Weidenfeld and
    Nicholson, 1974.

Woodham-Smith, Cecil. *Florence Nightingale, 1820–1910.* New York: McGraw-Hill,
    1951.

Journals of Dorothy Wordsworth. 2nd ed. Edited by Mary Moorman. New York: Oxford
    University Press, 1983.

Women Writers, Artists, and Musicians

Adburgham, Alison. *Women in Print: Writing Women and Women's Magazines from the
    Restoration to the Accession of Victoria.* London: George Allan & Unwin, 1972.

Basch, Françoise. *Relative Creatures: Victorian Women in Society and the Novel.* Trans-
    lated by Anthony Rudolf. New York: Schocken, 1974.

Beer, Patricia. *Reader, I Married Him: A Study of the Women Characters of Jane Austen,
    Charlotte Brontë, Elizabeth Gaskell and George Eliot.* New York: Harper & Row,
    1974.

Bowers, Jane, and Judith Tick, eds. *Women Making Music: The Western Art Tradition,
    1150–1950.* Chicago: University of Illinois Press, 1986.

Brinker-Gabler, Gisela, ed. *Deutsche Dichterinnen vom 16. Jahrhundert bis zur Gegen-
    wart.* Frankfurt am Main: Fischer, 1978.

Broude, Norma, and Mary D. Garrard, eds. *Feminism and Art History: Questioning the
    Litany.* New York: Harper & Row, 1982.

Brownstein, Rachel M. *Becoming a Heroine: Reading About Women in Novels.* New York:
    Penguin, 1984.

Byles, Joan Montgomery. *War, Women, and Poetry, 1914–1945: British and German
    Writers and Activists.* Newark, Del.: University of Delaware Press, 1995.

Colby, Vineta. *Yesterday's Woman: Domestic Realism in the English Novel.* Princeton,
    N.J.: Princeton University Press, 1974.

Crosland, Margaret. *Women of Iron and Velvet: French Women Writers After George Sand.*
    New York: Taplinger Publishing Company, 1976.

Datlof, Natalie, Jeanne Fuchs, and David A. Powell, eds. *The World of George Sand.* New
    York: Greenwood, 1991.

Dickenson, Donna. *George Sand: A Brave Man, The Most Womanly Woman.* New York:
    Berg, 1988.

Diethe, Carol. *Towards Emancipation: German Writers of the Nineteenth Century.* New York: Berghahn Books, 1998.

Felstiner, Mary Lowenthal. *To Paint Her Life: Charlotte Salomon in the Nazi Era.* New York: HarperCollins, 1994.

Greer, Germaine. *The Obstacle Race.* [Women artists] New York: Farrar, Straus & Giroux, 1979.

Harris, Ann Sutherland, and Linda Nochlin. *Women Artists, 1550–1950.* New York: Alfred A. Knopf, 1981.

Hedges, Elaine, and Ingrid Wendt. *In Her Own Image: Women Working in the Arts.* Old Westbury, N.Y.: Feminist Press, 1980.

Moers, Ellen. *Literary Women.* Garden City, N.Y.: Doubleday & Company, 1977.

Neuls-Bates, Carol, ed. *Women in Music: An Anthology of Source Readings from the Middle Ages to the Present.* New York: Harper & Row, 1982.

Orr, Clarissa Campbell, ed. *Women in the Victorian Art World.* Manchester, U.K.: Manchester University Press, 1995.

Petersen, Karen, and J. J. Wilson. *Women Artists: Recognition and Reappraisal from the Early Middle Ages to the Twentieth Century.* New York: Harper & Row, 1976.

Rogers, Katharine M. *Before Their Time: Six Women Writers of the Eighteenth Century.* New York: Frederick Unger, 1979.

Sheriff, Mary D. *The Exceptional Woman: Elisabeth Vigée-Lebrun and the Cultural Politics of Art.* Chicago: University of Chicago Press, 1996.

Showalter, Elaine. *A Literature of Their Own: British Women Novelists from Brontë to Lessing.* Princeton, N.J.: Princeton University Press, 1977.

Utter, Robert Palfrey, and Gwendolyn Bridges Needham. *Pamela's Daughters.* New York: Macmillan, 1936.

White, Cynthia. *Women's Magazines: 1693–1968.* [England] London: Michael Joseph, 1970.

Women As Salonières

Backer, Dorothy Anne Liot. *Precious Women: A Feminist Phenomenon in the Age of Louis XIV.* New York: Basic Books, 1974.

Bodek, Evelyn G. "Salonières and Bluestockings: Educated Obsolescence and Germinating Feminism," *Feminist Studies,* vol. III, no. 4 (Spring–Summer 1976), pp. 185–99.

Gelbert, Nina Rattner. *Feminine and Opposition Journalism in Old Regime France: Le Journal des Dames.* Berkeley: University of California Press, 1995.

Goldsmith, Elizabeth C., and Dena Goodman, eds. *Going Public: Women and Publishing in Early Modern France.* Ithaca, N.Y.: Cornell University Press, 1995.

Goodman, Dena. "Enlightenment Salons: The Convergence of Female and Philosophic Ambitions," *Eighteenth-Century Studies,* vol. 22 (Spring 1989), pp. 329–50.

Hargrave, Mary. *Some German Women and Their Salons.* London: T. Werner Laurie, 1912.

Herold, J. Christopher. *Love in Five Temperaments.* [Five eighteenth-century French salonières] New York: Athenaeum, 1961.

Hertz, Deborah. "Salonières and Literary Women in Late Eighteenth-Century Berlin," *New German Critique*, vol. 14 (Spring 1978), pp. 97–108.
———. *Jewish High Society in Old Regime Berlin*. New Haven: Yale University Press, 1987.
Hill, Bridget. *The Republican Virago: The Life and Times of Catherine Macauley, Historian*. Oxford: The Clarendon Press, 1992.
Latour, Anny. *Uncrowned Queens*. Translated by A. A. Dent. London: J. M. Dent, 1970.
Lougee, Carolyn C. *Le Paradis des Femmes: Women, Salons, and Social Stratification in Seventeenth-Century France*. Princeton, N.J.: Princeton University Press, 1976.
Scott, Walter S. *The Bluestocking Ladies*. London: John Green, 1947.
Spiel, Hilde. *Fanny von Arnstein: Daughter of the Enlightenment 1758–1818*. Translated by Christine Shuttleworth. New York: Berg, 1991.
Spencer, Samia I., ed. *French Women and the Age of Enlightenment*. Bloomington: Indiana University Press, 1984.
Steinbrügge, Lieselotte. *The Moral Sex: Woman's Nature in the French Enlightenment*. Translated by Pamela E. Selwyn. New York: Oxford University Press, 1995.
Trouille, Mary Seidman. *Sexual Politics in the Enlightenment: Women Writers Read Rousseau*. Albany: State University of New York Press, 1997.
Wallas, Ada. *Before the Bluestockings*. London: George Allan & Unwin, 1929.

Raising Families and Maintaining Households in the Nineteenth and Twentieth Centuries

Badinter, Elisabeth. *Mother Love: Myth and Reality–Motherhood in Modern History*. New York: Macmillan, 1981.
Banks, J. A., and Olive Banks. *Feminism and Family Planning in Victorian England*. New York: Schocken, 1964.
Beddoe, Deirdre. *Back to Home and Duty: Women between the Wars, 1918–1939*. [Britain] London: Pandora Press, 1989.
Bentham-Edwards, Matilda. *Home Life in France*. London: Methuen, 1905.
Bianquis, Geneviève. *Love in Germany*. Translated by James Cleugh. London: Frederick Muller, 1964.
Branca, Patricia. *Silent Sisterhood: Middle-Class Women in the Victorian Home*. Pittsburgh: Carnegie-Mellon University Press, 1975.
Conran, Shirley. *Superwoman*. New York: Bantam, 1978.
Crow, Duncan. *The Victorian Woman*. New York: Stein and Day, 1972.
Darrow, Margaret H. "French Noblewomen and the New Domesticity, 1750–1850," *Feminist Studies*, vol. V, no. 1 (Spring 1979), pp. 41–65.
Davidoff, Leonore. *Worlds Between: Historical Perspectives on Gender and Class*. Cambridge, U.K.: Polity Press, 1995.
Davidoff, Leonore, and Catherine Hall. *Family Fortunes: Men and Women of the English Middle Class, 1780–1850*. Chicago: University of Chicago Press, 1987.
Gavron, Hannah. *The Captive Wife: Conflicts of Housebound Mothers*. [Twentieth-century England] London: Routledge and Kegan Paul, 1966.
Hall, Catherine. *White, Male and Middle Class: Explorations in Feminism and History*. New York: Routledge, 1992.

Hall, Ruth, ed. *Dear Dr. Stopes: Sex in the 1920s*. Harmondsworth, U.K.: Penguin, 1981.

Hartman, Mary S. *Victorian Murderesses: A True History of 13 Respectable French and English Women Accused of Unspeakable Crimes*. New York: Schocken, 1977.

Hole, Christina. *English Home Life, 1500–1800*. London: B. T. Batsford, 1947.

Horn, Pamela. *Ladies of the Manor: Wives and Daughters in Country-House Society 1830–1914*. Stroud, U.K.: Sutton Publishers, 1997.

Kaplan, Marion A. *The Making of the Jewish Middle Class: Women, Family and Identity in Imperial Germany*. New York: Oxford University Press, 1991.

Koonz, Claudia. *Mothers in the Fatherland: Women, the Family and Nazi Politics*. New York: St. Martin's Press, 1987.

Loeb, Lori Anne. *Consuming Angels: Advertising and Victorian Women*. New York: Oxford University Press, 1994.

Lowry, Suzanne. *The Guilt Cage: Housewives and a Decade of Liberation*. London: Hamish Hamilton, 1980.

Malos, Ellen, ed. *The Politics of Housework*. London: Allison and Busby, 1980.

Mendus, Susan, and Jane Rendall, eds. *Sexuality and Subordination: Interdisciplinary Studies of Gender in the Nineteenth Century*. New York: Routledge, 1989.

Moore, Katherine. *Victorian Wives*. New York: St. Martin's Press, 1974.

Oakley, Ann. *The Sociology of Housework*. New York: Random House, 1974.

———. *Woman's Work: The Housewife, Past and Present*. New York: Random House, 1976.

Peterson, M. Jeanne. *Family, Love, and Work in the Lives of Victorian Gentlewomen*. Bloomington: University of Indiana Press, 1989.

Poovey, Mary. *Uneven Developments: The Ideological Work of Gender in Mid-Victorian England*. Chicago: University of Chicago Press, 1988.

Rees, Barbara. *The Victorian Lady*. London: Gordon and Cremonesi, 1979.

Ripa, Yannick. *Women and Madness: The Incarceration of Women in Nineteenth-Century France*. Translated by Catherine du Peloux Menage. Cambridge, U.K.: Polity Press, 1990.

Roberts, Mary Louise. *Civilization without Sexes: Reconstructing Gender in Postwar France, 1917–1927*. Chicago: University of Chicago Press, 1994.

Russett, Cynthia Eagle. *Sexual Science: The Victorian Construction of Womanhood*. Cambridge, Mass.: Harvard University Press, 1989.

Shanley, Mary Lyndon. *Feminism, Marriage, and the Law in Victorian England, 1850–1895*. London: I. B. Tauris & Co., 1989.

Sidgewick, Mrs. Alfred. *Home Life in Germany*. New York: Macmillan, 1909.

Smith, Bonnie G. *Ladies of the Leisure Class: The Bourgeoises of Northern France in the Nineteenth Century*. Princeton, N.J.: Princeton University Press, 1981.

Strumingher, Laura S. "'L'Ange de la Maison': Mothers and Daughters in Nineteenth-Century France," *International Journal of Women's Studies*, vol. II, no. 1 (January–February 1979), pp. 51–61.

Taylor, Gordon Rattray. *The Angel Makers: A Study in the Psychological Origins of Historical Change*. New York: E. P. Dutton, 1974 [1958].

Trumbach, Randolph. *The Rise of the Egalitarian Family: Aristocratic Kinship and Domestic Relations in Eighteenth-Century England*. New York: Harcourt Brace Jovanovich, 1978.

Vertinsky, Patricia Anne. *The Eternally Wounded Woman: Women, Doctors and Exercise in the Late Nineteenth Century*. Urbana: University of Illinois Press, 1994.

Walkowitz, Judith R. *City of Dreadful Delight: Narratives of Sexual Danger in Late-Victorian London*. Chicago: University of Chicago Press, 1992.

Wheeler-Bennett, Joan. *Women at the Top: Achievement and Family Life*. [Twentieth-century Britain] London: Peter Owens, 1977.

Charity Work, Education, and Employment

Bachrach, Susan. *Dames Employées: The Feminization of Postal Work in Nineteenth-Century France*. New York: Haworth Press, 1985.

Bernstein, George, and Lottelore Bernstein. "The Curriculum for German Girls' Schools, 1870–1914," *Paedagogica Historica: International Journal of the History of Education*, vol. XVIII, no. 2 (Ghent 1978), pp. 275–95.

———. "Attitudes Toward Women's Education in Germany, 1870–1914," *International Journal of Women's Studies*, vol. II, no. 5 (September–October 1979), pp. 473–88.

Black, J. L. "Educating Women in Eighteenth-Century Russia: Myths and Realities," *Canadian Slavonic Papers*, vol. XX (March 1978), pp. 23–43.

Bonner, Thomas N. *To the Ends of the Earth: Women's Search for Education in Medicine*. Cambridge, Mass.: Harvard University Press, 1992.

Burstyn, Joan. *Victorian Education and the Ideal of Womanhood*. New Brunswick, N.J.: Rutgers University Press, 1984.

Dyhouse, Carol. "Towards a 'Feminine' Curriculum for English Schoolgirls: The Demands of Ideology, 1870–1963," *Women's Studies International Quarterly*, vol. I, nos. 3–4 (Spring–Fall 1977), pp. 297–311.

Holcombe, Lee. *Victorian Ladies at Work: Middle-Class Working Women in England and Wales, 1850–1914*. Hamden, Conn.: Archon Books, 1973.

Jameson, Anna. *Sisters of Charity, Catholic and Protestant, and the Communion of Labour*. Boston: Ticknor and Fields, 1857.

Johanson, Christine. *Women's Struggle for Higher Education in Russia 1855–1900*. Kingston, Ont.: McGill-Queen's University Press, 1987.

Koonz, Claudia. "Conflicting Allegiances: Political Ideology and Women Legislators in Weimar Germany," *Signs*, vol. I, no. 3, part 1 (Spring 1976), pp. 663–84.

Margadant, Jo Burr. *Madame Le Professeur: Women Educators in the Third Republic*. Princeton, N.J.: Princeton University Press, 1990.

Parker, Julia R. *Women and Welfare: Ten Victorian Women in Public Service*. New York: Macmillan, 1988.

Petschauer, Peter. "Improving Educational Opportunities for Girls in Eighteenth-Century Germany," *Eighteenth-Century Life*, vol. III (December 1976), pp. 56–62.

Prelinger, Catherine M. "Religious Dissent, Women's Rights, and the Hamburger Hochschule für das weibliche Geschlecht in mid-nineteenth century Germany," *Church History*, vol. 45 (March 1976), pp. 42–55.

———. *Charity, Challenge and Change: Religious Dimensions of the Mid-Nineteenth-Century Women's Movements in Germany*. New York: Greenwood Press, 1987.

Prochaska, F. K. *Women and Philanthropy in Nineteenth-Century England*. Oxford: Clarendon Press, 1980.

Quartararo, Anne T. *Women Teachers and Popular Education in Nineteenth-Century France: Social Values and Corporate Identity at the Normal School Institution.* Newark, Del.: University of Delaware Press, 1995.

Shiman, Lilian Lewis. *Women and Leadership in Nineteenth-Century England.* New York: St. Martin's Press, 1992.

Vicinus, Martha, ed. *A Widening Sphere: Changing Roles of Victorian Women.* Bloomington: Indiana University Press, 1977.

———. *Independent Women: Work and Community for Single Women, 1850–1920.* London: Virago, 1985.

PART VIII   WOMEN OF THE CITIES

Autobiographies, Memoirs, and Oral Histories

Audoux, Marguerite. *Marie-Claire.* Translated by John Raphael. New York: George H. Doran, 1911.

Davies, Margaret Llewelyn, ed. *Life As We Have Known It By Co-Operative Working Women.* New York: W. W. Norton & Company, 1975 [1931].

Foakes, Grace. *My Part of the River.* London: Futura, 1976.

Hansson, Carola, and Karin Liden. *Moscow Women: Thirteen Interviews.* Translated by Gerry Bothmer, George Blecher, and Lone Blecher. New York: Pantheon, 1983.

Henderson, Kathy, ed. *My Song Is My Own: 100 Women's Songs.* London: Pluto Press, 1979.

Hiley, Michael. *Victorian Working Women: Portraits from Life.* Boston: David R. Godine, 1980.

Hobbs, May. *Born to Struggle.* Plainfield, Vt.: Daughters, 1975.

Klucsarits, Richard, and Friedrich G. Kürbisch eds. *Arbeiterinnen kämpfen um ihr Recht: Autobiographische Texte zum Kampf rechtloser und entrechteter "Frauenpersonen" in Deutschland, Osterreich, und der Schweiz des 19. and 20. Jahrhunderts.* Wuppertal, Ger.: Peter Hammer, n.d.

Luck, Lucy. "A Little of My Life," *The London Mercury,* vol. XIII (November 1925–April 1926), pp. 354–73.

Mayhew, Henry. *London Labour and the London Poor.* 4 vols. New York: Dover, 1968 [1861–1862].

McCrindle, Jean, and Sheila Rowbotham, eds. *Dutiful Daughters: Women Talk About Their Lives.* Austin: University of Texas Press, 1977.

Murray, Janet Horowitz, ed. *Strong-Minded Women and Other Lost Voices from Nineteenth-Century England.* New York: Pantheon, 1982.

Thompson, E. P., and Eileen Yeo, eds. *The Unknown Mayhew: Selections from the Morning Chronicle 1849–1850.* New York: Penguin, 1973.

Thompson, Flora. *Lark Rise to Candleford.* New York: Penguin, 1973 [1939].

Tristan, Flora. *The London Journal of Flora Tristan, 1842.* Translated by Jean Hawkes. London: Virago, 1982.

Family Life and the Rise of the Welfare State

Bock, Gisela, and Pat Thane, eds. *Maternity and Gender Politics: Women and the Rise of the European Welfare State, 1880s to 1950s.* London: Routledge, 1991.

Davies, Margaret Llewelyn, ed. *Maternity: Letters from Working Women*. New York: W. W. Norton, 1978 [1915].

Dufrancatel, Christine, et al. *L'histoire sans qualitées*. Paris: Galilée, 1979.

Evans, Richard J., and W. R. Lee, eds. *The German Family: Essays on the Social History of the Family in Nineteenth- and Twentieth-Century Germany*. London: Croom Helm, 1981.

Fuchs, Rachel. *Poor and Pregnant in Paris: Strategies for Survival in the Nineteenth Century*. New Brunswick, N.J.: Rutgers University Press, 1992.

Githens, Marianne, and Dorothy McBride Stetson. *Abortion Politics: Public Policy in Cross-Cultural Perspective*. New York: Routledge, 1996.

Harris, Henry J. *Maternity Benefit Systems in Certain Foreign Countries*. Washington, D.C.: General Printing Office, 1919.

Koven, Seth, and Sonya Michel, eds. *Mothers of a New World: Maternalist Politics and the Origins of Welfare States*. New York: Routledge, 1993.

Langer, William. "Infanticide: A Historical Survey," *History of Childhood Quarterly*, vol. I, no. 3 (Winter 1974), pp. 353–65.

Levine, David. *Family Formation in an Age of Nascent Capitalism*. New York: Harcourt Brace Jovanovich, 1977.

———. *Reproducing Families: The Political Economy of English Population History*. Cambridge, U.K.: Cambridge University Press, 1987.

Lewis, Jane. *The Politics of Motherhood*. London: Croom Helm, 1980.

McLaren, Angus. *Birth Control in Nineteenth-Century England*. New York: Holmes and Meier, 1978.

———. *Sexuality and the Social Order: The Debate over the Fertility of Women and Workers in France, 1770–1920*. New York: Holmes and Meier, 1983.

Moeller, Robert G. *Protecting Motherhood: Women and the Family in the Politics of Post-War West Germany*. Berkeley: University of California Press, 1993.

Pedersen, Susan. *Family, Dependence, and the Origins of the Welfare State, Britain and France, 1914–1945*. New York: Cambridge University Press, 1993.

Phayer, Michael J. "Lower-Class Morality: The Case of Bavaria," *Journal of Social History*, vol. 8 (Fall 1974), pp. 79–95.

———. *Sexual Liberation and Religion in Nineteenth-Century Europe*. London: Croom Helm, 1977.

Phillips, Roderick. "Women's Emancipation, the Family and Social Change in Eighteenth-Century France," *Journal of Social History*, vol. 12, no. 4 (Summer 1979), pp. 553–67.

Ransel, David. *Mothers of Misery: Child Abandonment in Russia*. Princeton, N.J.: Princeton University Press, 1988.

Reeves, Maud Pember. *Round About a Pound a Week*. London: Virago, 1979 [1913].

Rice, Margery Spring. *Working-Class Wives: Their Health and Conditions*. Harmondsworth, U.K.: Penguin, 1939.

Roberts, Elizabeth. *Women and Families: An Oral History, 1940–1970*. [England] Cambridge, U.K.: Blackwell, 1995.

Rosenberg, Charles E., ed. *The Family in History*. Philadelphia: University of Pennsylvania Press, 1975.

Ross, Ellen. " 'Fierce Questions and Taunts': Married Life in Working-Class London, 1870–1914," *Feminist Studies*, vol. 8, no. 3 (Fall 1982), pp. 575–602.

————. *Love and Toil: Motherhood in Outcast London, 1870–1918*. New York: Oxford University Press, 1993.

Sainsbury, Diane, ed. *Gendering Welfare States*. London: Sage, 1994.

Stewart, Mary Lynn. *Women, Work, and the French State: Labor Protection and Social Patriarchy, 1879–1919*. Montreal: McGill-Queens University Press, 1989.

Tomes, Nancy A. "A 'Torrent of Abuse': Crimes of Violence between Working-Class Men and Women in London, 1840–1875," *Journal of Social History*, vol. II, no. 3 (Spring 1978), pp. 328–45.

Earning Income: Domestic Service

Bridenthal, Renate. "Class Struggle Around the Hearth: Women and Domestic Service in the Weimar Republic," in Michael Dubkowski and Isidor Walliman, eds. *Towards the Holocaust: Anti-Semitism and Fascism in Weimar Germany*. Westport, Conn.: Greenwood Press, 1983.

Davidoff, Leonore. "Class and Gender in Victorian England: The Diaries of Arthur J. Munby and Hannah Cullwick," *Feminist Studies*, vol. 5, no. 1 (Spring 1979), pp. 84–141.

Dawes, Frank. *Not in Front of the Servants: Domestic Service in England, 1850–1939*. London: Wayland, 1973.

Fairchilds, Cissie. *Domestic Enemies: Servants and Their Masters in Old Regime France*. Baltimore: Johns Hopkins University Press, 1984.

Fraisse, Geneviève. *Femmes toutes mains: Essai sur le service domestique*. Paris: Seuil, 1979.

Gillis, John R. "Servants, Sexual Relations and the Risks of Illegitimacy in London, 1801–1900," *Feminist Studies*, vol. 5, no. 1 (Spring 1979), pp. 142–73.

Guiral, Pierre, and Guy Thuillier. *La Vie Quotidienne des Domestiques en France au XIXe Siècle*. Paris: Hachette, 1978.

Huggett, Frank E. *Life Below Stairs: Domestic Servants in England from Victorian Times*. London: Book Club Associates, 1977.

Maza, Sarah C. *Servants and Masters in Eighteenth-Century France: The Uses of Loyalty*. Princeton, N.J.: Princeton University Press, 1983.

McBride, Theresa M. *The Domestic Revolution: The Modernisation of Household Service in England and France, 1820–1920*. New York: Holmes and Meier, 1976.

————. "The Modernization of 'Woman's Work,'" *Journal of Modern History*, vol. 49, no. 2 (June 1977), pp. 231–45.

Earning Income: Women's Piecework and Domestic Industry

Berman, Sandra, ed. *Fit Work for Women*. New York: St. Martin's Press, 1979.

Coffin, Judith G. *The Politics of Women's Work: The Paris Garment Trade 1750–1915*. Princeton, N.J.: Princeton University Press, 1996.

Engel, Barbara. *Between the Field and the City: Women, Work and Family in Russia, 1860–1914*. Cambridge, U.K.: Cambridge University Press, 1994.

Fäy-Sallois, Fanny. *Les nourrices á Paris au XIXe Siècle*. Paris: Payot, 1980.

Gerhard, Ute. *Verhältnisse und Verhinderungen: Frauenarbeit, Familie und Rechte der Frauen im 19. Jahrhundert. Mit Dokumenten*. Frankfurt am Main: Suhrkamp, 1978.

Gordon, Eleanor, and Esther Beitenbach. *The World Is Ill Divided: Women's Work in Scotland in the Nineteenth and Twentieth Centuries.* Edinburgh: Edinburgh University Press, 1990.

Hausen, Karin. "Technischer Fortshritt und Frauenarbeit im 19. Jahrhundert. Zur Sozialgeschichte der Nähmaschine," *Geschichte und Gesellschaft,* vol. 4, no. 2 (Göttingen, 1978), pp. 148–69.

Herzog, Marianne. *From Hand to Mouth: Women and Piecework.* Translated by Stanley Mitchell. New York: Penguin, 1980.

Lewenhak, Sheila. *Women and Work.* New York: St. Martin's Press, 1980.

Mackie, Lindsay, and Polly Pattullo. *Women at Work.* London: Tavistock, 1977.

McDermid, Jane. *Women and Work in Russia, 1880–1930: A Study of Continuity through Change.* London: Longman, 1998.

Novarra, Virginia. *Women's Work, Men's Work: The Ambivalence of Equality.* London: Marion Boyars, 1980.

Quataert, Jean H. "The Shaping of Women's Work in Manufacturing: Guilds, Households, and the State in Central Europe, 1648–1870," *American Historical Review,* vol. 90, no. 5 (December 1985), pp. 1122–48.

Report of the Seminar on the Participation of Women in the Economic Evolution of the ECE Region and the Economic Role of Women in the ECE Region. A/CONF.94/14 (Copenhagen: World Conference of the United Nations Decade for Women, 1980).

Strumingher, Laura S. *Women and the Making of the Working Class: Lyon, 1830–1870.* St. Albans, Vt.: Eden Press Women's Publications, 1979.

Sullerot, Evelyne. *Historie et sociologie du travail féminin.* Paris: Editions Gonthier, 1968.

Sussman, George D. *Selling Mother's Milk: The Wet-Nursing Business in France, 1715–1914.* Urbana: University of Illinois Press, 1982.

Taylor, Barbara. " 'The Men Are As Bad As Their Masters' . . . : Socialism, Feminism and Sexual Antagonism in the London Tailoring Trade in the Early 1830s," *Feminist Studies,* vol. 5, no. 1 (Spring 1979), pp. 7–40.

Earning Income: Industrial Work and Trade Unionism

Beale, Jenny. *Getting It Together: Women as Trade Unionists.* London: Pluto Press, 1982.

Boxer, Marilyn J. "Foyer or Factory: Working Class Women in Nineteenth-Century France," *Western Society for French History Proceedings,* vol. II (November 1974), pp. 192–203.

Braybon, Gail. *Women Workers in the First World War: The British Experience.* London: Croom Helm, 1981.

Canning, Kathleen. *Languages of Labor and Gender: Female Factory Work in Germany 1850–1914.* Ithaca, N.Y.: Cornell University Press, 1996.

Cavendish, Ruth. *Women on the Line.* London: Routledge and Kegan Paul, 1982.

Downs, Laura Lee. *Manufacturing Inequality: Gender Division in the French and British Metalworking Industries, 1914–1939.* Ithaca, N.Y.: Cornell University Press, 1995.

Drake, Barbara. *Women in Trade Unions.* London: Virago, 1984 [1920].

Franzoi, Barbara. *At the Very Least She Pays the Rent: Women and German Industrialization, 1871–1914.* Westport, Conn.: Greenwood Press, 1985.

Glickman, Rose A. *Russian Factory Women: Workplace and Society, 1870–1914.* Berkeley: University of California Press, 1984.

Glucksmann, Miriam. *Women Assemble: Women Workers and the New Industries in Inter-War Britain.* London: Routledge, 1990.

Guilbert, Madeleine. *Les Femmes et l'organisation syndicale avant 1914.* Paris: Centre National de la Recherche Scientifique, 1966.

———. *Les fonctions des Femmes dans l'industrie.* Paris: Mouton, 1966.

Hewitt, Margaret. *Wives and Mothers in Victorian Industry.* Westport, Conn.: Greenwood Press, 1975 [1958].

Kamerman, Sheila B. "Work and Family in Industrialized Societies," *Signs,* vol. IV, no. 4 (Summer 1979), pp. 632–50.

Lewenhak, Sheila. *Women and Trade Unions: An Outline History of Women in the British Trade Union Movement.* New York: St. Martin's Press, 1977.

Messenger, Betty. *Picking Up the Linen Threads: A Study in Industrial Folklore.* Austin: University of Texas Press, 1980.

Neff, Wanda F. *Victorian Working Women: An Historical and Literary Study of Women in British Industries and Professions, 1832–1850.* London: Frank Cass, 1966 [1929].

Pinchbeck, Ivy. *Women Workers and the Industrial Revolution.* New York: F. S. Crofts, 1930.

Solden, Norbert C. *Women in British Trade Unions, 1874–1976.* Dublin: Gill and Macmillan, 1978.

Valenze, Deborah. *The First Industrial Woman.* New York: Oxford University Press, 1995.

Zappi, Elda Gentile. *If Eight Hours Seem Too Few: Mobilization of Women Workers in the Italian Rice Fields.* Albany, N.Y.: State University of New York Press, 1991.

Earning Income: Prostitution

Beattie, J. M. "The Criminality of Women in Eighteenth-Century England," *Journal of Social History,* vol. 8 (Summer 1975), pp. 80–116.

Bernstein, Laurie. *Sonia's Daughters: Prostitutes and Their Regulation in Imperial Russia.* Berkeley: University of California Press, 1995.

Christiane F.: Autobiography of a Girl of the Streets and Heroin Addict. Translated by Susanne Flatauer. New York: Bantam, 1982.

Corbin, Alain. *Women for Hire: Prostitution and Sexuality in France after 1850.* Translated by Alan Sheridan. Cambridge, Mass.: Harvard University Press, 1990.

Cordelier, Jeanne. *"The Life": Memoirs of a French Hooker.* New York: Avon, 1980.

Evans, Richard J. "Prostitution, State and Society in Imperial Germany," *Past and Present,* vol. 70 (February 1976), pp. 106–29.

Hersin, Jill. *Policing Prostitution in Nineteenth-Century Paris.* Princeton, N.J.: Princeton University Press, 1985.

Jaget, Claude, ed. *Prostitutes: Our Life.* Translated by Anna Furse, Suzie Fleming, and Ruth Hall. Bristol, U.K.: Falling Wall Press, 1980.

Mahood, Linda. *The Magdalenes: Prostitution in the Nineteenth Century.* [Scotland] New York: Routledge, 1991.

Richardson, Joanna. *The Courtesans: The Demi-Monde in Nineteenth-Century France.* Cleveland: World Publishing Co., 1967.

Walkowitz, Judith R. *Prostitution and Victorian Society: Women, Class and the State*. New York: Cambridge University Press, 1980.

Women in Revolutions and Politics

Abray, Jane. "Feminism in the French Revolution," *American Historical Review*, vol. 80, no. 1 (February 1975), pp. 43–62.

Applewhite, Harriet B., and Darline Gay Levy. *Women and Politics in the Age of the Democratic Revolution*. Ann Arbor: University of Michigan Press, 1990.

Barry, David. *Women and Political Insurgency: France in the Mid-Nineteenth Century*. New York: St. Martin's Press, 1996.

Clark, Anna. *The Struggle for the Breeches: Gender and the Making of the British Working Class*. Berkeley: University of California Press, 1995.

Clements, Barbara Evans. "Working-Class and Peasant Women in the Russian Revolution," *Signs*, vol. VIII, no. 2 (Winter 1982), pp. 215–35.

———. *Daughters of the Revolution: A History of Women in the USSR*. Arlington Heights, Ill.: Harlan Davidson, 1994.

Corbin, Alain, Jacqueline Lalouette, Michèle Riot-Sarcey, eds. *Femmes dans la Cité 1815–1871*. Paris: Créaphis, 1997.

Duchen, Claire. *Women's Rights and Women's Lives in France 1944–1968*. London: Routledge, 1994.

Fraisse, Geneviève. *Reason's Muse: Sexual Difference and the Birth of Democracy*. Translated by Jane Marie Todd. Chicago: University of Chicago Press, 1994.

George, Margaret. "The 'World Historical Defeat' of the Républicaines-Révolutionnaires," *Science and Society*, vol. 40, no. 4 (Winter 1976–1977), pp. 410–37.

Godineau, Dominique. *Citoyennes tricoteuses: The Women of Paris and Their Revolution*. Translated by Katherine Streip. Berkeley: University of California Press, 1998.

Grogan, Susan K. *French Socialism and Sexual Difference: Women and the New Society, 1803–1844*. Basingstoke, U.K.: Macmillan, 1992.

Gullickson, Gay L. *Unruly Women of Paris: Images of the Commune*. Ithaca, N.Y.: Cornell University Press, 1996.

Gutwirth, Madelyn. *The Twilight of the Goddesses: Women and Representation in the Era of the French Revolution*. New Brunswick, N.J.: Rutgers University Press, 1992.

Hilden, Patricia. *Working Women and Socialist Politics in France, 1850–1914*. Oxford: Oxford University Press, 1986.

———. *Women, Work and Politics: Belgium, 1830–1914*. Oxford: Oxford University Press, 1993.

Hill, Bridget. *Women, Work, and Sexual Politics in Eighteenth-Century England*. New York: Basil Blackwell, 1989.

Hufton, Olwen. *Women and the Limits of Citizenship in the French Revolution*. Toronto: University of Toronto Press, 1992.

Hunt, Lynn. *The Family Romance of the French Revolution*. Berkeley: University of California Press, 1992.

Johnson, Richard. "The Role of Women in the Russian Civil War, 1917–1921," *Conflict*, vol. II, no. 2 (1980), pp. 201–17.

Kaplan, Temma. "Female Consciousness and Collective Action: The Case of Barcelona, 1910–1918," *Signs*, vol. 7, no. 3 (Spring 1982), pp. 545–66.

Kapp, Yvonne. *Eleanor Marx.* 2 vols. New York: Pantheon, 1976.

Kelly, Gary, ed. *Women, Writing, and Revolution, 1790–1827.* New York: Oxford University Press, 1993.

Kent, Susan Kingsley. *Sex and Suffrage in Britain, 1860–1914.* Princeton, N.J.: Princeton University Press, 1990.

Landes, Joan. *Women and the Public Sphere in the Age of the French Revolution.* Ithaca, N.Y.: Cornell University Press, 1988.

Levy, Darline Gay, Harriet Branson Applewhite, and Mary Durham Johnson, eds. *Women in Revolutionary Paris: 1789–1795.* Urbana: University of Illinois Press, 1979.

Liddington, Jill, and Jill Norris. *One Hand Tied Behind Us: The Rise of the Women's Suffrage Movement.* London: Virago, 1978.

Maxwell, Margaret. *Narodniki Women: Russian Women Who Sacrificed Themselves for the Dream of Freedom.* New York: Pergamon, 1990.

Melzer, Sara E., and Leslie W. Rabine, eds. *Rebel Daughters: Women and the French Revolution.* New York: Oxford University Press, 1992.

Midgley, Clare. *Women Against Slavery: The British Campaigns, 1780–1870.* London: Routledge, 1992.

Moses, Claire Goldberg. *French Feminism in the Nineteenth Century.* Albany: State University of New York Press, 1984.

Otto, Louise. *"Dem Reich der Freiheit werb' ich Bürgerinnen": Die Frauen-Zeitung von Louise Otto.* Edited by Ute Gerhard, Elisabeth Hannover-Drück, and Romina Schmitter. Frankfurt am Main: Syndikat, 1980.

Pateman, Carole. *The Sexual Contract.* Stanford, Calif.: Stanford University Press, 1988.

Porter, Cathy. *Women in Revolutionary Russia.* New York: Cambridge University Press, 1987.

Rowbotham, Sheila. *Women, Resistance and Revolution: A History of Women and Revolution in the Modern World.* New York: Vintage, 1974.

Slaughter, Jane, and Robert Kern, eds. *European Women on the Left: Socialism, Feminism, and the Problems Faced by Political Women, 1880 to the Present.* Westport, Conn.: Greenwood Press, 1981.

Taylor, Barbara. *Eve and the New Jerusalem: Socialism and Feminism in the Nineteenth Century.* New York: Pantheon, 1983.

Thomas, Edith. *Les Femmes en 1848.* Paris: Presses Universitaires de France, 1948.

————. *The Women Incendiaries.* London: Secker and Warburg, 1967.

Tilly, Louise A. "Women's Collective Action and Feminism in France, 1870–1914," in Louise A. Tilly and Charles Tilly, eds. *Class Conflict and Collective Action.* Beverly Hills, Calif.: Sage Press, 1981.

Wood, Elizabeth A. *The Baba and the Comrade: Gender and Politics in Revolutionary Russia.* Bloomington: University of Indiana Press, 1997.

Yalom, Marilyn. *Blood Sisters: The French Revolution in Women's Memory.* New York: Basic Books, 1993.

Women, Fascism and World War II

Braybon, Gail, and Penny Summerfield. *Out of the Cage: Women's Experiences in Two World Wars.* [Britain] London: Routledge, 1987.

Bridenthal, Renate, Atina Grossmann, and Marion Kaplan, eds. *When Biology Became Destiny: Women in Weimar and Nazi Germany*. New York: Monthly Review Press, 1984.

Cassin-Scott, Jack. *Women at War, 1939–1945*. London: Osprey Publishing, 1980.

De Grazia, Victoria. *How Fascism Ruled Women, Italy, 1922–1945*. Berkeley: University of California Press, 1992.

Delbo, Charlotte. *None of Us Will Return*. Translated by John Githens. Boston: Beacon, 1968.

Evans, Richard J. "German Women and the Triumph of Hitler," *Journal of Modern History*, vol. 48, no. 1 (March 1976), pp. 123–75.

Fénelon, Fania. *Playing for Time*. Translated by Judith Landry. New York: Atheneum, 1979.

Anne Frank: The Diary of a Young Girl, A Definitive Edition. Edited by Otto H. Frank and Mirjam Tressler. New York: Doubleday, 1995.

Frauengruppe Faschismusforschung. *Mutterkreuz und Arbeitsbuch: Zur Geschichte der Frauen in der Weimarer Republik und im Nationalsozialismus*. Frankfurt am Main: Fischer, 1981.

Hart, Janet. *New Voices in the Nation: Women and the Greek Resistance, 1941–1964*. Ithaca, N.Y.: Cornell University Press, 1996.

Hawthorne, Melanie. *Gender and Fascism in Modern France*. Hanover, N.H.: New England University Press, 1997.

Higgonet, Margaret, et. al., eds. *Behind the Lines: Gender and the Two World Wars*. New Haven: Yale University Press, 1987.

Jancar-Webster, Barbara. *Women and Revolution in Yugoslavia, 1941–1945*. Denver, Co.: Arden Press, 1990.

Kaplan, Marion A. *Between Dignity and Despair: Jewish Life in Nazi Germany*. New York: Oxford University Press, 1998.

Ka-Tzetnik 135633. *House of Dolls*. Translated by Moshe M. Kohn. London: Granada, 1981 [1956].

Katz, Esther, and Joan Miriam Ringelheim, eds. *Women Surviving the Holocaust: Proceedings of the Conference*. New York: Occasional Papers from the Institute for Research in History, 1983.

Knef, Hildegard. *The Gift Horse: Report on a Life*. Translated by David Anthony Palastanga. New York: Dell, 1971.

Koonz, Claudia. *Mothers in the Fatherland: Women, the Family and Nazi Politics*. New York: St. Martin's Press, 1987.

Laska, Vera, ed. *Women in the Resistance and in the Holocaust: The Voices of Eyewitnesses*. Westport, Conn.: Greenwood Press, 1983.

Leitner, Isabella. *Fragments of Isabella*. New York: Dell, 1978.

Mangini, Shirley. *Memories of Resistance: Women's Voices from the Spanish Civil War*. New Haven: Yale University Press, 1995.

Martin, Elaine. *Gender, Patriarchy and Fascism in the Third Reich: The Response of Women Writers*. Detroit, Mich.: Wayne State University Press, 1993.

Mason, Tim. "Women in Germany, 1925–1940: Family, Welfare and Work," *History Workshop*, vol. I. Part I (Spring 1976), pp. 74–113; Part II (Autumn 1976), pp. 5–32.

Meed, Vladka. *On Both Sides of the Wall: Memoirs from the Warsaw Ghetto.* Translated by Steven Meed. New York: Holocaust Library, 1979 [1948].

Minns, Raynes. *Bombers and Mash: The Domestic Front, 1939–1945.* London: Virago, 1980.

Nash, Mary. *Defying Male Civilization: Women in the Spanish Civil War.* Denver: Arden Press, 1996.

Noggle, Anne. *A Dance with Death: Soviet Airwomen in World War II.* College Station: Texas A & M University Press, 1994.

Owings, Alison. *Frauen: German Women Recall the Third Reich.* New Brunswick, N.J.: Rutgers University Press, 1993.

Peterson, Brian. "The Politics of Working-Class Women in the Weimar Republic," *Central European History,* vol. X, no. 2 (June 1977), pp. 87–111.

Phayer, Michael. *Protestant and Catholic Women in Nazi Germany.* Detroit: Wayne State University Press, 1990.

Pickering-Iazzi, Robin, ed. *Mothers of Invention: Women, Italian Fascism, and Culture.* Minneapolis: University of Minnesota Press, 1995.

Pine, Lisa. *Nazi Family Policy, 1933–1945.* New York: Oxford University Press, 1997.

Rossiter, Margaret L. *Women in the Resistance.* New York: Praeger, 1986.

Rupp, Leila J. *Mobilizing Women for War: German and American Propaganda, 1939–1945.* Princeton, N.J.: Princeton University Press, 1978.

Hannah Senesh: Her Life and Diary. Translated by Marta Cohn. New York: Schocken, 1973.

Settle, Mary Lee. *All The Brave Promises: Memories of Aircraft Woman 2nd Class 2146391.* New York: Ballantine, 1980 [1960].

Slaughter, Jane. *Women and the Italian Resistance 1943–1945.* Denver, Co.: Arden, 1997.

Stephenson, Jill. *Women in Nazi Society.* New York: Harper & Row, 1975.

Stolzfuss, Nathan. *Resistance of the Heart: Intermarriage and the Rosenstrasse Protest in Nazi Germany.* New York: W. W. Norton, 1996.

Thomas, Theodore N. *Women Against Hitler: Christian Resistance in the Third Reich.* Westport, Conn.: Praeger, 1995.

Weitz, Margaret Collins. *Sisters in the Resistance: How Women Fought to Free France, 1940–1945.* New York: John Wiley, 1995.

Willson, Perry R. *The Clockwork Factory: Women and Work in Fascist Italy.* Oxford: Clarendon Press, 1993.

Zassenhaus, Hiltgunt. *Walls: Resisting the Third Reich—One Woman's Story.* Boston: Beacon, 1976.

PART IX TRADITIONS REJECTED

General Works on Feminism in Europe

Akkerman, Tjitske, and Siep Stuurman, eds. *Perspectives on Feminist Thought in European History: From the Middle Ages to the Present.* London: Routledge, 1998.

Albistur, Maïté, and Daniel Armogathe. *Histoire du féminisme français du moyen âge à nos jours.* 2 vols. Paris: Editions des femmes, 1977.

————. *Le grief des femmes: Anthologies du textes féministes du 2nd empire à nos jours.* Poitiers, France: Editions Hier et Demain, 1978.

Banks, Olive. *Faces of Feminism: A Study of Feminism as a Social Movement.* New York: St. Martin's Press, 1981.

Daley, Caroline, and Melanie Nolan, eds. *Suffrage and Beyond: International Feminist Perspectives.* New York: New York University Press, 1994.

Gordon, Felicia, and Maire Cross. *Early French Feminisms, 1830–1940: A Passion for Liberty.* Brookfield, Vt.: Edward Elgar, 1996.

Kelly, Joan. *Women, History, and Theory.* Chicago: University of Chicago Press, 1984.

Liddington, Jill. *The Road to Greenham Common: Feminism and Anti-Militarism in Britain since 1820.* Syracuse, N.Y.: Syracuse University Press, 1991.

Rabault, Jean. *Histoire des féminismes français.* Paris: Stock, 1978.

Riley, Denise. *"Am I That Name?" Feminism and the Category of "Woman" in History.* Minneapolis: University of Minnesota Press, 1988.

Rowbotham, Sheila. *Hidden from History: Rediscovering Women in History from the Seventeenth Century to the Present.* New York: Pantheon, 1974.

Rupp, Leila J. *Worlds of Women: The Making of an International Women's Movement.* Princeton, N.J.: Princeton University Press, 1997.

Scott, Joan Wallach. *Only Paradoxes to Offer: French Feminists and the Rights of Man.* Cambridge, Mass.: Harvard University Press, 1996.

Early Feminism in Europe

Ferguson, Moira, ed. *First Feminists: British Women Writers, 1578–1799.* Bloomington: Indiana University Press, 1985.

Flexner, Eleanor. *Mary Wollstonecraft.* Baltimore: Penguin, 1973.

Gagen, Jean Elisabeth. *The New Woman: Her Emergence in English Drama, 1600–1730.* New York: Twayne Publishers, 1954.

Janes, R. M. "On the Reception of Mary Wollstonecraft's *A Vindication of the Rights of Women,*" *Journal of the History of Ideas,* vol. XXXIX, no. 2 (April–June 1978), pp. 293–302.

Perry, Ruth. *The Celebrated Mary Astell: An Early English Feminist.* Chicago: University of Chicago Press, 1986.

Pizan, Christine de. *The Book of the City of Ladies.* Translated by Earl Jeffrey Richards. New York: Persea Books, 1982.

Rogers, Katharine M. *Feminism in Eighteenth-Century England.* Urbana: University of Illinois Press, 1982.

Smith, Hilda. *Reason's Disciples: Seventeenth-Century English Feminists.* Urbana: University of Illinois Press, 1982.

Tomalin, Clare. *The Life and Death of Mary Wollstonecraft.* London: Weidenfeld and Nicholson, 1974.

Wollstonecraft, Mary. *Maria, or the Wrongs of Woman.* New York: W. W. Norton & Company, 1975 [1797].

————. *Letters Written During a Short Residence in Sweden, Norway, and Denmark.* Edited by Carol H. Poston. Lincoln, Neb.: University of Nebraska Press, 1976 [1796].

―――. A Vindication of the Rights of Woman. New York: W. W. Norton & Company, 1977 [1792].

Equal Rights Feminism in Europe

Bard, Christine. Les filles de Marianne: Histoire des feminismes 1914–1940. Paris: Fayard, 1995.

Bidelman, Patrick Kay. Pariahs Stand Up! The Founding of the Liberal Feminist Movement in France, 1858–1889. Westport, Conn.: Greenwood Press, 1982.

Blom, Ida. "A Centenary of Organized Feminism in Norway," Women's Studies International Forum, vol. V, no. 6 (1982), pp. 569–74.

Bolt, Christine. The Women's Movements in the United States and Britain from the 1790s to the 1920s. Amherst: University of Massachusetts Press, 1993.

Bosch, Mineke, with Annemarie Kloosterman, eds. Politics and Friendship: Letters from the International Woman Suffrage Alliance 1902–1942. Columbus: Ohio State University Press, 1990.

Edmondson, Linda Harriet. Feminism in Russia, 1900–1917. Stanford, Calif.: Stanford University Press, 1984.

Evans, Richard J. The Feminist Movement in Germany, 1894–1933. London: Sage, 1976.

Faderman, Lillian, and Brigitte Eriksson, eds. Lesbian-Feminism in Turn-of-the-Century Germany. Weatherby Lake, Mo.: Naiad Press, 1981.

Fauré, Christine. Democracy Without Women: Feminism and the Rise of Liberal Individualism. Bloomington: University of Indiana Press, 1991.

Gleadle, Katherine. The Early Feminists: Radical Unitarians and the Emergence of the Women's Rights Movement, 1831–1851. [Britain] New York: St. Martin's Press, 1995.

Gordon, Felicia. The Integral Feminist: Madeleine Pelletier, 1874–1939. Minneapolis: University of Minnesota Press, 1990.

Gullickson, Gay L. "Feminists and Suffragists: The British and French Experiences," Feminist Studies, vol. 15 (1989), pp. 591–602.

Harrison, Brian. Separate Spheres: The Opposition to Women's Suffrage in Britain. New York: Holmes and Meier, 1978.

Hause, Steven C. Hubertine Auclert: The French Suffragette. New Haven, Conn.: Yale University Press, 1987.

Hause, Steven C., and Anne R. Kenney. "The Limits of Suffragist Behavior: Legalism and Militancy in France, 1876–1922," American Historical Review, vol. 86, no. 4 (October 1981), pp. 781–806.

Holcombe, Lee. Wives and Property: Reform of the Married Women's Property Law in Nineteenth-Century England. Toronto: University of Toronto Press, 1983.

Howard, Judith Jeffrey. The Woman Question in Italy, 1861–1880. Ph.D. dissertation, University of Connecticut, 1977.

Kaplan, Marion A. The Jewish Feminist Movement in Germany: The Campaigns of the Jüdischer Frauenbund, 1904–1933. Westport, Conn.: Greenwood Press, 1979.

Leneman, Leah. A Guid Cause: The Women's Suffrage Movement in Scotland. Aberdeen: Aberdeen University Press, 1991.

Liddington, Jill, and Jill Norris. *One Hand Tied Behind Us: The Rise of the Women's Suffrage Movement*. London: Virago, 1978.

Mackenzie, Midge. *Shoulder to Shoulder: A Documentary*. New York: Alfred A. Knopf, 1975.

Mill, John Stuart, and Harriet Taylor Mill. *Essays on Sex Equality*. Edited by Alice S. Rossi. Chicago: University of Chicago Press, 1970.

Murray, Janet Horowitz. *Strong-Minded Women and Other Lost Voices from Nineteenth-Century England*. New York: Pantheon, 1982.

Pugh, Martin. *Women and the Women's Movement in Britain 1914–1959*. New York: Paragon Press, 1992.

Reagin, Nancy R. *A German Women's Movement: Class and Gender in Hanover, 1880–1933*. Chapel Hill: University of North Carolina Press, 1995.

"Dem Reich der Freiheit werb' ich Bürgerinnen": Die Frauen-Zeitung von Louise Otto. Edited by Ute Gerhard, Elisabeth Hannover-Drück, and Romina Schmitter. Frankfurt am Main: Syndikat, 1979.

Rendall, Jane. *The Origins of Modern Feminism: Women in Britain, France and the United States*. New York: Schocken, 1984.

Spender, Dale. *There's Always Been a Women's Movement This Century*. London: Pandora, 1983.

Ward, Margaret. *Unmanageable Revolutionaries: Women and Irish Nationalism*. Dublin: Brandon, 1983.

Whittick, Arthur. *Woman into Citizen*. Santa Barbara, Calif.: ABC-Clio, 1979.

Feminist Socialism in Europe

Ackelsberg, Martha A. *Free Women of Spain: Anarchism and the Struggle for Emancipation of Women*. Bloomington: Indiana University Press, 1991.

Balabanoff, Angelica. *My Life As a Rebel*. Bloomington: Indiana University Press, 1973 [1938].

Bebel, August. *Women Under Socialism*. Translated by Daniel De Leon. New York: Schocken, 1971 [1879, 1883].

Boxer, Marilyn, and Jean H. Quataert, eds. *Socialist Women: European Socialist Feminism in the Nineteenth and Early Twentieth Centuries*. New York: Elsevier, 1978.

Brinker-Gabler, Gisela, ed. *Frauen gegen den Krieg*. Frankfurt am Main: Fischer, 1980.

Brion, Hélène. *La voie féministe*. Paris: Editions Syros, 1978 [1916].

Bussey, Gertrude, and Margaret Tims. *Pioneers for Peace: Women's International League for Peace and Freedom*. London: WILPF British Section, 1980.

Clements, Barbara Evans. *Bolshevik Feminist: The Life of Aleksandra Kollontai*. Bloomington: Indiana University Press, 1979.

Cross, Maire, and Tim Grey. *The Feminism of Flora Tristan*. Oxford: Berg, 1992.

Draper, Hal, and Anne G. Lipow. "Marxist Women versus Bourgeois Feminism," *The Socialist Register*, (1976), pp. 179–226.

Engel, Barbara Alpern, and Clifford N. Rosenthal, eds. *Five Sisters: Women Against The Tsar*. New York: Schocken, 1977.

———. *Mothers and Daughters: Women of the Intelligentsia in Nineteenth-Century Russia*. Cambridge: Cambridge University Press, 1983.

Evans, Richard. *Comrades and Sisters: Feminism, Socialism and Pacifism in Europe 1870–1945.* New York: St. Martin's Press, 1987.

Honeycutt, Karen. "Clara Zetkin: A Socialist Approach to the Problem of Woman's Oppression," *Feminist Studies,* vol. 3, nos. 3–4 (Spring–Summer 1976), pp. 131–44.

Hunt, Karen. *Equivocal Feminists: The Social Democratic Federation and the Woman Question 1884–1911.* New York: Cambridge University Press, 1996.

Kapp, Yvonne. *Eleanor Marx.* 2 vols. New York: Pantheon, 1976.

Klucsarits, Richard, and Friedrich G. Kürbisch, eds. *Arbeiterinnen kämpfen um ihr Recht: Autobiographische Texte zum Kampf rechtlose und entrechteter "Frauenpersonen" in Deutschland, Osterreich und der Schweiz des 19. und 20. Jahrhunderts.* Wuppertal: Peter Hammer, n.d.

Kollontai, Alexandra. *The Autobiography of a Sexually Emancipated Communist Woman.* Translated by Salvator Attanasio. New York: Herder and Herder, 1971 [1926].

———. *Love of Worker Bees.* Translated by Cathy Porter. Chicago: Cassandra Editions, 1978.

Selected Writings of Alexandra Kollontai. Edited and translated by Alix Holt. New York: W. W. Norton, 1977.

Kuhn, Annette, and AnnMarie Wolpe, eds. *Feminism and Materialism: Women and Modes of Production.* London: Routledge and Kegan Paul, 1978.

Lenin, V. I. *The Emancipation of Women.* New York: International Publishers, 1978. [Introduction by N. Krupskaya and includes Clara Zetkin's *Reminiscences of Lenin.*]

Meyer, Alfred G. *The Feminism and Socialism of Lily Braun.* Bloomington: Indiana University Press, 1985.

Michel, Louise. *The Red Virgin: Memoirs of Louise Michel.* Edited and translated by Bullitt Lowry and Elizabeth Ellington Gunter. University: University of Alabama Press, 1981.

Middleton, Lucy, ed. *Women in the Labour Movement: The British Experience.* London: Croom Helm, 1977.

Mitchell, Hannah. *The Hard Way Up: The Autobiography of Hannah Mitchell, Suffragette and Rebel.* Edited by Geoffrey Mitchell. London: Virago, 1977.

Moses, Claire Goldberg, and Leslie Wahl Rabine. *Feminism, Socialism and French Romanticism.* Bloomington: Indiana University Press, 1993.

Nettl, J. P. *Rosa Luxemburg.* Oxford: Oxford University Press, 1969.

Pore, Renate. *A Conflict of Interest: Women in German Social Democracy, 1919–1933.* Westport, Conn: Greenwood Press, 1981.

Quataert, Jean H. *Reluctant Feminists in German Social Democracy, 1885–1917.* Princeton, N.J.: Princeton University Press, 1979.

Scott, Hilda. *Women and Socialism: Experiences from Eastern Europe.* [Originally *Does Socialism Liberate Women?*] London: Alison and Busby, 1976.

Sowerwine, Charles. *Sisters or Citizens? Women and Socialism in France since 1876.* Cambridge: Cambridge University Press, 1982.

Stites, Richard. *The Women's Liberation Movement in Russia: Feminism, Nihilism, and Bolshevism, 1860–1930.* Princeton, N.J.: Princeton University Press, 1978.

Taylor, Barbara. *Eve and the New Jerusalem: Socialism and Feminism in the Nineteenth Century.* [England]. New York: Pantheon, 1983.

Thomas, Edith. *Pauline Roland: Socialisme et Féminisme au XIXe Siècle*. Paris: Marcel Rivière, 1956.

———. *Louise Michel*. Translated by Penelope Williams. Montreal: Black Rose Press, 1980.

Thönnessen, Werner. *The Emancipation of Women: The Rise and Decline of the Woman's Movement in German Social Democracy, 1863–1933*. Translated by Joris des Bres. Glasgow: Pluto Press, 1976.

Voilquin, Suzanne. *Mémoires d'une saint-simonienne en Russie (1839–1846)*. Edited by Maïté Albistur and Daniel Armogathe. Paris: Editions des femmes, 1977.

Woolf, Virginia. *A Room of One's Own*. New York: Harcourt, Brace and World, 1957 [1929].

———. *Three Guineas*. New York: Harcourt, Brace and World, 1966 [1938].

Clara Zetkin: Selected Writings. Edited by Philip Foner. New York: International Publishers, 1984.

Zylberberg-Hocquard, Marie-Hélène. *Femmes et féminisme dans le mouvement ouvrier français*. Paris: Éditions ouvrières, 1981.

The Women's Liberation Movement

Altbach, Edith Hoshino, Jeanette Clausen, Dagmar Schultz, and Naomi Stephan, eds. *German Feminism: Readings in Politics and Literature*. Albany, N.Y.: State University of New York Press, 1984.

Bair, Deirdre. *Simone de Beauvoir*. New York: Summit Books, 1990.

Barreno, Maria Isabel, Maria Teresa Horta, and Maria Velho da Costo. *The Three Marias: New Portuguese Letters*. Translated by Helen R. Lane. New York: Doubleday & Company, 1976.

de Beauvoir, Simone. *The Second Sex*. Translated by H. M. Parshley. New York: Bantam, 1961.

Birnbaum, Lucia Chiavola. *Liberazione della donna: feminism in Italy*. Middletown, Conn.: Wesleyan University Press, 1986.

Bouchier, David. *The Feminist Challenge: The Movement for Women's Liberation in Britain and the United States*. New York: Schocken, 1983.

Bradshaw, Jan, ed. *The Women's Liberation Movement: Europe and North America*. Oxford: Pergamon Press, 1982. [Originally vol. 4, no. 1 of the *Women's Studies International Forum*.]

Brøgger, Suzanne. *Deliver Us from Love*. Translated by Thomas Teal. New York: Dell, 1976.

Cambridge Women's Peace Collective. *My Country is the Whole World: An Anthology of Women's Work on Peace and War*. London: Pandora Press, 1984.

Coote, Anna, and Beatrix Campbell. *Sweet Freedom: The Struggle for Women's Liberation*. London: Pan Books, 1982.

Council of Europe Action for Equality between Women and Men. *Council of Europe Information Document Submitted to the World Conference on the UN Decade for Women*. Strasbourg, 1980.

Dahlerup, Drude, ed. *The New Women's Movement: Feminism and Political Power in Europe and the USA*. London: Sage, 1986.

Dalla Costa, Mariarosa, and Selma James. *The Power of Women and the Subversion of the Community*. Bristol, U.K.: Falling Wall Press, 1975.

Feminism and Non-Violence Study Group. *Piecing It Together: Feminism and Non-Violence*. Devon, U.K.: Feminism and Non-Violence Study Group, 1983.

Grossmann, Atina. *Reforming Sex: The German Movement for Birth Control and Abortion Reform 1920–1950*. New York: Oxford University Press, 1995.

Halimi, Gisèle. *La cause des femmes*. Paris: Grasset, 1973.

Iglitzin, Lynne B., and Ruth Ross, eds. *Women in the World: A Comparative Study*. Santa Barbara, Calif.: Clio Press, 1976.

Jones, Lynne, ed. *Keeping the Peace: A Woman's Peace Handbook*. London: Women's Press, 1983.

Kaplan, Gisela. *Contemporary Western European Feminism*. New York: New York University Press, 1992.

Katzenstein, Mary Fainsod, and Carol McClurg Mueller, eds. *The Women's Movements of the United States and Western Europe: Consciousness, Political Opportunity and Public Policy*. Philadelphia, Pa.: Temple University Press, 1987.

El Largo Camino Hacia La Igualdad: Feminismo en España 1975–1995. Madrid: Ministerio de Asuntos Sociales, Instituto de la Mujer, 1995.

Ledbetter, Rosanna. *A History of the Malthusian League: 1877–1927*. Columbus: Ohio State University Press, 1976.

Lovenduski, Joni. *Women and European Politics: Contemporary Feminism and Public Policy*. Amherst: University of Massachusetts Press, 1986.

Lovenduski, Joni, and Jill Hills, eds. *The Politics of the Second Electorate: Women and Public Participation*. London: Routledge and Kegan Paul, 1981.

Lovenduski, Joni, and Vicky Randall. *Contemporary Feminist Politics: Women and Power in Britain*. Oxford: Oxford University Press, 1993.

Marks, Elaine, and Isabelle de Courtivron, eds. *New French Feminisms*. New York: Schocken, 1981.

McLaren, Angus. *Sexuality and Social Order: The Debate over the Fertility of Women and Workers in France, 1770–1920*. New York: Holmes and Meier, 1983.

Meulenbelt, Ania. *The Shame Is Over: A Political Life Story*. Translated by Ann Oosthuizen. London: Women's Press, 1980.

Mohin, Lilian, ed. *One Foot on the Mountain: An Anthology of British Feminist Poetry 1969–1979*. London: Onlywoman Press, 1979.

Mohin, Lilian, and Anna Wilson. *Past Participants: a lesbian history diary for 1984*. London: Onlywomen Press, 1983.

Oakley, Ann, and Juliet Mitchell, eds. *Who's Afraid of Feminism?: Seeing Through the Backlash*. New York: New Press, 1997.

Potts, Malcolm, Peter Diggory, and John Peel. *Abortion*. Cambridge: Cambridge University Press, 1977.

Roussel, Nelly. *L'Eternelle Sacrifiée*. Edited by Maïté Albistur and Daniel Armogathe. Paris: Syros, 1979.

Rowbotham, Sheila. *A New World for Women: Stella Browne, Socialist Feminist*. London: Pluto Press, 1977.

———. *The Past Is Before Us: Feminism in Action Since the 1960s*. London: Penguin, 1989.

Rowe, Marsha, ed. *Spare Rib Reader*. Harmondsworth, U.K.: Penguin, 1982.

Russell, Diana E. H., and Nicole Van de Ven, eds. *Crimes Against Women: Proceedings of the International Tribunal*. Millbrae, Calif.: Les Femmes, 1976.

Sanger, Margaret. *My Fight for Birth Control*. New York: Farrar and Rinehart, 1931.

Sanger, Margaret, and Hanna Stone, eds. *The Practise of Contraception: An International Symposium and Survey*. Baltimore: Williams and Wilkins, 1931.

Scott, Hilda. *Sweden's "Right To Be Human"—Sex-Role Equality: The Goal and the Reality*. London: Allison and Busby, 1982.

Stefan, Verena. *Shedding*. Translated by Johanna Moore and Beth Weckmueller. New York: Daughters Publishing, 1978.

Vandelac, Louise. *L'italie au féminisme*. Paris: Éditions tierce, 1978.

Wittig, Monique. *Les Guérillères*. Translated by David Le Vay. New York: Avon, 1971.

Wolff, Charlotte. *Hindsight*. London: Quartet Books, 1980.

Women Against Rape, *Women at W.A.R.* Bristol, U.K.: Falling Wall Press, 1978.

The European Diaspora

Brettell, Caroline. *Men Who Migrate, Women Who Wait: Population and History in a Portuguese Parish*. Princeton, N.J.: Princeton University Press, 1986.

Burton, Antoinette. *Burdens of History: British Feminists, Indian Women and Imperial Culture, 1865–1915*. Chapel Hill: University of North Carolina Press, 1994.

Chaudhuri, Nupur, and Margaret Strobel, eds. *Western Women and Imperialism: Complicity and Resistance*. Bloomington: Indiana University Press, 1992.

Clancy-Smith, Julia, and Frances Gouda, eds. *Domesticating the Empire: Race, Gender, and Family Life in French and Dutch Colonialism*. Charlottesville: University Press of Virginia, 1998.

Dias, Joan Grant. *Women, Migration and Empire*. Stoke-on-Trent: Trentham Books, 1996.

Ferguson, Moira. *Subject to Others: British Women Writers and Colonial Slavery, 1670–1834*. London: Routledge, 1992.

Jayawardena, Kumari. *The White Woman's Other Burden: Western Women and South Asia During British Rule*. New York: Routledge, 1995.

Kranidis, Rita S., ed. *Imperial Objects: Essays of Victorian Women's Emigration and Unauthorized Imperial Experience*. New York: Twayne, 1998.

Melman, Billie. *Women's Orients: English Women and the Middle East, 1718–1918: Sexuality, Religion and Work*. Ann Arbor: University of Michigan Press, 1992.

Midgley, Clare, ed. *Gender and Imperialism*. New York: St. Martin's Press, 1998.

Quack, Sibylle, ed. *Between Sorrow and Strength: Women Refugees of the Nazi Period*. Cambridge, U.K.: Cambridge University Press, 1995.

Strobel, Margaret. *European Women and the Second British Empire*. Bloomington: Indiana University Press, 1991.

Ware, Vron. *Beyond the Pale: White Women, Racism and History*. London: Verso, 1992.

Epilogue

Allen, Beverly. *Rape Warfare: The Hidden Genocide in Bosnia-Herzegovina and Croatia*. Minneapolis: University of Minnesota Press, 1996.

Borchorst, Annette. *Family Policies in Western Europe: Convergence or Divergence?* Wellesley, Mass.: Center for Research on Women, Wellesley College, 1995.

Brill, Alida, ed. *A Rising Public Voice: Women in Politics Worldwide.* New York: The Feminist Press, 1995.

Buckley, Mary, ed. *Perestroika and Soviet Women.* Cambridge, U.K.: Cambridge University Press, 1992.

Einhorn, Barbara. *Cinderella Goes to Market: Citizenship, Gender and Women's Movements in East Central Europe.* London: Verso, 1993.

Einhorn, Barbara, and Eileen James Yeo, eds. *Women and Market Societies: Crisis and Opportunity.* Aldershot, U.K.: Edward Elgar, 1995.

Funk, Nanette, and Magda Mueller, eds. *Gender Politics and Post-Communism: Reflections from Eastern Europe and the Former Soviet Union.* New York: Routledge, 1993.

Garcia-Ramon, Maria Dolors, and Janice Monk, eds. *Women of the European Union: The Politics of Work and Daily Life.* New York: Routledge, 1996.

Hoerder, Dirk, and Leslie Page Moch, eds. *European Migrants: Global and Local Perspectives.* Boston: Northeastern University Press, 1996.

Lewis, Jane, ed. *Women and Social Policies in Europe: Work, Family and the State.* Aldershot, U.K.: Edward Elgar, 1993.

Long, Kristi S. *We All Fought for Freedom: Women in Poland's Solidarity Movement.* Boulder, Co.: Westview Press, 1996.

Marsh, Rosalind, ed. *Women in Russia and Ukraine.* Cambridge, U.K.: Cambridge University Press, 1996.

Noin, Daniel, and Robert Woods. *The Changing Population of Europe.* Oxford: Blackwell, 1993.

Stiglmayer, Alexandra, ed. *Mass Rape: The War Against Women in Bosnia-Herzegovina.* Lincoln: University of Nebraska Press, 1994.

Women's Studies International Forum. Special Issue: "Images from Women in a Changing Europe," nos. 2/3 (March/June 1994).

The World's Women 1995 Trends and Statistics. New York: United Nations, 1995.

INDEX

Illustration Credits

PART VI

1. Christine de Pizan, the British Library (Harl. MS 4431).
2. The Research Libraries, New York Public Library.
3. American Heritage Publishing Co., New York, N.Y.
4. National Trust Photographic Library, London, England (DRB 020/013P).
5. Engraving by Abraham Bosse.
6. Kunsthistorisches Museum, Vienna, Austria. Medal by J. G. Wächter.
7. Painting by Martin van der Meytens.
8. Alinari/Art Resource, NY
9. Portrait of Emilie du Châtelet.
10. "Die Familie Mozart" by J. N. della Croce. Internationale Stiftung Mozarteum, Salzburg, Austria.

PART VII

1. National Portrait Gallery, London, England (reg. no. 4501). Silhouette by A. Edouart.
2. National Portrait Gallery, London, England (reg. no. 1237).
3. Cliché Musée des Arts et Traditions populaires. Photograph by Cliché des Musées Nationaux, Paris, France.
4. Corbis. UUR 305.1
5. Musée Renan-Scheffer, Paris, France.
6. Bildarchiv Preussischer Kulturbesitz, Berlin, Federal Republic of Germany. Poster art by Jupp Wiertz.
7. Photo from *Madame Curie* by Eve Curie. Copyright © 1937 by Doubleday, a division of Bantam, Doubleday, Dell Publishing Group, Inc. Reprinted by permission of the publisher.
8. The 586th Fighter Regiment, U.S.S.R.
9. Marie Stopes Organisation, London, England.

PART VIII

1. Photograph attributed to Lord Somers, c. 1863–65.
2. Documentary Photography Archive, Manchester, England.
3. Bibliothèque Nationale, Paris, France.
4. Corbis. UNK 1.141
5. Photograph by Arkady Shaikhet.
6. From Maria-Antonietta Macciocchi, *La donna nero* (Milan: Feltrini, 1977).
7. International News Photo, New York, N.Y.
8. Nuremberg Trial Collection, National Archives, Washington, D.C. (238-NT-281).
9. New York Public Library, Picture Archives.

PART IX

1. The British Library, London, England (Har. MS. 4431).
2. Bibliothèque Nationale, Paris, France.
3. The London School of Economics, the British Library of Political and Economic Science, London, England.
4. Reprinted with permission from *Five Sisters: Women Against the Tsar*, edited by Barbara Alpern Engel and Clifford N. Rosenthal. (Boston: Allen and Unwin, 1987), p. 215.
5. Special issue of *Danish Journal*, 1980. Published by the Ministry of Foreign Affairs of Denmark.
6. By arrangement with International Publishers, New York, N.Y.
7. Hulton Deutsch/Corbis. UNK 2.59
8. Editions tierce, Paris, France.
9. Lynne Jones, ed., *Keeping the Peace* (London: The Women's Press, 1983).

CPSIA information can be obtained
at www.ICGtesting.com
Printed in the USA
BVOW03s1156040117
472502BV00001B/4/P